Urban Rainfall Runoff Pollution Control Technology and Application

Dezhi Sun · Fei Qi

Urban Rainfall Runoff Pollution Control Technology and Application

中国建筑工业出版社
CHINA ARCHITECTURE & BUILDING PRESS

Springer

Dezhi Sun
College of Environmental Science
and Engineering
Beijing Key Lab for Source Control
Technology of Water Pollution
Beijing Forestry University
Beijing, China

Fei Qi
College of Environmental Science
and Engineering
Beijing Key Lab for Source Control
Technology of Water Pollution
Beijing Forestry University
Beijing, China

ISBN 978-981-96-4173-4 ISBN 978-981-96-4174-1 (eBook)
https://doi.org/10.1007/978-981-96-4174-1

Jointly published with China Architecture Publishing & Media Co., Ltd.
The print edition is not for sale in China (Mainland). Customers from China (Mainland) please order the print book from: China Architecture Publishing & Media Co., Ltd.

This Springer imprint is published by the registered company Springer Nature Singapore Pte Ltd.
The registered company address is: 152 Beach Road, #21-01/04 Gateway East, Singapore 189721, Singapore

Preface

In recent years, China's rapid urbanization has led to dramatic changes in land use, with a significant increase in impermeable surfaces, which has notably reduced the permeability of urban land. As a result, the infiltration of urban rainfall runoff and the capacity for surface water storage have decreased substantially. This has caused a rise in runoff coefficients and a greater proportion of rainwater runoff in urban areas. This issue is further exacerbated by global climate change and the increasing frequency of extreme weather events, leading to higher peak flow rates and total flood volumes, which in turn heightens the risk of flooding. Additionally, the conflict between rapid urbanization and the need for effective urban management has become more pronounced. The rapid population growth, expansion of various industries, and busy transportation networks have led to an accumulation of contaminants on urban surfaces, placing additional strain on drainage systems. When rainfall flushes these contaminants into the already overburdened drainage system, it results in the contamination of initial rainwater and combined sewer overflows, contributing to water pollution, foul odors, and the deterioration of urban aquatic ecosystems.

Urban rainfall runoff pollution in China contributes significantly to overall urban water pollution, accounting for about 20% of the total pollution load. In major cities, this number can exceed 35%, making it a key source of urban water contamination. However, the scientific research and engineering practices aimed at mitigating urban rainfall runoff pollution in China have developed more recently and have faced a range of challenges. These efforts have progressed through various technological phases, including the construction of combined sewer systems, the development of separated sewer systems (rainwater and sewage separation), the elimination of combined sewer overflows, initial rainwater purification, and the implementation of sponge city initiatives.

With the support of the "the National Water Pollution Control and Treatment Science and Technology Major Project of China" (hereinafter referred to as "Water Project"), significant progress has been made in addressing urban runoff pollution in China since the Eleventh Five-Year Plan. This monograph, based on the research achievement of "Water Project" on urban rainfall runoff pollution control technology, alongside practical experience gained in China, integrated various technologies to develop a comprehensive approach to urban rainfall runoff pollution management. This approach includes several key areas: urban rainfall runoff pollution diagnosis and assessment, pollution source reduction, pollution process control, and pollution back-end treatment. The authors summarized the technology packages and application models of urban rainfall runoff pollution control, highlighted typical engineering cases, and evaluated the application performance of key technologies. The goal is to provide a technical guidance and scientific support for all stages of urban rainfall runoff control and urban water environment quality improvement, including technology selection, scheme design, project implementation, operation and maintenance, and other critical aspects.

This book comprises 10 chapters. Chapter 1 provides an overview of urban rainfall runoff pollution, including its definition, sources and types.It analyzes the current situation and reason of urban rainfall runoff pollution in China and introduces the development trends and main technical approaches for controlling this pollution. Chapter 2 presents methods for assessing the pollution load of urban rainfall runoff. Chapters 3–5 summarize the entire chain of urban rainfall runoff pollution control technologies, including source reduction, process control, and back-end treatment. Chapter 6 outlines the technology packages, technical models, and integrated approaches for urban rainfall runoff pollution control. Chapter 7 highlights the typical engineering cases of urban rainfall runoff pollution control in China. Finally, Chaps. 8–10 evaluate the typical projects, focusing on the application performance of key technologies and the outcomes of projects of various scales.

This book is edited by Sun Dezhi and Qi Fei, the editors also include: Wang Zhenbei, Wang Zheng, Chen Miao, Liu Longyan, Huang Chong, Liang Jiahui, Yang·Beibei, Zhou·Yang, Lin Bingquan, Zeng Lingwu, Fan Lu, Lai Junda, An Xinqi, Zhao Chen, Liu Yuxuan, Gao Yahong, Fang Zheng, Chen Yuan and Liu Yuxin.

This book has also been strongly supported by the urban theme expert group of "Water Project" and Ministry of Housing and Urban-Rural Development of China, for which I would like to express my heartfelt thanks.

Due to the editor's level of expertise and time constraints during the writing process, there may be some inevitable omissions and inaccuracies. I welcome any comments and feedback from readers.

Beijing, China
June 2024

Dezhi Sun
Fei Qi

Contents

Abbreviations

AHP	Analytic Hierarchy Process
ARM	Agricultural Runoff Management
BASINS	Better Assessment Science Integrating Point and Nonpoint Sources
BMPs	Best Management Practices
BOD_5	Biochemical Oxygen Demand
COD	Chemical Oxygen Demand
CREAMS	Chemicals, Runoff, and Erosion from Agricultural Management Systems
CSOs	Combined Sewer Overflows
D	Diameter
DDT	Dichlorodiphenyltrichloroethane
DEA	Data Envelopment Analysis
DO	Dissolved Oxygen
EMC	Event Mean Concentration
EPIC	Erosion-Productivity Impact Calculator
GLEAMS	Groundwater Loading Effects Model on Agricultural Ecosystems
HCB	Hexachlorobenzene
HCH	Hexachlorocyclohexanes
HRUs	Hydrological Response Units
HSP	Hydrological Simulation Program
HSPF	Hydrological Simulation Program-Fortran
LID	Low Impact Development
LIUDD	Low Impact Urban Design and Development
LUCC	Land Use/Cover Changes
MFF	Mass First Flush
N	Nitrogen
n_0	Interception ratio
NH_4^+-N	Ammonia nitrogen
NO_2-N	Nitrite nitrogen
NO_3-N	Nitrate nitrogen
NPDES	National Pollutant Discharge Elimination System

NPS	Nonpoint Source
NSE	Nash-Sutcliffe Efficiency
OCPs	Organochlorine Pesticides
P	Phosphorus
PA	Porous asphalt
PAC	Poly Aluminum Chloride
PAHs	Polycyclic Aromatic Hydrocarbons
PAM	Polyacrylamide
PC	Permeable Concrete
PCBs	Polychlorinated Biphenyls
PICP	Permeable Interlocking Concrete
POPs	Persistent Organic Pollutants
ROTO	Routing Output to Outlet Model
SMC	Site Mean Concentration
SS	Suspended Solids
SUDS	Sustainable Urban Drainage Systems
SUSTAIN	Stormwater Treatment and Analysis Integration
SWAT	Soil and Water Assessment Tool
SWC	Storm Water Calculator
SWMM	Storm Water Management Model
TBM	Tunnel Boring Machines
TN	Total Nitrogen
TOPSIS	Technique for Order Preference by Similarity to Ideal Solution
TP	Total Phosphorus
WSUD	Water Sensitive Urban Design

Chapter 1
Introduction

1.1 Development Background of Urban Rainfall Runoff Pollution Control Technology

1.1.1 Definition of Urban Rainfall Runoff

Urban runoff pollution, also known as nonpoint source pollution (NPS pollution), can be categorized broadly and narrowly.

Broadly, it refers to urban surface and runoff pollution caused by urban rainfall runoff, atmospheric deposition, flushing of pipe sediments by rainwater, combined sewer overflow, and urban soil erosion.

Narrowly, it specifically refers to the contamination of urban water bodies caused by various pollutants entering the rainwater runoff through the washing and scouring action of urban rainfall on urban atmosphere and underlying surfaces (such as residential areas, commercial areas, parking lots, and roads), resulting in varying degrees of pollution and ecological damage to the water bodies.

In recent years, the impacts of point source pollution on urban water bodies have been effectively controlled. However, the overall water quality has not improved due to the increasingly prominent issues of urban runoff pollution. With the accelerated process of urbanization, there have been significant changes in land use types, leading to an increase in impermeable surfaces. This has greatly reduced the infiltration and retention capacity of urban rainfall, accelerating surface runoff and causing significant impacts on the urban water environment. Urban runoff pollution, carried by urban rainfall runoff, has become a crucial factor limiting the sustainable development of urban areas. According to statistics, the pollution load generated by urban rainfall runoff pollution each year is equivalent to the pollution load discharged by wastewater treatment plants. In terms of pollutant concentration, many pollutants in urban runoff pollution are comparable to or even higher than untreated urban sewage. The harmfulness of urban runoff pollution to the urban water environment

D. Sun and F. Qi, *Urban Rainfall Runoff Pollution Control Technology and Application*,
https://doi.org/10.1007/978-981-96-4174-1_1

has gained widespread attention from many countries worldwide and has become a forefront and hot topic in global environmental research. The US Environmental Protection Agency (EPA) has listed urban runoff pollution as the third largest source of pollution in rivers and lakes. Countries like the United States and European countries have conducted extensive research on urban runoff pollution since the 1970s, while research in China started in the 1980s. In the twenty-first century, China has gradually accelerated its research on urban runoff pollution control, especially after 2015. Significant progress has been made in technological research and engineering applications in this field.

1.1.2 Sources, Types, and Loads of Urban Rainfall Runoff Pollution

a. **Sources of Urban Rainfall Runoff Pollution**

Urban rainfall runoff pollution comes from multiple and complex sources, primarily from three aspects: atmospheric wet deposition (precipitation), atmospheric dry deposition (surface wash-off), and stormwater drainage systems.

i. Atmospheric Wet Deposition

Pollutants in the air are deposited to the ground with urban precipitation, known as atmospheric wet deposition. The pollutant load in this part mainly depends on the local air pollution conditions. The more severe the air pollution, the higher the concentration of pollutants brought by urban rainfall.

ii. Atmospheric Dry Deposition

Atmospheric dry deposition refers to the phenomenon where aerosols and other pollutants in the atmosphere on dry days are adsorbed by plants or deposited to the ground by gravity. During dry days, these pollutants accumulate on urban surfaces (such as residential areas, commercial areas, parking lots, and roads, etc.); during rainfall, the deposited pollutants are washed and flushed on the ground, forming runoff pollution.

In areas with inadequate drainage systems, the surface-deposited pollutants can be washed into urban waterways as surface runoff. In these areas, poor drainage leads to surface water accumulation, which immerses road pollutants, domestic waste, and organic waste from plants and animals. This immersion dissolves more pollutant components, increasing the pollutant load in stormwater. Especially when organic matter such as domestic waste is soaked in rainwater for a long time, it significantly increases COD, TN, and TP in the runoff.

iii. Stormwater Drainage Systems

The impact of stormwater runoff on water quality through stormwater drainage systems can be divided into three parts. One part is the stormwater and wastewater overflow from combined sewer systems. Another part is the entry of garbage and wastewater into stormwater in separate sewer systems. The third part is the sediment in the pipes being washed out by stormwater runoff and entering the water body, causing impact pollution.

In combined sewer systems, domestic sewage (possibly including a small amount of industrial wastewater) and urban stormwater are conveyed together in the same pipe system to wastewater treatment plants. When the total flow of stormwater and wastewater exceeds the conveying and treatment capacity of the pipe system, sewer overflow occurs. Without storage facilities, the overflow of mixed stormwater and wastewater is discharged directly into receiving water bodies.

Another important source of rainfall runoff pollution is the mixing of garbage and wastewater in stormwater in separate sewer systems. These waste and wastewater will directly discharge into receiving water bodies along with the stormwater, causing runoff pollution.

The third source of rainfall runoff pollution is pipe sediments. Whether it is a combined sewer system or a separate sewer system, pipe sediments are commonly present. When the flow velocity of stormwater runoff is high, pipe sediments are resuspended and discharged into waterways with the runoff. This is often the main reason for the sudden change in water quality in receiving water bodies during rainfall, and the occurrence of unpleasant odors during heavy rain.

b. **Types of Urban Rainfall Runoff Pollution**

The pollutants carried by urban stormwater runoff can be classified into three categories: physical, chemical, and biological. These pollutants have complex sources and are widely distributed. They mainly include corrosive substances from construction materials, sediments and sludge from construction sites, sand and debris from road surfaces, tire abrasion particles, oil leaks from vehicles, heavy metals in vehicle emissions, atmospheric dry and wet deposition, organic waste from animals and plants, pesticides sprayed in urban parks, and other dispersed industrial and urban pollution sources. These pollutants accumulate in streets, drains, and impervious surfaces in various forms. Under heavy rainfall, they enter receiving water bodies through different pathways and have significant negative impacts on the water ecosystem quality.

Commonly studied pollutants in urban stormwater runoff are suspended particles, organic matter, nutrients, heavy metals, pesticides, and emerging pollutants.

i. Suspended Particles

The most basic pollutants in urban stormwater runoff are suspended particles with particle sizes ranging from 1.0 (colloidal) to 10,000 μm (sand grains). These particles primarily come from soil particles, atmospheric deposition, tire and road wear particles, and road deicing agents. These particles have a large specific surface area and can adsorb various pollutants, accounting for over 80% of the pollutants in urban stormwater runoff. Additionally, some particles themselves contain pollutants. For example, particles generated from tire wear contain heavy metals, and it is estimated that 15–60% of Zn in urban stormwater runoff is derived from tire wear particles.

During heavy rainfall, a large amount of suspended particles is carried in stormwater runoff. When the flow velocity of stormwater runoff is low, these particles settle and become surface runoff deposits. These pollutant-carrying surface runoff deposits have different properties from soil and riverbed sediments and pose potential environmental risks as they serve as food sources for benthic organisms.

ii. Organic Matter

Organic matter in urban stormwater runoff primarily originates from soil, plant debris, bird excrement, urban waste, and petroleum byproducts. It can cause an increase in water color, COD, and BOD_5, while decreasing the dissolved oxygen content. Organic matter in urban stormwater runoff can also act as a carrier for various pollutants, allowing them to combine and coexist.

iii. Nutrients

Nutrients such as N and P in urban stormwater runoff mainly come from atmospheric deposition, urban green spaces, and waste disposal. The entry of nutrients into water bodies can stimulate rapid proliferation of algae and other zooplankton, leading to a decrease in dissolved oxygen and causing significant mortality among aquatic organisms such as fish. It results in eutrophication of the water body. Studies have shown that the annual output of TN and TP from urban surface runoff is 10 kg/ha and 1.0 kg/ha, respectively. Research has reported that urban road runoff contributes as high as 83.1% of TP to the receiving water bodies in Beijing. Urban stormwater runoff has become a significant contributor to nutrient pollution in urban rivers and lakes.

iv. Heavy Metals

Heavy metals are the most common pollutants in urban stormwater runoff. Among them, Cu, Zn, and Pb are the most prevalent heavy metals. These metals have toxic effects on aquatic organisms, fish, and birds, and can bioaccumulate in the food chain, eventually leading to neurological disorders, cancer, and the occurrence of birth defects in humans. Among the numerous heavy metal pollutants in urban stormwater runoff, Cu, Zn, Pb, and Mn mainly come from vehicular emissions. These metals exist in stormwater runoff in soluble salt form, chelated form, or in association with particulate matter, with varying proportions in different particle size fractions. For example, Cu and Zn exist in dissolved form in stormwater runoff in quantities ranging from 20 to 100%, and in association with particulate matter in quantities above 70%. Pb and Cr are only present on particles with a size greater than 5 μm. Fe, Al, and Si predominantly exist on larger particles, with more than 70% present, while a small portion is associated with particles ranging from 0.45 to 5 μm in size [1].

v. Pesticides

Pesticides are one of the commonly concerned pollutants in surface runoff near urban green spaces and residential areas. It has been reported that neonicotinoid pesticides are gradually replacing organophosphate pesticides as the main pesticide pollutants in urban stormwater runoff [2]. Various neonicotinoid pesticides have been detected in surface runoff surrounding residential areas in the capital city of Sacramento, California, USA. Among them, the concentration of imidacloprid in surface runoff is 73 mg/L, and the concentration in suspended particulate matter is 1.211 mg/kg. The pollutant load of this substance in stormwater runoff from the area, over a duration of 3.0 h, is equivalent to the pollutant load from six months of agricultural irrigation runoff. Thus, the pesticide pollution load in surface runoff is significant. Along highways, herbicides are frequently applied to suppress weed growth. Investigation and research on California highways have detected all the herbicides used in surface runoff, including aminosulfonyl chloride, diuron, glyphosate, picloram, and 2,4-D. Among them, the highest concentrations of glyphosate and picloram reach 10 mg/L, and the highest concentration of aminosulfonyl chloride reaches 200 mg/L. Liu Zhigang et al. conducted a study on the pollution characteristics of organochlorine pesticides (OCPs) in surface runoff from urban roads in Beijing. The geometric mean concentrations of hexachlorocyclohexanes (HCH), dichlorodiphenyltrichloroethane (DDT), and hexachlorobenzene (HCB) were found to be 25.1 ng/L, 10.6 ng/L, and 5.91 ng/L, respectively [2].

vi. Emerging Contaminants

Emerging contaminants refer to toxic and harmful chemicals that possess characteristics such as biotoxicity, environmental persistence, and bioaccumulation. These chemicals pose significant risks to the ecological environment or human health but have not yet been included in environmental management or have insufficient management measures in place. Currently, widely recognized emerging contaminants at the international level include persistent organic pollutants, endocrine-disrupting chemicals, antibiotics, and microplastics.

(i) *Persistent Organic Pollutants*

Persistent organic pollutants (POPs) are characterized by their long-term persistence and semi-volatility, allowing them to be volatilized from water and soil into the air, attaching to airborne particles and entering surface runoff with rainfall. In rainfall runoff, the main POPs are polycyclic aromatic hydrocarbons (PAHs) and polychlorinated biphenyls (PCBs).

PAHs are mainly generated from asphalt road wear and gasoline combustion and are prone to exposure in surface runoff. Reports have indicated that the content of PAHs in surface runoff sediments and particulate matter ranges from 0.09 to 80 mg/kg. Besides the influence of water body topography, hydrological conditions, and climate, the content and size distribution of particulate matter in the water, as well as the molecular weight of PAHs, are significant factors affecting the presence and migration of PAHs in surface runoff. Research on the pollution of PAHs in stormwater runoff in Zhejiang New City revealed that 40% of street dust particles had a diameter below 250 nm, with up to 55% of PAHs being attached to their surfaces. Due to the larger surface area of fine particles, they are more prone to PAH deposition and adhesion. Therefore, the content of PAHs in street dust particles increases with decreasing particle size. Additionally, high molecular weight PAHs tend to bind with particulate matter, while low molecular weight PAHs have weaker interactions with particles and are more easily transported. Wang et al. studied the pollution characteristics and flushing patterns of PAHs in rainfall runoff from a typical overpass bridge in Beijing, and the results showed that the mean concentrations of $\sum_{16}$PAHs and $\sum_{6}$PAHs in urban rainfall runoff were 1923.7 ng/L and 460.2 ng/L, respectively [3].

PCBs primarily originate from urban industrial production, the combustion of fuel used in electrical equipment, and fossil fuels. Research has reported the detection of PCBs in sediment samples from 68 rainwater retention wells in Bergen, Norway, with concentration ranges from 0.0004 to 0.704 mg/kg. Among them, PCB concentrations at 14 sampling points exceeded 0.1 mg/kg. The types and concentration distribution of PCBs in this rainwater retention well sediments corresponded to the distribution of PCBs adhering to various buildings in the surface area, confirming that the pollution of PCBs in the region mainly originates from atmospheric deposition. Han et al. studied the concentrations of 14 PCBs in rainfall runoff samples from different urban functional areas, including residential roofs, traffic intersections, confluences, residential roads, and parking lots. The results showed that the concentrations and composition of particulate PCBs varied slightly but differed among different urban functional areas, with the following sequence: residential roofs > traffic intersections > confluences > parking lots > residential roads (337.9, 306.3, 240.1, 193.2, 172.7 ng/L) [4].

(ii) *Endocrine-Disrupting Chemicals*

Endocrine-disrupting chemicals (EDCs), also known as environmental hormones, are a type of chemical substance that exerts external interference on the endocrine system. They do not directly act as toxic agents to cause abnormal effects on organisms but rather mimic the effects of estrogens on organisms. They exhibit low-dose effects, meaning that even at ng/L levels, they can disrupt the endocrine system of organisms, leading to abnormal effects and even cancer occurrence. There is a wide variety of chemical substances in the environment that have endocrine-disrupting effects, and they have diverse sources. The classification and major substances of EDCs are shown in Table 1.1, these substances exhibit rich chemical structural diversity, with most EDCs demonstrating lipophilicity, hydrophobicity, and chemical stability. They also have long half-lives, making them difficult to degrade and remove.

Stormwater runoff is one of the significant sources of phenolic EDCs in urban water environments. Bisphenol A (BPA) is a raw material used in the synthesis of phenol formaldehyde resins, which are widely installed in the outer walls of urban buildings, particularly in large public structures. Ethoxylated alkylphenols (APE) have been used as detergents in cleaning private cars, building walls, glassware, and other items. Therefore, phenolic compounds such as BPA and alkylphenols may be present in stormwater runoff. Zhao et al. investigated the load of three types of EDCs, namely BPA, NP, and 4-tert-octylphenol (4-t-OP), in stormwater runoff and point source pollution in Huicheng District, Huizhou City, within the Dongjiang River Basin. The results showed that the maximum concentrations of BPA in urban stormwater runoff samples reached 5873 ng/L and 2397 ng/L at two sampling points, respectively, averaging several tens of times higher than the concentrations of NP and

Table 1.1 Classification of endocrine disrupting chemicals (EDCs) [5]

Categories	Endocrine disrupting chemicals	
Natural endocrine disrupting chemicals	Steroid hormone	Estrone (E1), estradiol (E2), estriol (E3)
	NON-steroid hormone	Phytohormone
Synthetic endocrine disrupting chemicals	Steroid hormone	Ethinylestradiol (EE)
	Polyhalogenated compounds	Polybrominated biphenyls, polybrominated diphenyl ethers (PBDE), perfluorooctane, alkylsulfonic acids, perfluorooctanoic acid
	Phenol	Bisphenol A (BPA), nonyl phenol, octyl phenol
	Phthalates	Bis(2-ethylhexyl) phthalate, di-iso-nonyl phthalate (DINP), phthalates (PAEs), etc.
	Pesticides	Organochloride (dichlorodiphenyltrichloroethane, lindane, etc.), organophosphates (chlorpyrifos, etc.), carbamate (carbaryl, etc.)
	Pharmaceutical and personal care products (PPCPs)	Anti-inflammatory drugs, antihistamine, antipsychotic, hormone drugs, etc.

4-t-OP. The load of phenolic EDCs in urban stormwater runoff was 3–62 times higher than that in wastewater treatment plant effluent, indicating that stormwater runoff is a significant source of phenolic EDCs in receiving water bodies [6, 7]. Fan Chongyang et al. analyzed the seasonal distribution characteristics of PAHs, PAEs, BPA, E1, E2, and EE2, major endocrine disruptors, in the Shiyixi River Basin in Shenzhen City. The results indicated that urban stormwater runoff pollution influenced the seasonal distribution of these pollutants, with the concentrations of major endocrine disruptors during the rainy season being 0.15–7.9 times higher than during the dry season, particularly for BPA and most PAHs [8].

(iii) *Antibiotics*

Different environmental conditions, population densities, and human activities in various functional areas of urban settings result in variations in the types of antibiotics generated and accumulated. Luo Lichan analyzed the levels of dissolved antibiotics in road dust and adsorbed antibiotics on particulate matter in different functional areas of Shenzhen City from the perspective of antibiotic pollutants. The study identified the typical antibiotic pollutants in urban stormwater runoff contamination [9].

Table 1.2 presents the levels of dissolved antibiotics in five functional areas of the city. It can be observed that two types of antibiotics, namely norfloxacin and Anhydroerythromycin, were detected in dissolved samples from different functional areas. In the commercial area, norfloxacin was detected with a concentration of 5.2 ng/m^2. In the hospital surroundings, norfloxacin and Anhydroerythromycin were detected with concentrations of 4.7 ng/m^2 and 1.9 ng/m^2, respectively. No dissolved antibiotics were detected in residential, industrial, and landfill areas.

Table 1.3 displays the concentrations of antibiotics adsorbed onto particulate matter in five different functional zones within urban areas. The results reveal the presence of eight antibiotics within the particulate matter samples, namely lincomycin, norfloxacin, ofloxacin, ciprofloxacin, anhydroerythromycin, erythromycin, clarithromycin, and cefradine. Of these antibiotics, six were exclusively detected in the particulate matter samples as opposed to the dissolved state samples, specifically lincomycin, ofloxacin, ciprofloxacin, erythromycin, clarithromycin, and cefradine. Concentrations of the detected antibiotics ranged from ND to 882.2 ng/g.

Three types of antibiotics adsorbed on particulate matter were detected in the commercial area, with the antibiotic concentrations in the following order: norfloxacin > ofloxacin > anhydroerythromycin. In the residential area, seven types of antibiotics adsorbed on particulate matter were detected, with the antibiotic concentrations in the following order: erythromycin > ofloxacin > norfloxacin > clarithromycin > ciprofloxacin > anhydroerythromycin > cefradine. In the industrial area, only one type of antibiotic, anhydroerythromycin, adsorbed on particulate

Table 1.2 Dissolved antibiotic concentrations (ng/m^2) in five functional zones of urban areas

Chemical compound	Commercial area	Residential area	Industry area	Landfill area	Hospital surroundings
Norfloxacin	5.2	ND	ND	ND	4.7
Anhydroerythromycin	ND	ND	ND	ND	1.9

Note ND indicates non-detection

Table 1.3 Concentrations of antibiotic adsorbed on particulate matter in five functional zones of the urban area (ng/m^2)

Chemical compound	Commercial area	Residential area	Industry area	Landfill area	Hospital surroundings
Lincomycin	ND	ND	ND	188.2	ND
Norfloxacin	281.5	611.7	ND	429.6	94.1
Ofloxacin	76.6	735.4	ND	50.2	148.9
Ciprofloxacin	ND	294.9	ND	ND	57.0
Anhydroerythromycin	14.4	230.4	118.6	ND	25.3
Erythromycin	ND	882.2	ND	121.6	39.3
Clarithromycin	ND	350.4	ND	ND	18.7
Cefradine	ND	39.6	ND	2.3	ND

Note ND indicates non-detection

matter was detected. In the landfill area, five types of antibiotics adsorbed on particulate matter were detected, with the antibiotic concentrations in the following order: norfloxacin > lincomycin > erythromycin > ofloxacin > cefradine. In the hospital surroundings, six types of antibiotics adsorbed on particulate matter were detected, with the antibiotic concentrations in the following order: ofloxacin > norfloxacin > ciprofloxacin > erythromycin > anhydroerythromycin > clarithromycin.

(iv) *Microplastics*

Due to their tiny size, microplastics can easily be absorbed by aquatic organisms. Apart from their inherent toxicity, microplastics can also adsorb other toxic substances and transfer them together, leading to continuous accumulation of these toxic substances through the food chain and posing a threat to the survival environment of other species.

Wang investigated the occurrence and release patterns of microplastics in areas including plastic tracks, artificial turf, tennis courts, basketball courts, and badminton courts. The study revealed that microplastics were only detected in the runoff from artificial turf and plastic tracks, with concentrations of 9,820 particles/L and 6,910 particles/L, respectively [8, 10].

c. **Pollutant Load in Urban Rainfall Runoff**

Through literature research, water quality data from some countries or cities worldwide were collected and summarized, as shown in Table 1.4. In urban rainfall runoff, the United States has a wide range of pollutants, while Germany's urban rainfall runoff is limited to conventional pollutants, including COD, NH_3–N, TN, and TP, with low frequencies and quantities of heavy metal detections.

Similar to the United States, the types of pollutants in urban runoff from Beijing are diverse, with slightly higher concentrations. Apart from Beijing, cities in China such as Shanghai, Nanjing, and Kunming have fewer types of pollutants in urban runoff, all of which are conventional pollutants including SS, COD, TN, and TP. Except for TN, which has higher concentrations compared to the United States and Germany, the concentrations of other pollutants are lower than those in the United States and Germany.

Table 1.4 Water quality of urban stormwater runoff in domestic and foreign cities (mg/L)

Urban	SS	COD	NH_3–N	TN	TP	Pb	Zn	Cu
America	100	65	–	1.5	0.33	0.14	0.16	0.034
Germany	–	84.5	0.75	2.1	0.35	–	–	–
Beijing	734	582	2.4	11.2	1.74	0.1	1.23	–
Shanghai	76.9	42.6	–	4.8	0.14	–	–	–
Macau	–	2–49	–	2.1–2.7	–	–	–	–
Nanjing	47–77	40–51	–	5.8–7.4	0.19–0.27	–	–	–
Kunming	27.2	4.92	–	4.86	0.25	–	–	–

Table 1.5 Concentrations of pollutants in different types of stormwater runoff (mg/L)

Types of stormwater and sewage	SS	COD	TN	TP
Natural stormwater	<1.0	9–16	–	0.02–0.15
Runoff stormwater	67–101	40–73	0.4–1.0	0.7–1.7
Overflow sewage	150–400	260–480	3.0–24	1.0–10
Domestic sewage	100–350	260–900	20–85	4.0–15
Effluent from wastewater treatment plants	<5–30	–	15–25	<1–5

Through literature research, it can be observed that there are significant differences in the pollution load of natural rainfall, runoff rainwater, overflow sewage, domestic sewage, and wastewater treatment plant effluent, as shown in Table 1.5.

From Table 1.5, it can be concluded that the pollution load of runoff rainwater is significantly higher than that of natural rainfall. This is because natural rainfall only dissolves a small amount of air pollutants, while runoff rainwater not only dissolves a small amount of air pollutants but also washes away and carries a large amount of surface pollutants. A comparison between overflow sewage and domestic sewage reveals that the pollution load of overflow sewage is higher than that of runoff rainwater but lower than that of domestic sewage. In combined sewer systems, large amounts of runoff rainwater flow into the pipes during rainy days, diluting the domestic sewage and reducing some of the pollution load. However, once the pipe network becomes clogged, the pollution load significantly increases. Obviously, the SS load of runoff rainwater and overflow sewage far exceeds the SS load of wastewater treatment plant effluent. Untreated runoff rainwater and overflow sewage have a severe impact on the water environment governance of the receiving water body and are one of the important limiting factors for improving urban water environment. If the control of runoff rainwater and overflow sewage is not implemented, they will inevitably cause serious harm to the water bodies.

1.2 Status and Causes of Rainfall Runoff Pollution in China's Cities and Towns

1.2.1 Pollution Load of Rainfall Runoff from Cities and Towns in China

a. **Urban Rainfall Runoff Pollution Characteristic Analysis**

Based on a comprehensive review of nearly 300 relevant literature on urban rainfall runoff pollution, 100 articles were selected that provided monitoring data on rainfall runoff pollution loads. The monitoring data were extracted and organized according to the literature [11–110], as shown in Fig. 1.1.

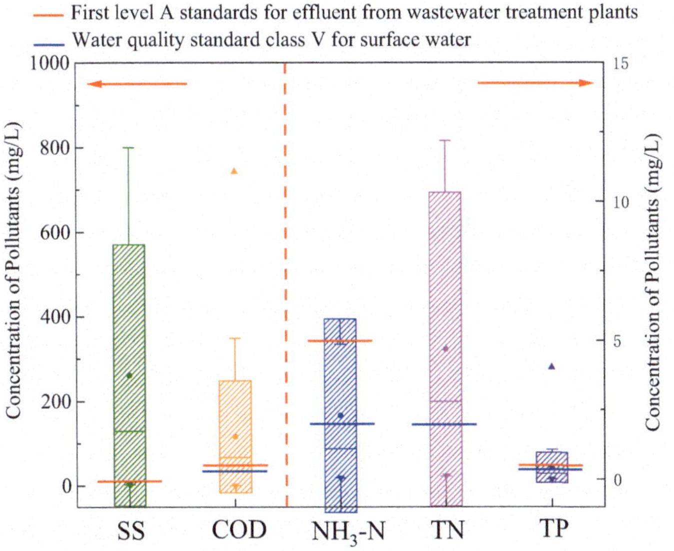

Fig. 1.1 Surface runoff pollution characteristics

As illustrated in Fig. 1.1, the average concentrations of SS, COD, NH_3–N, TN, and TP in the rainfall runoff were 261.83 mg/L, 116.71 mg/L, 2.30 mg/L, 4.69 mg/L, and 0.42 mg/L, respectively. Since specific standards for rainfall runoff water quality have not been enacted in our country yet, the evaluation of the impact of rainfall runoff on natural water bodies in this section was based on the Class V standards in "*Environmental Quality Standards for Surface Water*" (GB 3838-2002). However, since this standard does not provide a specific limit for SS, the evaluation of SS was based on the first level A standard in "*Discharge Standard of Pollutants for Municipal Wastewater Treatment Plant*" (GB 18918-2002). The specific evaluation criteria can be found in Table 1.6.

Table 1.6 Water quality evaluation standards (mg/L)

Reference standards	SS	COD	NH_3–N	TN	TP
The class V standards in "Environmental Quality Standards for Surface Water" (GB 3838-2002)	–	≤40	≤2	≤2	≤0.4
The first level A standard of "Discharge Standard of Pollutants for Municipal Wastewater Treatment Plant" (GB 18918-2002)	≤10	≤50	≤5	≤15	≤0.5

Compared with the above standards, the average concentration of SS in stormwater runoff is 26.18 times the concentration limit of the first level A standard for wastewater treatment plant effluent in China. Additionally, the average concentrations of COD, NH_3–N, TN, and TP in the rainfall runoff were 2.91 times, 1.15 times, 2.35 times, and 1.05 times higher than the concentration limits of Class V standard for surface water in China, respectively. These results indicate that the current rainfall runoff pollution in our country is severe. If left untreated and directly discharged into receiving water bodies, it could lead to eutrophication and the occurrence of water body odors.

b. **Characteristics of Combined Sewer Overflow Pollution**

Based on a review of nearly 200 relevant literature on combined sewer overflow (CSO) pollution, 19 articles were selected that provided monitoring data on CSO pollution loads. The monitoring data were extracted and organized according to the literature [111–129], as shown in Fig. 1.2.

As illustrated in Fig. 1.2, the average concentrations of SS, COD, NH_3–N, TN, and TP in CSO were 463.56 mg/L, 309.98 mg/L, 15.65 mg/L, 20.65 mg/L, and 2.76 mg/L, respectively. Comparing SS with the Class A standard for "*Discharge Standard of Pollutants for Municipal Wastewater Treatment Plant*" (GB 18918-2002), the average concentration of SS in CSO was 46.36 times higher than the concentration limit of the Class A standard. Comparing COD, NH_3–N, TN, and TP with the Class V standard for "*Environmental Quality Standards for Surface Water*" (GB 3838-2002), the average concentrations of COD, NH_3–N, TN, and TP in CSO were 7.75 times, 7.83 times, 10.33 times, and 6.90 times higher than the concentration limits of the Class V standard, respectively. It is evident that the degree of CSO pollution is higher than that of stormwater runoff pollution, making it a significant factor influencing the water environment quality in urban areas in our country.

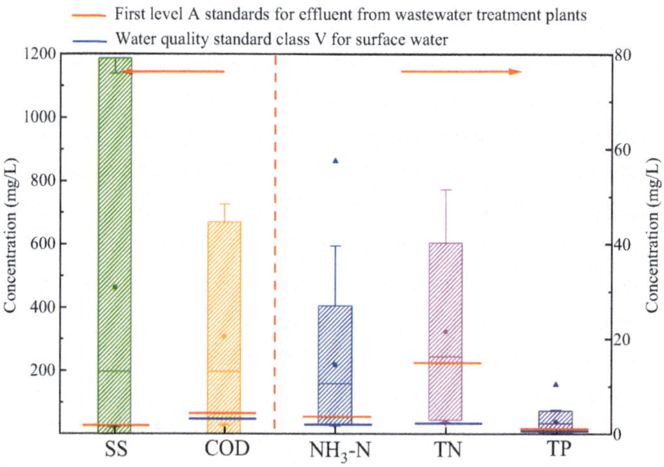

Fig. 1.2 Characteristics of overflow pollution in a combined system

1.2.2 Analysis of the Causes of Urban Rainfall Runoff Pollution

a. **Influence of Rainfall Type on Runoff Pollution Characteristics**

According to the classification standards of meteorological departments, rainfall types are categorized as light rain, moderate rain, heavy rain, rainstorm, heavy rainstorm, and super heavy rainstorm based on the amount of precipitation within 24 h. Due to the limited number of available samples for rainstorm, heavy rainstorm, and super heavy rainstorm extracted from the literature, this section focuses only on the study of light rain, moderate rain, and heavy rain. The classification method for rainfall types can be seen in Table 1.7.

Through a review of the literature, the impact of different rainfall types on runoff pollution characteristics was analyzed according to the aforementioned classification, as shown in Fig. 1.3. From Fig. 1.3, it can be observed that during heavy rain, the average concentrations of SS, COD, NH_3–N, TN, and TP were 547.07 mg/L, 133.24 mg/L, 1.63 mg/L, 3.22 mg/L, and 0.29 mg/L, respectively. During moderate rain, the average concentrations of SS, COD, NH_3–N, TN, and TP were 162.36 mg/L, 96.21 mg/L, 2.17 mg/L, 4.26 mg/L, and 0.41 mg/L, respectively. During light rain, the average concentrations of SS, COD, NH_3–N, TN, and TP were 108.72 mg/L, 92.69 mg/L, 2.80 mg/L, 4.67 mg/L, and 0.56 mg/L, respectively. Comparatively, the concentrations of SS and COD showed a positive correlation with rainfall intensity, with higher concentrations associated with greater rainfall. On the other hand, the concentrations of NH_3–N, TN, and TP exhibited a negative correlation with rainfall intensity, with higher rainfall resulting in lower concentrations. This can be attributed to the higher accumulated pollutant load of SS and COD, which are more influenced by the wash-off effect of rainfall and less influenced by dilution, compared to the lower accumulated pollutant load of NH_3–N, TN, and TP, which are more easily diluted by rainfall, leading to lower concentrations.

b. **Influence of Urban Areas on the Characteristics of Rainfall Runoff Pollution**

The urban area can be divided into the new urban area and the old urban area according to the construction time, as shown in Fig. 1.4. The influence of urban areas on the characteristics of rainfall runoff pollution is illustrated in Fig. 1.4.

According to Fig. 1.4, the average concentrations of SS, COD, NH_3–N, TN, and TP in rainfall runoff in the old urban area are 387 mg/L, 151 mg/L, 3.5 mg/L, 3.9 mg/L, and 0.4 mg/L respectively. In the new urban area, the average concentrations of

Table 1.7 Classification of rainfall types

Rainfall types	Rainfall characteristics
Light rain	Rainfall of less than 10 mm in 24 h
Moderate rain	Rainfall between 10.0 and 24.9 mm in a 24 h period
Heavy rain	Rainfall between 25.0 and 49.9 mm in a 24 h period

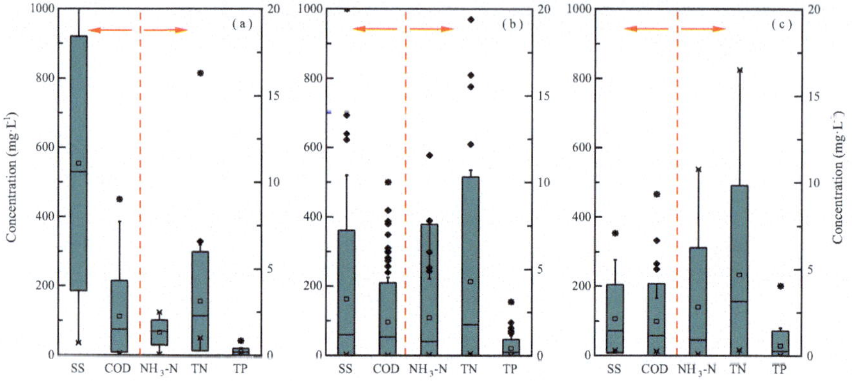

Fig. 1.3 Effect of rainfall type on surface runoff pollution characteristics. **a** Heavy rain; **b** moderate rain; **c** light rain

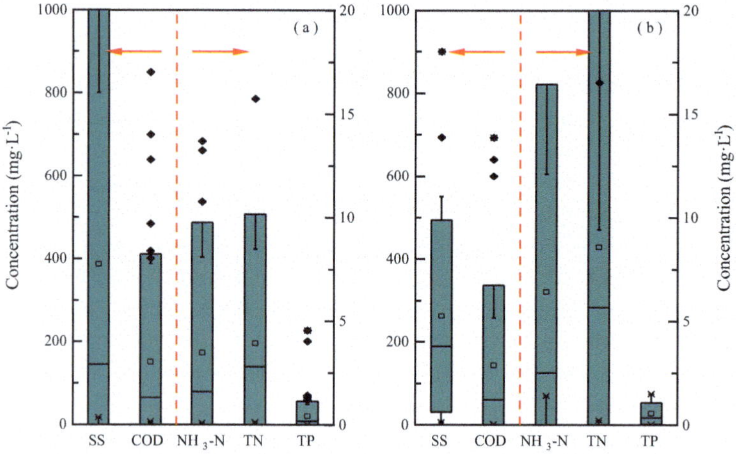

Fig. 1.4 Pollution characteristics of rainfall runoff in urban areas. **a** The old urban area; **b** the new urban area

SS, COD, NH_3–N, TN, and TP in rainfall runoff are 262 mg/L, 143 mg/L, 6.4 mg/L, 8.6 mg/L, and 0.6 mg/L respectively. The average concentrations of SS and COD in rainfall runoff are higher in the old urban area than in the new urban area. This is mainly because the old urban area is usually densely built with low green coverage, mostly consisting of tree pits, street flower beds, etc., lacking large-scale green spaces. The high dust content in the air and the presence of surface deposits lead to higher concentrations of SS in rainfall runoff. In addition, the old urban area has a dense population and thriving commercial activities, resulting in increased pollution such as road garbage and oil, contributing to higher concentrations of COD in rainfall runoff. The dense road network and narrow roads in the old urban area make it inconvenient for cleaning vehicles to enter, resulting in incomplete manual cleaning and the accumulation of particles and organic matter, leading to higher concentrations of SS and COD in rainfall runoff.

The average concentrations of NH_3–N, TN, and TP in rainfall runoff are higher in the new urban area than in the old urban area. This is mainly because the new urban area has convenient transportation and a high volume of traffic, leading to a large amount of ammonia oxidation and deposition from vehicle exhaust. The high green coverage in the new urban area requires fertilization and regular maintenance, resulting in significantly increased concentrations of ammonia and phosphorus in rainfall runoff. In addition, the new urban area is subject to industrial pollution, and atmospheric wet and dry deposition lead to higher pollutant concentrations in rainfall runoff.

c. **Influence of Typical Functional Areas on the Characteristics of Rainfall Runoff Pollution**

In order to comprehensively reflect the characteristics of urban rainfall runoff pollution, representative functional areas were selected as research areas based on the analysis of urban land use and functional zoning. These areas include residential areas, commercial areas, industrial areas, transportation areas, and educational areas, as shown in Fig. 1.5. The influence of typical functional areas on the characteristics of rainfall runoff pollution is analyzed in Fig. 1.5.

According to Fig. 1.5, different functional areas exhibit varying pollutant concentrations in rainfall runoff due to differences in production structure and lifestyle.

The order of SS concentrations in rainfall runoff is as follows: transportation area > commercial area > residential area > industrial area > educational area. This is mainly because the transportation area is the most frequented place for human activities in the city, and in addition to the large amount of waste generated by humans, solid particles generated by vehicle movement and friction between tires and roads are also major sources of SS. Therefore, the transportation area has the highest average concentration of SS in rainfall runoff. In the educational area, there are fewer heavily-polluting industries nearby, the area has a large green coverage, and the roads are generally well-maintained, resulting in a reduction in SS concentration.

Since most COD in rainfall runoff is adsorbed on solid particles, the highest concentrations of COD occur in the transportation area, while the lowest concentrations occur in the residential area. The order of COD concentrations is as follows: transportation area > commercial area > educational area > industrial area > residential area. The transportation area is densely populated by vehicles, and COD mainly comes from the friction between vehicle tires and roads as well as the deposition of particles. Residential areas have fewer vehicles and are regularly cleaned, resulting in relatively lower COD concentrations.

The order of NH_3–N concentrations in rainfall runoff is as follows: transportation area > educational area > residential area > industrial area > commercial area. During rainfall and runoff formation, nitrogen not only comes from atmospheric wet and

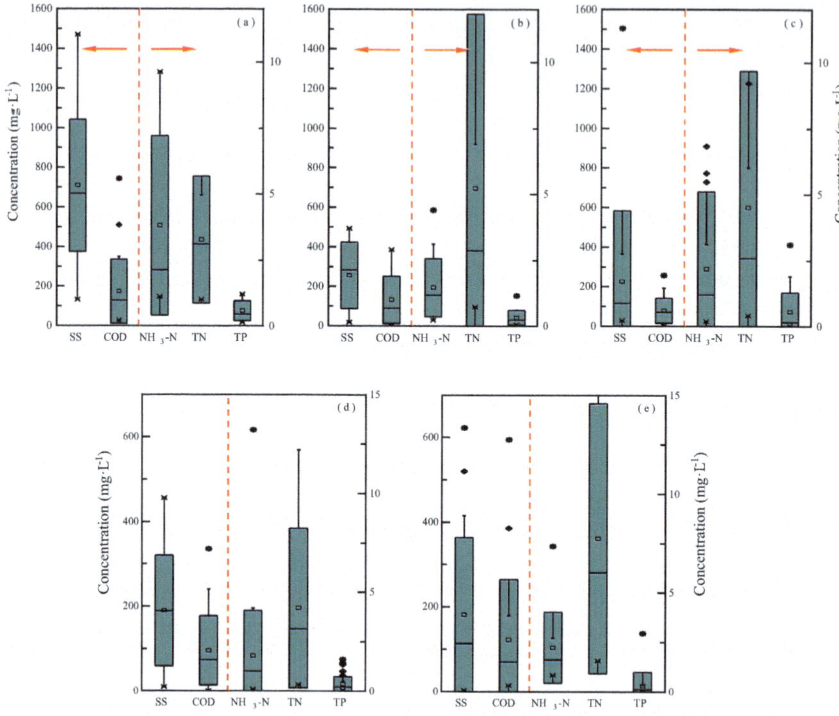

Fig. 1.5 Impact of typical functional areas on pollution characteristics of rainfall runoff. **a** Transportation area; **b** commercial area; **c** residential area; **d** industrial area; **e** educational area

dry deposition, but also from fertilization in green belts and emissions from motor vehicles.

The order of TN concentrations in rainfall runoff is as follows: educational area > commercial area > residential area > industrial area > transportation area. In the educational area, washing and nitrogen-containing detergents used in restaurants, as well as fertilization and maintenance of flowers and plants, are important sources of TN in rainfall runoff.

The TP concentration in rainfall runoff also varies among the five functional zones, in the following order of magnitude: transportation zone > residential zone > T industry zone > commercial zone > educational zone. In addition to atmospheric deposition, phosphorus in the transportation zone also comes from fertilization and maintenance of green belts, which accumulate on road surfaces and are then transported into runoff with rainwater.

Overall, the average concentrations of SS in rainfall runoff exceed the first level A standard of the "*Discharge Standard of Pollutants for Municipal Wastewater Treatment Plant*" GB 18918-2002, and the average concentrations of COD and TN exceed the Class V standard of the "*Environmental Quality Standards for Surface Water*"

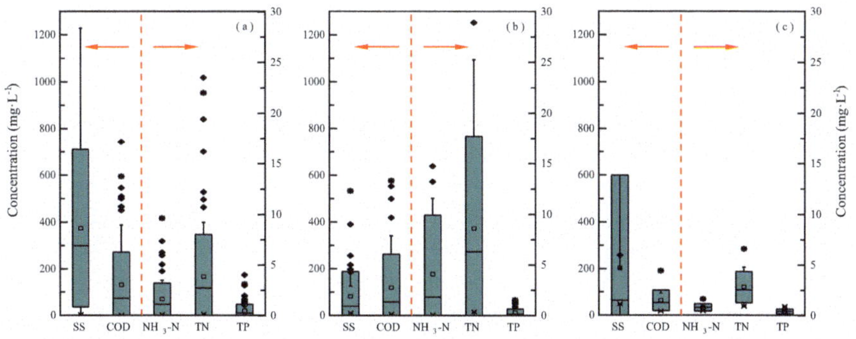

Fig. 1.6 Influence of typical town subsurface on pollution characteristics of rainfall runoff. **a** Road surface; **b** roof; **c** green space

GB 3838-2002. It is evident that SS, COD, and TN are the main pollutants in rainfall runoff from each functional zone. In comparison, the transportation zone and commercial zone exhibit the most severe pollution in rainfall runoff and require focused control measures.

d. **Influence of Typical Urban Underlying Surface on Rainfall Runoff Pollution Characteristics**

Road surface, roof, and green space are the main types of underlying surfaces in urban areas. When rainfall runoff occurs, a large amount of pollutants on the underlying surfaces is washed away by rainwater and enters receiving water bodies, posing a serious threat to the urban ecological environment. Figure 1.6 shows the impact of typical urban underlying surfaces on rainfall runoff pollution characteristics.

From Fig. 1.6, it can be observed that the average concentrations of COD and TN in the rainfall runoff from the three types of underlying surfaces all exceed the Class V standards of the "*Environmental Quality Standards for Surface Water*" (GB 3838–2002); the SS indicator is compared to the first level A standard of "*Discharge Standard of Pollutants for Municipal Wastewater Treatment Plant*" (GB 18918-2002), and the average concentration of SS in the rainfall runoff from various underlying surfaces exceeds this standard. Clearly, SS, COD, and TN are the main pollutants in typical urban underlying surfaces.

In comparison, due to the influence of human activities such as traffic, road runoff pollution is mainly affected by complex pollution sources, including atmospheric dry and wet deposition, tire and road surface friction, vehicle structure wear, exhaust emissions from vehicles, and carried substances from vehicles and pedestrians. Therefore, the SS, COD, and TP indicators are particularly prominent.

The concentrations of NH_3–N and TN are highest in roof runoff, while COD concentration is relatively high. The pollutants in roof runoff are closely related to atmospheric dry and wet deposition and roof materials. On one hand, nitrogen-containing pollutants in the atmosphere form high concentrations of nitrogen pollution in the runoff through rain flushing. On the other hand, there are various types of roofs, including asphalt roofs, tar roofs, concrete roofs, and tile roofs, among which asphalt and tar roofs release nitrogen-containing organic substances during the aging and decomposition process.

Compared to the other two types of underlying surfaces, the concentration of pollutants in green space runoff is relatively low. This is mainly because green spaces can capture a large amount of SS, COD, and NH_3–N in surface runoff through plant interception. In addition, vegetation in green spaces can also absorb some COD and NH_3–N in surface runoff. However, due to the need for regular fertilization and pesticide spraying, the TP concentration in green spaces is higher than that of roofs.

1.3 Development History and Technical Components of Urban Rainfall Runoff Pollution Control

1.3.1 Development History of Urban Rainfall Runoff Pollution Control at Home and Abroad

In the process of urbanization, natural land is being converted into residential land or commercial areas, resulting in more impervious surfaces in urban areas. This leads to increased uncontrolled surface runoff, causing more flooding disasters for urban residents and introducing a certain burden of runoff pollution to urban water environments, thus affecting the quality of urban water environments.

Traditional urban drainage systems have fallen behind in addressing rainfall runoff pollution associated with climate change and urban development. The unreasonable urban planning has posed a serious threat to the environment due to rainfall runoff pollution in urban areas. Various concepts for rainwater resource management have been proposed globally, including the Best Management Practices and Low Impact Development techniques in the United States, sustainable urban drainage systems in the United Kingdom, water-sensitive urban design in Australia, low-impact urban design and development in New Zealand, water resource policy guidelines in Germany, the "ABC Water Program" in Singapore, and the construction of sponge cities in China.

a. **Best Management Practices and Low Impact Development Techniques in the United States**

The United States has developed Best Management Practices (BMPs) for stormwater management, which can be categorized into engineering measures and non-engineering measures. Engineering measures primarily involve the construction of green infrastructure, permeable filtration systems, dry/wet ponds, and rainwater wetlands, aiming to control rainfall runoff pollution. The practical application of engineering measures requires comprehensive consideration of various factors, such as natural geography and climate conditions, in order to make appropriate choices. Non-engineering measures refer to intensified management approaches, such as restricting the use of de-icing salts, regular road cleaning, periodic maintenance of drainage networks, etc. These measures aim to control pollution.

As ongoing exploration and development of BMPs continued, Low Impact Development (LID) emerged as a new concept in stormwater management in the 1990s. The LID research center was established in the United States in 1998, leading to the widespread adoption and gradual improvement of LID techniques.

LID mainly utilizes various rainwater control technologies, including storage, infiltration, evaporation, filtration, purification, and retention, to restore post-development stormwater discharge conditions to a level close to pre-development conditions. The main measures include permeable paving, green roofs, depressed green spaces, rain gardens, vegetated swales, rainwater wetlands, rainwater tanks, and cisterns, as well as non-engineering measures such as alternative road designs, reducing impervious areas, minimizing pollution sources, and increasing public awareness. Extensive research has shown that areas that implement LID measures have significantly improved control of runoff during small rainfall events compared to areas without such measures. Additionally, LID measures also have positive effects on water quality purification. These typical engineering and non-engineering measures have been widely implemented in both domestic and international engineering practices, including the United Kingdom, Australia, New Zealand, and China.

b. **Sustainable Urban Drainage Systems in the United Kingdom**

Sustainable Urban Drainage Systems (SUDS) were introduced in the late twentieth century. They are defined as systems that not only fulfill the basic functions of urban drainage but also adapt well to urban development while maintaining the integrity of urban ecology and the environment. SUDS can be understood as engineering techniques based on the concept of low impact development, aiming to achieve the circulation and efficient utilization of urban water resources without significantly disrupting the natural drainage system.

The United Kingdom was the first to apply the concept of sustainable urban drainage systems in practice. The primary objectives of initially constructing such systems in the UK were to optimize the utilization of rainwater resources, reduce urban flood hazards during heavy rainfall, and maintain a benign water circulation system in urban areas.

c. **Water Sensitive Urban Design in Australia**

Water Sensitive Urban Design (WSUD) in Australia elevates the optimization and management of urban water systems to the fundamental principles of urban planning. It emphasizes reducing negative impacts on water systems through an integrated planning and design system during the initial stages of urban development, guiding sustainable urban development through top-level strategies. This reflects a fundamental improvement in traditional urban development practices.

The definition of WSUD by the International Water Association is a comprehensive design approach that combines urban water cycle management, urban planning and design, and environmental protection while maintaining the natural water cycle.

WSUD emphasizes the sustainability of water resources in urban design, focusing on comprehensive management of the urban water cycle as a whole. It also highlights the preservation of the natural ecological environment, the prohibition of untreated wastewater discharge, and the integration of stormwater runoff and sewage treatment facilities with landscape construction.

d. **Low Impact Urban Design and Development in New Zealand**

Low Impact Urban Design and Development (LIUDD) is a novel urban stormwater management strategy developed in New Zealand based on the concepts of LID and WSUD. LIUDD emphasizes the application of localized vegetation in low impact urban design and its organic integration with urban blue-green spaces, enabling urban greenery to play a crucial role in urban development. Unlike the water-centric approach of WSUD, LIUDD considers water cycle as a branch of the entire urban development and design system, placing greater emphasis on the broader environment beyond the site of low impact development design. It aims to fully utilize any available land while minimizing damage to urban ecosystems, ensuring the efficient utilization of rainwater resources.

e. **German Water Resource Policy Guidelines**

Directive 2000/60/EC (2000), issued by the European Union in 2000, highlights that participating countries should employ various measures in their comprehensive planning to achieve the good status of water bodies, meeting the requirements of the community. These measures include adopting integrated management approaches (preventing diffusion and controlling point sources) to prevent pollution and achieve environmental standards and limits. The directive also prohibits the discharge of pollutants and toxic substances. The water resource policy, known as the German Water Resource Policy Guidelines, effectively guides the construction and management of drainage systems in European cities for at least the next 20 years, if not longer, that have an impact on the entire European continent.

f. **The "ABC Water Programme" in Singapore**

In order to address water environmental issues and improve water resource management, the Singaporean government introduced the "ABC Water Programme" in 2007. This program advocates for the integration of society, environment, and humanities, combining urban water elements such as reservoirs and channels with the urban environment to create a water-friendly and nature-oriented urban space, thus enhancing quality of life. The "ABC" in this program represents "Active," which promotes the integration of human activities with water by creating livable community spaces that encompass living, activities, and leisure; "Beautiful," which focuses on beautifying the landscapes along the rivers; and "Clean," which employs vegetation and other methods to treat pollutants in rainfall runoff, preventing them from entering water bodies and promoting rainwater reuse.

g. **Sponge City Construction in China**

In order to effectively alleviate urban water environmental problems caused by rainfall runoff pollution and improve rainwater utilization, China introduced the concept of sponge cities in December 2013. In 2014, the Ministry of Housing and Urban–Rural Development issued the "*Technical Guidelines for Sponge City Construction—Low Impact Development Rainwater System Construction (Trial).*" This guideline provides a clear definition of sponge cities as, "Cities that, like sponges, break through the traditional concept of urban rainwater management, where drainage is the main approach, and instead use urban infrastructure such as buildings, roads, and green spaces as carriers to absorb, store, infiltrate, and purify rainwater during rainfall events and release the stored rainwater for utilization when needed." The guideline also outlines the content, requirements, and methods of constructing sponge city systems from the aspects of planning, design, engineering construction, and maintenance management, and provides some practical cases from China. With the large-scale construction of sponge cities, domestic scholars have also conducted research and engineering applications related to the control of combined sewer overflow pollution, rainfall runoff regulation, and rainwater reuse.

1.3.2 Composition of Urban Stormwater Runoff Pollution Control Technologies

In the process of urban stormwater runoff pollution control, various technologies have been developed. Based on their characteristics and applicable stages, these technologies can be divided into diagnostic and evaluation techniques, source reduction techniques, process control techniques, and end-of-pipe treatment techniques, as shown in Fig. 1.7.

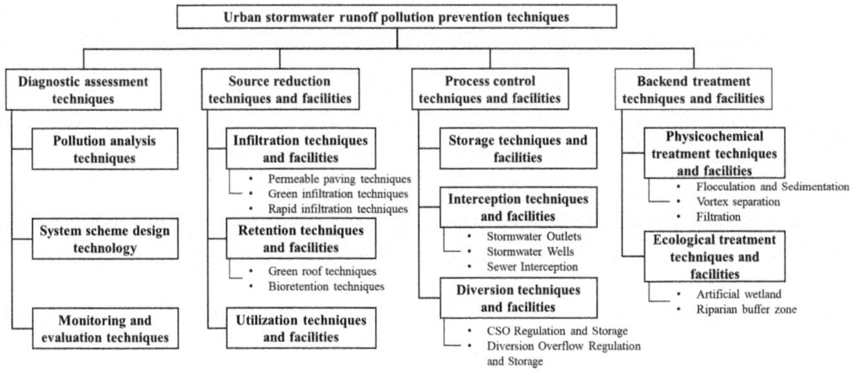

Fig. 1.7 Pollution control technology components for urban rainfall runoff

In order to address the current lack of foundational big data support for urban stormwater runoff pollution control in China and the serious deficiencies in technical parameter selection, technical integration, and the lack of standardization of existing technologies, it is necessary to deeply analyze the relevant research literature on urban stormwater runoff pollution control technologies both domestically and internationally. Through literature research, the development process of urban stormwater runoff pollution control technologies can be effectively reviewed, the research hotspots can be identified, and the technological gaps between China and other countries can be compared. This analysis can also help determine the future development direction of urban stormwater runoff pollution control technologies in China and assess the recent development status of these technologies. Ultimately, this will provide a basis and reference for the subsequent evaluation of urban stormwater runoff pollution control technologies.

a. **Diagnostic Assessment Technology**

The development process of urban stormwater runoff diagnostic assessment technology is presented in Table 1.8. Both domestic and international diagnostic assessment technologies for urban stormwater runoff primarily rely on models to simulate and assess rainfall runoff quantity, runoff water quality, and runoff pollution load. Throughout the development of this technology, three stages have been identified: model establishment and calibration, model coupling and integration, and model-based evaluation of technological performance.

Researchers both domestically and internationally have primarily focused on the development of new rainfall-runoff models, model calibration, the coupling and integration of multiple models, using models to predict rainfall-runoff water quality and quantity, and using models to evaluate technological performance to optimize and improve parameters for controlling rainfall-runoff pollution. The development of diagnostic assessment technology has been more advanced overseas compared to domestic research. In the early stages, foreign researchers mainly focused on constructing simple models, optimizing model prediction accuracy, and applying them in engineering. The emphasis was on the development of scouring models and pollution load models. With the widespread use of internet technology and the introduction of advanced algorithms, foreign scholars integrated these advanced algorithms into the establishment and parameter optimization of related models, improving the accuracy of predicting rainfall-runoff quantity, runoff water quality, and runoff pollution load. Since 2010, there has been increasing attention to the development and establishment of composite models that couple and integrate different models. The development of composite models is important for leveraging the advantages of each model, expanding the scope, spatial scale, accuracy, and operational convenience of their applications. Evaluating the effectiveness of technology through models can further guide the selection and optimization of technological parameters. In the future, overseas urban rainfall-runoff diagnostic assessment technology will continue to evolve towards intelligent and automated online monitoring system models.

Table 1.8 Development process of urban stormwater runoff diagnosis and assessment technologies

Year	China				Water major project		Foreign			
	Model development	Model calibration	Model coupling	Model evaluation	Model coupling	Model evaluation	Model development	Model calibration	Model coupling	Model evaluation
1995–1999	Simulation for impervious rainfall runoff volume calculation						Runoff scouring model			
2000–2004		Quasi-Newton methods					Water quality model			
2005–2009	Runoff and confluence calculation simulation	Simplex algorithm					Hydrological impact assessment system	GA	Coupling of SWMM and GIS	
2010–2014	Runoff model; confluence model	Pattern search algorithm	Coupling of SWMM and GIS	SWMM evaluation of LID	Coupling of SWMM and GIS	SWMM and LID, combined use of storage tanks	Hybrid intelligent model for runoff prediction	ANNs	Meteorological model and rainfall-runoff model	Integration of model and LID
2015–present	Rainwater quality and quantity online monitoring system	GA; ANNs	Integration and utilization of multiple model	SWMM model and sponge city construction			Meteorological rainfall-runoff forecasting model; satellite rainfall runoff model		Coupling of satellite remote sensing and rainfall-runoff model	

In the early stages, domestic development has been comparatively slower, mainly drawing on advanced foreign experience and utilizing mature model tools to address issues related to controlling rainfall-runoff pollution in China. In recent years, with the gradual improvement in understanding urban waterlogging and rainfall runoff pollution, China has begun to calibrate parameters in model tools to better suit local conditions and has also conducted algorithm corrections to improve prediction accuracy. In the current stage, China mainly focuses on researching the SWMM model, advanced algorithms, the coupling and integration of multiple models, as well as local calibration of these model parameters. These approaches have been applied in numerous urban rainfall-runoff pollution load diagnostics and facility planning and design.

Overall, the development of urban rainfall-runoff diagnostic assessment technology in China lags behind that of foreign countries. In the future, it is expected that the development trend of China's urban rainfall-runoff diagnostic assessment technology will involve further promoting the use of the SWMM model, and developing diagnostic assessment models suitable for different characteristics of Chinese cities.

b. **Source Reduction Technologies**

Source reduction technologies for urban rainfall runoff pollution focus on the infiltration and purification of runoff rainwater. They primarily aim to slow down the runoff velocity, increase infiltration, and reduce the volume of runoff by utilizing low-impact development techniques and rainwater bioretention facilities. As researchers gained a deeper understanding of runoff, they discovered that the composition and complexity of pollutant components in initial rainfall runoff posed significant challenges for treatment. Therefore, in addition to rainwater infiltration, further research has been conducted on effective reduction and removal of pollutants in rainfall runoff through source reduction technologies. The typical principle behind this technology involves the selection and application of matrix fillers with high permeability, strength, and pollutant removal efficiency, enabling simultaneous reduction of pollutants during the process of rainfall runoff infiltration. The development history of source reduction technologies for urban rainfall runoff pollution is presented in Table 1.9.

As early as the early 1990s, foreign countries had started the research and engineering applications of Low Impact Development (LID) technologies based on rainwater infiltration, gradually progressing towards the development and application of media fillers with pollutant interception capabilities. The focus was on increasing rainwater infiltration through the use of bioretention measures, reducing runoff volume, and highlighting key technologies such as large-scale permeable pavement and ecological grass ditch techniques. With the continuous advancement of technology, foreign researchers began to focus on the removal of pollutants in rainfall runoff. They improved the technical performance by optimizing technical parameters, selecting modified fillers as matrix materials, and using a combination of techniques. In order to effectively address the randomness of rainfall runoff pollution and the difficulty in monitoring the effectiveness of pollution control facilities, foreign countries introduced mathematical/numerical simulation and hydraulic

Table 1.9 Development process of source reduction technologies for urban stormwater runoff pollution

Year	China			Water major project			Foreign		
	Rainwater infiltration	Rainwater purification	Landscape construction	Rainwater infiltration	Rainwater purification	Landscape construction	Rainwater infiltration	Rainwater purification	Landscape construction
1995–1999							Grass swales; permeable pavement; bioretention technology		
2000–2004	Traditional bioretention technology								
2005–2009	Rain gardens; permeable pavement; sunken greenbelt; grass swales; green roofs	High-efficiency media screening	Plant selection				Development and promotion of LID technologies	High-efficiency media screening	Plant selection and landscape architecture

(continued)

Table 1.9 (continued)

Year	China			Water major project			Foreign		
	Rainwater infiltration	Rainwater purification	Landscape construction	Rainwater infiltration	Rainwater purification	Landscape construction	Rainwater infiltration	Rainwater purification	Landscape construction
2010–2014	Application of LID technology and optimization improvement of technical parameters	Adsorbent for phosphate removal	Plant combinations and site planning	Rain gardens; permeable pavement; sunken greenbelt; grass swales; green roofs; Dry stream	High-efficiency media screening	Landscape rain gardens; plant combinations and site planning	Optimization and improvement of LID technology parameters; integration of LID with drainage systems	Modification and application of purification media; improvement of purification performance of permeable concrete	Facilities landscape maintenance and management
2015–present	Guide LID construction with SWMM and GIS; integrated use of LID and stormwater drainage technology; improvements in permeable concrete clogging	Media modification and structural design					Integration of reflective pavement and permeable pavement		

models into the planning design and effectiveness evaluation of LID engineering measures around the year 2000. By integrating advanced management systems such as the Storm Water Management Model (SWMM) and GIS, they achieved the integration of urban rainfall runoff pollution control and urban planning and construction. It is expected that foreign source reduction technologies for urban rainfall runoff pollution will further focus on the development of media materials with high infiltration and pollutant removal efficiency and will carry out intelligent monitoring and management of related facilities in the future.

The development of source reduction technologies for urban rainfall runoff pollution in China mainly involves the introduction of related LID facilities and rainwater management technologies from foreign countries, combined with the characteristics of Chinese cities, to further develop rainwater control measures suitable for different cities and typical underlying surfaces in China. Ecological retention and detention techniques are commonly used in this regard. Additionally, the coordination of LID technologies with detention facilities, such as models and pipe networks, is being gradually studied. In order to enhance urban aesthetics, China also integrates landscape planning and design with source control facilities for rainfall runoff pollution. The future development trend of source reduction technologies for urban rainfall runoff pollution in China is expected to focus on the optimization and management of technical parameters in low-impact development facilities for pollutant control, as well as the development of efficient, low-cost, and low-carbon media that enable infiltration, retention, detention, and purification functionalities. Furthermore, facility monitoring and maintenance technologies oriented towards intelligent systems will be a major focus of future development.

c. **Process Control Technologies**

Process control technologies for urban rainfall runoff pollution mainly involve the control of urban drainage pipe inlets, rainwater storage and detention of rainfall runoff, overflows of stormwater and sewage, and purification of initial rainfall runoff. The development history of process control technologies for urban stormwater runoff pollution is presented in Table 1.10.

In the field of urban stormwater runoff pollution control technology, foreign countries first established corresponding overflow stormwater wastewater models to guide the design of detention reservoirs through simulating and predicting overflow volume. Preliminary treatment of the inflowing rainwater through vortex separation technology was employed to remove high concentrations of SS and prevent deposition and blockage in the pipes. Secondly, by optimizing the design of rainwater outlets and integrating them with other types of urban non-point source and stormwater pollution control technologies, effective removal of SS from stormwater runoff was achieved. Due to the difficulty in reducing pollutants during the runoff confluence process, the current focus is on detention and SS removal. Therefore, it is expected that the future development trend of this type of technology would be the coordinated removal of various pollutants during the runoff confluence process and the efficient

Table 1.10 Development process of process control technologies for urban stormwater runoff pollution

Year	China		Water major project		Foreign	
	Rainwater diversion and storage	Interception and purification of runoff	Rainwater diversion and storage	Interception and purification of runoff	Rainwater diversion and storage	Interception and purification of runoff
1995–1999		Design technology for interception-type rainwater inlets; design technology for rainwater wells; design technology for hydrocyclone separation			Construction and design technology of rainwater storage tank; hydrocyclone separation technology	Design technology for rainwater inlets; design technology for rainwater wells
2000–2004	Runoff erosion calculation; runoff concentration prediction					
2005–2009	Design technology for rainwater storage tanks	Optimized design of environmentally-friendly rainwater inlets			Model-based design technology for stormwater storage tanks; model-based integrated technology for rainwater storage tanks and storm sewers	Stormwater manhole; dynamic water quality monitoring technology for rainwater inlets

(continued)

Table 1.10 (continued)

Year	China		Water major project		Foreign	
	Rainwater diversion and storage	Interception and purification of runoff	Rainwater diversion and storage	Interception and purification of runoff	Rainwater diversion and storage	Interception and purification of runoff
2010–2014	Design of stormwater drainage system; integrated technology of storm sewers and LID; digital rainwater storage and control technology	Design technology for interception-type rainwater inlets	Design technology for rainwater storage tanks; numerical management technology for rainwater storage tanks; integrated technology of rainwater detention tanks with stormwater networks and wastewater treatment plants	Optimized design techniques for rainwater outlets; automatic purification technology for stormwater manhole	Dynamic monitoring technology for water quality and quantity in rainwater storage tanks; optimization and scheduling integration technology for rainwater storage tanks; integrated technology of rainwater storage tanks and LID; maintenance, management, and evaluation technology for rainwater storage tanks	Design technology of intelligent purification rainwater inlet; high-efficiency purification technology of stormwater manhole
2015–present	Integrated technology of water quality simulation platform and stormwater system; integrated technology of stormwater networks and rainwater storage tanks	Integrated technology of rainwater inlets with LID and storm sewers				

integration with other types of urban stormwater runoff pollution control technologies, thus developing key technologies capable of achieving efficient runoff control while simultaneously addressing multiple runoff pollution control issues.

In this field, the overall development trend in China is similar to that of foreign countries. The difference lies in the faster development of China in the design and construction of SWMM models combined with stormwater detention reservoirs and urban drainage systems. Through the combination with SWMM models, it effectively guides the construction of stormwater detention reservoirs and urban drainage systems. Additionally, the development of efficient pollutant interception and purification-type rainwater outlet technology has achieved reduction of SS and other pollutants during the runoff confluence process. It is expected that the future development trend of urban stormwater runoff pollution control technology in China would be the optimized design of efficient pollutant interception and purification-type rainwater drainage pipeline systems, as well as the design of water quality purification-type multi-stage stormwater detention reservoirs.

d. **Backend Treatment Technology**

Backend treatment technology for urban stormwater runoff pollution mainly relies on artificial rainwater wetlands, ecological buffer zones, and physicochemical measures to achieve efficient reduction of pollutant load before entering the river. These technologies mainly include biological, ecological, and physicochemical treatments, with an emphasis on ecological/biological treatment. The development process of backend treatment technology for urban stormwater runoff pollution is shown in Table 1.11.

The post-treatment of rainfall runoff pollution both domestically and internationally focuses on runoff ecological/biological treatment to achieve efficient treatment of rainfall runoff pollution and create hydrophilic landscapes through techniques such as rainwater artificial wetlands and ecological buffer zones. In this research field, a large amount of research work has been carried out both domestically and internationally on the scientific construction of rainwater artificial wetlands and ecological buffer zones, with foreign research leading the way. This is mainly reflected in the function of pollutant interception and purification media in rainwater artificial wetlands and ecological buffer zones, as well as optimized vegetation configuration. At the current stage, ecological/biological treatment techniques combined with hydrophilic landscapes are mainly utilized to construct urban rainwater wetlands and buffer zones with runoff pollution reduction effects, and in-depth research is conducted on the optimization of their operation and maintenance management techniques. With the construction and use of a large number of bio/ecological treatment facilities, the development of efficient, low-cost, and intelligent operation and maintenance techniques is crucial and will also become one of the important directions for future development. In the context of the "dual carbon" targets, targeted allocation of green vegetation and efficient application of high-biomass water purification materials will also be one of the important directions for future development, while achieving pollutant reduction and improving the carbon sequestration capacity of ecological/biological treatment technologies.

Table 1.11 Development process of back-end governance technologies for urban stormwater runoff pollution

Year	China		Water major project		Foreign	
	Ecological treatment technology	Physicochemical treatment technology	Ecological treatment technology	Physicochemical treatment technology	Ecological treatment technology	Physicochemical treatment technology
1995–1999					Research and selection of filler for buffer strip and stormwater wetlands; research and selection of plants in buffer strips and rainwater wetlands	Stormwater filtration tanks technology
2000–2004	Design technology of buffer strips; design technology of constructed wetlands	Screening of efficient coagulants				
2005–2009	Technology for the selection of plants in wetland and buffer strip	Design technology for efficient stormwater sedimentation			Integrated technology of stormwater wetlands and storage tanks; integration technology of models and stormwater wetlands	Fixed rainwater filtration technology
2010–2014	Type selection of constructed wetlands; selection of high efficient fillers	Coupled application of coagulants, rainwater storage tanks and rainwater sedimentation tanks	Model-based rainwater wetland design technology; screening of efficient media for stormwater wetlands	Screening of efficient coagulants; rainwater diatom earth treatment technology; rainwater inclined tube precipitation technology	Management technology of constructed wetland; evaluation model of constructed wetland	Design technology for rainwater purification and reuse systems
2015–present	Construction technology of wetland and buffer strip based on GIS and SWMM; construction technology of rainwater wetland and sponge city	Technology for automatic treatment of rainwater coagulation and sedimentation				

In addition to ecological/biological treatment techniques, physicochemical treatment techniques still play an important role in removing suspended solids (SS) from rainfall runoff, mainly including coagulation and sedimentation, flocculation and sedimentation, centrifugal separation, rapid filtration, ultrafiltration membrane separation, and other technical measures. In China, these techniques are mainly applied to localized storage treatment and comprehensive utilization of rainwater.

In terms of backend treatment techniques and facilities, in recent years, the focus of domestic research and development has been on the performance of different types of artificial wetlands in reducing pollutants in rainfall runoff and the screening and modification of high-quality wetland fillers. Equipment for flocculation and sedimentation, centrifugal separation, and rapid filtration has been developed and extensively used in engineering applications. Miniaturization, integration, automation, and intelligent physicochemical treatment techniques and equipment have the advantages of saving space, easy operation, and maintenance and will be one of the important directions for future development. In addition, the coupling of backend physicochemical and bio/ecological rainfall runoff pollution control techniques, targeting efficient removal of new pollutants, serves as the last line of defense to safeguard the ecological risks of receiving water bodies and is an important direction for development.

1.4 Urban Rainfall Runoff Pollution Control Objectives and Technical Routes

1.4.1 Control Objectives

After nearly 30 years of rapid development, urban sewage treatment systems in China have become increasingly improved, effectively controlling point source pollution in urban water environments. However, the issue of rainwater runoff pollution caused by urbanization has become increasingly prominent. Rainwater runoff often contains high concentrations of SS and COD, with average concentrations exceeding Class V standards of the "*Environmental Quality Standards for Surface Water*" (GB 3838–2002) of China. Rainwater runoff pollution load now accounts for nearly 20% of urban pollution load, exceeding 35% in some large cities, gradually becoming one of the major sources of pollution in urban water environments. The control objectives for urban rainwater runoff pollution include reducing the volume and pollution load of runoff while increasing rainwater reuse. To achieve this, a combination of measures, such as source reduction, process control, and end treatment, are adopted to minimize the entry of rainwater runoff into receiving water bodies, consolidate the effectiveness of treating black and odorous water bodies, and improve the urban water environment.

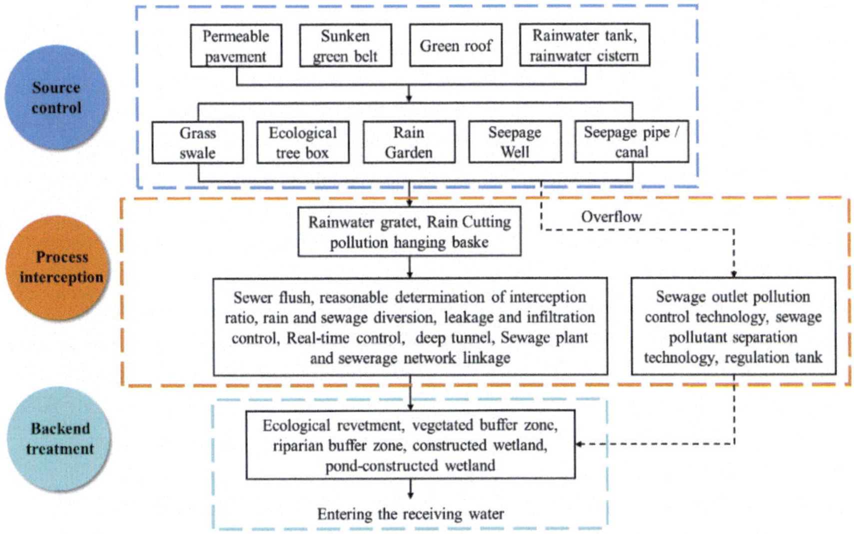

Fig. 1.8 The control approach for urban rainwater runoff pollution is illustrated

1.4.2 Technical Approach

Urban rainwater runoff pollution is formed through the interaction between rainwater runoff and surface pollutants. The process of urban rainwater runoff pollution involves the dissolution and scouring of surface pollutants by rainwater and resulting runoff, ultimately discharging into receiving water bodies.

Based on different underlying surfaces, the initial runoff-forming areas in urban areas mainly include roofs, roads, squares, and green spaces. These four types of runoff-forming areas are considered sources for reducing urban rainwater runoff pollution. The area where rainwater runoff enters the drainage network is considered the process control zone, and the area between the drainage network and the receiving water body is considered the end treatment zone. Finally, by considering these three zones, various integrated control technologies are employed to effectively manage urban rainwater runoff pollution. The control approach for urban rainwater runoff pollution is illustrated in Fig. 1.8.

References

1. Tuccillo, M. (2006). Size fractionation of metals in runoff from residential and highway storm sewers. *Science of the Total Environment, 355*(1–3), 288–300.
2. Liu, Z., O, L., Hu, D., et al. (2012). Contamination of organochlorine pesticides in rain, canopy throughfall and runoff water in Beijing urban area. *Chinese Journal of Environmental Engineering, 6*(03), 804–810 (in Chinese).

3. Wang, J., Xia, X., & Feng, W. (2020). Pollution characteristics of polycyclic aromatic hydrocarbons in urban viaduct stormwater runoff based on event rainfall process. *Environmental Chemistry, 39*(07), 1832–1838 (in Chinese).
4. Han, J., Bi, C., Chen, Z., et al. (2013). Pollution characteristics and toxicological assessment of PCBs in urban runoff. *China Environmental Science, 33*(03), 546–552 (in Chinese).
5. Yang, Q., Mei, X., Sun, J., et al. (2018). Sources, distribution and major transformation process of typical endocrine disruptors in the environment. *Asian Journal of Ecotoxicology, 13*(03), 42–55 (in Chinese).
6. Zhao, J., Huang, Z., Zhang, Q., et al. (2021). Distribution and mass loads of xenoestrogens bisphenol A, 4-nonylphenol, and 4-tert-octylphenol in rainfall runoff from highly urbanized regions: A comparison with point sources of wastewater. *Journal of Hazardous Materials, 401*, 123747.
7. Lei, K., Pan, H., Zhu, Y., et al. (2021). Pollution characteristics and mixture risk prediction of phenolic environmental estrogens in rivers of the Beijing–Tianjin–Hebei urban agglomeration, China. *Science of the Total Environment, 787*, 147646.
8. Fan, C., Qin, H., & Xu, N. (2014). Seasonal distribution of endocrine disruptors and their initial scouring effects in the Shiyan River Basin, Shenzhen, China. In *12th National conference on water treatment chemistry 2014* (in Chinese).
9. Luo, L. (2017). *Pollution characteristics and control of PPCPs in rainfall runoff in urban environment*. Tinghua University (in Chinese).
10. Wang, Y. (2021). *Research on the occurrence and risk of typical pollutants and microplastics in athletic field runoff*. Beijing University of Civil Engineering and Architecture (in Chinese).
11. Cao, H., Huang, S., Li, J., et al. (2011). Research on pollution characteristics and initial effect of surface runoff in Lingang new city of Shanghai. *Journal of Water Resources & Water Engineering, 22*(06), 66–71 (in Chinese).
12. Chen, S. (2016). *Study on characteristics of urban storm runoff in Yixing City*. Xi'an University of Architecture and Technology (in Chinese).
13. Chen, T. (2019). *Monitoring and control of rainfall runoff pollution of small watershed in southern city*. Harbin Institute of Technology (in Chinese).
14. Chen, W., Pan, Z., & Xing, Y. (2018). Preliminary analysis of rainfall pollution characteristics of urban roofs in Shijiazhuang city. In *2018 CSES annual conference on environmental science and technology* (in Chinese).
15. Chen, W., Pei, S., & Pan, Z. (2020). Observation and experimental analysis of rainfall runoff from rainwater wetland treated roofs. In *(8th) China water ecological summit 2020* (in Chinese).
16. Chu, X. (2020). *Characteristics of rainwater pollution and environmental risk control in Shanghai Chemical Industry Park*. East China Normal University (in Chinese).
17. Chu, J., Cai, Y., Wu, C., et al. (2012). Analysis on characterization of Zhenjiang old city roadway runoff. *Industrial Safety and Environmental Protection, 38*(12), 58–61 (in Chinese).
18. Di, W. (2014). *Research on infiltration and reuse of road runoff by artificial soil technology in Jiaxing City*. Beijing University of Civil Engineering and Architecture (in Chinese).
19. Ding, Q., Wang, Q., Zhang, Q., et al. (2017). Analysis of runoff pollution from the main road in Yixing City, Taihu Lake Basin. *Acta Scientiae Circumstantiae, 37*(09), 3456–3463 (in Chinese).
20. Dong, Z. (2018). *Study on rainwater runoff pollution control in Yixing new urban district and Beijing Liangshui river basin*. Beijing Forestry University (in Chinese).
21. Dong, Z., Qu, D., & Sun, D. (2018). Study on pollution control of stormwater runoff from new urbanized areas in Yixing. In *2018 CSES annual conference on environmental science and technology* (in Chinese).
22. Dou, Y., Wu, J., Chen, Z., et al. (2017). Spring rain runoff pollution process and its characteristics analysis in Hefei City. *Water Resources Protection, 33*(05), 86–90 (in Chinese).
23. Du, Y. (2018). *Study on pollution characteristics of initial rainwater and its intercepted storage at Hefei economic and technological development area*. Hefei University of Technology (in Chinese).

24. Feng, C., Mi, N., Wang, X., et al. (2015). Analysis of road runoff pollutants in northern city based on the typical rainfall. *Ecology and Environmental Sciences, 24*(03), 418–426 (in Chinese).
25. Fu, J., Chen, Z., You, K., et al. (2016). Feature analysis on pollution of runoff in north cold region. *Journal of Shenyang Jianzhu University (Natural Science), 32*(03), 553–559 (in Chinese).
26. Gan, C., Zhu, Z., Liu, M., et al. (2020). Comprehensive assessment of urban stormwater runoff pollution in Xiushan county, Chongqing. *Ecology and Environmental Monitoring of Three Gorges, 5*(04), 73–81 (in Chinese).
27. Gao, B., Xu, Y., Lu, M., et al. (2020). Analysis of rainfall runoff pollution and pollution load estimation for urban communities in a highly urbanized region. *Environmental Science, 41*(08), 3657–3664 (in Chinese).
28. Gong, M., Cai, C., Miao, H., et al. (2019). Pollution characteristic and source apportionment of road rainfall runoff in different functional zones of Lin'an. *The Administration and Technique of Environmental Monitoring, 31*(04), 18–22 (in Chinese).
29. Gu, M., Chang, S., Xu, W., et al. (2018). Classification of rainwater runoff pollution indexes and analysis of pollution sources in Tianjin urban area. *South-to-North Water Transfers and Water Science & Technology, 16*(05), 85–92+101 (in Chinese).
30. Gu, Y., Wang, X., & Fang, Z. (2020). Investigation and analysis of pollution characteristics of rainfall runoff in typical cities. *China Rural Water and Hydropower*, (06), 46–50+57 (in Chinese).
31. Gu, Y., & Zhang, N. (2013). Characterization of urban surface runoff in main districts of Kunming. *Chinese Journal of Environmental Engineering, 7*(07), 2587–2595 (in Chinese).
32. Guo, F., Guo, J., Liu, Q., et al. (2012). Characteristics of water quality change in different under-laying surface rainfall runoff of Handan city. *Water Sciences and Engineering Technology, 01*, 85–87 (in Chinese).
33. He, G. (2015). *Reaserch on the characteristics of different urban roads pollution—The case of Hangzhou.* Zhejiang University of Technology (in Chinese).
34. He, S., Chen, W., Liang, D., et al. (2021). Rapid estimation method of pollution load from each rainfall-runoff in urban area. *Journal of Lake Sciences, 33*(01), 138–147 (in Chinese).
35. He, T., Yang, L., Wang, D., et al. (2014). Pollution characteristics and discharge rules in road runoff of culture and education area, park district in the rapid city area. *Pearl River, 35*(03), 95–98 (in Chinese).
36. Hong, G., Yuan, M., & You, Z. (2019). Effects of rainfall characteristics on runoff pollution in urban area of Wuxi. *Journal of China Hydrology, 39*(02), 33–38+66 (in Chinese).
37. Hu, M., Liu, X., Yan, Y., et al. (2021). Recognition of rainfall runoff pollution characteristics of different river drainage outlets. *Acta Scientiae Circumstantiae, 41*(01), 164–173 (in Chinese).
38. Hua, L., Zou, B., Lu, H., et al. (2012). Study on the variation of urban roof-runoff pollution. *Environmental Monitoring in China, 28*(05), 109–115 (in Chinese).
39. Huang, S., Zeng, Y., Hong, W., et al. (2012). Characteristics of rainfall runoff pollution in city center area of Fuzhou. *Journal of Subtropical Resources and Environment, 7*(03), 26–31 (in Chinese).
40. Jiang, Y. (2015). *Urban catchment rainfall runoff pollution characteristics and control simulation.* Nanchang University (in Chinese).
41. Kong, Y., Feng, H., & Nie, J. (2018). Research on pollution characteristics and emission load of urban rainfall runoff in the Fuxian Lake Watershed, Yunnan. *Environmental Science Survey, 37*(05), 26–32 (in Chinese).
42. Lai, X., Zhao, J., Li, D., et al. (2015). Pollution characteristics of runoff on different underlying surfaces in industrial area of Taiyuan City. *Bulletin of Soil and Water Conservation, 35*(06), 97–100+105 (in Chinese).
43. Li, C., Liu, M., Hu, Y., et al. (2014). Characteristics of pollutants in urban rainfall runoff in Shenyang City. *Chinese Journal of Ecology, 33*(05), 1327–1336 (in Chinese).
44. Li, C., Liu, K., Lin, J., et al. (2013). Analysis on water-runoff pollutions of different underlying urban surfaces in Haikou City. *Environmental Monitoring in China, 29*(05), 80–83 (in Chinese).

45. Li, H., Li, T., & Li, C. (2008). Research on the characteristics of urban roof runoff in Shanghai cultural and educational area. *Environmental Science, 01*, 47–51 (in Chinese).
46. Li, H., Huang, W., Yang, G., et al. (2006). Non-point source pollutant concentration in typical towns of Taihu Upriver Region. *Journal of Agro-Environment Science, 06*, 1598–1602 (in Chinese).
47. Li, J., Cao, H., Wang, J., et al. (2011). Research on the pollution characteristics of surface runoff in Lingang New City. *Journal of Shanghai Ocean University, 20*(04), 594–599 (in Chinese).
48. Li, M., Qu, Z., Liu, P., et al. (2020). Characteristics of urban road runoff pollution in northern city based on artificial rainfall test. *China Water & Wastewater, 36*(21), 110–114 (in Chinese).
49. Li, Q. (2018). *Research on pollution characteristics and treatment countermeasures of runoff based on Liuzhen Expressway in Guizhou Province.* Beijing University of Chemical Technology (in Chinese).
50. Li, S., Xiang, M., & Chen, Q. (2012). Pollution characteristics of runoff in different function areas of Hangzhou city. *Journal of Zhejiang University of Science and Technology, 24*(06), 475–480 (in Chinese).
51. Liu, D., Li, Q., Li, T., et al. (2015). Research on pollution characteristics of rainfall runoff in Tianjin City. *China Water & Wastewater, 31*(11), 116–119 (in Chinese).
52. Liu, G., & Yi, L. (2012). Research on initial abandonment control technology for stormwater runoff pollution in Huzhou. *Science & Technology Information, 29*, 91–93 (in Chinese).
53. Liu, Q. (2020). *Study on rainfall runoff characteristics and initial pollution control in Xi'an city.* Xi'an University of Technology (in Chinese).
54. Long, J., Li, X., Wang, S., et al. (2014). Water quality of stormwater runoff from different roofs in urban region. *Chinese Journal of Environmental Engineering, 8*(07), 2895–2900 (in Chinese).
55. Ma, H. (2016). *Study on the characteristics of rainfall runoff pollution in Beijing urban city.* Henan University (in Chinese).
56. Pan, L. (2018). *Study on rainfall runoff pollution characteristics and pollution load in typical campus in Wuhan.* Hubei University of Technology (in Chinese).
57. Pan, L., Liu, D., & Liu, R. (2018). Characteristics of rainfall runoff pollution in a campus parking lot in Wuhan during rainy season. *Journal of Hubei University of Technology, 33*(04), 94–99 (in Chinese).
58. Qi, X. (2020). Process and characterization of water quality pollution under summer rainfall runoff in Yingkou area. *Ground Water, 42*(02), 78–79 (in Chinese).
59. Qin, Y. (2018). *Research on the characters of rainwater runoff pollution in Nanning city.* Guangxi University (in Chinese).
60. Qiu, L. (2012). Characterization of nitrogen and phosphorus pollution in urban rainfall runoff in Huizhou city, China. *Environment, S2*, 8–9 (in Chinese).
61. Ren, H., Li, Z., Wang, X., et al. (2014). Study on the characteristics of pollution in precipitation runoff in Wuxi University. *Journal of Qilu University of Technology, 28*(02), 15–20 (in Chinese).
62. Sheng, J., Zeng, P., Zhang, C., et al. (2011). Characteristics of pollution in precipitation runoff in Zhenjiang urban area. *Journal of Jiangsu University of Science and Technology (Natural Science Edition), 25*(05), 496–499 (in Chinese).
63. Shi, S. (2021). Study on characteristics of rainfall runoff pollution in a district of Shenzhen. *Guangdong Water Resources and Hydropower, 05*, 57–61 (in Chinese).
64. Sun, Z. (2017). *Study on non-point source pollution reduction technology in typical cities of Taihu Lake Basin.* Xi'an University of Architecture and Technology (in Chinese).
65. Sun, Z., Zhang, J., Wang, Q., et al. (2018). Pollution characteristics of road runoff of municipal solid waste transfer stations. *Environmental Engineering, 36*(04), 88–92 (in Chinese).
66. Teng, J., Yin, Q., Li, F., et al. (2014). Characteristics of water quality and load of pollution for highway rainfall runoff in Shanghai. *Water Purification Technology, 33*(03), 18–21+47 (in Chinese).

67. Tian, Y., Li, T., He, S., et al. (2009). Characteristics of non-point pollution of Fengqiao Industrial Park in Suzhou. *China Water & Wastewater*, *25*(13), 89–91+94 (in Chinese).
68. Tong, Z., Fu, L., & Li, J. (2021). Research on a southern college urban rainfall-runoff non-point source pollution. *Journal of East China Jiaotong University, 38*(02), 136–141 (in Chinese).
69. Tu, J., & Ai, N. (2018). Analysis on the pollution characteristics of urban rainwater runoff—Taking Foshan new town as an example. *Guangdong Water Resources and Hydropower, 09*, 33–39 (in Chinese).
70. Wan, Y., Zhang, B., & Wu, C. (2011). Analysis on the pollution characteristics of surface runoff in Zhenjiang city. *Journal of Anhui Agricultural Sciences*, 39 (27), 16716–16717+16787 (in Chinese).
71. Wang, H., Cao, X., Ren, Y., et al. (2019). Analysis on pollution load of rainfall runoff in Beijing urban arterial roads. *Acta Scientiae Circumstantiae, 39*(06), 1860–1867 (in Chinese).
72. Wang, H. (2015). *Pollution characteristics analysis and pollution load estimation in rainfall runoff of different underlying surface in Wuhan city*. Huazhong University of Science and Technology (in Chinese).
73. Wang, J., Jing, H., Wang, H., et al. (2011). Pollution characterization and analysis of rainfall runoff in Beijing urban area. *Water & Wastewater Engineering, 47*(S1), 135–139 (in Chinese).
74. Wang, X., Lai, Q., Du, J., et al. (2016). Analysis on water-runoff pollutions of different underlying surfaces in Ningbo. *Environmental Engineering, 34*(S1), 312–316 (in Chinese).
75. Wang, X. (2021). *Non-point source of rainfall and runoff around Fu River in Baoding city pollution characteristics research*. Hebei Agricultural University (in Chinese).
76. Wang, X., Wu, W., & Li, H. (2019). Analysis of pavement rainfall runoff in various functional areas of Suzhou Ancient City. *Jiangsu Water Resources, 07*, 9–14 (in Chinese).
77. Wang, X. (2018). *Study on urban runoff contaminants characteristics and a prediction model*. Wuhan University (in Chinese).
78. Wang, X., Yan, J., He, J., et al. (2019). Study of the road runoff pollution at different sections of Qionghai lakeside road at Xichang. *Highway, 64*(01), 265–271 (in Chinese).
79. Wang, Y., Yang, X., Hu, R., et al. (2017). Analysis of rainfall-runoff pollution load of Hutang textile industrial park in Changzhou city. *Water Resources Protection, 33*(03), 68–73 (in Chinese).
80. Wang, Z. (2016). *Study on the discharge characteristics of road runoff pollution of Xi'an city*. Chang'an University (in Chinese).
81. Wei, X. (2015). *Research on pavement runoff pollution characteristics and treatment measures of Changsha*. Changsha University of Science & Technology (in Chinese).
82. Wu, M. (2020). *Study on runoff pollution characteristics of Tianjin Binhai Lingang Industrial Park*. Hebei University of Engineering (in Chinese).
83. Wu, M., Li, S., Zhang, W., et al. (2020). Polltion characteristics and control strategies of runoff in Tianjin Binhai Lingang Industrial Park. *Chinese Journal of Environmental Engineering, 14*(12), 3435–3446+3240 (in Chinese).
84. Wu, Y., Chen, Y., Chen, W., et al. (2018). Characteristics of runoff pollution of different underlying surfaces in typical cultural and educational area of Xi'an. *China Environmental Science, 38*(08), 3104–3112 (in Chinese).
85. Xia, Z., Zhang, J., Yi, Q., et al. (2020). Analysis of pollution characteristics of rainfall runoff process of underlying surface of Yantai cultural and educational area. *Journal of Yantai University (Natural Science and Engineering Edition), 33*(02), 238–245 (in Chinese).
86. Yan, J. (2014). *Research on the characteristics of pavement runoff pollution and control programs in Hohhot city*. Inner Mongolia University (in Chinese).
87. Yan, Z., Liu, H., Sun, H., et al. (2012). Study on the polluted characteristics of the surface runoff in the different functional districts of Wenzhou. *Environmental Science & Technology, 35*(S1), 203–208 (in Chinese).
88. Yuan, Y., Li, X., & Fu, J. (2014). Study on pollution characteristics of urban roadway runoff in Suzhou city. *Environmental Science and Management, 39*(12), 68–73 (in Chinese).
89. Zhang, H., & Chen, F. (2010). Non-point pollution statistics and control measures in Taihu Basin. *Water Resources Protection, 26*(03), 87–90 (in Chinese).

90. Zhang, L., & Xu, H. (2014). Analysis of pollution characteristics of rainfall runoff in Wuhan. *Journal of Municipal Technology, 32*(05), 90–93 (in Chinese).
91. Zhang, S. (2019). *Study on rainfall runoff pollution characteristics and regional generalization model in urban industrial area.* Qingdao University of Technology (in Chinese)
92. Zhang, S., Jiao, C., Lv, M., et al. (2020). Study on pollution characteristics of rainfall runoff in Licang Industrial Park in Qingdao city. *Pearl River, 41*(03), 103–108 (in Chinese).
93. Zhang, S., & Li, X. (2008). Pollute character of road earth's surface radial flow in Tianjin Heartland. *Environmental Monitoring in China, 03*, 65–69 (in Chinese).
94. Zhang, W., Luo, Y., Zhong, X., et al. (2019). Characteristics of runoff pollution in asphalt roof and metal roof in downtown Beijing. *Science Technology and Engineering, 19*(23), 358–365 (in Chinese).
95. Zhang, X. (2020). *Study on characteristics of non-point source pollution on roads in Longgang district, SZ.* Shenyang Jianzhu University (in Chinese).
96. Zhang, Y. (2014). *Study of runoff rules and pollution load in Dalian urban.* Huazhong University of Science and Technology (in Chinese).
97. Zhang, Z., Meng, Q., & Ma, Z. (2016). Pollution characterization of urban surface source pollution. *Water & Wastewater Engineering, 52*(S1), 163–167 (in Chinese).
98. Zhang, Z. (2019). Characterization of rainfall runoff pollution in plains cities. *Shandong Chemical Industry, 48*(08), 170–171 (in Chinese).
99. Zhao, D., Xu, Z., Bian, Z., et al. (2020). An analysis of the variation characteristics of rainfall runoff water quality in different underlying surfaces in Jinan. *China Rural Water and Hydropower, 09*, 177–181 (in Chinese).
100. Zhao, K. (2018). *Study on characteristics of storm water runoff pollution in typical cities.* Chongqing Jiaotong University (in Chinese).
101. Zhao, X., Wang, S., Yu, H., et al. (2019). Analysis on characteristics of rainfall runoff pollutions flush of typical underlying surfaces in Tianjin central city. *Environmental Engineering, 37*(07), 34–38+87 (in Chinese).
102. Zhao, Y., & Mei, S. (2019). Analysis of urban surface runoff pollutant concentration and pollution characteristics in Tai Hu Basin. *Environmental Science and Technology, 32*(04), 52–59 (in Chinese).
103. Zhou, F. (2014). Research progress on urban rainfall runoff pollution and its control. *Construction Science and Technology, 12*, 68–71 (in Chinese).
104. Zhou, F., Cao, M., Ke, F., et al. (2017). Analysis of pollution characteristics and first flush effect in rainfall runoff in separate storm sewer system in upstream of Tangxi river, Chaohu catchment. *Journal of Lake Sciences, 29*(02), 285–296 (in Chinese).
105. Zhou, G., Lv, M., Zhang, Z., et al. (2021). Analysis of runoff pollution characteristics of typical industrial parks in Qingdao city based on SWMM model. *Water Resources and Power, 39*(06), 42–45 (in Chinese).
106. Zhou, M., Wang, J., Ge, Y., et al. (2020). Research and analysis of water pollution load in a river network area in Suzhou city. *Environment and Development, 32*(12), 177–179 (in Chinese).
107. Zhou, Z., Chen, T., Yang, S., et al. (2020). Study on characteristics of urban rainfall runoff pollution in southern typical small watersheds based on rainfall patterns. *Tianjin Science & Technology, 47*(09), 86–92 (in Chinese).
108. Zhu, H. (2016). *Research of intercepting and treatment technology for first flush rainwater from urban road.* Tsinghua University (in Chinese).
109. Zhu, Z. (2019). *Analysis of factors affecting storm runoff pollution of Wanzhou.* Chongqing Jiaotong University (in Chinese).
110. Zhuang, J., Luo, H., & Lei, Y. (2017). Analysis of road runoff pollution load at southern city based on the type of rainfall. *Journal of Water Resources & Water Engineering, 28*(02), 110–114 (in Chinese).
111. Luo, H., Chen, X., Wu, Q., et al. (2020). Spatio-temporal characteristics of non-point source and sewer overflow pollution and its impacts on water environment in Shenzhen Bay Basin. *Journal of Natural Resources, 35*(12), 3018–3028 (in Chinese).

112. Bian, Z., Cai, T., & Dai, H. (2020). Assessment and treatment of overflow pollution load in combined partition drainage system. *China Water & Wastewater, 36*(21), 115–120 (in Chinese).
113. Xie, B., Zhang, J., Wu, X., et al. (2019). Study on the influence of overflow pollution from sewage interception project of tributary of drinking water source reservoir. In *2019 CSES annual conference on environmental science and technology* (in Chinese).
114. Wang, Q., Quan, C., Ye, W., et al. (2018). Study on pollution control countermeasures for rainwater and sewage combined drainage overflow in the old city of Beijing. In *Annual national planning conference 2018* (in Chinese).
115. Zhou, M., Chen, J., Dong, L., et al. (2015). Pollution characteristics of combined sewer overflows (CSOs) in old urban area in South Jiangsu. *Chinese Journal of Environmental Engineering, 9*(07), 3159–3164 (in Chinese).
116. Li, H., Xu, S., Huang, Y., et al. (2013). Pollution loading of overflow in combined drainage channels during rainy season. *Acta Scientiae Circumstantiae, 33*(09), 2522–2530 (in Chinese).
117. Wu, C., Zhang, B., Ren, Y., et al. (2011). Analysis of combined overflow pipe pollution of Zhenjiang. *Environmental Science & Technology, 34*(10), 182–185 (in Chinese).
118. Duan, Z., Chen, S., Yao, J., et al. (2020). Wastewater quantity and quality characteristics of combined sewer system in urban village in Shallow Hilly region of Zhuhai city. *China Water & Wastewater, 36*(13), 101–105 (in Chinese).
119. Yin, H., Huang, Y., Chen, L., et al. (2021). Case analysis on treatment project of black and odorous water body in Xiaoqiong lake of Yuanjiang city. *China Resources Comprehensive Utilization, 39*(03), 201–204 (in Chinese).
120. Li, H., & Li, T. (2006). Study on the characteristics of combined sewer overflow from the high density residential area in Shanghai. *Environmental Science, 08*, 1565–1569 (in Chinese).
121. Li, T., Dai, M., Zhang, W., et al. (2013). Pollutant source apportionment of combined sewer overflows for a pump lifting drainage system. *Journal of Tongji University (Natural Science), 41*(10), 1513–1518+1525 (in Chinese).
122. Wang, L., Duan, B., Zhao, J., et al. (2015). Pollution characteristics of surface runoff of typical town in Chongqing city. *Environmental Science, 36*(08), 2809–2816 (in Chinese).
123. Liu, C., Che, W., & Dong, C. (2007). Sewer water quality comparison of separate discharge and combined overflow. *Water & Wastewater Engineering, 04*, 51–55 (in Chinese).
124. Li, S., Guan, Y., Chen, J., et al. (2015). Water quality characteristics of combined sewer overflows in southern region of China. *Water & Wastewater Engineering, 51*(S1), 344–348 (in Chinese).
125. He, Q., Li, L., Kong, L., et al. (2005). Characteristics of storm runoff pollution in Hanyang district of Wuhan city. *China Water & Wastewater, 02*, 101–103 (in Chinese).
126. Han, Y., Peng, D., Xu, W., et al. (2007). Research on characteristics of combined sewer overflows. *Journal of Xi'an University of Architecture & Technology (Natural Science Edition), 06*, 834–838 (in Chinese).
127. Yin, H., Huang, Y., Liu, L., et al. (2021). Case analysis on the treatment of black and odorous water body in Shiji lake of Yuanjiang city. *Journal of Hunan City University (Natural Science), 30*(04), 15–18 (in Chinese).
128. Hai, Y., Dian, L., Yu, D., et al. (2020). Effect of rainfall on the pollution characteristics of combined sewer overflows. *Chinese Journal of Environmental Engineering, 14*(11), 3082–3091 (in Chinese).
129. Shen, J., Li, T., Qian, J., et al. (2012). Types of detention tank for Xinghua drainage system in Hefei city. *China Water & Wastewater, 28*(17), 40–43+48 (in Chinese).

Chapter 2
Methods for Assessing Pollution Loads from Rainfall Runoff in Towns and Cities

In recent years, with the continuous progress of urbanization in China, urban flood and waterlogging disasters have occurred frequently. The pollution caused by rainfall-induced runoff has become an important factor constraining the improvement of urban black and odorous water bodies and the quality of urban water environment. In order to improve urban resilience and effectively control the runoff pollution problem induced by urban flood and waterlogging, higher requirements have been proposed for urban stormwater management. However, due to the significant randomness and complexity of urban rainfall and its induced runoff pollution, direct monitoring is difficult, which poses great challenges for timely analysis of rainfall quantity and water quality. Therefore, both domestic and foreign research have developed theoretical calculation and model simulation-based approaches for diagnosing and assessing urban rainfall runoff quantity and water quality. These approaches aim to express the "source-sink-flow" relationship of rainfall runoff pollution, so as to effectively guide the planning, construction, and management of rainfall runoff pollution control facilities, and achieve efficient utilization and purification of urban rainfall.

This chapter discusses the calculation methods of urban rainfall runoff pollution load. It reviews the calculation process and methods, as well as commonly used commercial models and software, for rainfall infiltration, surface runoff, pollutant accumulation, and wash-off in urban areas. These include the SWMM, urban drainage network model (InforWorks CS), Hydrological Simulation Program-Fortran (HSPF), Soil and Water Assessment Tool (SWAT), and MIKE (hydrodynamic calculation model). These methods, commercial models, and software for calculating urban rainfall runoff pollution load can be used to analyze the runoff quantity and pollution load of urban rainfall in different scales and scenarios, and evaluate and predict the reduction effects of green infrastructure, initial stormwater purification facilities, and overflow pollution control facilities on runoff pollution.

D. Sun and F. Qi, *Urban Rainfall Runoff Pollution Control Technology and Application*,
https://doi.org/10.1007/978-981-96-4174-1_2

2.1 Calculation of Urban Rainfall Runoff Pollution Load

Typically, the relationship and corresponding expressions between pollution load and influencing factors are established by monitoring and calculating the deposition of pollutants on sunny days, the infiltration and runoff volume on rainy days, as well as water quality. This indirect approach allows obtaining the urban rainfall runoff pollution load. Alternatively, the pollution load of different rainfall durations can be obtained by synchronously observing the runoff volume and water quality after the formation of rainfall runoff. This section will introduce various aspects, including rainfall infiltration, surface runoff, pollutant accumulation, and rainfall runoff scouring.

2.1.1 *Rainfall Infiltration*

Rainfall infiltration refers to the amount of rainwater that falls onto the ground and infiltrates through unsaturated soil into the underground.

The calculation of rainfall infiltration is usually done using the Green-Ampt equation, which is applicable not only to cases where rainfall intensity exceeds the infiltration capacity but also to cases where rainfall intensity is initially smaller than the infiltration capacity during heavy rain.

Rainfall infiltration is generally calculated in two situations: when the cumulative infiltration is less than the saturated cumulative infiltration, and when the cumulative infiltration is equal to or greater than the saturated cumulative infiltration.

(1) When the cumulative infiltration is less than the saturated cumulative infiltration, i.e., $F < F_s$, the equation for rainfall infiltration is $f = i$.

$$\begin{aligned} F_s &= \frac{S \cdot IMD}{i/K_s - 1} \quad (i > K_s) \\ F_s &= 0 \quad (i \le K_s) \end{aligned} \tag{2.1}$$

(2) When the cumulative infiltration is equal to or greater than the saturated cumulative infiltration, i.e., $F \geq F$, the equation for rainfall infiltration is $f = f_p$.

$$f_p = K_s\left(1 + \frac{S_{wf} \cdot IMD}{F}\right) \tag{2.2}$$

where, F—cumulative infiltration, mm;
F_s—saturated cumulative infiltration, mm;
S—capillary suction at the wetting front of the soil, mm;
f—infiltration rate, mm/s;
i—rainfall intensity, mm/s;
S_{wf}—capillary suction at the wetting front, mm;

IMD—initial soil unsaturated degree, mm/mm;
K_s—saturated hydraulic conductivity of the soil, mm/s;
f_p—steady-state infiltration rate, mm/s.

2.1.2 Surface Runoff

Surface runoff refers to the amount of rainfall that falls on the ground and forms runoff along the surface. The calculation of surface runoff is usually done using nonlinear storage equations. Depending on the characteristics of the region, it is divided into several sub-catchments (including road networks and channels), which can be generalized into three types of units: impervious storage unit A1, pervious storage unit A2, and impervious non-storage unit A3, as shown in Fig. 2.1. The runoff from each unit only enters the catchment outlet or channel, and the runoff from one unit does not pass through another unit.

The nonlinear storage equation for surface runoff is established through the continuity equation combined with the Manning equation. The continuity equation for each unit is given in Eq. (2.3).

$$\frac{\mathrm{d}V}{\mathrm{d}t} = A_n \cdot \frac{\mathrm{d}h}{\mathrm{d}t} = A_n I^* - Q_{wn} \tag{2.3}$$

where, V—runoff volume in a specific unit, m^3, $V = A_n h$;

A_n—surface area of the n-th unit, m^2;

h—water depth (m); t represents the runoff time, s;

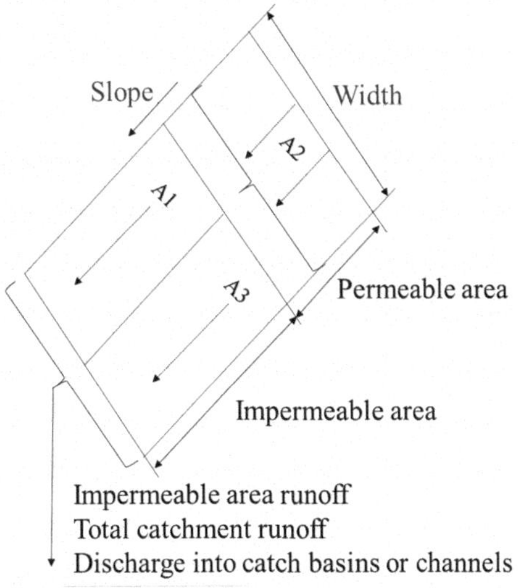

Fig. 2.1 Diagram of the 3 types of unit divisions in the catchment area

I^*—net rainfall, m/s, net rainfall = the rainfall intensity-vaporization-seepage, (note: the infiltration rate can be obtained using Eqs. (2.1) and (2.2));

Q_{wn}—outflow rate of the nth unit, m^3/s.

The outflow rate Q_{wn} is expressed with the Manning formula, as shown in Eq. (2.4).

$$Q_{wn} = W \cdot \frac{1.49}{n}(h - h_p)^{5/3} slp^{1/2} \tag{2.4}$$

where, W—width of the sub-catchment, m;

n—roughness coefficient;

h_p—depth of the groundwater table, m;

slp—slope of the sub-catchment, %.

2.1.3 *Accumulation of Pollutants*

The process of pollutant accumulation refers to the cumulative changes of pollutants on the surface of the watershed during sunny weather, i.e., the accumulation of pollutants over time. The accumulation of pollutants on the surface typically occurs during sunny periods and can be calculated using models such as linear accumulation model, power-law accumulation model, logarithmic accumulation model, progressive accumulation model, and other models.

a. **Linear Accumulation Model**

The accumulation of pollutants is calculated by using the equivalent number of sunny days and the daily load. The equivalent number of sunny days is determined by the number of sunny days and the road cleaning conditions. This model has a simple structure and is easy to understand. However, as the number of sunny days increases, the accumulation will show unlimited growth, and this model does not consider the accumulation rate of pollutants during sunny days. Therefore, its use is limited. The expression of the model is shown in Eq. (2.5).

$$\begin{aligned} P_t &= t_e \cdot Y(S)_u \\ t_e &= (y - t_s)(1 - \varepsilon_S) + t_s \end{aligned} \tag{2.5}$$

where, P_t—pollutant accumulation per unit area on sunny days t days after the last rainfall, mg/m^3;

t_e—equivalent number of sunny days, day;

$Y(s)_u$—solid daily load per unit area of the underlying surface, $mg/(m^2 \cdot day)$;

y—number of days since the last rainfall event, day;

t_s—number of days since the last street cleaning event, day;

ε_s—street cleaning frequency.

b. **Cumulative Power Law Model for Pollutant Accumulation**

The accumulation process of pollutants can also be represented in the form of a power law based on the number of clear days, as shown in Eqs. (2.6) and (2.7). In Eq. (2.6), it is assumed that the initial cumulative surface pollutant residue is zero, indicating that a single rainfall event completely flushes the surface clean. Equation (2.7), on the other hand, accounts for the presence of an initial pollutant load, where the surface pollutant residue is not zero. It can be observed that the rate of pollutant accumulation decreases accordingly with the initial residual load.

$$P_t = P_m\left(1 - e^{-k_1 t_q}\right) \tag{2.6}$$

$$P_t = P_s + (P_m - P_s)\left(1 - e^{-k_1 t_q}\right) \tag{2.7}$$

where, P_m—maximum cumulative pollutant concentration within the watershed (mg/m^2);

P_s—surface residual pollutant load at the end of the previous rainfall event, which is the initial pollutant load on clear days (mg/m^2);

k_1—accumulation coefficient (day^{-1});

t_q—number of clear days elapsed since the most recent rainfall event (day).

c. **Logarithmic Model for Pollutant Accumulation**

Based on the assumption that the rate of pollutant accumulation decreases with time and approaches a maximum value, the accumulation model can be expressed as a logarithmic equation, as shown in Eq. (2.8).

$$P_t = a \cdot \ln t + c \tag{2.8}$$

where, a, c—constant coefficients.

d. **Progressive Accumulation Model for Pollutants**

The progressive model (also known as the saturation model) was proposed by Sartor and Boyd (1972) and later by Roesner (1982) [1]:

$$P_t = \frac{P_m t}{N + t} \tag{2.9}$$

where, N—half-saturation constant;

P_t—the time elapsed when the cumulative concentration reaches half of the maximum concentration P_m, day.

The widely-used SWMM model currently includes the progressive model as an option due to its satisfactory simulation performance [2].

e. **Other Models for Pollutant Accumulation**

The accumulation of pollutants on clear days can also be calculated using Eq. (2.10):

$$\frac{dL_{si}}{dt} = k_i - k_{2i}L_{si} \tag{2.10}$$

where, i—the i-th type of surface feature;

L_{si}—deposition amount of fallout per unit area, g/m^2);

k_i—deposition rate of fallout, g/(m^2·day);

k_{2i}—consumption rate of fallout (caused by wind, traffic, and biodegradation), day.

For a certain pollutant, the accumulation amount L_{ij} is directly proportional to the fallout accumulation amount L_{si}:

$$P_t = L_{ij} = f_{ij} \cdot L_{si} \tag{2.11}$$

where, f_{ij}—proportion coefficient of the j-th pollutant on the i-th surface feature, mg/g;

L_{ij}—cumulative amount of the j-th pollutant per unit area on the i-th surface feature at the beginning of rainfall, mg/m^2.

Different surface features, such as rooftops, road surfaces, and grasslands, have varying levels of fallout residue. For rooftops, only the fallout residue after rainfall runoff is considered, as cleaning is not taken into account. For road surfaces, street sweeping needs to be taken into consideration. As for grasslands, cleaning is not considered, but the effective portion of fallout contributing to the initial accumulation amount is taken into account in the calculations.

2.1.4 Rainfall Runoff Erosion

Rainfall runoff erosion refers to the impact of surface runoff generated by rainfall or snowmelt on the surface, where it washes away pollutants deposited on the surface, dissolves them, and forms runoff pollution.

a. **Initial Flushing Effect**

Under the wash of rainfall, accumulated pollutants on the surface are transferred from the surface to the runoff water. It is commonly observed that the concentration of pollutants in the initial runoff is higher than that in later stages [3], known as the first flush phenomena. Non-permeable areas such as urban parking lots and asphalt pavements are the main regions where the initial flushing effect occurs.

Different definitions have been proposed for the initial flushing effect of pollutants in rainfall runoff and are usually represented by dimensionless pollutant transport functions. When the dimensionless pollutant transport rate is higher than the dimensionless rainfall runoff transport rate, the initial flushing effect is observed.

Through the study of initial rainfall runoff, it has been found that within the initial 20–30% of the runoff volume, the transport rate of pollutants is higher than the transport rate of the runoff volume, indicating the presence of over 50% of the runoff water pollutants and exhibiting a significant initial flushing effect [4, 5].

The initial flushing rate, quantitatively described as mass first flush (MFF), is commonly used to represent the initial flushing effect. It is the ratio of standardized pollutant mass to runoff volume.

$$MFF = \frac{\int_0^t C(t)Q(t)dt/M_f}{\int_0^t Q(t)dt/V} = \frac{\int_0^t C(t)Q(t)dt/\int_0^t Q(t)dt}{M_f/V} \tag{2.12}$$

where, MFF—initial flushing rate;

$C(t)$—pollutant concentration in milligrams per liter, mg/L;

$Q(t)$—rainfall runoff volume in liters per second, L/s;

M_f—the total amount of flushed pollutants in milligrams, mg;

V—volume of runoff in liters, L.

Ma et al. [6] proposed the use of the Mass First Flush (MFF) as a quantitative measure to describe the initial flushing effect of rainfall runoff, as depicted in Fig. 2.2. The MFF curve lies above the straight line with a slope of 1.0, indicating that pollutant transport in rainfall runoff is faster than the transport of runoff volume, demonstrating the presence of the initial flushing effect. The ratio of pollutant transport rate to runoff volume transport rate, at m% of runoff volume transport, can be described as MFF_n.

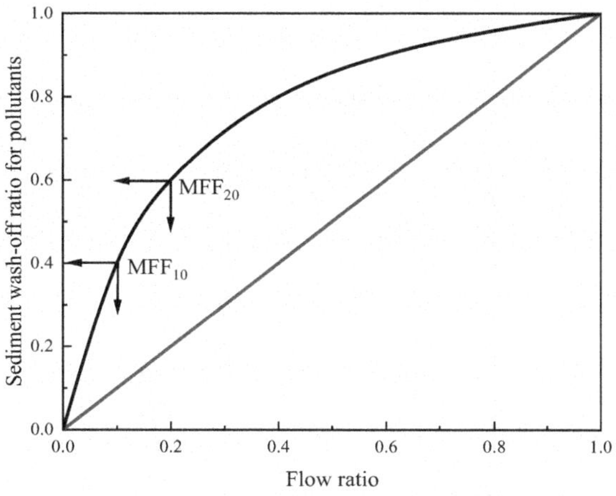

Fig. 2.2 Schematic diagram for quantitative description of scouring effect

The initial wash-off effect of rainfall-runoff pollutants is influenced by various factors, including rainfall intensity, catchment area, the proportion of impervious surfaces, and the type of pollutants. As rainfall intensity increases, the accumulated surface pollutants are more prone to wash-off, thus intensifying the initial wash-off effect, albeit with a relatively modest rate of increase. The catchment area exerts a significant influence on the initial wash-off effect, featuring a threshold value beyond which its impact diminishes. When the catchment area is smaller than this threshold, insufficient wash-off occurs, leading to a reduction in the strength of the initial wash-off effect. Conversely, when the catchment area exceeds this threshold, the prolonged conveyance time results in a diminished initial wash-off effect. Notably, the increase in impervious surfaces amplifies the initial wash-off effect due to the greater transfer of pollutants from impermeable surfaces to rainfall-runoff. Moreover, the type of pollutants also impacts the initial wash-off effect; soluble pollutants, being more prone to transfer into rainfall-runoff, tend to exhibit a more pronounced trend in the initial wash-off effect, emphasizing its significance in the early transport of pollutants.

Due to the relatively higher concentration of pollutants in the initial stages of rainfall runoff compared to later stages, the primary focus of rainfall runoff pollution control lies in mitigating pollutants during the early runoff phase.

b. **Empirical Model for Pollutant Washoff**

Common methods for calculating pollutant wash-off in rainfall runoff include the cumulative quantity approach and empirical techniques.

(1) Cumulative Quantity Method

The formula for the Cumulative Quantity Method is depicted in Eq. (2.13):

$$\Delta L_{ij} = L_{ij}\left[1 - \exp(-c_{3i} \cdot P)\right] \tag{2.13}$$

where, ΔL_{ij}—the amount of the j-th pollutant flushed from the ith characteristic unit area of subsurface by the rainfall process, mg/m^2;

c_{3i}—the flushing coefficient of the i-th characteristic surface, mm^{-1};

P—the amount of rainfall, mm.

(2) Empirical Method

The empirical method's calculation formula is depicted in Eq. (2.14).

$$\frac{dM_{t_F}}{dt_F} = -c \cdot \overline{I} \cdot M_{t_F} \tag{2.14}$$

where, M_{t_F}—the amount of pollutants per unit area of the substrate at the moment of t_F, mg/m^2;

t_F—flushing time, s;

c—pollutant flushing coefficient, mm^{-1};

$\overline{I}$—average rainfall intensity, mm/s.

Integrating Eq. (2.15) results in:

$$M_{t_F} = P_t e^{-c \cdot \overline{I} \cdot t_F} \tag{2.15}$$

The rate at which accumulated pollutants are transferred to rainwater runoff is expressed as follows:

$$R_M = -\frac{dM_{t_F}}{dt} = c \cdot \overline{I} \cdot M_{t_F} \tag{2.16}$$

where, R_M—the scour rate per unit area of underlayment, mg/(m^2 min).

However, research has shown that the flushing coefficient "c" is not a constant, but rather a characteristic value that is intimately linked to surface and contaminant characteristics [7]. Scientists such as Ian Brodie [8] and Prasanna Egodawatta [9] have argued that the amount of pollutant subject to flushing is less than the total amount accumulated on the surface. This flushing process is related to the kinetic energy of the raindrops and the intensity of the rainfall, resulting in non-constant values. Therefore, an adjustment was made to Eq. (2.16) as shown in Eq. (2.17).

$$R_M = -\frac{dM_{t_F}}{dt_F} = c \cdot \overline{I} \cdot c_F \cdot P_t \cdot e^{-c \cdot i \cdot t_F} \tag{2.17}$$

where, c_F—the correction factor, which has a value less than 1.0 and increases with increasing rainfall intensity.

The wash-off coefficient of pollutants is closely tied to the nature of the pollutants. Typically, soluble pollutants tend to have higher wash-off coefficients, whereas non-soluble particulate pollutants exhibit lower coefficients [10, 11]. The reason for this lower coefficient in particulate pollutants lies in the simultaneous sedimentation process during the wash-off [12]. When the wash-off quantity surpasses the deposition, it indicates positive wash-off; otherwise, it is termed negative wash-off. Consequently, the surface transport of particulate pollutants can be described using the model presented in Eq. (2.18):

$$\frac{\partial M_s}{\partial t_F} + \frac{\partial V M_s}{\partial L} = v_e - v_h \tag{2.18}$$

where, M_s—suspended particulate matter transported per unit area of underlayment, mg/m^2;

L—the length of the slope, m;

V—runoff volume, L;

v_e—pollutant scouring velocity, mg/(m^2·s);

v_h—pollutant settling velocity, mg/(m^2·s).

The wash-off effect of particulate pollutants over the entire rainwater runoff within the land surface can be represented using Eq. (2.19).

$$\frac{dM_g}{dt_F} = v_h - v_e \tag{2.19}$$

where M_g—the amount of particulate matter per unit area of underlayment, mg/m^2.

The variable 'v_e' can be described using a stochastic wash-off model [13, 14], as depicted in Eq. (2.20).

$$v_e = c\bar{I}M_g \tag{2.20}$$

The settling of particulate pollutants in surface runoff is expressed using the shallow flow settling model as defined in Eq. (2.21) [12]:

$$v_h = \frac{\alpha v_{set} M_s}{h} \tag{2.21}$$

where, α—the concentration change coefficient;

v_{set}—settling velocity, m/s;

h—represents runoff depth, m.

c. **Simulation and Calculation of Rainfall Runoff Processes**

Rainfall, after accounting for rainfall losses, transforms into surface runoff. The hydraulic model for surface runoff can be mathematically represented by the complete one-dimensional Saint–Venant equation system.

The Saint–Venant equation system comprises momentum equations, as shown in Eq. (2.22), and continuity equations, as depicted in Eq. (2.23).

$$u\frac{\partial u}{\partial t} + \frac{\partial u}{\partial t} + g\frac{\partial h}{\partial t} = g\left(slp_o - slp_f\right) - \frac{uI}{h} \tag{2.22}$$

$$\frac{\partial q}{\partial x} + \frac{\partial h}{\partial t} = I \tag{2.23}$$

where, x—the distance in the direction of water flow, m;

q—the single width flow rate through the corresponding section, m^3/s;

u—the flow rate in the x direction section, m/s;

slp_o—the surface slope, %;

slp_f—the resistance slope, %;

I—net rain intensity, mm/s;

g—gravitational acceleration, m/s^2.

When the convective terms and pressure gradients in the Saint–Venant equation system are disregarded, we obtain the kinematic wave equation. Similarly, omitting the convective and inertial terms results in the diffusion wave equation [15]. In most scenarios, the kinematic wave equation serves as a robust model for simulating overland flow processes and offers computational simplification.

Rainfall runoff within stormwater pipes can be computed using the Saint–Venant equations, albeit with a higher parameter complexity and computational demands. In contrast, the Muskingum method offers an elegant and highly accurate alternative for quantifying pipe flow [16–18], making it a suitable replacement for the Saint–Venant equations.

The Muskingum method employs a water balance equation expressed in the form of a continuity equation, replacing the momentum equation with the storage-discharge equation, as depicted in Eq. (2.24).

$$\begin{cases} Q_{In} - Q_{Out} = \frac{dS_w}{dt_F} \\ S_w = K_w\left[\mu I + (1-\mu)Q\right] \end{cases} \tag{2.24}$$

where, Q_{In} and Q_{Out}—the inlet and outlet flows, m^3/s;

S_w—the storage volume of the river reach, m^3;

t_F—the scour time, s;

K_w—the travel time of the steady state river segment under the corresponding river segment storage volume (S_w), s;

μ—the parameter also known as the flow weight factor.

The differential decomposition of Eq. (2.25) is:

$$\begin{cases} Q_{t_F+\Delta t_F} = Q_{I,t_F} + c_2 Q_{I,(t_F+\Delta t_F)} + c_3 Q_{t_F} \\ c_1 = (\Delta t_F + 2K\mu)/[\Delta t_F + 2K(1-\mu)] \\ c_2 = (\Delta t_F - 2K\mu)/[\Delta t_F + 2K(1-\mu)] \\ c_3 = [2K(1-\mu) - \Delta t_F]/[\Delta t_F + 2K(1-\mu)] \end{cases} \tag{2.25}$$

where, Q_{I,t_F} and $Q_{I,(t_F+\Delta t_F)}$—the inlet flow at the beginning and end of time period Δt, m^3/s;

Q_{t_F} and $Q_{t_F+\Delta t_F}$—the outgoing flow at the beginning and end of time period Δt, m^3/s;

Δt_F—calculation time, s;

c_1, c_2 and c_3—coefficients, $c_1 + c_2 + c_3 = 1$.

The lower flow can be calculated time by time if the upper inlet flow and the lower inlet flow at the start time are known.

i. Calculate the Water Quality of Fully Mixed Rainfall Runoff.

In the calculation of water quality during rainfall runoff, it is generally assumed that the rainfall runoff and pollutants are thoroughly mixed within the unit. This assumption implies that the pollutant concentration within the unit is considered to be equal to the concentration in the runoff, as shown in Eq. (2.26).

$$\frac{dVC_{ij}}{dt} = V\frac{dC_{ij}}{dt} + C_{ij}\frac{dV}{dt} = Q_I C_{Iij} - QC_{ij} - k_1 C_{ij} V + S_c \tag{2.26}$$

where, C_{ij}—the concentration of the j-th contaminant in the i-th subsurface, mg/L;

V—outflow volume, m^3;

Q_I—influent flow rate, m^3/s;

C_{Iij}—the concentration of the j-th pollutant in the i-th type of subsurface, mg/L.

Q—effluent flow rate, m^3/L.

k_1—the first attenuation coefficient, day^{-1};

S_c—the source or leakage, g/s.

ii. Calculation of Water Quality of Stormwater Runoff

During rainfall events, the subsequent runoff effectively flushes away residual pollutants from the previous rainfall event and any pollutants that have accumulated on the ground surface during periods of clear weather. This process results in the incorporation of surface contaminants into the runoff, thereby contributing to the contamination of stormwater runoff.

Commonly used methods for assessing the concentration of pollutants in rainfall runoff include determining the average concentration of runoff from individual rainfall events, calculating the average concentration of runoff from multiple rainfall events, and estimating the average concentration of runoff from multiple rainfall events representative of a specific geographic location.

Commonly used methods for assessing the concentration of pollutants in rainfall runoff include determining the average concentration of runoff from individual rainfall events, calculating the average concentration of runoff from multiple rainfall events, and estimating the average concentration of runoff from multiple rainfall events representative of a specific geographic location.

(i) *Event Mean Concentration*

Within the stormwater runoff resulting from a single rainfall event, pollutant concentrations exhibit significant temporal variation, which varies considerably (by orders of magnitude) due to the stochastic nature of rainfall intensity. As a result, in both national and international practice, the concept of the 'average concentration of runoff from the second rainfall' is commonly used to represent the typical concentration of a particular pollutant discharged during the entire surface runoff process following a rainfall event, as shown in Eq. (2.27).

$$EMC = \frac{M}{V} = \frac{\int_0^T C_t Q_t dt}{\int_0^T Q_t dt} \tag{2.27}$$

where, *EMC*—the average concentration of a pollutant in rainfall runoff from a site, mg/L;

M—the total amount of a pollutant in rainfall runoff from a particular site, g;

V—the volume of runoff caused by rainfall at a particular site, m^3;

C_t—the instantaneous concentration of a pollutant at time t, mg/L;

Q_t—the flow rate of surface runoff at time t, m^3/s;

T—the total duration of a particular rainfall, s.

(ii) *Average Event Mean Concentration*

The stochastic nature of rainfall runoff pollution is manifested not only in the unpredictable variation of pollutant concentrations within specific rainfall runoff events, but also in the random variation of Event Mean Concentration (EMC) values between different rainfall events. Extensive field studies have consistently shown different EMC values for different rainfall runoff events. The United States Environmental Protection Agency has observed that EMC values over different rainfall events follow a lognormal distribution. Typically, the magnitude of pollutant concentrations is represented by the arithmetic mean of EMC values from multiple rainfall events, referred to as the average EMC or EMC mean.

(iii) *Site Mean Concentration*

The Site Mean Concentration (SMC) is the average concentration of runoff from multiple rainfall events at a given site. It is determined through systematic testing, data collection and calculations that allow it to approximate the true average concentration of runoff from multiple rainfall events at the site. The accuracy of the SMC value in representing the true average runoff concentration at the site increases as the number of runoff fields tested increases. Typically, the SMC is expressed as the median of the EMC values obtained from different rainfall fields.

(iv) Water Quality Modelling of Stormwater Runoff

The volume of rainfall runoff flushed is influenced by factors such as rainfall intensity, duration and various other parameters. The volume of surface rainfall runoff flushing can typically be determined using simple first-order kinetic equations, as shown in Eq. (2.28).

$$\begin{gathered} \frac{dM}{dt} = -kM_f \\ k = cr \end{gathered} \tag{2.28}$$

integral result: $M_t = P_0 e^{-cR_t}$

Where, M_f—the amount of pollutant that can be washed off the impervious surface, kg;

M_f—the amount of pollutant remaining on the impervious surface after rainfall runoff begins at time t, kg;

P_0—the amount of pollutants accumulated in the watershed after 0 d of sunny weather since the last rainfall, kg;

R_t—the cumulative runoff volume after the start of rainfall at time t, mm;

k—attenuation coefficient, s^{-1};

r—runoff velocity, mm/s.

Then the amount of urban surface pollutants discharged by a rainfall flush is:

$$M' = P_0\left(1 - e^{-cR_T}\right) \tag{2.29}$$

where, M'—the amount of surface pollutants discharged from a single rainfall event, kg;

R_T—the total runoff volume of the sub-rainfall, mm.

Assuming that the catchment area is A, the pollutant concentration C_t in the rainfall runoff can be expressed as

$$C_t = -\frac{1}{A}\frac{dM_t}{dR_t} = \frac{cP_0}{A}e^{-cR_t} \tag{2.30}$$

Therefore, the concentration of pollutants at the beginning of rainfall runoff can be expressed as:

$$C_0 = \frac{cP_0}{A} \tag{2.31}$$

The development of this rainfall-runoff water quality model is based on a specific theoretical framework that assumes an exponential correlation between the removal process of surface pollutants. The formulation of the rainfall-runoff water quality calculation model is achieved through a quantitative parameter-based linkage that relates the accumulation of surface pollutants to their flushing efficiency.

(v) Empirical Calculation Method

The empirical calculation method is based on data collected in the Washington area as part of the National Urban Rainfall Runoff Pollution Study Programme, as shown in Eq. (2.32).

$$m_t = [C_F \times \psi \times A \times P \times C] \times 0.01 \tag{2.32}$$

where, m_t—the pollutant load discharged by rainfall runoff in time period t, kg;

C_F—the coefficient of the proportion of the total number of rainfall events in a year that produce rainfall runoff;

ψ—average runoff coefficient;

A—the total area of the runoff area, ha;

P—annual precipitation, mm;

C—the weighted average concentration of the pollutant runoff, mg/L;

0.01—the unit conversion factor.

This method has been successfully applied to study the pollution load from roads and bridges in the sub-basins of lakes in Jiuzhaigou, Sichuan, China [19]. In addition, Zhang et al. [20] conducted a comprehensive study of surface runoff pollution in Shanghai using this method, which provided valuable insights.

2.2 Conventional Commercial Models for Estimating Pollution Loads from Urban Stormwater Runoff

Traditional mathematical equations used to calculate pollution loads from rainfall runoff are complex and require significant human and material resources. To improve computational efficiency and better manage the effects of rainfall runoff and associated pollution, researchers have developed various urban runoff models that integrate and optimize these expressions within computer systems. These models are enhanced by models that accumulate pollutants, washout runoff, quantify water, and measure water quality, to provide a comprehensive simulation of pollutant transport within the subsurface, alongside rainfall runoff, accounting for processes like sedimentation and chemical reactions occurring internally. Simulating urban rainfall runoff, tracking the migration and transformation of pollutants within the runoff, and evaluating the effectiveness of runoff pollution control technology and the feasibility of stormwater management programs are crucial steps in achieving the ultimate goal of flood prevention and drainage, overflow mitigation, and runoff pollution control.

Currently, various models are available for calculating and predicting urban rainfall runoff pollution loads. Among this assortment, commonly used models are the SWMM, HSPF, SWAT, Infoworks CS, and MIKE models. A summary of the unique features and suitability of each model is presented in Table 2.1.

Table 2.1 Frequently utilized commercial models for estimating pollution loads in urban rainfall runoff

Model	SWMM model	HSPF model	SWAT model	Infoworks CS model	MIKE model
Input parameter	Characteristic width of overland flow, average slope, percent imperviousness area, manning's roughness coefficient, depth of depression storage, surface permeability	Hydrology and meteorology, land use, DEM, buildup and washoff coefficient, pollutant decay coefficient, characteristics of receiving waters, etc.	Digital elevation maps, land use raster imagery, soil type maps, soil type tables, land use tables, groundwater depth data, stormwater pipe network layouts, precipitation data, meteorological data, and natural hydrologic data	Runoff depth, rain depth, runoff coefficient, infiltration rate, catchment area, rainfall intensity, flow, manning's coefficient, slope and depression storage, etc.	Land use, digital elevation maps, infiltration conditions, evaporation, initial soil moisture content, stormwater pipe layouts, rainfall data, meteorological data, hydrological parameters, pollutant accumulation, and erosion parameters
Hydraulic simulation function	Using the dynamic wave model to solve the Saint Venant equations, analyzing the flow conditions in the drainage network, applied to the design and optimization of the system	The river and reservoir hydrological and water quality process simulation module can effectively simulate hydraulic behaviors	Utilizing simplified Saint Venant equations for grid cell runoff calculations	Applying dynamic wave models, solving the Saint Venant equations, analyzing flow conditions in the pipe network for system design and optimization	By applying dynamic wave models and solving the Saint Venant equations, this study analyzes flow conditions in the pipe network to optimize system design

(continued)

Table 2.1 (continued)

Model	SWMM model	HSPF model	SWAT model	Infoworks CS model	MIKE model
Hydrologic simulation function	Use of groundwater infiltration models to simulate the effect of groundwater strata on infiltration flow, capable of evaluating any infrastructure	The module 'Pervious Lot Hydrologic and Water Quality Processes Simulation' enables the simulation of hydrologic processes in rainfall runoff	Effectively assess non-point source pollution at multiple scales, identify critical pollution areas, and create different management scenarios to evaluate the effectiveness of different management actions	Various models exist for generating runoff, including fixed runoff coefficients, SCS, Green-Ampt, Horton, and other models. Additionally, there are routing models like Large Catch Wallingford	Able to realistically simulate the complete process from rainfall to runoff, in line with the principle of runoff formation
LID facility evaluation function	Capable of simulating various types of LID facilities on developed and applied low impact development underlay surfaces	LID construction can be guided by the permeable underlayment module and the impermeable underlayment module	There are few applications in urban areas in China, mainly in watershed hydrologic studies and nonpoint source pollution modeling	Including hydrological and hydraulic simulation methods, it can be applied in bulk to watersheds or used to simulate individual LID facilities in detail, aiding in design processes	Developed and applied LID modules, capable of simulating various types of low-impact development facilities
Simulation computing capability	Capable of simulating both single event and continuous rainfall, the model provides stable calculations and operates at a relatively high speed	Surface runoff calculation employs the Philip equation and the Chezy-Manning formula. In impervious areas, a linear function accumulation model and scour model are employed to simulate the runoff process	Using hydrological response units (HRUs) as the calculation unit, the calculation is efficient and does not require unnecessary manual input	"Utilizing a stable computation engine with variable step sizes, accompanied by graphical and reporting components, including prompts and data management tools; Capable of parallel computing, using dedicated graphics cards, supporting multitasking and multithreading, and able to leverage hardware to increase computing speed	Capable of simulating both single event and continuous rainfall, the model provides stable calculations and operates at a relatively high speed

(continued)

Table 2.1 (continued)

Model	SWMM model	HSPF model	SWAT model	Infoworks CS model	MIKE model
Simulation of calibration capabilities	Provides the ability to import and export simulation and monitoring data, facilitating calibration and verification of model parameters	The HSPF model is calibrated and optimized using methods such as orthogonal collocation, manual calibration, and response surface methodology.。	Improvement of simulation accuracy through precise spatial parameters (thematic maps, hydrological and meteorological monitoring, etc.) and fitting with measured data; long-term runoff simulation results are more accurate than short-term ones, but the accuracy of daily simulation results needs further improvement	Flow rate and velocity are adjusted and matched using predicted and observed curves	Provides the ability to import and export simulation and monitoring data facilitating calibration and validation of model parameters
Ease of operation	The model interface is user-friendly, accompanied by comprehensive operation manuals and case studies, facilitating the utilization by technical professionals	The model includes numerous parameters with complex interactions among them, requiring training to become proficient in its use	The model interface is easy to use and comes with detailed operating instructions and case studies, making it easy for technical staff to use	Proficiency in ArcGIS is essential for using this software. With numerous software modules, training is required	The model interface is easy to use and comes with detailed operating instructions and case studies, making it easy for technical staff to use
Degree of difficulty in promotion	Open-source software, facilitating widespread adoption	This is a commercial software that requires purchase for use	Open-source software, facilitating widespread adoption	This is a commercial software that requires purchase for use	This is a commercial software that requires purchase for use

2.2.1 SWMM Model

a. **Principles of SWMM Modelling**

The United States Environmental Protection Agency (USEPA) developed SWMM in 1971 as a dynamic model for predicting and managing water quality and quantity in rainfall-runoff processes. SWMM is an efficient tool for researching urban stormwater resourcing by simulating the entire urban rainfall-runoff process, inclusive of the calculation of yield-sink flow and pollutant concentrations during cumulative flushing. Its primary uses consist of simulating distributed surface runoff, performing quantitative analysis of regional water quality and sewage discharge. The model is suitable for use in drainage planning, performance analysis, design, and management measure evaluation.

The SWMM model comprises four computational modules and one service module. The computational modules are Runoff, Transport, Extran, and Storage/Treatment. The service module provides statistical analysis and mapping functions.

Input information for the SWMM model includes hydrometeorological data, land use characteristics, recharge and flushing coefficients, and drainage network parameters. Output information includes pollutant loads, pipe overflows, and assessments of the effectiveness of best management practices at each location within the simulated area.

Through the above modules and input information, the model can calculate and simulate the runoff volume and pollutant transport in the urban area, including: time-varying rainfall, surface water evaporation, surface runoff from impervious areas, soil erosion and infiltration in permeable areas, pipe overflow, and water quality changes in receiving water bodies. The pollutant indicators simulated by the model mainly include 10 kinds of SS, BOD5, COD, TN, TP and E. coli, etc., and user-defined pollutants can also be simulated.

SWMM, widely used and recognised as the most comprehensive and sophisticated model, is particularly well suited to simulating runoff in extensive urban areas characterised by multiple subsurface components. Its approach of evaluating each sub-watershed run-by-run makes it an ideal choice. In addition, SWMM offers flexibility with no predefined input time interval. Compared to other models, such as the HSPF model, SWMM's simulation results closely match the measured values, with the added advantage of achieving the shortest time required for simulated runoff to reach its peak.

The limitations and drawbacks of SWMM are notable, including its demanding requirements for input data and parameters, its reliance on a single input format, and the lack of a direct facility for importing pipe network maps in GIS or CAD formats, which limits the versatility of the model. In addition, SWMM lacks the ability to simulate biochemical reactions and inter-transformation of pollutants, and provides limited representation of sediment movement within pipelines, a critical aspect that is closely linked to water quality.

b. **SWMM Model Calculations**

i. Generalization of the Study Area

The study area for the surface yield and watershed model is shown in Fig. 2.3. In SWMM modeling, the study area is first divided into several sub-areas based on the type of surface and the direction of rainfall flow. Each subarea is further subdivided into permeable area S1, impervious area S2 with detention capacity, and impervious area S3 without detention capacity. The surface of each subarea is represented as a nonlinear reservoir with an inflow term that includes precipitation and runoff from the upstream subarea. Meanwhile, the discharge term includes infiltration, evapotranspiration, and runoff. As shown in Fig. 2.3, the characteristic width of S1 is equal to the width of the entire catchment area (L1), while the characteristic widths of S2 and S3, denoted as L2 and L3, are determined using Eq. (2.33). The runoff production and pollutant loads of each subarea are calculated individually, and their sum provides the runoff process line for the study area, which enables the estimation of pollutant concentrations at each location.

$$L_2 = \frac{S_2}{S_2 + S_3} \times L_1; L_3 = \frac{S_3}{S_2 + S_3} \times L_1 \tag{2.33}$$

where, L_1—the characteristic width of subregion S_1, m;

L_2—the feature width of subregion S_2, m;

L_3—the characteristic width of sub-area S_3, m;

S_2—the impervious area with stagnant storage capacity, m^2;

S_3—the impervious area without stagnant storage capacity, m^2.

ii. Calculation of Runoff Yield

Surface yield is the amount of runoff remaining after various losses, commonly referred to as rainfall losses, have been deducted from the initial rainfall. When

Fig. 2.3 Schematic diagram of study area generalization

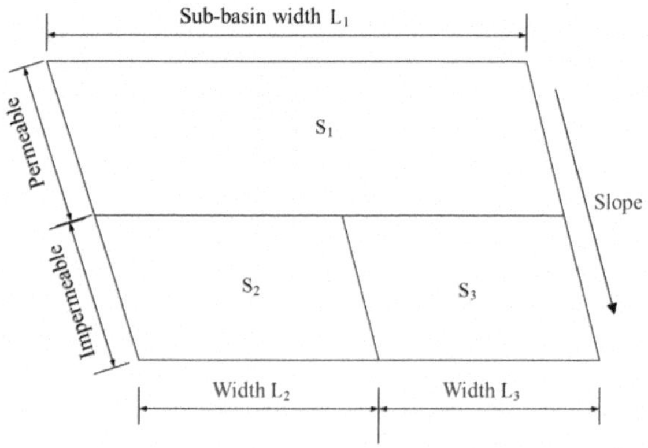

rainfall intensity exceeds the infiltration capacity of the soil (if any), water begins to accumulate on the surface, resulting in surface runoff.

(i) On impervious surfaces without ponding, the dominant source of rainfall loss is evapotranspiration. The calculation of the generated runoff is detailed in Eq. (2.34).

$$R_1 = P - E \tag{2.34}$$

where, R_l—the amount of runoff produced by the un-lying impervious surface, mm;

P—the rainfall, mm; E represents evapotranspiration, mm.

(ii) Rainfall losses on ponded impervious surfaces are primarily pond filling and the flow yield calculations are shown in Eq. (2.35).

$$R_2 = P - D \tag{2.35}$$

where, R_2—the flow rate produced by a puddled impervious surface, mm;

D—the volume of the puddle storage, mm.

(iii) Rainfall losses on permeable surfaces are mainly due to ponding and infiltration. The equation for calculating flow production is shown in Eq. (2.36).

$$R_3 = (I - f) \cdot T - D \tag{2.36}$$

where, R_3—the amount of flow produced by the permeable surface, mm;

I—rainfall intensity, mm/h;

f—infiltration intensity, mm/h;

T—rainfall time, h.

iii. Infiltration Calculation During Runoff

The SWMM model presents three distinct techniques to compute the quantity of infiltration during flow production: the Horton model, the Green-Ampt model, and the SCS-CN model (Curve Number Method, or Runoff Curve Number Method).

(i) *Horton Model*

The Horton model, proposed by Horton in 1933, is an empirical model.

$$f_t = f_c + (f_0 - f_c)e^{-kt} \tag{2.37}$$

where, f_t—the infiltration rate at time t, mm/h;

f_c—the stabilized infiltration rate of the soil, mm/h;

f_0—the initial infiltration rate, mm/h;

k—the infiltration attenuation coefficient related to the physical properties of the soil, h^{-1}.

(ii) *Green-Ampt Model*

The Green-Ampt model, proposed in 1911 by Green and Ampt, is a physically grounded theoretical model rooted in Darcy's theorem for water flow in porous media.

$$\mathrm{F} = \frac{K_s S_w(\theta_s - \theta_i)}{i - K_s} \tag{2.38}$$

where, F—the cumulative infiltration depth of rainfall, mm;

θ_s and θ_i—the water content per unit volume at saturation and at initial time, respectively;

S_w—the water absorption capacity of the soil at the infiltration surface, mm;

K_s—the saturated hydraulic conductivity, mm/h.

(iii) *SCS-CN Model*

The SCS-CN model is an empirical model originally developed by the U.S. Soil and Water Conservation Service in the 1950s. Its original purpose was to estimate 24 h rainfall in agricultural regions. Over time, however, it has found applications in the analysis of flood flow processes in urbanized watersheds. The model calculates net precipitation, which is equivalent to surface runoff, as shown in Eq. (2.39).

$$P_e = \frac{(P - J_a)^2}{P - J_a + S} \tag{2.39}$$

where, P_e—accumulated net rainfall (effective rainfall or surface runoff), mm;

P—cumulative rainfall, mm;

J_a—initial loss (including surface depression storage, interception and infiltration before runoff formation), mm;

S—potential maximum infiltration, mm.

In Eq. (2.39), the relationship between J_a and S is determined using empirical Eq. (2.40).

$$J_a = \alpha \cdot S \tag{2.40}$$

The calibration of the coefficient α can be determined from observed runoff data and is typically set to $\alpha = 0.2$.

By analyzing runoff data from different watersheds, a series of runoff curves can be derived. Thus, Eq. (2.39) can be transformed into Eq. (2.41).

$$P_e = \frac{(P - 0.2S)^2}{P + 0.8S} \tag{2.41}$$

According to Eq. (2.39), the amount of runoff in the watershed depends primarily on two key parameters: cumulative rainfall and potential maximum infiltration. The SCS infiltration equation calculates the potential maximum infiltration, denoted as S, based primarily on the parameter *CN*, which encapsulates the comprehensive characteristics of the watershed.

$$S = 25.4\left(\frac{1000}{CN} - 10\right) \tag{2.42}$$

where, *CN* is the runoff curve coefficient for the day, a parameter related to the Antecedent Moisture Condition (AMC), slope, vegetation, and land use practices of the watershed. It provides a comprehensive assessment of watershed characteristics prior to precipitation. A higher *CN* value indicates a greater propensity for runoff generation, while a lower value indicates reduced runoff potential.

SCS classifies Antecedent Moisture Condition (AMC) into three categories: I (dry condition, wilting point), II (moderate moisture condition), and III (wet condition, field water holding capacity condition). The *CN* value for condition I represents the minimum value on the daily dryness curve. The *CN* value for condition II can be obtained by referring to the table (Appendix 1). The *CN* values for Condition I and III can be derived from the *CN* value for Condition II using Eqs. (2.43) and (2.44).

$$CN_1 = CN_2 - \frac{20 \times (100 - CN_2)}{100 - CN_2 + \exp[2.533 - 0.0636 \times (100 - CN_2)]} \tag{2.43}$$

$$CN_3 = CN_2 \times \exp[0.00673 \times (100 - CN_2)] \tag{2.44}$$

where CN_1, CN_2, and CN_3 represent the values of the curves corresponding to dry, average humidity, and wet conditions, respectively.

Table 2.2 compares the characteristics, advantages, and application scopes of the three infiltration models. It is worth noting that the Horton model is highly suitable for urban areas and is extensively used in multiple urban rainfall runoff models for infiltration calculations.

iv. Surface Runoff Calculation

The objective of the surface runoff calculation task is to simulate stormwater collection from each sub-basin and its eventual discharge at the control cross-section of the urban river and stormwater network outlet. In the SWMM model, this is achieved by approximating the three components of the sub-basin (permeable area with puddle storage, impermeable area with puddle storage, and impermeable area without puddle storage) as nonlinear reservoirs. This requires a unified solution to Manning's equation and the continuity equations.

Table 2.2 Comparison of three SWMM infiltration models

Model	Features and Benefits	Scope of application
Horton model	This model assumes that the intensity of rainfall always exceeds the rate of infiltration. It describes the connection between the infiltration rate and the length of rainfall without taking into account variables like soil type, soil moisture content, or the amount of infiltration. Rather, it functions as a dependent variable of soil water storage	Fewer parameters need to be calibrated, making it suitable for small-scale simulations
Green-Ampt model	Assuming that there is a drastically changing soil-wet/dry interface in the soil layer, the intensity of rainfall at the beginning of the rainfall period can be less than the infiltration rate; the infiltration rate can be calculated over time, taking into account that the infiltration of rainfall causes the subsurface to change from unsaturated to saturated, and dividing the infiltration process into a soil-unsaturated stage and a soil-saturated stage, respectively, for the purpose of the calculations	The requirements for soil data are high
SCS-CN model	Infiltration calculations were performed using a composite parameter, CN, that summarizes the characteristics of the watershed. The CN value was obtained by analyzing precipitation in relation to runoff probabilities on a daily basis. Next, this value was correlated with factors including soil type, land use, vegetation, initial soil saturation, and current soil conditions The method provides a complete flow hydrograph by accounting for rainfall losses through the calculation of soil moisture absorption capacity, without reflecting the influence of rainfall processes (rainfall intensity) on runoff generation; underestimation of smaller flood flows	Suitable for modeling large areas (as large as 50 km^2) and larger design storm intensities

Figure 2.4 presents a schematic depiction of the catchment subregion utilizing the nonlinear reservoir approach. Rainfall (I) accounts for the inflow into the nonlinear reservoir while infiltration and surface runoff serve as the outflow (Q). The model proposes that the surface runoff at the subregion's outlet is a uniform flow with a water depth identified as (y—yd). Furthermore, it is assumed that the flow rate from the reservoir is a non-linear function of the water depth within the reservoir.

The Eq. (2.45) demonstrates the continuity equation for a linear reservoir.

$$\frac{dV}{dt} = A\frac{dh}{dt} = AI^* - Q \tag{2.45}$$

where, I^*—net rain intensity, mm/s.

The rate of outflow is calculated by utilizing Manning's equation in conjunction with Eq. (2.4) and the coupled Eq. (2.45), which is formulated as a nonlinear

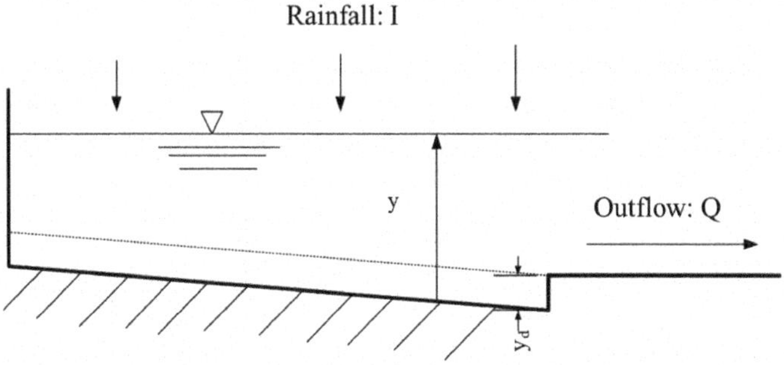

Fig. 2.4 Schematic representation of catchment subregion generalization through the nonlinear reservoir approach

differential equation for the purpose of solving for the variable 'y.'

$$\frac{dh}{dt} = i^* - \frac{1.49W}{An}(h - h_p)^{\frac{5}{3}} slp^{\frac{1}{2}} = i^* + WCON(h - h_p)^{5/3} \tag{2.46}$$

where, WCON—the flow algorithm parameter, which consists of area, width, slope, and roughness, and is equivalent to $-\frac{1.49W}{An} slp^{1/2}$.

During each time step, the Eq. (2.46) is solved using the finite difference method. The right-hand side of the equation takes into account the net inflow and outflow as time-step averages, while the net rainfall intensity value I* is treated as a time-step average in the computations. This leads to the transformation of Eq. (2.46) into (2.47).

$$\frac{h_2 - h_1}{t} = I^* + WCON\left[h_1 - \frac{1}{2}(h_2 - h_1) - h_p\right]^{5/3} \tag{2.47}$$

where, h_1—the initial value of water depth in the time period, m;

h_2—the last value of water depth in the time period, m.

The model involves a three-step process. (1) The average potential infiltration rate during the rainfall period can be calculated using either the Green-Ampt or Horton infiltration equations. (2) The water depth at the end of the period, h_2, can be determined by iterative solution using the Newton–Raphson iterative method. (3) The instantaneous discharge rate, Q_2, at the end of the period for the catchment subregion can be calculated by substituting h_2 into the Manning equation.

For an impermeable area without standing water storage, both the infiltration rate f and the standing water storage yd are assumed to be zero. For an impermeable area with standing water storage, the infiltration rate f is assumed to be zero.

v. Cumulative Surface pollutant Model

The accumulation process pertains to the gradual alteration of pollutants on the catchment area surface during sunny days, involving the buildup of pollutants over time. The Revised Soil Loss Equation (RUSLE) is employed in the permeable zone, while in the impermeable zone, it can be described by four cumulative equations: linear, power, exponential, and saturation functions. As a result, the accumulation of pollutants is dependent on time in these four equations, with pollutant accumulation ultimately reaching a maximum cumulative amount before ceasing.

(1) Linear Function Cumulative Equation:

$$P_t = \mathrm{Min}(P_m \times C_1 \times t_e) \tag{2.48}$$

(2) Power Function Cumulative Equation:

$$P_t = \mathrm{Min}\left(P_m \times C_1 \times t_e^{C_2}\right) \tag{2.49}$$

(3) Exponential Function Cumulative Equation:

$$P_t = P_m\left(1 - e^{-C_1 t_e}\right) \tag{2.50}$$

(4) Saturation Function Cumulative Equation:

$$P_t = \frac{P_m t_e}{N + t_e} \tag{2.51}$$

where, P_t—the cumulative amount per unit of subsoil, kg/m^2;

P_m—the maximum accumulation per unit of subsoil, kg/m^2;

C_1—the accumulation rate constant, d^{-1};

C_2—the time index, when $C_2 = 1$ power function cumulative equation into a linear function cumulative equation of the special case;

N—the half-saturation constant, the number of days when half of the maximum accumulation is reached.

vi. Surface Pollutant Washout Model

The process of pollutant flushing involves the erosion and dissolution of pollutants on the surface after rainwater runoff forms. Three primary flushing models exist in SWMM: the exponential equation, the flow-characterized flushing curve, and the average concentration curve of the second rainfall.

(i) *Exponential Equation*

The exponential equation shows that the quantity of pollutants flushed away is directly proportional to the amount of pollutants remaining on the surface and is exponentially related to the runoff volume.

$$P_{off} = \frac{-dP_p}{dt} = C_1 r^n P_p \quad (2.52)$$

where, P_{off}—the amount of pollutants flushed by runoff at moment t, kg/s or kg/h, which has a certain exponential relationship with the runoff volume and is proportional to the amount of remaining surface pollutants;

C_1—flushing coefficient;

P_p—the amount of remaining pollutant factor per unit of subsurface at moment t, kg/h m^2 or kg/m^2;

r—the runoff rate per unit area of the sub-basin at moment t, mm/h;

n—scour index.

(ii) *Flow Characteristic Flush Curve*

This method assumes that flushing is a simple function of runoff volume and that the flushing load of pollutants is unrelated to the total amount of pollutants accumulated at the surface.

$$P_{off} = C_1 r^n \quad (2.53)$$

(iii) *EMC Curve*

This technique is a unique instance of a flushing curve's flow characteristics. When the flushing index is 1.0, the coefficient C1 denotes the typical pollutant concentration (g/L)—also referred to as EMC—computed as per Eq. (2.27). In the three models described, the process of flushing comes to a halt only after all surface contaminants have been eliminated.

c. **SWMM Model Application Examples**

The SWMM model has found extensive application in the assessment and forecasting of surface runoff dynamics, runoff volumes, and pollutant loads during urban storm events, both domestically and internationally. Furthermore, it plays a pivotal role in the strategic planning, analysis, and design of integrated and diversionary drainage networks, sewage pipelines, and various other drainage systems. The following describes the application and research progress of some representative SWMM models for surface runoff in recent years.

i. Program Evaluation and Impact Analysis

(i) Evaluation of Planning Schemes.

Dong et al. [21] take the drainage planning of Shenzhen Hewan area as an example, refer to the drainage situation in typical rainfall years, quantitatively analyze the drainage situation of the area with SWMM model, and evaluate the environmental impact of the "layout planning scheme" of the area in the near future and in the long term, and the results show that SWMM is one of the effective tools for calculating regional drainage analysis.

Zhou [22] established a SWMM model for the Songjiang International Eco-Business District in Shanghai to evaluate the effect of sponge city construction on rainfall runoff control and pollution load control in the region, and the results showed that with the increase of the reproduction period, the total runoff reduction rate of sponge city construction showed a decreasing trend, and the peak runoff reduction showed an increasing trend, compared with the traditional development mode.

(ii) Drainage Network Design Assessment

Chen et al. [23] took the new and old urban areas of Dongguan City as a pilot, and established a SWMM-based hydraulic model of the drainage pipe network to evaluate the drainage pipe carrying capacity in this urban area; to solve the urban flooding, they proposed to increase the pipe diameter, and simulated the optimized design scheme to improve the pipe fullness and reduce the waterlogging points.

Cheng et al. [24] used the SWMM model to compare and analyze the changes in node water accumulation and pipe overflow capacity of the vertical and horizontal pipe arrangements in flat areas under actual rainfall conditions with the increase in return period. The results show that with the improvement of the return period, the vertical pipe arrangement in flat areas can effectively reduce the number of inspection wells under heavy rainfall conditions, which is more advantageous than the horizontal arrangement.

Ma et al. [25] applied the SWMM model to simulate the operating condition of a combined sewer system in a northern county at different time periods, and compared that the main reason for overflow is the pipe diameter is too small and the surface impervious area is too large. After increasing the downstream pipe diameter, the overflow duration of the simulated node was reduced from 70 to 8 min, and the optimization effect was significant.

Li et al. [26] used the SWMM model to simulate the current operation of the drainage system in the selected area, analyzed its "bottleneck constraints", and proposed two pipe network improvement schemes, namely, changing the node elevation and increasing the pipe diameter.

Li et al. [27] analyzed the role of construction measures in runoff control in a sponge pilot area in Xiamen City using the SWMM model. The results showed that the arrangement of low-impact development measures can effectively relieve the drainage pressure of the stormwater pipe network, and most of the outlet discharge is less than the corresponding volume of the storage wet pond, which can realize the runoff control requirements.

(iii) Modeling the Impact of Urbanization on Precipitation

Camorani et al. [28] conducted an analysis of the impact of urbanization on rainfall with different return periods using SWMM model simulations. The results showed that urbanization significantly increases the occurrence of urban rainfall events characterized by high frequency and low flow rates, while its influence on rainfall events with longer return periods is comparatively less significant.

Liang et al. [29] used the SWMM model to simulate the runoff processes under different scenarios: pre-urbanization (in its natural state), post-development without regulatory measures (using ordinary concrete pavement), and post-development with the implementation of permeable pavement to manage 2 year, 10 year, and 100 year design storms, respectively. The results indicate that the urbanization process results in a significant increase in peak flow rates and a reduction in runoff duration. In addition, the integration of permeable pavement plays a critical role in mitigating peak flow rates and reducing flood volumes.

Wang et al. [30] used the SWMM model to simulate urban hydrologic effects before and after urbanization and under different scenarios, including the incorporation of two LID measures: permeable brick and depressed green space. The results showed that the peak flood flow increased significantly, the peak flood time advanced, and the runoff coefficient became larger after urbanization. The combination of permeable brick and recessed green space effectively relieved the drainage pressure on the stormwater network, reduced the peak flow, and decreased the runoff coefficient. The synergistic use of these two measures demonstrated superior flow control and improved utilization of stormwater resources.

(iv) Analysis of the Effectiveness of Source Reduction Measures

Zhang [31] used the SWMM model to simulate the effects of source reduction measures such as green roofs on flood flow reduction, peak flow delay, and runoff coefficient reduction. The results showed that the flood peak flow was reduced by 53.0% and the peak flow occurrence time was delayed from 52 to 133 min after the application of source reduction measures compared to conventional development; the runoff coefficient was reduced by 0.22.

Wu et al. [32] used the SWMM model to simulate stormwater runoff and water quality in a representative area surrounding Qing Lake, an inland river in the urban landscape of Shenzhen. Their study investigated the temporal relationship between stormwater runoff and water quality under different return periods and different permeable surface conditions.

Xiong et al. [33] used the SWMM model to evaluate the effectiveness of three strategies for managing surface runoff and controlling watersheds in urban areas: green roofs, infilled green spaces, and permeable pavement. The results showed that the combined implementation of these measures in piped residential neighborhoods within the watershed resulted in a significant improvement. The total annual runoff control rate increased to 69% and the total annual suspended solids (SS) discharge decreased by 41%.

Wang et al. [34] used the SWMM model to evaluate the impact of stormwater regulation before and after the implementation of a sponge cell retrofit. The study found that sponge cell retrofits effectively reduce peak flows and delay their onset during shorter return periods. This not only increases drainage capacity, but also reduces external discharge and total runoff. However, as the return period increases, the increase in stormwater runoff control after sponge cell retrofitting gradually decreases, along with a decrease in the peak flow reduction rate. When the return period exceeds 5 years, both the increase in total runoff control and the peak flow reduction begin to decrease.

Xiong et al. [35] developed a SWMM model in Guangming New District, Shenzhen, China, to evaluate the effects of sponge city engineering measures on stormwater runoff. The results showed that infilled green spaces had more pronounced benefits in reducing surface runoff, while permeable pavement had benefits in mitigating flood flows.

ii. SWMM Model Calibration and Refinement

(i) Verification of Drainage Capacity

Zhong [36] used the SWMM model to calibrate the drainage system's surface water removal capacity in a Shanghai municipality, considering a 3 year return period. The results show that SWMM provides a more accurate representation of urban drainage system performance under different operational scenarios.

Cong et al. [37] conducted an analysis of a representative neighborhood in Beijing using the SWMM model. They evaluated the drainage performance, water accumulation, and street slope flow of the neighborhood under different design storm frequencies. The results showed that during a 10 year rainfall event, the flat and concave green spaces had 10 and 36% higher infiltration, 30 and 53% lower runoff volume, and 10 and 35% lower peak flood flow, respectively, than the convex green spaces.

Liu et al. [38] used the SWMM model to simulate the drainage and surface flooding processes in the city of Shanghai. The comparison of inundation results obtained from model simulations using recorded rainfall data with the observed field conditions generally agrees with the inundation trends at certain nodes, confirming the accuracy and reliability of the SWMM model.

(ii) Verification of Pollution Load Simulation Errors

Ma et al. [39] used the SWMM model to calculate the runoff pollution load in a typical residential area in Wenzhou City. They calibrated the model parameters using local observed data and analyzed the pollution load and cumulative changes of SS, COD, TN, and TP under four different rainfall scenarios. The results showed that the relative errors in simulating the four pollutants by the SWMM model were all less than 10%.

(iii) Improvin'g Accuracy

Park et al. [40] explored the influence of sub-catchment delineation methods and resolution on SWMM model simulations and runoff pollution predictions. Their research indicated a reduction in the overall pollutant load when merging sub-catchments with different land use types. Consequently, to ensure simulation precision, it is advisable to maintain uniform land use types within sub-catchments.

(iv) Multi-Method Coupled Simulation

Barco et al. [41] applied SWMM to a large urban area (217 km^2) in Southern California. The results showed that the coupled application of SWMM, GIS, and optimization methods is an effective tool for modeling nonpoint source pollution in large urban areas.

Zheng et al. [42] used the processing and analysis function of GIS on spatial data, combined with the SWMM model to establish a GIS system for stormwater network planning, realized the integrated management of attribute data and spatial data of stormwater network equipment, and provided a means of visualization and analysis for the calculation of urban storm water accumulation.

2.2.2 HSPF Model

a. **Principles of the HSPF Model**

The HSPF model, developed by the US Environmental Protection Agency in the late 1970s using the Fortran programming language, is based on the Stanford Watershed Model. It integrates features from several models, including HSP (Hydrological Simulation Program), ARM (Agricultural Runoff Management), NPS, among others. The HSPF model aims to replicate hydrological and water quality processes in vast watersheds, accounting for urban, forested, and rural areas. This model is especially efficient for assessing the enduring hydrologic effects in urban areas.

The HSPF model is known for its versatility since it can be applied to a wide range of watersheds and climate zones. It includes arid terrains in Saudi Arabia, the humid environments of Eastern United States and Europe, and the ice-covered expanses in Eastern Canada. It effectively simulates flood flows on various temporal scales, including hourly, daily, or minute-by-minute. Moreover, the software's adaptable structure streamlines the implementation of modifications and expansions. Functioning as a semi-distributed model for hydrological and hydraulic quality, it presents significant benefits in tackling all-encompassing concerns pertaining to water quality and quantity on a larger scale.

The HSPF model can be applied to diverse watersheds and various climatic zones, such as the arid regions of Saudi Arabia, the humid regions of the eastern United States and Europe, and the ice-covered regions of eastern Canada. Moreover, the model can simulate flood flows at different time scales—minute, hourly, or daily. Additionally, the program's structural features facilitate easy changes and extensions. The HSPF model is a semi-distributed hydrological and hydraulic quality model that provides significant benefits in resolving larger-scale water-quality and water-quantity issues.

The HSPF model necessitates extensive spatial and attribute information input. The model inputs encompass a comprehensive array of data, comprising hydrometeorological parameters, land-use classifications, elevation data, accumulation and washout coefficients, pollutant attenuation coefficients, and the specific attributes of the receiving water body. The output data of the model includes time-series records of surface runoff, profiles of pollutant loading, changes in water quantity and quality at various watershed sites, and the effects of pollutants on the water bodies that receive them.

HSPF provides comprehensive simulation of surface runoff, soil erosion, and water flow in canals and channels. It allows continuous modeling of the complex interactions and transport of pollutants, including sediment, BOD_5, DO, N, P, pesticides, and E. coli, among others. The HSPF model is widely used in flood control planning and management, nonpoint and point source pollution assessment, soil erosion and sediment transport research, and evaluation of urban best management practices (BMPs) and their associated impacts.

The major drawbacks and limitations of the HSPF model include:

(1) The complex calculations of water flow within drainage pipes cannot be performed, making it unsuitable for simulating extreme rainfall events.
(2) Its applicability in urban areas is considerably limited, with non-unique parameter values during model calibration.
(3) Simulation accuracy is constrained by spatial and attribute data limitations, which place high demands on data input and require continuous hydrologic and water quality monitoring data for model calibration.
(4) Assuming that the Stanford watershed model is universally applicable to all regions, the simulations are limited to homogeneously mixed rivers, reservoirs, and one-dimensional water bodies. Therefore, for the study of complex watersheds or water bodies, it is necessary to combine it with other models.

HSPF consists of three application modules: PERLND, IMPLSD, and RCHRES, which simulate hydrology and water quality for permeable areas, impervious areas, rivers, and mixed reservoirs, respectively. These three main application modules can be further subdivided into several submodules, as shown in Figs. 2.5, 2.6, 2.7 and 2.8.

b. **HSPF Model Calculations**

i. Hydrological Flow Computation Simulation

The primary objective of the hydrologic flow calculation simulation is to replicate the changes in surface precipitation runoff within a given area. In permeable areas, the runoff volume is determined using the Stanford model with parameter adjustments. The runoff processes are simulated using three different methods that include surface runoff, interflow, and groundwater flow, with surface runoff calculations using the Philip equation and the Chezy-Manning formula. In impervious areas, runoff processes are simulated using a linear function accumulation model and an erosion model.

When simulating the hydrological processes of precipitation, the calculation method used for permeable surfaces is based on the Stanford Model VI, with modifications in some of its parameters. The process of surface runoff generation can be summarized as follows: during rainfall or snowfall events, a portion is intercepted by the ground surface and then subtracted for surface depression storage, infiltration, and evaporation, resulting in surface runoff. Surface vegetation interception and depression storage, as well as the water-holding capacity of the soil layers (top, middle, and bottom), vary with soil characteristics, time, crop types, and agricultural practices. For impervious surfaces, the primary hydrologic process is rainfall and snowmelt runoff, after accounting for interception by roofs, evaporation, and vegetative cover, resulting in surface runoff.

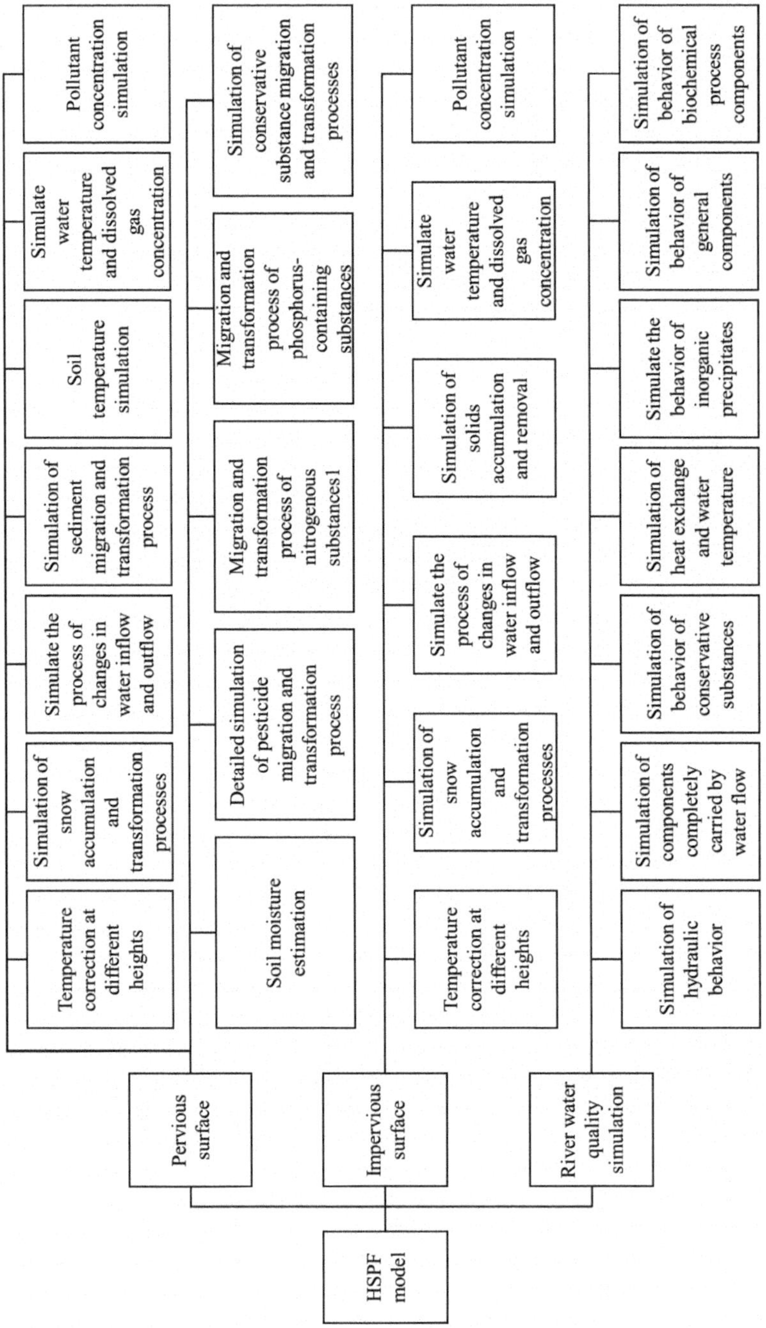

Fig. 2.5 Structure of the HSPF model

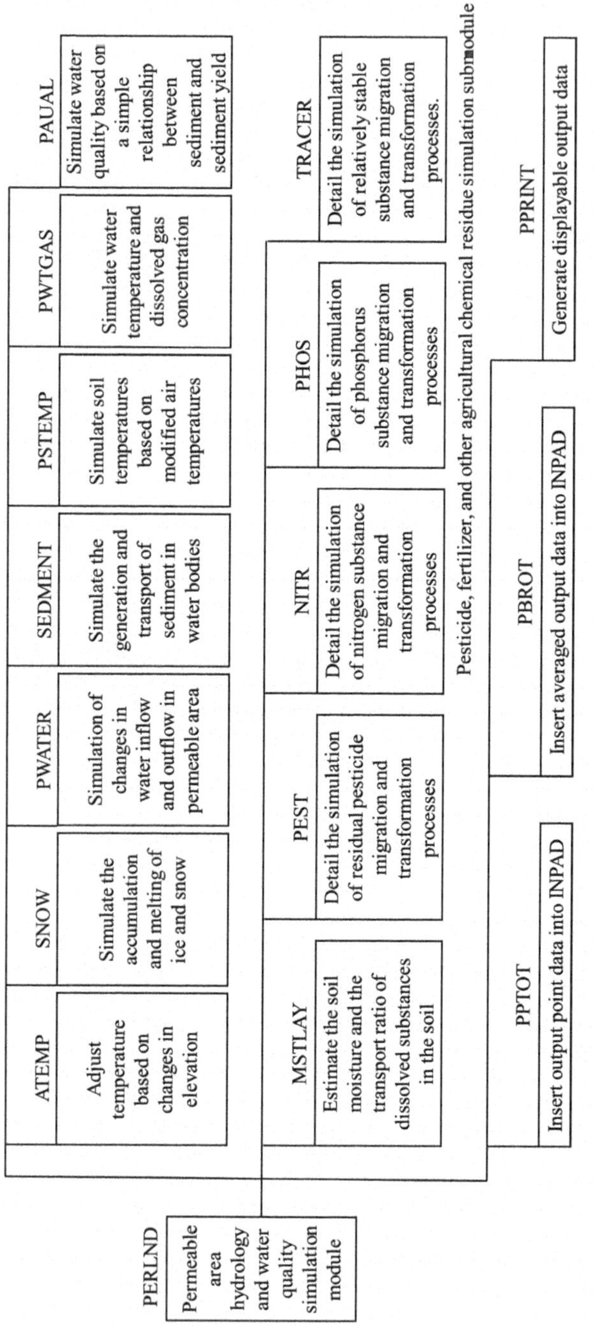

Fig. 2.6 Structure of the PERLND module

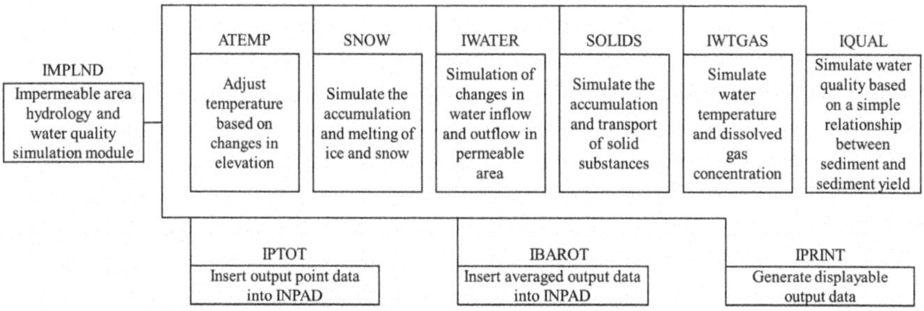

Fig. 2.7 Structure of the IMPLND Module

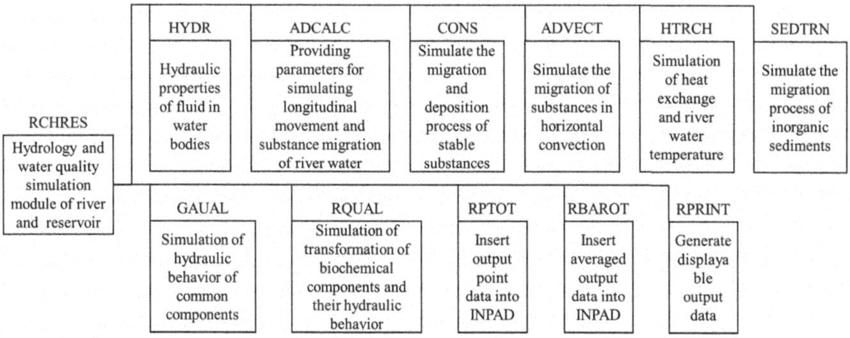

Fig. 2.8 Structure of the RCHRES module

ii. Water Quality Computation Simulation

The HSPF model can simulate several water quality constituents, including sediment, DO, BOD_5, pesticides, TN, NO_2–N, NO_3–N, TP, and chloride. Of these, sediment, N, and P are the most important pollutants in non-point source pollution. The transport and transformation of pollutants involves complex equilibrium processes for substances such as nitrogen, phosphorus, and pesticides.

(i) *Sediment Simulation*

In the generation of stream sediment, soil erosion is divided into several sub-processes, including raindrop splash erosion, runoff detachment erosion, and runoff transport. These processes are calculated using soil surface rainfall erosion models.

(ii) *Simulation of Nitrogen and Phosphorus*

The HSPF model can simulate several water quality constituents, including the ability to simulate the migration and transformation of inorganic N and inorganic phosphorus in both soil and water. The processes of N and P migration and transformation in soils include plant fixation and uptake, mineralization, adsorption, desorption, and more. In water bodies, N migration and transformation include vertical convection of dissolved nitrate, nitrite, and ammonia; ammonia release at the bottom of the water body; ammonium ionization; ammonium volatilization; nitrification of ammonia and nitrite; denitrification of nitrate; ammonification from BOD decay; sediment adsorption and desorption of ammonia in water; and deposition, flushing, and vertical convection of adsorbed ammonia. The migration and transformation of P in water bodies includes vertical convection of dissolved phosphate, release of phosphate at the bottom of the water body, sediment adsorption and desorption of phosphate in the water body, and deposition, scouring, and vertical convection of adsorbed phosphate.

2.2.3 *SWAT Model*

a. **SWAT Model Principles**

The SWAT model was developed by the United States Department of Agriculture's Agricultural Research Service in the early 1990s. It is a watershed-scale distributed model based on physical mechanisms. As of 2011, the SWAT model has undergone seven versions. This model is not only capable of simulating hydrological processes such as surface runoff, infiltration, groundwater flow, and return flow but can also investigate the impacts of environmental factors, including soil temperature, humidity, and atmospheric conditions. Additionally, it has the capability to simulate changes in water quality, such as sediment transport, nutrient substances, and more. Currently, the SWAT model is applicable to watershed scales ranging from approximately 0.004 to $491,665\ km^2$.

The SWAT model includes essential elements of non-point source pollution models at the field level, such as CREAMS (Chemicals, Runoff, and Erosion from Agricultural Management Systems), the groundwater loading effects model on agricultural ecosystems (GLEAMS), the erosion-productivity impact calculator (EPIC), and the routing output to outlet model (ROTO). Its comprehensive nature assimilates the primary features of these models.

The SWAT model has been the subject of various adaptations tailored to the characteristics of specific simulation targets. These include improved models such as SWIM, SWATMOD, SWAT-G, AVSWAT, ArcSWAT, and E-SWAT. In addition, the SWAT model has been integrated into the BASINS (Better Assessment Science Integrating Point and Nonpoint Sources) system.

The SWAT model is widely used worldwide and offers advantages such as minimal observed monitoring data requirements, high computational efficiency, and free access to model source code, theoretical documents, and user manuals. On the one hand, SWAT, based on physical mechanisms and linked to HSPF in terms of data requirements, can continuously simulate changes in water quality and quantity even in the absence of certain observed monitoring data, with a daily time step. SWAT, on the other hand, is a distributed model. Because the SWAT model uses Hydrological Response Units (HRUs) as its computational units, it calculates each HRUs region individually based on specific land use and soil type combinations and overlays the results. Therefore, when simulating large areas or regions with multiple management decisions, SWAT demonstrates high computational efficiency with minimal time and financial investment.

The combination of SWAT with GIS and BASINS is an existing successful example. GIS supports the data input of SWAT model in terms of topography, land use, soil and other spatial information, which greatly improves the efficiency of SWAT model in terms of data management, parameter extraction and result expression.

Generally, the SWAT model provides more accurate results for long-term runoff simulations compared to short-term simulations, and there is room for improvement in the accuracy of daily-scale simulation results. To achieve more precise simulation outcomes and broaden its application scope, the SWAT model has several aspects that require further refinement, primarily in the following five areas:

(1) Enhance pollutant simulation capabilities. Presently, the SWAT model mainly concentrates on simulating nitrogen and phosphorus nutrients for non-point source pollution, and the migration simulation module for bacteria and other microorganisms needs improvement. Besides, the simulation abilities need to expand to cover different pollutants such as petroleum, heavy metals, endocrine-disrupting substances, and more.
(2) The establishment of a China-adapted database and substitution procedure module is critical. Due to the significant differences between the soil, terrain, runoff characteristics, and default soil and vegetation databases in the SWAT model and those in China, the efficiency and accuracy of the model are severely affected.
(3) Acquiring more precise spatial parameters. Utilizing extensive measured data, including professional maps, hydrological and meteorological monitoring, to enhance model calibration methods and improve simulation accuracy.
(4) The model needs to be used in conjunction to harness the capabilities of the SWAT model. Depending on regional characteristics and management requirements, developing SWAT models under different conditions has become a new demand.
(5) The hydrological response simulation of the SWAT model to urbanization requires enhancement. Human activities have an increasingly significant impact on natural cycles, significantly affecting watershed rainfall, floods, and water quality. Currently, the simulation of the hydrological response to urbanization with the SWAT model is not flawless. Enhancements to the model include more realistic simulation of urban landscapes and the use of shorter time steps.

b. **SWAT Model Computation**

i. Basic Computational Framework of SWAT Model

The simulation of the SWAT model includes the road section of the water cycle (i.e., the production and slope flow section) and the water section (i.e., the river network flow section). Urban surface runoff only considers the road section. In the road section of the water cycle, the SWAT model takes into account various factors such as climate, hydrology, vegetation cover, and underlying surface types. When calculating evapotranspiration, three types are considered: water surface evaporation, bare soil evaporation, and vegetation transpiration, each simulated separately.

The SWAT model, based on water balance, treats the unsaturated zone above the impermeable layer and the shallow groundwater as a whole. It simulates both surface runoff and peak flow for each HRU. The water balance equation for simulating runoff components is shown in Eq. (2.54).

$$SW_t = SW_0 + \sum_{i=1}^{t}\left(R_{day} - R_{surf} - E_a - W_{seep} - R_{gw}\right) \tag{2.54}$$

where, SW_t—the soil water content at the end of the calculation, mm;

SW_0—the soil water content at the beginning of calculation, mm;

R_{day}—the precipitation on day i, mm;

R_{surf}—the surface runoff volume on day i, mm;

E_a—the evapotranspiration on day i, mm;

W_{seep}—the seepage and lateral flow from the bottom of the soil profile on day i, mm;

R_{gw}—the base flow on day i, mm.

ii. Runoff Calculation

In general, the runoff components simulated by the SWAT model include surface runoff, interflow, shallow groundwater flow, and deep groundwater flow. The model provides the SCS runoff curve number method and the Green-Ampt infiltration method to calculate surface runoff. The Green-Ampt method requires rainfall data in hourly units, but its practical application is limited. The SCS runoff curve number method is discussed in Sect. 2.1 of this chapter.

iii. Concentration Calculation

(i) Time of Concentration

The time of concentration is the amount of time from the beginning of a rainfall event until the entire subbasin area is contributing to flow at the outlet. The time of concentration is calculated by summing the overland flow time (the time it takes for flow from the remotest point in the subbasin to reach the channel) and the channel flow time (the time it takes for flow in the upstream channels to reach the outlet):

$$t_{conc} = t_{\mathrm{ov}} + t_{\mathrm{ch}} \tag{2.55}$$

$$t_{ov} = \frac{L_{slp}}{3600 \cdot u_{ov}}, \quad t_{ch} = \frac{L_c}{3.6 u_c} \tag{2.56}$$

where, t_{conc}—the time of concentration for a subbasin, h;

t_{ov}—the time of concentration for overland flow, h;

t_{ch}—the time of concentration for channel flow, h;

L_{slp}—slope length of the subbasin, m;

u_{ov}—overland flow velocity, m/s;

u_c—the average channel flow velocity, m/s;

L_c—the average channel flow length of the subbasin, km.

The L_c can be approximated using Eq. (2.57):

$$L_c = \sqrt{L \cdot L_{cen}} \tag{2.57}$$

where, L—the distance from the furthest point within the subbasin to the outlet, km;

L_{cen}—the distance between the river channel and the centroid of the subbasin, km.

It is generally approximated that $L_{cen} = 0.5L$, which implies $L_c = 0.71L$.

u_{ov} and u_c can be obtained using the Manning formula, as shown in Eqs. (2.58) and (2.59).

$$u_{ov} = \frac{q_{ov}^{0.4} \cdot slp^{0.3}}{n^{0.6}} \tag{2.58}$$

$$u_c = \frac{0.489 \cdot q_{ch}^{0.25} \cdot slp_{ch}^{0.375}}{n^{0.75}} \tag{2.59}$$

where, q_{ov}—the average overland flow rate, m^3/s,

q_{ch}—the average channel flow rate, m^3/s;

n—Manning's roughness coefficient for the subbasin;

slp—the average slope in the subbasin, m/m;

slp_{ch}—channel slope, m/m.

In summary, Eq. (2.55) can be transformed into Eq. (2.60).

$$t_{conc} = \frac{L_{slp} \cdot n^{0.6}}{3600 \cdot q_{ov}^{0.4} \cdot slp^{0.3}} + \frac{0.403 \cdot L \cdot n^{0.75}}{q_{ch}^{0.25} \cdot slp_{ch}^{0.375}} \tag{2.60}$$

(ii) Concentration Volume of Overland Flow

The SWAT model accounts for the time delay of overland flow, while the runoff routing equation for surface flow, calculated using the SCS runoff curve number method, is presented below:

$$R_{surf} = \left(R'_{surf} + R_{sotr,i-1}\right)\left[1 - \exp\left(-\frac{k_{surlag}}{t}\right)\right] \tag{2.61}$$

where, R_{surf}—the amount of surface runoff discharged to the main channel on a given day, m^3/d;

R'_{surf}—the amount of surface runoff generated in the subbasin on a given day, m^3/d;

$R_{sort,i-1}$—surface runoff stored or lagged from the previous day, m^3/d;

k_{surlag}—surface runoff lag coefficient.

Given the specified time of concentration, a larger surface runoff detention coefficient indicates less detained water in the sub-watershed.

c. **SWAT Model Application Cases**

The SWAT model, with its sophisticated model structure and powerful modeling capabilities, has been widely utilized in watershed hydrological process simulation, runoff pollution simulation, and water resources management. Its applications include watershed water balance, long-term surface runoff, average runoff simulation, sediment yield, pesticide transport, non-point source pollution, and more. Additionally, the SWAT model can assess the impact of existing and potential future factors on large-scale water resources, such as human activities like land management practices, BMPs, and natural changes like climate warming.

i. Hydrological Process Simulation and Non-Point Source Pollution Analysis

Yang et al. [43] pointed out through their research that the SWAT model can effectively assess non-point source pollution at different scales and identify critical pollution areas. It can establish different management scenarios to evaluate the effectiveness of various management measures.

Chiang et al. [44] evaluated the impacts of land use changes and grazing management on sediment yield and nitrogen and phosphorus nutrient losses in the basin. The results indicated that an increase in nutrient application in grazing areas led to an increase in total nitrogen loss in the Beatty basin. Urban land use resulted in increased sediment yield and nitrogen loss in the Moore Creek sub-basin.

Tang et al. [45] analyzed the temporal and spatial characteristics of non-point source pollution in the Wenyu River basin within Beijing. The results indicated that non-point source outputs accounted for over 90% during the flood season throughout the year. The pollution load modulus was highest in rain-fed areas with predominant crop rotation of wheat and corn, while it was relatively smaller in forested and orchard areas.

Zhu et al. [46] utilized the SWAT model to investigate the Hongmen River basin in the mid-scale Miyun County, Beijing. The results indicated that the relative error of monthly flow simulation was less than 15% during both the calibration and validation periods. The determination coefficient (R^2) was greater than 0.8, and the Nash coefficient was higher than 0.7, indicating that the SWAT model provided good simulation results for runoff in the Hongmen River basin. This confirms that the sensitivity analysis module in the SWAT model can be applied to relatively small and mid-scale basin.

ii. Hydrological Response under Environmental Changes

Applying the SWAT model to simulate hydrological conditions under diverse environmental changes, including land use/cover changes (LUCC) and climate variations, enables investigating hydrological responses to environmental variations. This approach aims to identify ideal land use/land cover scenarios for different climatic conditions within the watershed.

Wang et al. [47] used the SWAT model to simulate non-point source pollution in the Qinhuai River Basin. The study indicated that, under rainfall conditions, the runoff coefficient increased by less than 0.1 when land cover changed from forest to cropland or paddy field to cropland. However, the conversion from cropland or paddy field to developed land resulted in an increase in the runoff coefficient by 0.2–0.3, and the conversion from forest to developed land led to an increase in the runoff coefficient of over 0.3.

Based on the simulation results of the Rock Creek watershed in the Portland area of the United States using the SWAT model, Franczyk et al. [48] proposed that highly urbanized watersheds are more sensitive to climate change in terms of runoff response. In the case of a climate change scenario with a 1.20 °C increase in average annual temperature and a 2% increase in average annual precipitation, coupled with intensified urban development, it is projected that by the year 2040, the annual average runoff in the Rock Creek watershed will increase by at least 5.2%.

iii. Localized Version Enhancement

The wide applicability of the SWAT model in the international arena is manifested in terms of the model's reliable simulation at different spatial and temporal scales and under different terrain conditions, as well as its continuous improvement of simulation accuracy. For different geographic features around the world, the SWAT model has been improved in various corresponding ways, and has been better applied in various types of areas, such as low mountains, forests, cities, and wetlands.

After Sang et al. [49] improved the irrigation module and water use model of the SWAT model, they applied it in Tianjin City. This addressed limitations in the SWAT model for areas with intense human activity. The Nash–Sutcliffe efficiency (NSE) coefficient and correlation coefficient demonstrated substantial improvements.

Gassman et al. [50] summarized 37 studies that applied the SWAT model for simulating pollutant loss and evaluated the determination coefficients and NSE coefficients based on 113 studies using the SWAT model for calibration and validation results.

iv. Adaptation Studies of the Model in China

The research and application of the SWAT model in urban areas in China are limited, mainly applied in watershed hydrological studies and non-point source pollution simulations. The theoretical research of SWAT model mainly includes model improvement to enhance its convenience and effectiveness. However, there is a lack of research in areas such as management measures, crop yield, and model comparison and integration.

The application research of the SWAT model in China has covered all primary water resource regions, with study watershed sizes ranging from approximately 0.7 thousand to 428 thousand square kilometers. The research is predominantly concentrated in the three major river basins of the Yangtze River, Haihe River, and Yellow River. It involves studies on the applicability in different topographical environments, including arid and semi-arid regions, semi-arid and semi-humid regions, northwest cold regions, hilly areas, wetlands, lakes, reservoirs, dam group regions, irrigation areas, and areas with intensive human activities.

Xue et al. [51] applied the SWAT model to simulate nonpoint source pollution in the Chishui River basin. The results indicate that the SWAT model exhibits a certain degree of applicability in this basin, with favorable performance in simulating runoff and pollution load.

2.2.4 *Infoworks CS Model*

The Infoworks CS model is an improved and integrated version based on the Wallingford program developed by the British company Wallingford. The main computational units of the model include runoff calculation, flow routing calculation, pipeline hydraulic calculation, and runoff hydrograph, enabling it to be applied to runoff calculation and water quality simulation.

Input data for the InfoWorks CS model includes hydro-meteorological information, land use data, accumulation and wash-off coefficients, and parameters related to the layout and size of the pipe network. The InfoWorks CS model is integrated with GIS functionality, has a user-friendly interface, and uses a linked database with graphical analysis capabilities. In addition, the model can predict and compare the actual effects of different scenarios.

a. **Runoff Yield and Concentration Calculation**

The Infoworks CS model is based on detailed subbasins partitioning, uses a distributed model to simulate rainfall runoff processes, and calculates runoff based on the runoff characteristics of different land types.

Infoworks CS provides a variety of general and proprietary rainfall-runoff models. The main computational units include initial loss, runoff yield models, and flow accumulation models. Runoff yield models include the Fixed Percentage Runoff Model, Wallingford Procedure (Fixed) PR Model, New UK (Variable) PR Model, US SCS Model, Green-Ampt Infiltration Model, Horton Infiltration Model, and Fixed Infiltration Model. Concentration calculation models include the Bilinear Reservoir Model, Large Catch Model, Desbordes Runoff Model, and SWMM Model. A comparison of the various modules of Infoworks CS runoff yield and concentration models is shown in Table 2.3.

Table 2.3 Comparison of various modules of infoworks CS runoff yield models

Model name	Introduction	Applicability
Fixed percentage runoff model	You supply the figure for the percentage of rainfall that actually runs off into the system	Catchments where you a good estimate of the percentage runoff can be made
Wallingford procedure (fixed) PR model	The default model, which is calibrated for the UK	Urban catchments in or close to the UK
New UK (variable) PR model	This is a UK model, which represents the condition of the catchment changing throughout the simulation for a pervious surface	UK pervious catchments where it is important to take account of the change in catchment wetness during long storms
US SCS model	A rural catchment model	Rural catchments and pervious surfaces within a catchment
Green-Ampt model	Permeation model for pervious and semi-pervious surfaces	Rural surfaces and pervious surfaces within a catchment. This model is associated in the US in conjunction with the storm water management model (SWMM) runoff routing model
Horton model	Permeation model for pervious and semi-pervious surfaces	Rural surfaces and pervious surfaces within a catchment. This model can be used with all routing models
Bilinear reservoir model	The bilinear reservoir model uses linear reservoirs to simulate hillslope runoff. Each node corresponds to the net precipitation generated by a sub-watershed and is converted to an inflow hydrograph. Two linear reservoirs in series are used to simulate the storage capacity of the water body, each with a relatively fixed storage and discharge relationship. The routing parameters of this model depend on rainfall intensity, watershed area, and slope	Calculation of subarea runoff applicable to watersheds with subarea sizes smaller than 1 h m^2

(continued)

Table 2.3 (continued)

Model name	Introduction	Applicability
Large catch model	The model takes into account the characteristics of the watershed by assuming that the pipes have a detention effect, which aligns their runoff process line with reality. To accurately reflect flow characteristics, a lag factor is used to indicate that the peak flow will lag slightly behind the peak rainfall	Calculation of subarea runoff applicable to watersheds with subarea sizes smaller than 100 h m^2
Desbordes runoff model	The model assumes a certain parameter to reflect the proportionality between the watershed's outlet flow and the rainfall in the watershed	The model is suitable for continuous simulation
SWMM model	Typically coupled with Horton or Green-Ampt infiltration models and solved simultaneously by the continuity equation and Manning's equation	Use the nonlinear reservoir method to simulate surface runoff

b. **Calculation of Water Quality Simulation**

The Infoworks CS model can simulate the process of water quality changes, including the accumulation and transport of sediment and pollutants. It can model various water quality parameters, such as SS, BOD_5, COD, NH_3–N, TN and TP.

When simulating the accumulation and erosion of surface sediments in the Infoworks CS model, the sub-watershed can be divided into different urban functional zones, such as residential, commercial, industrial, and mixed-use areas. Alternatively, it can be divided into different land use types, such as streets, rooftops, and green spaces. Different urban functional zones or land use types can have different pollutant accumulation and erosion parameters to enhance the model's ability to more realistically simulate the surface conditions of the watershed.

c. **Surface Sediment Accumulation Model**

The Infoworks CS model defaults to the following equation to describe the deposition process of surface sediments. The mathematical expression for pollutant deposition is given by Eq. (2.62).

$$M_0 = M_d \cdot e^{-k_1 NJ} + \frac{P_s}{k_1}\left(1 - e^{-k_1 NJ}\right) \tag{2.62}$$

where, M_0—the maximum accumulated sediment or the pollutant deposition per time period, kg/ha;

M_d—the initial sediment amount, kg/ha;

NJ—dry day count, d;

P_s—cumulative rate, kg/(ha·d);

k_1—decay coefficient, d^{-1}; as the surface sediment increases, the sediment accumulation rate will decay.

d. **Surface Sediment Erosion Simulation**

Erosion refers to the process of surface erosion and the dissolution of pollutants during runoff. InfoWorks CS provides two erosion models for simulating sediment scouring: the Desbordes Model and the hydraulic runoff model.

(1) At the beginning of the Desbordes Model, the total sediment accumulation at the surface from the dry weather accumulation model. At the beginning of the rainfall runoff, the model calculates the storm erosion coefficient at the rainfall intensity to obtain the amount of sediment flushed into the pipe network, and then the total amount of pollutants flushed is calculated from the pollutant attachment coefficient.

$$K_a t = C_1 \cdot I(t)^{C2} - C_3 \cdot I(t) \tag{2.63}$$

where, $K_a(t)$—storm water erosion coefficient;

$I(t)$—effective rainfall, m/s;

C_1, C_2 and C_3—coefficients.

$$\frac{dMe}{dt} = K_a Mt - f_t \tag{2.64}$$

where M_e—the sediment load entering the pipe network, kg/(ha·s);

$M(t)$—the surface accumulated sediment load, kg/(ha·s);

K_a—the erosion coefficient corresponding to rainfall intensity;

f_t—the same as in Eq. (2.37).

(2) The pollutant attachment coefficient K_{pn} in the hydraulic flow model is used to characterize the relationship between the amount of sediment and the amount of pollutant attached. In other words, the pollutant mass is equal to the sediment mass multiplied by the pollutant attachment coefficient as expressed in Eq. (2.65).

$$K_{pn} = C_1 (I_{MKP} - C_2)^{C_3} + C_4 \tag{2.65}$$

where, C_1, C_2, C_3, and C_4—coefficients;

I_{MKP}—maximum rainfall intensity, mm/h.

The formula for calculating the amount of pollutants due to erosion is as follows:

$$M_n t = K_{pn}(i) \cdot f_n(t) \tag{2.66}$$

where, $M_n(t)$—pollutant load, kg/(ha·s);

K_{pn}—attachment coefficient;

$f_{\text{n}}(t)$—TSS load, kg/(ha·s).

2.2.5 MIKE Model

The MIKE model is a product developed by the Danish Hydraulic Institute (DHI) and has been widely used in various fields, including urban flooding and watershed pollution simulation. The model series software includes MIKE 11, MIKE 21, MIKE FLOOD, MIKE URBAN, MIKE BAISIN, MIKE SHE, etc. Among them, MIKE 11 is a one-dimensional water modeling software widely utilized in issues related to water quality, flow, and sediment transport; MIKE 21 is a two-dimensional water modeling software commonly used to simulate sediment, water flow and environmental conditions in rivers, estuaries and oceans, providing the necessary design conditions and parameters for engineering applications and planning; MIKE URBAN is a software for the simulation of surface runoff and pipe networks in urban areas. It includes comprehensive models for both supply and drainage pipe networks, capable of calculating flow conditions in both pressurized and non-pressurized pipes; MIKE FLOOD incorporates a comprehensive one-dimensional and two-dimensional flood simulation engine. Built on the FLOOD platform, it allows coupling of three models: MIKE 11, MIKE URBAN, and MIKE 21. This integration enables the simulation of water flow in urban drainage systems, both within pipe networks and on the surface where flooding may occur. It also facilitates simulations of issues such as floods, ocean storms, and dam breaches; MIKE BASIN is designed for basin or regional scale applications and serves as a GIS-based tool for water resources planning and management, addressing issues related to surface water runoff and water quality simulation. MIKE SHE can simulate many processes in the hydrologic cycle and is widely used in watershed management, floodplain studies, environmental assessment, and surface water-groundwater interaction.

Liang et al. [52] used the MIKE 11 software to construct a watershed rainfall-runoff model for a typical region in Beijing. The simulation results show that from the late 1990s to 2005, as the level of urbanization increased significantly, the initial rainfall losses and alongflow losses during the runoff process decreased. With the further improvement of the urban drainage network system, the duration of precipitation runoff was significantly shortened, and the rates of runoff generation and flow accumulation accelerated. Concurrently, there was a significant increase in the total volume of floodwater generated during the same period.

Yang et al. [53] took a residential community in Shenzhen as an example and, based on MIKE FLOOD, developed a coupled model of the drainage network and surface with a precision of 1.5 m × 1.5 m. The model simulated the runoff and peak flow in the area and assessed the risk of urban flooding. Sun et al. [54] used the MIKE FLOOD model to simulate the sponge city transformation effects in an old residential community in Taiyuan, Shanxi. Dong et al. [55] used MIKE FLOOD to analyze the effects of sponge city transformation in the pilot old urban area of Pingxiang. Under the conditions of a 30 year return period and a 2 h rainfall with a volume of 86.6 mm, the annual runoff control rate reached 75.8%, and the TSS reduction rate was 68.9%. Zhang et al. [56] used MIKE URBAN to construct a drainage pipe network model for the central urban area of Xi'an. Under different rainfall return periods, over 90% of the pipes in the study area were in full flow and over 60% of the manholes were in overflow. The number of full-flow pipe sections and overflow manholes increased with rainfall frequency, but the rate of increase decreased relatively.

Tian et al. [57] used the MIKE SHE model to simulate runoff in the Bahe River basin. The results showed good performance in annual runoff simulation, confirming the applicability of the MIKE SHE model in the Bahe River basin. Zhang et al. [58] conducted a simulation study on the hydrodynamics and water quality of the Shunyi section of the Chaobai River in Beijing using the MIKE 21 model. COD, TP, and NH3–N were selected as characteristic indicators.

2.2.6 Other Models

a. **SWC Model**

The Storm Water Calculator (SWC) program is a computer model developed by the US Environmental Protection Agency in 2014 for managing urban stormwater. It employs the Runoff, Infiltration, and LID submodels from the widely used SWMM 5 as its underlying calculation engine. The model utilizes long-term data on meteorology, hydrology, and LID facilities in the United States to calculate and simulate the runoff volume in a specified area. The model is primarily used to analyze rainfall runoff generation, retention volume, and the types and areas of LID facilities in the simulated region. It can also intuitively display the proportional relationship between rainfall and runoff in the simulated area. Each type of control facility comes with specific design parameters. At the same time, users have the flexibility to adjust the default parameters of the desired facilities based on actual conditions. The SWC model is suitable for environments with uniform soils and relatively small site scales. The simulated hydrologic processes include evaporation from the vegetation surface, evaporation from ponded rainfall in depressions, infiltration loss through the soil, and surface runoff.

b. **DRAINMOD Model**

The DRAINMOD model, developed by Professor Skaggs at North Carolina State University in the 1970s, is a computer simulation model for long-term agricultural drainage. Primarily used in agriculture, the model has been used since its development in a variety of applications including drainage control, irrigation, wetland hydrology, nitrogen dynamics, onsite wastewater treatment, forest hydrology, and other agricultural drainage systems. In recent years, the DRAINMOD model has been gradually applied to simulate the regulatory effects of LID facilities, including the validation and calibration of hydrological characteristics for bioretention cells.

c. **HYDRUS Model**

The HYDRUS model, developed by the United States Salinity Laboratory, is used to calculate salt transport processes and water movement in the vadose zone. This model is capable of analyzing not only water flow in saturated and unsaturated porous media, but also the transport of various solutes. In addition, it can simulate water flow regions in non-uniform soil. In recent years, the model has found wide application in agriculture and indoor simulation experiments.

d. **RECARGA Model**

The RECARGA model, developed by the University of Wisconsin, is capable of analyzing the hydrologic performance of bioretention cells under various design elements. This provides a theoretical basis for the rational design of bioretention cells. The RECARGA model uses the TR-55CN program to separately simulate runoff from permeable and impermeable areas in the study area. It uses the Green-Ampt equation to simulate infiltration from the storage layer to the medium layer of soil and uses the van Genuchten nonlinear equation to simulate water movement within the controlling soil layer (from the medium layer to the gravel layer) and from the gravel layer to the natural soil. The RECARGA model can be used to iteratively design and simulate various elements of a bioretention cell, such as area and root zone soil characteristics, to achieve specific performance goals.

e. **SUSTAIN Model**

The System for Urban Stormwater Treatment and Analysis Integration (SUSTAIN) model is a decision support system used for site selection, design, simulation, and optimization of LID/BMPs in urban development areas. The SUSTAIN model is built upon ArcGIS 9.3 and integrates hydrological, hydraulic, and water quality analysis models. It also utilizes cost management and optimization analysis techniques to evaluate the economic and effective aspects of stormwater management schemes in basins of varying scales. The SUSTAIN model utilizes ArcGIS 9.3 as its foundational platform, integrating hydrological, hydraulic, and water quality analysis models. Simultaneously, it incorporates cost management and optimization analysis techniques to assess and analyze the economic and effective aspects of stormwater management schemes in different-scale basins.

Appendix

See Table 2.4.

Table 2.4 CN value for urban areas

Land cover description				Hydrological soil group			
Land cover and hydrological conditions			Impervious surface ratio[a]	A	B	C	D
Fully developed urban areas (vegetation identified)							
Open space	Lawn, park, golf course, cemetery, etc.[b]	Poor (grass coverage ratio < 50%)		68	79	86	89
		Medium (grass coverage ratio 50–75%)		49	69	79	84
		Good (grass coverage ratio > 50%)		39	61	74	80
Impervious surface	Parking lots, roofs, driveways, etc. (including highways)			98	98	98	98
	Streets, highways	Paved with drainage systems (including highways)		98	98	98	98
		Paved with open channels (including highways)		83	89	92	93
		Gravel layer (including highways)		76	85	89	91
		Soil (including highways)		72	82	87	89
Western desert urban area	Natural desert landscapesc[c]			63	77	85	88
	Artificial landscape			96	96	96	96
Urban area	Commercial area		85	89	92	94	95
	Industrial area		72	81	88	91	93
	Residential area	Less than 1/8 acre	65	77	85	90	92
		1/4 acre	38	61	75	83	87
		1/3 acre	30	57	72	81	86
		1/2 acre	25	54	70	80	85
		1 acre	20	51	68	79	84
		2 acre	12	46	65	77	82

(continued)

Table 2.4 (continued)

Land cover description		Hydrological soil group			
Land cover and hydrological conditions	Impervious surface ratio[a]	A	B	C	D
Developing urban areas					
Reclassified permeable zones (without vegetation)[d]		77	86	91	94

[a] The impervious surface ratio is determined by assuming that the impervious surface is directly connected to the drainage system. The *CN* value for impervious surfaces is taken as 98, while pervious surfaces are considered equivalent to open spaces in good hydrologic condition. Multiple hydrologic conditions require the calculation of a composite *CN* value

[b] The *CN* value also applies to pasture land, but the method of calculating the composite *CN* value is different

[c] In calculating the composite *CN* value for natural desert landscapes, the *CN* value for impervious surfaces is assumed to be 98, while the *CN* value for permeable surfaces is assumed to be equivalent to that of desert shrubs under poor hydrological conditions

[d] Determination of the composite *CN* value in developing areas requires consideration of the development intensity (impervious surface ratio) and the *CN* values of the reclassified impervious areas. Note: 1 acre $\approx$ 4047 m^2

References

1. Roesner, L. A., Aldrich, J. A., & Dickinson, R. E. (1988). *Storm water management model*, User's Manual, Version 4: Addendum I, EXTRAN. EPA.
2. Zhang, W., & Zhou, Y. (2005). Pollution load calculation of city rainwater runoff and its evaluation model. *Journal of Hunan City University (Natural Science), 14*(3), 27–29. (in Chinese).
3. Deletic, A., & Maksumovic, C. T. (1998). Evaluation of water quality factors in storm water from paved areas. *Jouranl of Environmental Engineering, 124*(9), 869–879.
4. Wang, B., Huang, T., Cheng, H., et al. (2010). Small area stormwater runoff pollutant transport study. *Water & Wastewater Engineering, 36*(3), 128–131. (in Chinese).
5. Sansalone, J. J., & Cristina, C. M. (2004). First flush concepts for suspended and dissolved solids in small impervious watershed. *Jouranl of Environmental Engineering, 130*(11), 1301–1314.
6. Ma, M., Khan, S., Li, S., et al. (2002). Fist flush phenomena for highways: How it can meaningfully define. In *Proceedings of 9th international conference on urban drainage.*
7. Charbeneau, R. J., & Barretti, M. (1998). Evaluation of methods for estimating stormwater pollutant loads. *Water Environment Research, 70*(7), 1295–1302.
8. Brodie, I., & Rosewell, C. (2007). Theoretical relationships between rainfall intensity and kinetic energy associated with stormwater particle washoff. *Journal of Hydrology, 340*, 40–47.
9. Egodawatta, P., Thomas, E., & Goonetilleke, A. (2007). Mathematical interpretation of pollutant wash-off from urban road surfaces using simulated rainfall. *Water Research, 41*(13), 3025–3031.
10. Wüst, W., Kern, U., & Hermann, R. (1994). Street wash-off behaviour of heavy metals, polyaromatic hydrocarbons and nitrophenols. *Science of the Total Environment, 146*(147), 457–463.
11. Pariente, S. (2001). Soluble salts dynamics in the soil under different climatic conditions. *CATENA, 43*, 307–321.
12. Hairsine, P. B., & Rose, C. W. (1991). Rainfall detachment and deposition: Sediment transport in the absence of flow-driven processes. *Soil Science Society of America Journal, 55*, 320–324.
13. Lisle, I. G., Rose, C. W., Hogarth, W. L., et al. (1998). Stochastic sediment transport in soil erosion. *Journal of Hydrology, 204*, 217–230.

14. Gao, B., Walter, M. T., Steenhuis, T. S., et al. (2003). Investigating ponding depth and soil detachability for a mechanistic erosion model using a simple experiment. *Journal of Hydrology, 277*(1–2), 116–124.
15. Zoppou, C. (2001). Review of urban storm water models. *Environment Modeling and Software, 16*(3), 195–231.
16. Cen, G., Shen, J., & Fan, R. (1995). The application of Muskingum method to storm sewers flow routing. *Journal of Xi'an University of Technology, 11*(4), 275–279. (in Chinese).
17. Wang, G., Yao, C. M., Okoren, C., et al. (2006). 4-Point FDF of Muskingum method based on the complete St Venant equations. *Journal of Hydrology, 324*(1–4), 339–349.
18. Ren, B. (2004). *Study on the urban design rainstorm and rainfall runoff calculation model.* Chongqing University (in Chinese).
19. Zhang, D. (2007). *The annual pollution carry model of highway and the wood-pavement in the tour area.* Southwest Jiaotong University (in Chinese).
20. Zhang, S., Li, T., & Gao, T. (2006). Study on pollution load of urban surface runoff in Shanghai. *China Water & Wastewater*, (21), 57–60+63 (in Chinese).
21. Dong, X., Chen, J., & Zhao, D. (2006). Application of SWMM modeling in urban drainage planning. *Water & Wastewater Engineering, 05*, 106–109. (in Chinese).
22. Zhou, Z. (2020). Effect assessment of sponge city construction in the international eco-business district in Songjiang district in Shanghai based on SWMM. *Environmental Engineering, 38*(08), 167–173. (in Chinese).
23. Chen, M., Huang, P., Wu, F., et al. (2014). The carrying capacity evaluation and optimization research of urban drainage network based on SWMM. *Bulletin of Surveying and Mapping*, (03), 54–57+62 (in Chinese).
24. Cheng, W., Wang, H., & Chen, Z. (2013). Optimization of rainwater drainage system based on SWMM in the flat area. *Shanxi Architecture, 39*(25), 135–137. (in Chinese).
25. Ma, J., Li, J., Xu, Y., et al. (2012). Solution to the problem on urban drainage system overflow in rainy season with storm water management model (SWMM). *Water Purification Technology, 31*(03), 10–15+19 (in Chinese).
26. Li, Y., You, X., Ji, M., et al. (2010). Optimization of rainwater drainage system based on SWMM mode. *China Water & Wastewater, 26*(23), 40–43. (in Chinese).
27. Li, P., He, J., Wu, Z., et al. (2019). Application of SWMM model in runoff control simulation of sponge city construction. *Urban Roads Bridges & Flood Control*, (11), 69–72+80+12 (in Chinese).
28. Camorani, G., Castellarin, A., & Brath, A. (2005). Effects of land-use changes on the hydrologic response of reclamation systems. *Physics and Chemistry of the Earth, Parts A/B/C, 30*(8–10), 561–574.
29. Liang, C., & Sun, Y. (2012). Hydrological effect of porous pavement based on SWMM. *Shanxi Hydrotechnics*, (03), 6–7+27 (in Chinese).
30. Wang, W., Zhao, Z., & Qin, H. (2012). Hydrological effect assessment of low impact development for urbanized area based on SWMM. *Acta Scientiarum Naturalium Universitatis Pekinensis, 48*(02), 303–309. (in Chinese).
31. Zhang, S. (2013). A simulation study of the effectiveness of low impact development measures using the stormwater management model (SWMM). *Informatization of China Construction, 19*, 76–78. (in Chinese).
32. Wu, J., Sun, F., Dong, W., et al. (2012). SWMM simulation analysis based on rainfall runoff and water quality within urban inland river area. *Water Resources and Hydropower Engineering, 43*(08), 90–94. (in Chinese).
33. Xiong, Y., Li, Z., Hu, A., et al. (2015). Simulation of effect of LID facilities on water quantity and quality in residential area based on SWMM. *China Water & Wastewater, 31*(17), 100–103. (in Chinese).
34. Wang, J., Tu, N., Xi, G., et al. (2017). Discussion of residential district renovation combined with sponge city development. *China Water & Wastewater, 33*(18), 1–8. (in Chinese).
35. Xiong, X., Na, J., Pan, X., et al. (2018). Simulation research on effects of sponge city measures based on SWMM: A case study of Guangming new district in Shenzhen. *Water & Wastewater Engineering, 54*(04), 129–133. (in Chinese).

36. Zhong, L. (2014). Checking and verification on drainage capacity of ground waterlogging in an urban community of Shanghai based on SWMM. *Urban Roads Bridges & Flood Control*, (01), 78–80+84+9–10 (in Chinese).
37. Cong, X., Ni, G., Hui, S., et al. (2006). Simulative analysis on storm flood in typical urban region of Beijing based on SWMM. *Water Resources and Hydropower Engineering, 04*, 64–67. (in Chinese).
38. Liu, J., Guo, L., Zhang, J., et al. (2006). Study on simulation of drainage and flooding in urban areas of shanghai based on improved SWMM. *China Water & Wastewater*, (21), 64–66+70 (in Chinese).
39. Ma, X., Zhu, Y., Mei, K., et al. (2012). Application of SWMM in the simulation of non-point source pollution load in urban residential area. *Research of Environmental Sciences, 25*(01), 95–102. (in Chinese).
40. Park, S. Y., Lee, K. W., Park, I. H., et al. (2008). Effect of the aggregation level of surface runoff fields and sewer network for a SWMM simulation. *Desalination, 226*(1–3), 328–337.
41. Barco, J., Wong, K. M., & Stenstrom, M. K. (2008). Automatic calibration of the U.S. EPA SWMM model for a large urban catchment. *Journal of Hydraulic Engineering, 134*(4), 466–474.
42. Zhen, L., Yang, X., & Ge, Y. (2013). SWMM model and GIS system integration. *China Science and Technology Information, 03*, 76–77. (in Chinese).
43. Yang, J., & Zhang, W. (2009). Research progress on SWAT model and its application on water environmental nonpoint source pollution. *Research of Soil and Water Conservation, 16*(05), 260–266. (in Chinese).
44. Chiang, L., Chaubey, I., Gitau, M. W., et al. (2010). Differentiating impacts of land use changes from pasture management in a CEAP watershed using the SWAT model. *Transactions of the ASABE, 53*(5), 1569–1584.
45. Tang, L., Lin, W., Zhang, S., et al. (2010). Simulation and analysis of non-point source pollution based on SWAT model for the Wenyu river basin. *Journal of Hydroelectric Engineering, 29*(04), 6–13. (in Chinese).
46. Zhu, Li., Qin, F., Yao, Y., et al. (2011). Research of sensitivity analysis module of SWAT model in middle-scale watershed—A case study of Hongmenchuan watershed in Miyun county. *Research of Soil and Water Conservation, 18*(01), 161–165+275 (in Chinese).
47. Wang, Y., Lv, H., Shi, Y., et al. (2009). Impacts of land use Changeson hydrological processes in an urbanized basin—A case study in the Qinhuai river basin. *Journal of Natural Resources, 24*(01), 30–36. (in Chinese).
48. Franczyk, J., & Chang, H. (2009). The effects of climate change and urbanization on the runoff of the Rock Creek basin in the Portland metropolitan area, Oregon, USA. *Hydrological Processes: An International Journal, 23*(6), 805–815.
49. Sang, X., Zhou, Z., Wang, H., et al. (2010). Development of soil and water assessment tool model on human water use and application in the area of high human activities, Tianjin, China. *Journal of Irrigation & Drainage Engineering, 136*(1), 23–30.
50. Gassman, P. W., Reyes, M. R., Green, C. H., et al. (2007). The soil and water assessment tool: Historical development, applications, and future research directions. *Transactions of the ASABE, 50*(4), 1211–1250.
51. Xue, Y., Sun, W., Zou, C., et al. (2020). Study on non-point source pollution in Chishui river watershed based on SWAT model. *Journal of Subtropical Resources and Environment, 15*(03), 17–23. (in Chinese).
52. Liang, L., Yang, Z., & Liu, C. (2012). Rainfall runoff simulation of Beijing typical area based on MIKE11. *Journal of China Hydrology, 32*(01), 39–42+28 (in Chinese).
53. Yang, J., Hong, D., & Zhang, B. (2019). Rainwater system evaluation and analysis of waterlogging in residential area based on high precision MIKE model. *Journal of Water Resources and Architectural Engineering, 17*(03), 236–241. (in Chinese).
54. Sun, N., Fan, X., & Guo, M. (2020). Study on the sponge transformation of old residential area based on mike flood model. *Shanxi Hydrotechnics*, (03), 1–5+10 (in Chinese).

55. Dong, L., & Gao, Z. (2019). Exploration and oractice of the sponge city construction in Obsolete region. *Environmental Engineering, 37*(07), 13–17. (in Chinese).
56. Zhang, X., Li, Z., He, W., et al. (2019). Simulation of flood process in the central urban area of XI'AN based on MIKE URBAN. *Journal of Water Resources & Water Engineering, 30*(06), 157–163. (in Chinese).
57. Tian, K., Shen, B., & Jia, X. (2016). Application of MIKE SHE model in runoff simulation of Bahe river basin. *Journal of Water Resources & Water Engineering, 27*(01), 91–95. (in Chinese).
58. Zhang, Y., Meng, D., Yu, Z., et al. (2020). Analysis of urban river water quality improvement and compliance based on MIKE21. *Water Resources and Power, 38*(09), 48–52. (in Chinese).

Chapter 3
Source Reduction Technologies for Urban Rainfall Runoff Pollution

The source reduction technology of urban rainfall runoff pollution mainly refers to the engineering and non-engineering technical measures, which taken in the surface before the rainfall forms runoff into the drainage network. It includes the interception and infiltration, collection and storage, evaporation and transpiration of rainwater, and also includes the precipitation, cohesion, absorption, adsorption, filtration, biodegradation and other roles of the pollutants in the rainwater. The ultimate goal is to realize the reduction of the volume and pollution load of the rainfall running off into the drainage network. Source reduction technology is one of the most effective and economical control measures in the overall control of rainfall runoff pollution.

This chapter describes the source reduction technologies for urban rainfall runoff pollution control, including: interception and infiltration technologies, storage technologies, transmission control technologies, and their combined technologies and related equipment and materials.

3.1 Interception and Infiltration Technologies

Interception and infiltration technologies for urban rainfall runoff pollution are usually achieved through facilities with interception or infiltration functions such as green roof, sunken green space, rain garden, and permeable pavement.

D. Sun and F. Qi, *Urban Rainfall Runoff Pollution Control Technology and Application*,
https://doi.org/10.1007/978-981-96-4174-1_3

Table 3.1 Comparison of green roof features

Item	Ecological green roof	Garden-type green roof
Maintenance	Easy maintenance with drought-tolerant plants	High maintenance requirements
Plant species	Turf, lichens, herbs, etc.	Trees, perennials, etc.
Total thickness	60~300 mm	300~500 mm
Load-bearing requirements	60~150 kg/m^2	180~500 kg/m^2
Construction cost	Low	High
Main application	Ecology, stormwater management	Rooftop parks, recreational areas

3.1.1 Green Roof

3.1.1.1 Technical Overview and Principles

Green roof, also referred to as planted roof, are cultivated on sturdy concrete rooftop. Through the infiltration and accumulation of vegetation layer, water storage and drainage layer, filter layer and substrate layer, the purpose of regulating roof rainwater runoff and intercepting pollutants is achieved.

Some green roofs are exclusively covered with turf and flower beds, capable of partially stagnating runoff and retaining pollutants in rainwater. These roofs feature a relatively small total thickness, low load requirements for the roof, and simple maintenance, categorizing them as ecological green roof. In contrast, some green roofs are adorned with trees, offering not only runoff regulation and pollution purification but also creating rooftop gardens and providing recreational spaces. However, due to their substantial thickness and high load-bearing demands, these are known as garden-type green roof. Table 3.1 outlines the primary characteristics of these two types of green roofs.

3.1.1.2 Structural Characteristics and Design Parameters

i. Structure

The green roof consists of a multi-layer structure, and its basic construction is shown in Fig. 3.1. According to the climatic characteristics, roof form, plant species and other factors of the construction site, the composition level of the green roof can be appropriately increased or decreased.

ii. Structural Design Requirements

Structural design requirements for green roof include: slope requirements, impermeable layer requirements, drainage layer requirements, geotextile requirements, soil layer requirements, and overflow design requirements.

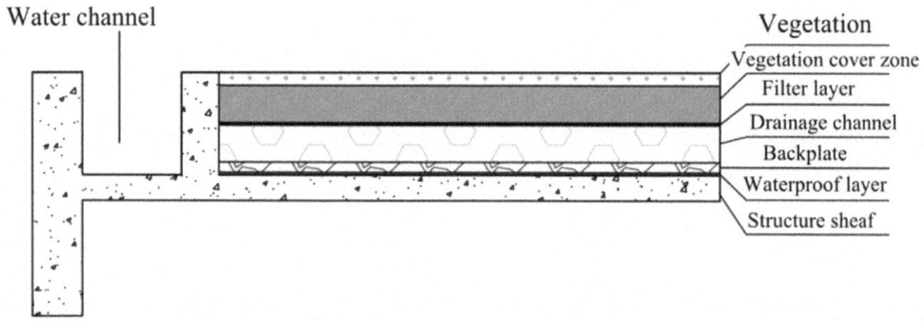

Fig. 3.1 Schematic diagram of common green roof structure

(i) *Slope Requirements for Green Roof*

The slope of the roof should be more than 2% (or 1°) and less than 15% (or 8°). The minimum slope requirement for green roof is for smooth drainage; if the slope is less than 2%, cement plastering should be used to make the slope in the direction of rainwater drainage meet the requirements. The maximum slope requirement for green roof is to maintain the stability of the roof; if the slope is greater than 15%, structures such as braces can be used to achieve it.

The maximum slope for building green roof varies from region to region, for example, Washington State in the United States stipulates that the maximum slope for building green roof is 36% (20°), and Michigan stipulates that the maximum slope for building green roof is 18% (10°). According to the information, the maximum slope of green roof constructed in the United States is 100% (45°), and the total load of green roof is guaranteed to be 60–150 kg/m^2.

(ii) *Impermeable Layer Requirements for Green Roof.*

(1) Fiberglass, PVC, HDPE and EPDM and other impermeable materials can be used;
(2) The thickness of impermeable layer should be greater than 60 mm.

The soil layer of the green roof is very thin and the plant roots can easily reach the impermeable layer after a long time. If the impermeable layer cannot be guaranteed not to be pierced by plant roots, it is recommended that a protective layer be used, and the protective layer should meet the following requirements:

(1) The protective layer may be thermoplastic or other protective film that meets the requirements;
(2) The thickness of the protective layer should be more than 30 mm.

(iii) *Drainage Layer Requirements for Green Roof*

(1) Drainage layer can be made of finished water transfer plate, gravel, ceramic granule or other materials that meet the requirements;
(2) Meet the load-bearing requirements;

(3) Thickness of the drainage layer is more than 30 mm;
(4) Maximum drainage capacity is more than 4 L/s.

(iv) *Geotextile Requirements for Green Roof*

Green roof geotextile should be non-woven needle punched geotextile, and should meet the following requirements:

(1) Puncture strength is greater than 10 kg, to prevent geotextile breakage;
(2) Usually, the green roof soil layer permeability coefficient in the 1×10^{-5}–1×10^{-4} m/s, so it is stipulated that the permeability coefficient of the geotextile is greater than 1×10^{-4} m/s;
(3) Geotextile aperture requirements are determined by the planting soil, and it is required to planting soil through the geotextile shall not exceed 7%.

(v) *Soil Layer Requirements for Green Roof*

(1) The thickness of soil layer should be determined in accordance with the requirements of planting plants, and its suitable thickness is 100–250 mm.

Other requirements for the thickness of the soil layer: satisfy the cultivation of growing plants; reduce the load-bearing capacity of the roof; the greater the thickness of the soil layer, the greater the amount of rainwater it will retain. Table 3.2 illustrates that as the thickness of the soil layer increases, the runoff factor decreases, but the greater the roof bearing weight, the more an appropriate balance needs to be found.

Drought and flood tolerant turf, lichens, and herbs are preferred for green roof. If some species of plants require a thicker thickness of planting soil, instead of raising the thickness of the soil layer as a whole, the drainage layer in these areas can be partitioned off and planting soil can be used. This usually increases the thickness of the planting soil by 5 to 10 cm. The specific construction is shown in Fig. 3.2.

(2) Field water-holding point humidity is greater than 10%.

The field water-holding point humidity is essential to meet the growth needs of plants and minimize the need for frequent watering. Additionally, stipulated maximum porosity and permeability coefficients serve the purpose of storing significant rainwater within the soil and and at the same time, it can meet the rate of rainwater infiltration into the drainage layer and ensure the removal of pollutants.

(3) Clay content less than 1%.
(4) Maximum porosity greater than 25%.
(5) The coefficient of permeability is greater than 1×10^{-5} m/s and less than 1×10^{-4} m/s; pH value should be between 5.5 and 7.9.

Table 3.2 Relationship between runoff factors and soil layer thickness in green roof

Soil layer thickness (mm)	50	75	100	150	200
Runoff factors	94	92	88	85	77

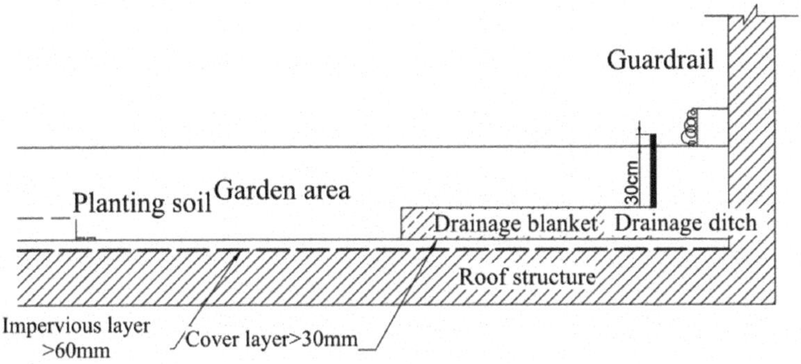

Fig. 3.2 Treatment of garden-type green roof

(vi) *Overflow Design Requirements for Green Roof*

In situations where rainwater from the platform splashes directly onto the green roof, it is advisable to install cobblestone or gravel troughs within the splash area. This measure helps prevent rainwater erosion of the planting soil. The arrangement of gravel or cobblestone troughs is illustrated in Fig. 3.3.

Overflow mechanisms for green roof can take the form of deflector covers, cobble/gravel troughs, and cobble/gravel channels. When gravel or cobble channels are incorporated into green roof, they can be connected to the drainage layer. The channel's perimeter is isolated from the planting soil layer by an impermeable barrier, and the channel sits 2 to 5 cm above the planting soil's surface, allowing it to function as an overflow facility. Figure 3.4 illustrates the implementation of a cobblestone channel as an overflow facility.

Drip or micro-sprinkler irrigation systems are desirable for green roof. Green roof uses flammable materials such as various organic synthetic materials and plantings, which require the installation of fire-fighting measures.

iii. Green Roof Construction Process

The green roof shall be constructed in accordance with the procedures in Fig. 3.5.

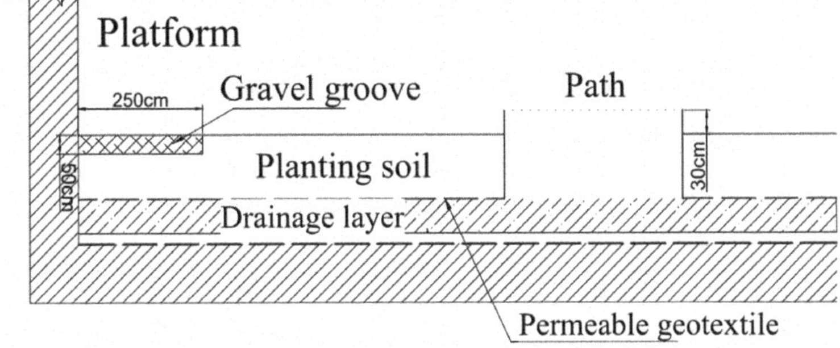

Fig. 3.3 The resolution of green roof deck falling water

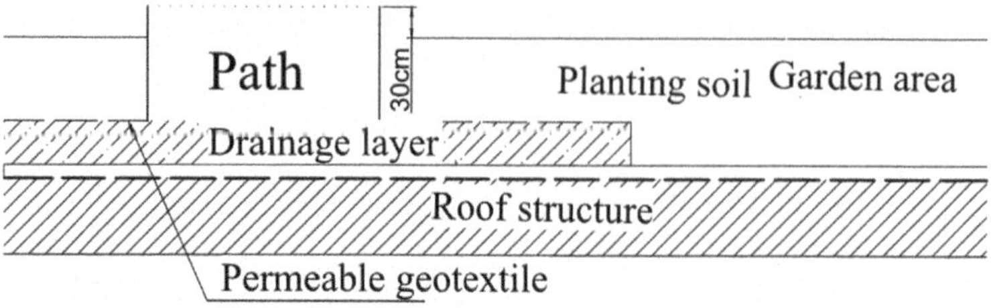

Fig. 3.4 The resolution of overflow rainwater of green roof with cobblestone paths

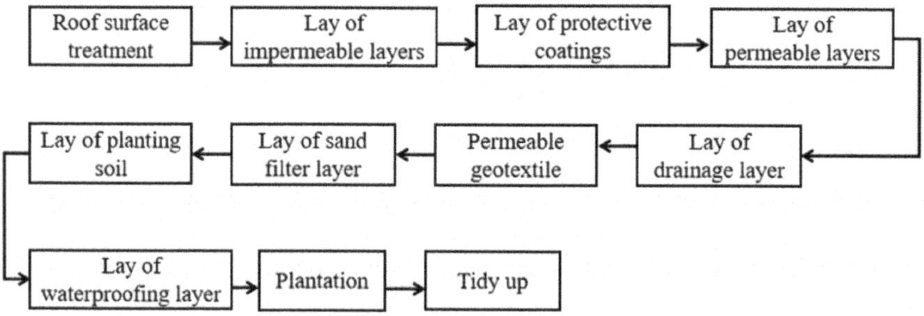

Fig. 3.5 Construction process of green roof

Prior to installation, the planting soil for green roofs must undergo testing to ensure it aligns with the design specifications. Before laying, the soil should be sieved using the geotextile designated for green roofs. The percentage of particles passing through the sieve should not exceed 7%. It is essential to lay the soil evenly, maintaining its natural state without compacting it.

Selection of plant species: (1) low herbaceous ground cover, shrubs and climbing plants, small trees can be selected. It is not appropriate to choose large trees; (2) it is appropriate to choose the weak root piercing ability, wind resistance, drought resistance, high temperature resistant plants; (3) it is appropriate to choose easy transplanting, pruning resistance, sloppy management, slow-growing plants; (4) it is appropriate to choose plants with strong anti-pollution ability, tolerate, retention of harmful pollutant gases.

3.1.1.3 Scope of Application and Advantages and Disadvantages

Green roofs are suitable for flat roof buildings that meet the conditions of roof loading and waterproofing and sloped roof buildings with a slope of $\leq 15°$. It have the following three advantages:

(1) It can not only reducing rainwater runoff generated by the roof, reduce the peak flow and purify the quality of rainwater runoff, but also contribute to diminishing the greenhouse effect, purifying the air, and mitigating noise pollution;

(2) Enhance the roof's insulation properties and improve the local microclimate;
(3) Do not require additional urban land, and the presence of green vegetation increases urban green spaces, enhancing the overall urban landscape environment. Moreover, they mitigate the urban heat island effect and effectively protect rooftops, thereby extending the lifespan of buildings.

The technique also suffers from the following four disadvantages:

(1) Unstable efficiency of pollutant removal;
(2) Higher cost of construction and maintenance than ordinary roofs;
(3) Constrained by roof loads;
(4) Higher requirements on the level of design and construction, requiring regular inspection, maintenance or replacement of waterproofing materials, etc.

3.1.1.4 Operation and Maintenance Requirements

By utilizing technical and management tools, green roof maintenance can ensure good performance and reduce overall costs during the service period. Typical green roof maintenance includes maintenance of greening and drainage facilities.

i. Greening Maintenance

Green roofs cannot effectively fulfill their runoff pollution control role without the healthy growth of plants, and plant maintenance should be a priority. Generally grown plants require management measures such as watering, fertilizing, pruning, weeding and pest control.

Weeding on green roofs holds significant importance as invasive plant roots can harm the waterproofing layer. It is crucial to refrain from using chemical agents, as they expedite the aging of the waterproofing layer. Additionally, sections of the roof exposed to intense sunlight and reflected heat require increased watering for the plants in those areas. The use of non-polluting and odorless fertilizers is encouraged, and over-fertilization should be avoided to minimize the entry of nitrogen and phosphorus elements into the drainage system, preventing potential secondary pollution.

Smaller green roofs, which are challenging to maintain manually, often do not pose significant management issues due to their limited size. Conversely, larger rooftop gardens, which can be managed manually, present greater operational challenges and require a substantial workload. Consequently, it is advisable that extensive rooftop gardens, especially those intended for public use, be entrusted to dedicated full-time staff with expertise in landscaping, planting, and management.

ii. Maintenance of Drainage Facilities

Apart from the maintenance of greenery, roof drainage facilities play a crucial role. Therefore, it is essential to conduct regular inspections and maintenance of drainage facilities such as drainage pipelines, gullies, and drainage observation wells on green roofs. Regular cleaning is necessary to prevent the accumulation of debris, which can block drainage outlets and channels. This preventive measure is vital to avoid rainwater and its contaminants from eroding the roof surface, leading to leaks. Such leaks can significantly impact both the building's lifespan and that of the green roofs. Additionally, during winter, it is imperative to ensure timely water drainage in the irrigation system to prevent freezing and cracking of water pipes. In the event of heavy snowfall or adverse weather conditions, organizing personnel to promptly remove snow can help reduce the load on the roof.

3.1.1.5 Technology Effectiveness and Costs

Numerous studies have demonstrated the substantial regulatory impact of green roofs on runoff. While green roofs can effectively reduce roof runoff, their interception effectiveness varies considerably. It is primarily influenced by factors such as substrate thickness, slope, rainfall intensity, seasonal variations, vegetation density, and the duration of green roof operation. De Nardo et al. [1] reported an interception effectiveness of 45% for green roofs. Van Woert et al. [2] demonstrated that during light rainfalls, the interception effectiveness of three typical roofs (substrate-only, gravel-layered, and green roofs) was 99.3%, 79.9%, and 96.2%, respectively. Additionally, green roofs exhibited 48% higher interception effectiveness than gravel-layered roofs during medium rainfalls. The interception efficiencies of these three roof types were 38.9%, 22.2%, and 52.4%, respectively. Carter et al. [3] conducted a pilot study in Georgia [4] and found that the interception capacity of green roofs ranged from 39 to 100%, with an average capacity of 78%. Teemusk et al. [5] demonstrated that green roofs could intercept 87% of roof runoff when the rainfall intensity was 21 mm. However, their effectiveness diminished significantly during heavy rainfall. Gregoire et al. [6] found that the average annual interception effectiveness of green roofs in Europe and the United States was 56%. In a comprehensive study analyzing 28 years of rainfall data, Jarrett et al. [7] revealed that intensive green roofs reduced runoff volume by 45% to 55%.

There are more research results about the control of green roof runoff volume, while there are fewer research results about the control of water quality aspects. Wang Shumin et al. [8, 9] discovered a distinction in pH levels between runoff from bare roofs and green roofs. Runoff from bare roofs had a pH below 7, whereas green roof runoff exhibited an increase, reaching approximately 8.0. This indicates that green roofs can effectively regulate runoff pH [5, 10]. In addition, NH_4^+-N, TN, and TP concentrations in green roof runoff were lower than those of bare roofs, and seasonal differences were evident (lower pollutant concentrations in summer and higher pollutant concentrations in spring). Berndtsson et al. [11], in their research conducted in Malmö and Lund, Sweden, demonstrated the effective reduction of NH_4^+-N and TP concentrations in runoff due to green roofs. Generally, green roofs exhibit substantial purification effects on runoff water quality, consistently removing NH_4^+-N, SS, COD, and heavy metals. Notably, COD removal rates could exceed 70%, while the removal of heavy metals like Cu, Zn, and Cd varied with the seasons. The stability of TN and NO_3^--N removal appeared less consistent, possibly due to microbial denitrification processes. A comprehensive investigation into the microorganisms and enzymes within the substrate could provide valuable insights. However, the removal of TP was found to be relatively poor.

The components of the cost of a green roof mainly include: waterproofing layer and protection layer, drainage layer, landscaping, green sprinkler system and ancillary facilities. The change of its cost mainly lies in whether it is necessary to set up additional waterproof layer and the selection of ancillary facilities. Typically, the waterproofing and protection layer costs range between 200 to 300 CNY/m^2, the drainage layer costs around 200 to 300 CNY/m^2, landscaping expenses are approximately 100 CNY/m^2, and the combined cost of green irrigation and related facilities falls between 100 to 200 CNY/m^2.

3.1.1.6 Application Examples

Green roof technology has been implemented on the podiums of Buildings 1 to 5 in Wanzhangpo, Guangming New District, Shenzhen city, China, covering a total area of approximately 6380 m^2 [12]. These podiums, functioning as flat roofs, facilitate partial infiltration of natural rainfall through the green roofs, directing it to discharge through the drainage boards. The surplus rainfall overflows from the roof surfaces, with the resulting water collected through rainwater hoppers positioned on the roofs and subsequently discharged outward. The actual project and structure are shown in Figs. 3.6 and 3.7 respectively.

This technology was also implemented on the podium of Building A and the roofs of A6, B3, and B4 within the Guangming Science and Technology Park in Guangming New District, Shenzhen city, China, covering a total area of approximately 6,070 m^2 [12]. During rainfall, rainwater permeates down from the roofs and is collected by drainage boards, gradually flowing into the drainage ditch due to the slope. Overflow rainwater directly enters the drainage ditch through surface runoff. The cover of the

Fig. 3.6 Actual picture of green roof in Wanzhangpo residential district, Guangming New District, Shenzhen

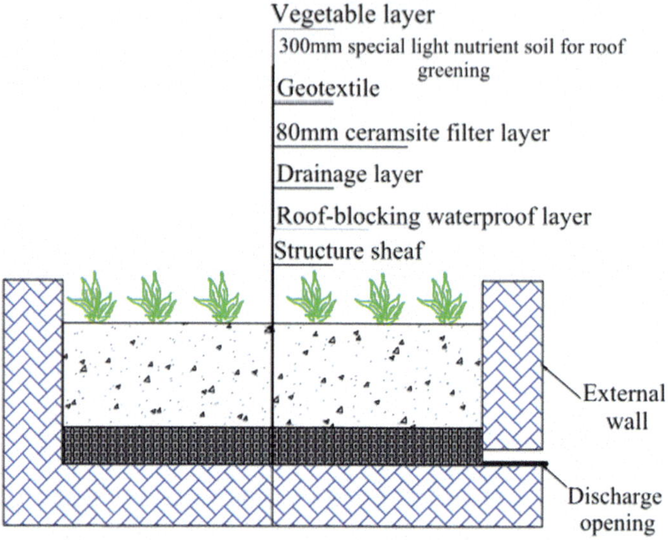

Fig. 3.7 Green Roof Structure of Wanzhangpo residential district, Guangming New District, Shenzhen

drainage ditch is surfaced with pebbles, providing a certain degree of purification as runoff seeps through. The water in the drainage ditch ultimately enters the rainwater riser through the rainwater hopper and is discharged. The actual project and structure are shown in Fig. 3.8.

3.1.2 Sunken Green Space

3.1.2.1 Technical Overview and Principles

Sunken green spaces are intentionally designed with a lower elevation than their surrounding ground, so as to facilitate the inflow of rainwater runoff. This design aims to promote the infiltration of rainwater runoff and enable the retention of pollutants through vegetation.

Sunken green spaces are classified into two distinct concepts: narrow and broad. Narrow sunken green spaces refer to areas where the green space elevation is approximately 200 mm lower than the surrounding impermeable surfaces or ground level. This design facilitates the flow of rainfall runoff from adjacent impermeable surfaces into the green space, enabling purification and infiltration processes, thereby reducing urban rainfall runoff pollution. In addition to the narrowly defined sunken green spaces, the broadly defined category also encompasses facilities with specific storage capacities, including rain garden, rainwater wetland, rainwater pond, depression, and other ecological rainwater management facilities.

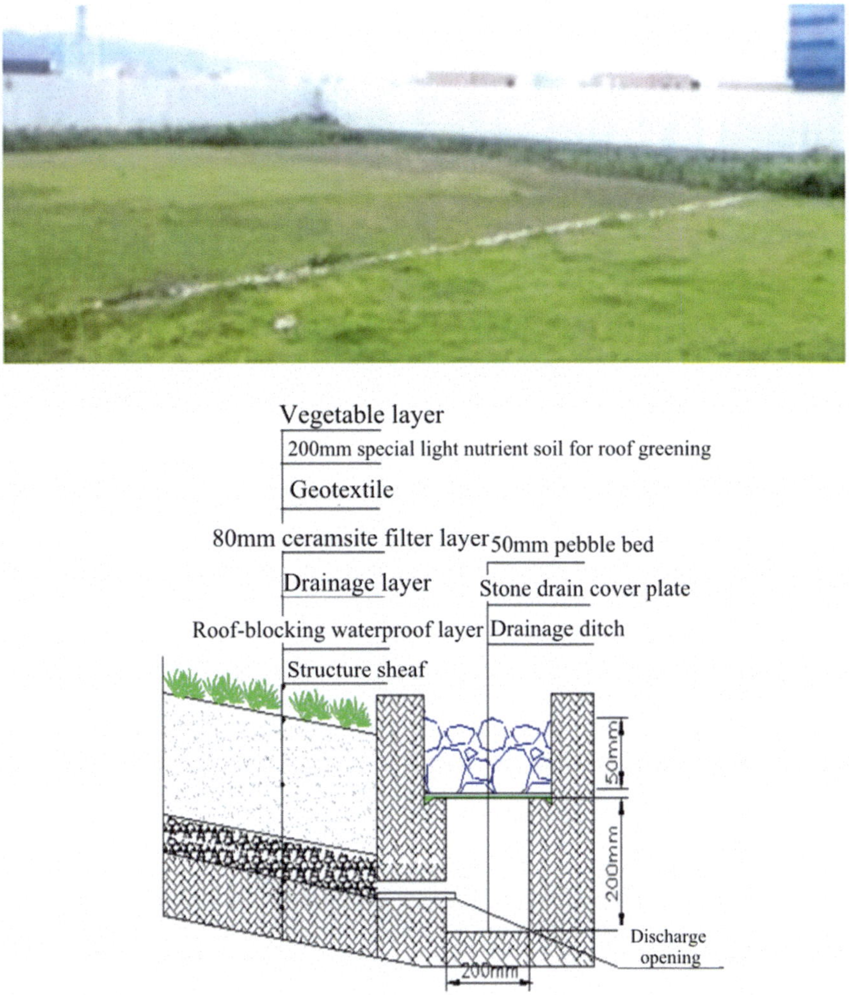

Fig. 3.8 Real scene and structural diagram of green roof of Guangming Science Park of China Merchants Bureau in Guangming New District, Shenzhen

3.1.2.2 Structural Characteristics and Design Parameters

i. Structure

The fundamental structure of a sunken green space consists of several key components: rainwater conditioning space, water storage layer, planting soil layer, gravel drainage layer, and composite geotextile, as illustrated in Fig. 3.9. Runoff from the surrounding park roads and green squares is collected via linear drainage ditches into the sunken green area. Rainwater runoff from municipal roads is intercepted by the interception device in the inlet and directed into the sunken green area. At the

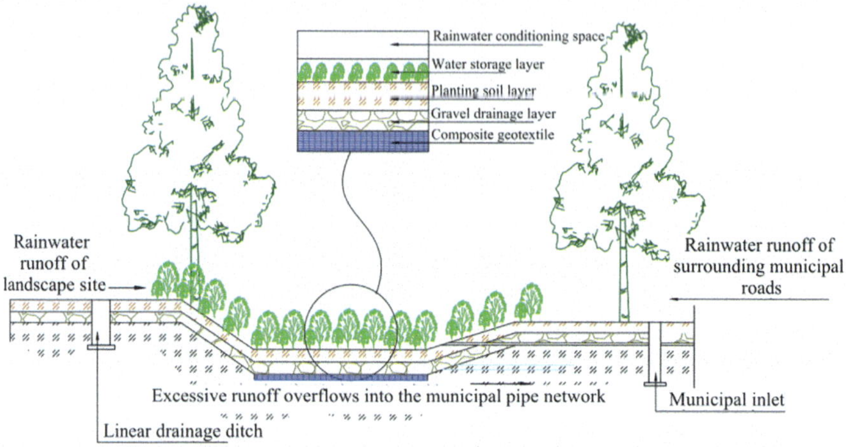

Fig. 3.9 Schematic diagram of sunken green space structure

base of the facility, a gravel drainage layer is installed. Seepage runoff is channeled through the drainage layer, and infiltration runoff is swiftly discharged via blind pipes within the drainage layer. The infiltration runoff collected by the blind drainage pipe can be utilized for watering the surrounding greenery or directly released into the municipal pipeline network. In the event of heavy rainfall, if the runoff cannot infiltrate promptly, it is discharged directly into the nearby municipal pipeline network through the 500 mm × 500 mm overflow outlet situated above the green area.

ii. Structural Design Requirements

Sunken green space is typically constructed in low-lying areas of the catchment surface, allowing rainwater from the road surface to naturally diffuse into the green space. This design reduces the need for extensive rainwater conveyance systems like pipes and ditches, leading to cost savings in construction. Rainwater from the surrounding areas should be directed into the sunken green space, and when it is concentrated, buffer measures should be implemented at the entrance. In cases where the green space has a substantial width, it is advisable to create a buffer area in front of the sunken green space and connect it via a gentle slope transition.

(1) The depth of the sunken green space should be determined according to the plant flooding resistance and soil permeability, generally 100–200 mm.
(2) Sunken green space should generally be set up in the overflow outlet (such as inlet), to ensure that the overflow discharge of runoff during heavy rainfall, the top elevation of the overflow outlet should generally be higher than the green space of 50–100 mm.
(3) The design thickness and material of the structural layer of the sunken green space is shown in Table 3.3.

Table 3.3 Design thickness and material of structural layer for sunken green space

Structural layer	Thickness (mm)	Material
Ultra High/Aquifer	100~200	–
Overflow outlet	50 above design level	Overflow pipe, drainage grate and other devices
Planting soil	200~300	Permeability coefficient $\geq 10^{-5}$ m/s
Raw soil	The top of the original soil should be more than 1 m above the water table	Permeability coefficient $\geq 10^{-6}$ m/s

(4) Horizontal design

The horizontal design of sunken green spaces allows for flexible and decentralized arrangements. It is advisable to avoid designing a single group of sunken green spaces with excessively large scales; the design area should be determined based on the service catchment area of the sunken green space. The ratio between the sunken green area and the catchment area is typically calculated with the goal of controlling the total annual runoff, usually ranging from 10 to 20%. However, when the sunken green is designed with water purification as the control objective, the ratio between its area and the catchment area should align with the upper limit value. Moreover, the horizontal distance between sunken green spaces and building or road foundations should exceed 3 m to prevent potential secondary disasters. If this distance is not met, necessary measures must be implemented to mitigate such risks.

1) Determine the volume of treated runoff

$$V = 0.001P \cdot \psi \cdot A \tag{3.1}$$

where, V-the volume of rainwater runoff entering the sunken green space to be treated, m^3;

P-precipitation, mm;

ψ-combined runoff coefficient at the catchment surface, if the surface type is single, it can be selected according to "*Code for Design of Outdoor Wastewater*" (GB50014). If it is a variety of ground types, it can be obtained by weighted average of different ground types according to the specifications;

A-catchment area, m^2.

2) Determine the amount of rainwater infiltration

First, the construction area of the sunken green space was determined based on the planning objectives of green space system construction and site conditions. Second, the rainwater infiltration volume F was calculated based on the measured soil infiltration conditions on site.

$$F = k \cdot J \cdot A_s \cdot t \tag{3.2}$$

where, F-the amount of rainwater infiltration in the sunken green space during the calculation period, m^3;

k-soil permeability coefficient, m/s;

J-hydraulic gradient, which can generally be taken as 1;

A_s-area of sunken green space, m^2;

t-infiltration time, s, refers to the infiltration ephemeris of the facility during rainfall, which can generally be taken as 2 h, or the local average rainfall ephemeris can be selected empirically.

3) Determine the depth of sinking of green areas

$$h = \frac{V - F}{A_s} \tag{3.3}$$

where, h-submergence depth, m.

4) Calibration of flooding time

The flooding time of the green space is related to the sinking depth and soil infiltration performance, and is generally set as the time required for all rainwater to infiltrate when the green space is full of rainwater. When calibrating, the flooding time should be controlled according to the type of green space plants not more than 1–3 d, generally 48 h.

$$t_d = \frac{h}{k} \tag{3.4}$$

where, t_d-flooding time, h.

iii. Plant Configuration

Sunken green spaces typically integrate a diverse array of community structures, including arbor, shrub and herbs, to create a green space with rich landscape levels. Priority is given to selecting native plants that exhibit a certain degree of waterlogging tolerance, as listed in Table 3.4. To enhance the aesthetic appeal and ecological functions of sunken green spaces, they can also be integrated with other rainwater collection facilities.

3.1.2.3 Scope of Application and Advantages and Disadvantages

Sunken green space can be widely used in all types of green space except road isolation zones. It can be implemented in urban buildings and communities, green spaces, squares and green spaces along the street except sidewalks on both sides of roads, and has the following three advantages:

Table 3.4 Recommended plant species and advantages and disadvantages for sunken green space

Name	Families genera	Advantages	Disadvantages
Bermuda grass	Poaceae Cynodon	Well-developed rhizomes, rapid reproduction, flood-tolerant	–
Paspalum	Poaceae Paspalum	Common grass species in wetlands, flood tolerant	–
Dichondra micrantha	Convolvulaceae Dichondra micrantha	Tolerant of shade, moisture, slightly drought tolerant	Intolerant of trampling
Miscanthus sinensis Variegatus	Poaceae Hypocreeae	Light-loving, cold-tolerant, drought-tolerant, flood-tolerant	–
Miscanthus sinensis	Poaceae Hypocreeae	Light-loving, cold-tolerant, drought-tolerant, flood-tolerant, ornamental	–
Arrhenatherum elatius cv Variegatum	Poaceae Arrhenatherum	Light-loving, cold-tolerant, drought-tolerant, water-wet tolerant	–
Pampasgrass	Poaceae Cortaderia	Evergreen, cold-tolerant, drought-tolerant, ornamental	–
Stipa lessingiana	Poaceae Stipa	Evergreen, slender leaves, light-loving, ornamental	–
Gold leaf sedge	Cyperaceae Carex	Evergreen, golden leaves, light-loving, ornamental	Intolerant of floods
Brown leaved sedge	Cyperaceae Carex	Evergreen, brownish-yellow leaves, light-loving, ornamental	Intolerant of floods

(1) Utilizing the concave space to fully accumulate rainwater, which can significantly increase the infiltration time of rainwater;
(2) The initial purification of the accumulated rainwater can be carried out through the green space, which has a great effect on the reduction of pollutants;
(3) The infiltration water can also play a role in replenishing groundwater.

This technology also suffers from 3 disadvantages:

(1) The edge of the sunken green space is lower than the surrounding road surface, which is easy to create a sense of insecurity;
(2) If the supporting drainage facilities in the area are not perfect, it is easy to cause dirty sewage to flow directly into the sunken green space, which increases the difficulty of management;
(3) When the sunken green space is used on a large scale, the sunken green space is limited by the slope and the vertical condition of the catchment surface, and the actual storage volume is often small.

3.1.2.4 Operation and Maintenance Requirements

(1) Replanting and pruning of plants, and removal of weeds should be carried out in a timely manner;
(2) When the water inlet cannot effectively collect runoff rainwater from the catchment surface, the size of the water inlet should be increased or localized undercutting treatment should be carried out, etc.;
(3) When water and soil erosion occur at the water inlet or overflow port due to scouring, it is necessary to promptly install gravel buffers or take other anti-erosion measures;
(4) When the water inlet or overflow port is blocked or silted up, resulting in poor water flow, it is necessary to promptly clean up garbage and sediment;
(5) When the storage space has insufficient storage capacity due to siltation of sediments, the sediments should be cleared in time;
(6) When slope protection collapse occurs, reinforcement should be done promptly;
(7) When the storage capacity of the storage space is insufficient due to the slope, additional water-blocking weir should be installed or the elevation of the water-blocking weir and overflow port should be raised;
(8) When the emptying time of rainwater in the storage space exceeds 36 h, the bark cover layer or top layer of planting soil should be replaced in time;
(9) The filler should be replaced when the water quality does not meet the design requirements.

3.1.2.5 Technology Effectiveness and Costs

Numerous scholars have conducted extensive research on the runoff storage capabilities of sunken green spaces using experimental devices, field observations, and model simulations. For instance, Cong et al. [13] employed the SWMM model to analyze a neighborhood in Beijing. Their findings indicated that during a 10-year rainstorm recurrence period, sunken green spaces exhibited a 36% increase in infiltration compared to convex green spaces. Additionally, runoff volume was decreased by 53% and peak flow decreased by 35%. Tian et al. [14] investigated the runoff reduction effects of sunken green spaces in urban areas of Jinan under various design rainfall frequencies. Their study concluded that the rainfall storage effect of sunken green spaces was prominent, with runoff reduction rates of 76.55% and 63.45% for rainfall return periods of 1 year and 3 years, respectively. Qu Chan [15] utilized the SWMM model to simulate sunken green spaces in Chang'an Campus of Northwest University. The study revealed that with a sunken depth of 10 cm and a greening rate of 40%, the campus's sunken green spaces could absorb 88.54 m^3 of runoff generated from surrounding 4.7 hm^2 impermeable areas. Additionally, under a rainfall intensity of 9.69 mm/h and a precipitation of 27.5 mm, the campus's sunken green spaces reduced runoff by 10.17% and delayed the flood peak time by 0.16 h. Song Zhaofeng et al. [16] calculated the flood reduction effect of sunken green space under different design storm conditions, showing that the runoff reduction rate rises with the increase of the depth of the sinking, and decreases with the increase of the rainfall return period, and the reduction rate of sunken green space on rainfall runoff with a return period of less than 4 years is as high as 100%.

Sunken green space not only possess rainwater runoff storage capabilities but also contributes to rainwater purification to a certain extent. Several studies have demonstrated the effectiveness of small natural-type sunken green spaces in reducing pollutants. On average, these spaces have shown reduction rates of 52.21% for COD, 48.98% for NH_4^+-N, and 47.35% for TP. Yang et al. [17] conducted research in Yixing City and reported significant purification effects on rainfall in green spaces. They observed annual reductions of 233.6 × 10^3 kg for COD, 70.9 × 10^3 kg for TN, 12.6 × 10^3 kg for NH_4^+-N, and 1.7 × 10^3 kg for TP. Another study by Fan Qunjie [18] simulated green spaces and found that COD exhibited the highest reduction effect, with a maximum reduction rate of 83.08% and an average rate of 56.87%. The reduction rates for NH_4^+-N, TN, and TP averaged 70.11%, 39.70%, and 40.81%, respectively. These studies indicate that sunken green spaces play a vital role in purifying rainwater runoff. However, the reduction rates for each pollutant vary. Sunken green spaces exhibit the most effective reduction for COD and SS, followed by TP and heavy metals. NH_4^+-N and TN exhibit relatively lower reduction rates compared to other pollutants.

Generally, sunken green space offers several advantages, including rainwater storage and infiltration, peak flow reduction, rainwater purification, and prevention of soil erosion, and low investment. Moreover, it is a new rainwater utilization measure that can meet the multiple effects of environment, ecology and economy at the same time. Its application can effectively mitigate urban water scarcity, reduce rainfall runoff pollution, and to some extent, prevent urban flooding.

3.1.2.6 Application Examples

In the Binshui park of Future Science and Technology City, Beijing, a sunken green space covering an area of 900 m^2 has been established. This space has a catchment area of 1914 m^2 and collects runoff from the main road, steep slopes, adjacent green areas, and permeable pavements. The site layout is depicted in Fig. 3.10 [19]. Along the steep slope, a standard runoff cell measuring 20 m × 5 m has been installed, with a 1 m^3 catchment basin below it for monitoring rainfall runoff. This sunken green space achieved a reduction rate of more than 55% for SS concentration. Additionally, it exhibited a COD reduction rate ranging from 24.28% to 36.35%. However, the removal rate for nitrogen and phosphorus concentrations was relatively low. It was observed that the reduction rate of pollutant concentration decreased with higher influent pollutant loads under certain rainfall intensities. Similarly, when the influent pollutant load was constant, higher rainfall intensities led to lower reduction rates of pollutant concentration.

The sunken green space experienced outflow solely during the rainfall events on September 26 and October 7, 2016. However, for all other conditions, it effectively reduced both runoff and pollutant concentrations. During natural rainfall, pollutant concentrations were significantly low. In contrast, the initial runoff pollutant concentrations from the runoff plots were considerably high, reaching 373 mg/L and 141.40 mg/L for SS and COD, and up to 1.503 mg/L and 11.46 mg/L for TP and TN. This observation suggests that the primary source of rainfall runoff pollution is the erosion of subsurface materials by rainfall, leading to the loss of soil particles, road deposits, and associated pollutants. Overall, the sunken green space effectively retained almost all the rainfall runoff and reduced the total pollutant load by more than 99%.

Fig. 3.10 Diagram of sunken green space runoff plot installation in Binshui park of Future Science and Technology City, Beijing **a** Runoff plot; **b** Outlet collecting barrel of runoff plot; **c** Green space drain collecting bucket

Table 3.5 Rain Garden Forms and Characteristics

Form	Characteristics
Infiltration rain garden	Pollution control and rainwater infiltration can be realized at the same time; the structure is simple and easy to construct and maintain; the groundwater level and the depth of the impermeable layer need to be greater than 1.2 m
Filtration rain garden	It can be used as a pre-treatment facility for rainwater collection and reuse; it can be designed in conjunction with the superstructure of the roof slab of an underground building; the water table and the depth of the impermeable layer need to be greater than 0.7 m
Bioretention	It can be used as a pre-treatment facility for rainwater collection and reuse; it is not restricted by groundwater level and depth of impermeable layer; it has a wide range of applications and can be used for sidewalks, plazas and so on

3.1.3 Rain Garden

3.1.3.1 Technical Overview and Principles

Rain garden is shallow green area constructed in low-lying terrain, designed to mitigate runoff and retain pollutants. This area utilizes plants, soil, and microorganisms to store and infiltrate rainwater, effectively purifying pollutants. Surface runoff gradually permeates the soil, replenishing groundwater, or it can be redirected for urban purposes like landscaping, toilets, and other municipal water uses. Rain garden serves as ecologically sustainable facilities, effectively reducing the sources of runoff pollution.

Rain garden can be categorized into infiltration rain garden, filtration rain garden, and bioretention based on topographic and geological conditions, uses, and landscape effects, with their respective characteristics shown in Table 3.5.

3.1.3.2 Structural Characteristics and Design Parameters

i. Structure

Rain garden is mainly composed of water intake facilities, water storage area, mulch layer, soil layer, planting, sand filter layer (only for rainwater runoff of Class III and IV water quality class), underground drainage layer and overflow facilities, and typical construction facilities are shown in Fig. 3.11.

ii. Design Parameters

(i) *General Requirements*

The catchment area of a single rain garden should not be greater than 0.5 hm^2, for the following reasons: (1) single occupancy area is too large, and the landscape effect is not coordinated; (2) prolong the catchment time, increasing the amount of runoff pollution control.

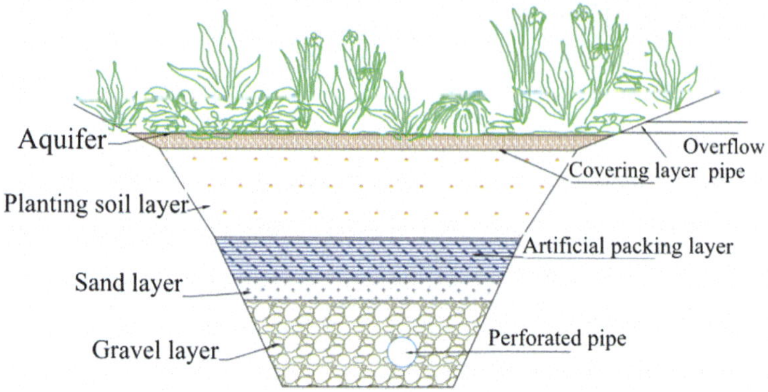

Fig. 3.11 Rain garden construction

Rainwater from infiltration rain garden needs to be infiltrated, and requirements need to be made for the depth of site soil, water table and impermeable layer, with a bottom soil permeability coefficient of greater than 4×10^{-6} m/s, and a water table and depth of impermeable layer of greater than 1.2 m.

Filtration rain gardens only require the depth of water table and impermeable layer, depth of the water table and the impermeable layer at the bottom of the rain garden should be greater than 0.7 m. The soil of the rain garden itself needs to be configured to ensure that its permeability coefficient meets the requirements.

According to the rainwater runoff water quality and rainwater destination, rain gardens can be set up as in-line or off-line, and the combination form of Fig. 3.12 can be selected. Rainwater quality according to China's "*Environmental Quality Standards for Surface Water*" (GB 3838-2002) is divided into five categories, I–III for clean water quality rainwater; IV for rainwater with a certain degree of pollution water quality; V for more serious pollution water quality rainwater.

(ii) *Water Intake Requirements*

Rain garden, designed as primary rainwater runoff treatment facilities, function to discharge, infiltrate, or collect initial rainwater runoff for subsequent reuse post-treatment. It's important to note that Category III and IV water quality rainwater should not be collected for reuse. Class IV water quality rainwater runoff demands more frequent maintenance owing to its inferior water quality. Moreover, there's a risk of infiltrated rainwater contaminating groundwater, necessitating the incorporation of pre-treatment facilities. These pre-treatment facilities for rain gardens can take the form of sedimentation groove, filtration systems, and similar structures.

(1) Rainwater from roofs is usually connected directly to rain gardens using down-pipes, and rainwater from squares and roads flows directly into rain gardens. While rainwater along the road can enter the rain garden by using curb, vertical grate or other means.

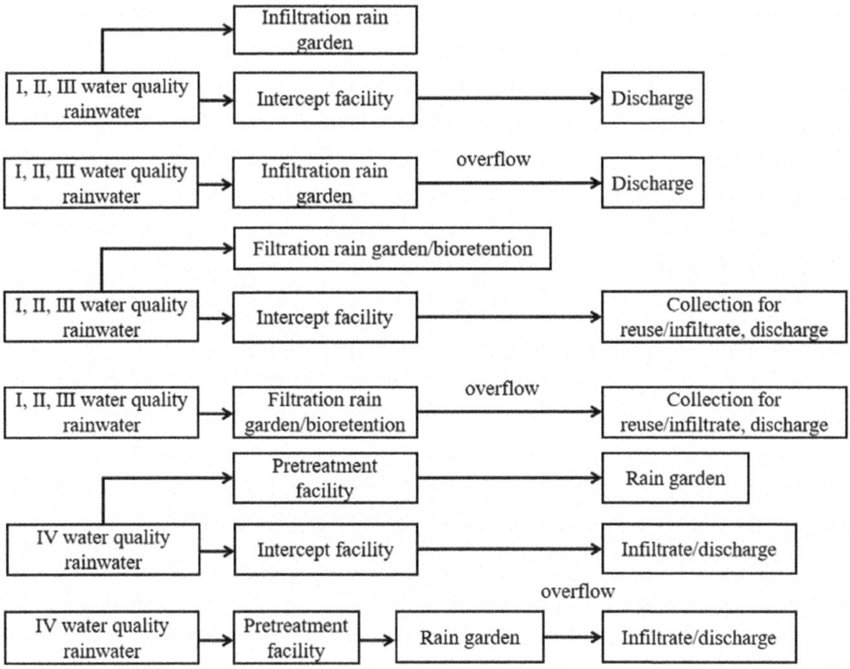

Fig. 3.12 Rain garden design methods for different water qualities

(2) In order for rain gardens to work well, rainwater should enter the rain garden smoothly and evenly. If the catchment area has a certain slope, the rainwater runoff will erode the rain garden, and it is necessary to use water distribution facilities to make the water enter the rain garden uniformly and not erode the rain garden. The maximum storage depth of rain gardens should be set at 10–30 cm, and slopes greater than 2:1 (H:V) should be set around the storage area of infiltration and filtration rain gardens.

(3) The rain garden's effectiveness in retaining rainwater increases with a larger maximum water storage depth, promoting the achievement of low-impact development goals. However, an excessively large storage depth presents two challenges. Firstly, it impacts the aesthetic appeal of the landscape. Secondly, it prolongs drainage time, affecting plant growth and potentially leading to mosquito breeding during summer. To balance both aesthetic and functional aspects, a storage depth of 10 to 30 cm is recommended. This depth ensures not only the preservation of the landscape but also allows stagnant rainwater to infiltrate or be discharged within 36 h. Additionally, a 5 to 10 cm mulch layer, comprising dead tree bark and leaves, should be incorporated into rain garden designs.

(4) When the water quality of rainwater runoff is class III or IV, it is appropriate to set up a sand filter layer of 15–30 cm under the soil layer. When the water

quality of rainwater runoff is class I or II, the sand filter layer can be replaced by a 5–10 cm layer of pea gravel.

(iii) *Mulch Layer Design*

The main effects of mulch layer on rain gardens are as follows:

(1) Maintain soil moisture in the rain garden, which is conducive to the growth of species of plants;
(2) Prevent rainwater erosion of the soil layer;
(3) Provide a microbial environment conducive to the removal of pollutants in rainwater;
(4) Sediment is wrapped in bark and leaves, which is conducive to rain garden maintenance.

(iv) *Area Design*

The formula for calculating rain garden area is derived from Darcy's Law. The permeability coefficient for infiltration rain gardens uses the permeability coefficient of the subsoil, while the permeability coefficient for filtration rain gardens and bioretention is the permeability coefficient of the soil configured in them; the average depth of water storage is calculated as 0.5 times the maximum depth of water storage; and the rainwater emptying time for rain gardens is usually 48 h or 72 h.

The infiltration rain garden area design is calculated according to Eq. (3.5).

$$A_f = \frac{V_{WQ} \times d_f}{i \times \left(h_f + d_f\right) \times t_f \times 3600} \tag{3.5}$$

where, A_f-rain garden area, m^2;

V_{WQ}-rain Garden Runoff Pollution Control Volume, m^3;

df-planting soil layer thickness, m;

i-rain garden subsoil permeability coefficient, m/s;

h_f-average depth of water storage in rain gardens, m, $h_f = 0.5 \times$ Maximum water depth;

t_f-rain garden rainwater emptying time, h, Generally designed for 36 h.

The filtration and bioretention area design is calculated as Eq. (3.6).

$$\mathrm{Af} = \frac{V_{WQ} \times d_f}{k \times (h_f + d_f) \times t_f \times 3600} \tag{3.6}$$

where, k-configured soil permeability coefficient in rain gardens, m/s;

(v) *Planting Requirements*

Native species of plants should be used in rain gardens, and their flooding tolerance or survival time should be greater than 36 h. The arrangement of species of plants in rain gardens should be designed in conjunction with the landscape profession.

Planting requirements for rain gardens are listed below:

(1) Arbors have strong water retaining ability, so they should be planted around the rain garden, which is conducive to maintaining the humidity of the soil layer in the rain garden. This practice fosters the growth of seed plants and sustains the rain garden's effectiveness in removing pollutants. Arbors can't be planted at the water inlet of the rain garden. If planted inside the rain garden, it will easily lead to poor water flow and affect the normal operation of the rain garden.
(2) The surface of the bioretention is not much space for planting, and all other space is covered with a cover, so it is recommended to plant 1–2 shrubs according to the requirements of landscape design.

(vi) *Soil layer design*

The soil layer in the rain garden should have a thickness ranging from 40 to 80 cm and Using the configuration soil, the configuration soil shall meet the requirements in Table 3.6. The soil should be free from any weeds, plant seeds, gravel, concrete blocks, bricks, or other debris. Additionally, the soil's porosity and permeability coefficient must be maintained by controlling the volume percentages of Clay, silty soil and sandy soil. This ensures that the organic matter content in the soil not only facilitates plant growth within the rain garden but also aids in the absorption of pollutants by the soil and plant root system.

The permeability coefficients in Table 3.6 were determined based on the following:

(1) When the maximum storage depth of the rain garden is 30 cm, the soil permeability coefficient required to empty the rainwater for 36 h is 3×10^{-6} m/s;
(2) Excessively high soil permeability coefficients can lead to a decrease in the pollution removal effect of rain gardens.

The results of the study showed that pollutants such as N and P were most easily adsorbed by soil and plant roots when the pH was between 5.5 and 6.5.

(vii) *Sand Filter Layer Design*

The purposes for setting up the sand filter layer: (1) to strengthen the rainwater filtration and improve the pollutant removal effect; (2) laying pea gravel can separate the geotextile and the soil layer to prevent the clay from clogging the geotextile.

The permeability coefficient of the sand filter layer must be at least ten times greater than that of the soil layer to ensure that the installation of the sand filter layer does not impede the rainwater drainage time of the entire rain garden.

Table 3.6 Rain garden configuration soil characteristics

Parameters	Permeability coefficient(m/s)	pH	Organic matter content(%)	Clay(%)	Sandy soil(%)	Silty soil(%)
Requirement	3×10^{-6}~1×10^{-5}	5.5~6.5	3~5	< 10	30~55	35~60

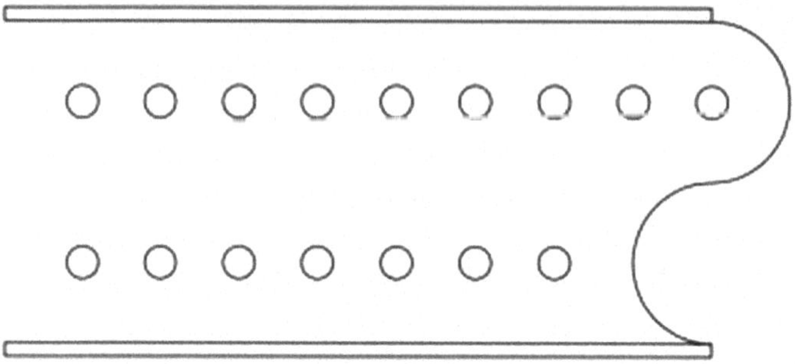

Fig. 3.13 Underground perforated pipe design for rain gardens

(viii) *Underground Drainage Layer Design*

(1) Perforated pipes are set up as shown in Fig. 3.13. One perforated pipe should be installed every 3 m in the width direction, with a diameter of not less than 150 mm, and the drainage capacity should not be less than the maximum infiltration capacity of the rain garden.
(2) The perforated pipe is encased in gravel layer, with the washed gravel in the layer serving the dual purpose of water storage and drainage. Additionally, it promotes the cultivation of biofilm, enhancing nitrogen removal. It's important to note that cobbles cannot substitute for washed gravel in this context.
(3) The gravel layer should be covered with permeable geotextile. Non-woven geotextile should be selected as permeable geotextile, and its permeability should be greater than the maximum seepage requirement of rain garden to meet the requirements of soil conservation, water permeability and anti-blocking.

Short fiber needle-punched geotextile stands out as one of the most widely utilized non-woven geotextiles available in the market. The manufacturing process involves opening and blending fibers, carding (or employing airflow) to create a network, followed by laying, stretching, needling, and cementation processes, ultimately resulting in the final product. The entanglement strength achieved through needling is sufficient to withstand tensile stress during installation, preventing tearing and surface rupture. Due to its thickness, fluffy structure, and the presence of three-dimensional fiber channels, this geotextile offers high filtration efficiency and excellent drainage performance, boasting a permeability coefficient ranging from 0.01 to 0.1 m/s. short fiber needle-punched geotextile is convenient to install, cost-effective, and provides moderate reinforcement and isolation. Hence, it is particularly suitable for applications involving reverse filtration and drainage.

(4) Over time, sediment accumulation may occur in the underground perforated pipes of the rain garden. To address this issue, dredging can be done by pouring water into the riser, or by overflowing rainwater from the riser. The diameter of the dredging riser should be 100~150 mm.

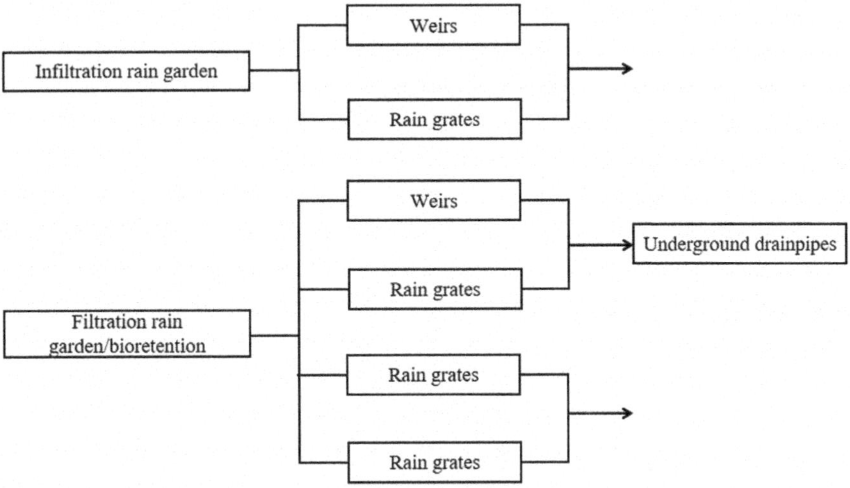

Fig. 3.14 Different types of rain garden overflow forms

(ix) *Overflow Design*

Different types of rain garden overflow forms are shown in Fig. 3.14. Rain gardens can choose the following overflow forms based on their type and where the rainwater goes.

(1) Rain gardens can discharge rainwater downstream through weirs or rain grates for infiltration, collection for reuse, retention or discharge.
(2) Filtration rain gardens and bioretention can overflow rainwater through their underground drainpipes. These two forms of overflow are shown in Figs. 3.15 and 3.16.

Regardless of the type of overflow system employed, the elevation of the overflow inlet should align with the maximum storage depth line of the rain garden. Failing to do so can diminish the rain garden's capacity for rainwater runoff pollution treatment and effect the effectiveness of its low-impact development. The drainage capacity of rain garden overflow systems must not be lower than the designed influent flow. When underground drainpipes are used to overflow the filtration, rain gardens and bioretention, the drainage capacity of underground drainpipes should be designed according to the designed influent flow, not according to its maximum infiltration capacity.

iii. Selection of Plants

Plants play a pivotal role in rain gardens, primarily removing pollutants through absorption and transpiration. When selecting plants, attention should be given to the following criteria: (1) Choose plants with both decontamination abilities and ornamental value. (2) Select plants with wet or semi-wet characteristics and tolerance to waterlogging. (3) Opt for plants with well-developed root systems, robust vitality,

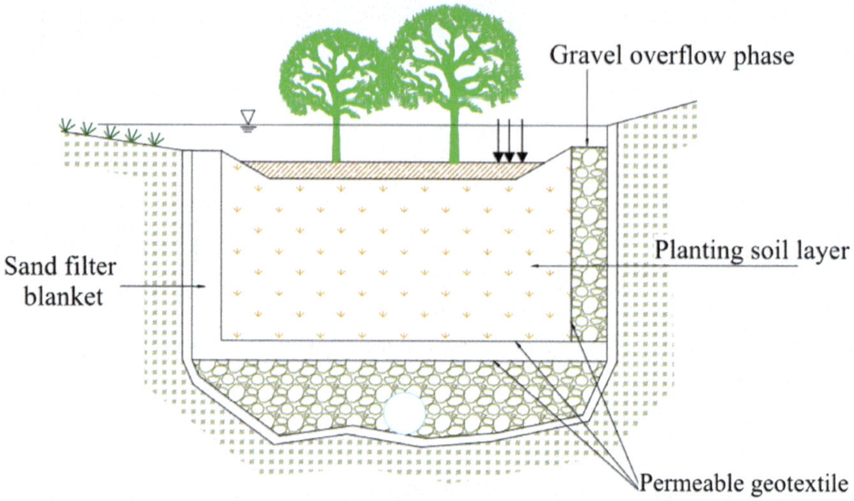

Fig. 3.15 Rain garden uses gravel tanks to overflow rainwater

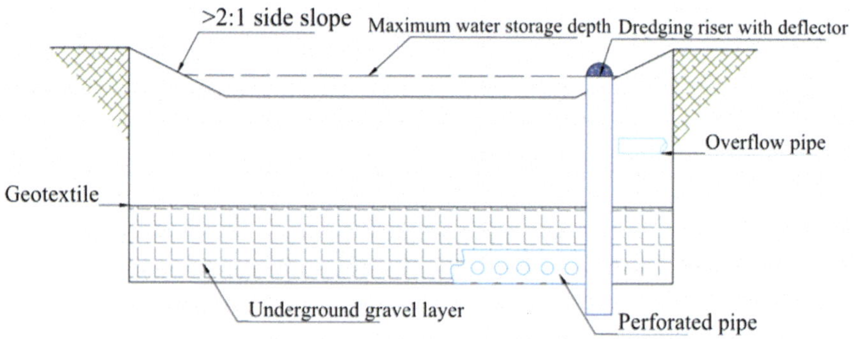

Fig. 3.16 Rain garden uses overflow rainwater from dredging risers with deflectors

high resilience to adverse conditions, resistance to pollution, pests, and diseases. (4) Choose plants that can be planted together to establish a stable ecosystem. (5) Prioritize native plants with superior adaptability to local climate, soil, and environmental conditions. Exotic species should be limited in quantity. Table 3.7 provides a list of recommended plant species.

iv. Construction Process for Rain Gardens

There are three types of rain gardens, which are structurally different and have somewhat different construction processes. Filtration rain gardens and bioretention usually use underground perforated pipes as overflow facility; infiltration rain gardens usually use weir to overflow, and the construction of their overflow facilities is included in the construction of pretreatment facilities.

Table 3.7 Recommended plant species for use in rain gardens and their advantages and disadvantages

Name	Genus	Advantage	Disadvantage
Reed	Poaceae Phragmites genus	Well-developed root system with superior oxygen transfer performance, favorable to COD degradation, adaptability and resistance to adversity	Plants are tall and spread too quickly for small rain gardens
Arundo donax	Poaceae and Asparagus	High biomass, drought tolerant	Tall plants, not suitable for small rain gardens
Vetiver	Gramineae Vanilla	Well-developed root system, resistant to drought and flooding, cold and heat, Resistant to acid and alkali, obvious effect on nitrogen and phosphorus removal	Tall plants, reproduce quickly, not suitable for small rain gardens
Cattail	Typha genus in the family Typha	Well-developed root system, high production, COD and ammonia nitrogen removal	Tall plants, reproduce quickly, not suitable for small rain gardens
Canna	Canna family, Canna genus	Removal of COD and ammonia nitrogen was effective	Shallow root system
Hydrocotyle vulgaris	Umbelliferae Tianhu Genus	Light-loving, can be planted on land and in shallow water, and has a high capacity for integrated pollutant uptake	Intolerant of cold
Garland-flower	Ginger family, ginger flower genus	High biomass, high nitrogen uptake and ornamental	Intolerant of drought and cold
Zizania latifolia	Gramineae, Zizania genus	Enrichment of Mn, Zn and other metals, high removal rate of BOD5, edible	Intolerant of cold
Arrowhead	Alismaceae Sagittaria	The foliage is oddly shaped, ornamental, has a high removal rate of BOD5, and is edible	Shallow root system
Bog rush	Lamiaceae and Lamiaceae	Drought-tolerant, well-developed root system, good water purification	–
Acorus tatarinowii	Araceae Acorus genus	Evergreen, rhizomatous, stems cross-traveling and much-branched	Intolerant of drought
Cyperus alternifolius	Cyperaceae	Evergreen, stems erect, cespitose unbranched	Intolerant of cold
Stripe moss	Cyperaceae moss genus	Evergreen, light-loving, moisture-loving, cold-tolerant	–

(continued)

Table 3.7 (continued)

Name	Genus	Advantage	Disadvantage
Loosestrife	Chrysanthemum genus in the family Chrysanthemaceae	Drought tolerant and ornamental	Inadequate removal of pollutants
Yellow iris	Iridaceae and Iris genus	Drought tolerant and ornamental	–
Rhizoma alismatis	Alismaceae and Alisma genus	Cold hardy, drought tolerant, ornamental	–
Alternanthera paronychioides	Amaranthus genus	Drought tolerant, reddish foliage all year round, ornamental	–
Saururus chinensis	Sanbaicao family Sanbaicao genus	Drought tolerant and ornamental	–

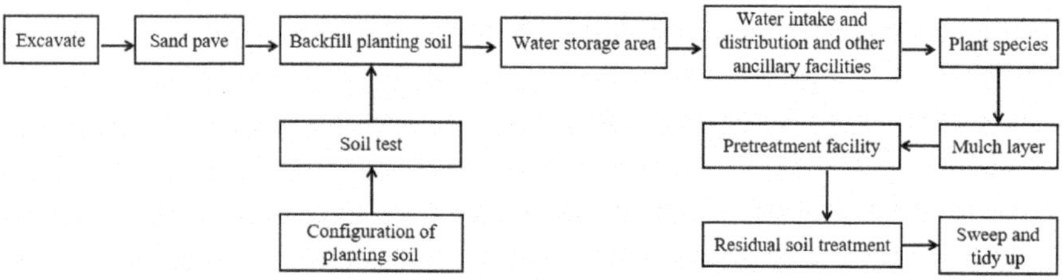

Fig. 3.17 Construction process for infiltration rain gardens

Excavation for infiltration and filtration rain gardens can be carried out manually or using small machinery. It is crucial that the soil at the bottom is not compacted, while the bottom of the bioretention should be compacted. In cases where the bottom soil is found to be overly compacted after excavation, it can be over-excavated by 30 cm and then backfilled with the over-excavated soil mixed evenly with 5 cm thick fine construction sand. Following the completion of earth excavation, sand must be immediately laid according to the design specifications. Mechanical crushing should not be used after sand paving. To prevent direct contact between the underground drainage layer gravel and the bottom soil layer, geotextile should be placed as an isolation barrier. Additionally, the excavation surface should be convenient for the construction and secure fixation of the geotextile.

Infiltration and filtration rain gardens can be constructed by referring to Figs. 3.17 and 3.18.

3.1.3.3 Scope of Application and Advantages and Disadvantages

Rain gardens find extensive applications in various settings, including residential areas, commercial zones, industrial sectors, urban roads, and parking lots. They can also used in the surrounding green spaces of buildings and communities, roads and

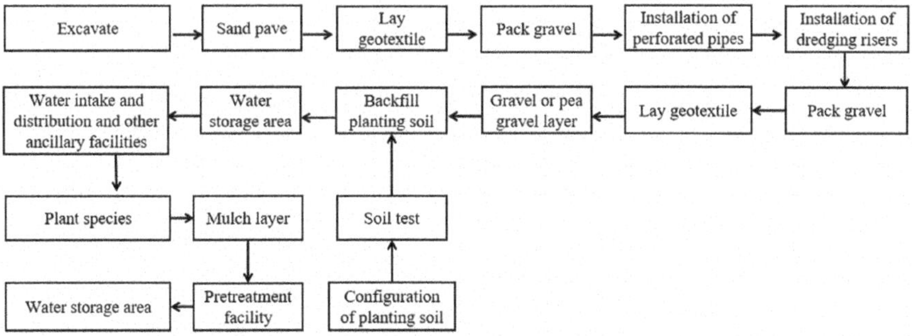

Fig. 3.18 Construction process for filtration rain garden

parking lots, and urban green spaces such as urban road green belts. However, if the quality of rainwater falls below class IV surface water quality, pre-treatment is necessary before the rainwater enters the rain garden.

Rain gardens have the following 4 advantages:

(1) They can reduce the amount of rainwater runoff, delay the appearance of flood peaks, and reduce the occurrence of floods;
(2) They can effectively remove pollutants, especially for suspended particulate matter, heavy metals, nitrogen, phosphorus and pathogens;
(3) Infiltration of rainwater can be recharged to groundwater;
(4) Lower construction and maintenance costs, high integration of natural landscape, and good landscape effect.

The technology also has the following four disadvantages:

(1) The large footprint and the fact that each individual unit can generally only treat rainfall runoff within a smaller catchment area;
(2) Requires high frequency of removal of floating debris;
(3) The top soil is prone to sloughing, requiring regular top soil repair and replacement, and high maintenance frequency requirements;
(4) High requirements for soil drainage performance.

3.1.3.4 Operation and Maintenance Requirements

Rain gardens tend to operate with clogging, soil sloughing, overgrown plants, drought and frost damage. Only targeted maintenance measures can ensure continued long-lasting operation. Table 3.8 lists rain garden maintenance component, maintenance priorities and objectives, and maintenance cycle.

Table 3.8 Rain garden maintenance requirements

Maintenance component	Maintenance priorities and objectives	Maintenance cycle
Maintenance of Planting	Replanting plants; fertilizing; removing weeds	Twice a year; determine based on inspection results
Plant pruning	Pruning of plantings	Three times a year; meets landscaping requirements
Debris and Litter Removal	Water intake and distribution facilities; water storage areas; overflow facilities	Four times in a year; determine based on inspection results
Mulch layer	Replace mulch	Once a year; determine based on inspection results
Top layer of planting soil	Replace top layer of planting soil	Once a year; when new plantings are needed; to meet landscaping requirements
Underground Drainage Layer	Clean out underground perforated pipes using dredging risers	Once a year; determine based on inspection results
Planting soil layer	Replace the top planting soil layer or sand filter layer	Inspection results show that the filtration layer and underground drainage layer lose their effectiveness, usually after 5 to 10 years of use
Geotextile	Geotextile is damaged, replace with new geotextile	

Notes 1. Replacement of mulch layer should not be done mechanically; manual methods should be used so that debris is not cleaned out of the mulch layer

2. Resurfacing of mulch layer should be based on what was required at the time of initial construction

3.1.3.5 Technology Effectiveness and Costs

Vegetation transpiration and rainwater evaporation within rain garden play a vital role in regulating the temperature and humidity of the surrounding environment, mitigating the heat island effect. Additionally, rain gardens facilitate soil infiltration and vegetation interception, impeding rainwater runoff, reducing flow rates, replenishing groundwater sources, and minimizing the risk of floods. Furthermore, rain garden is effective in removing various pollutants from runoff, including SS, TN, TP, heavy metals, and pathogens. Their removal rates typically exceed 80% for SS, surpass 50% for TN, exceed 60% for TP, range from 45 to 95% for heavy metals, and reach 70% to 100% for pathogens.

The economic value of a rain garden primarily stems from its ecological attributes. Rain garden enhances the surrounding environment significantly, which, in turn, elevates the value of the properties nearby. Homes featuring rain gardens tend to have higher market value, while pedestrianized shopping streets adorned with these green spaces attract more visitors. Moreover, areas with large rain garden often become focal points for regional activities and community engagement. Additionally, rain garden serves as reservoirs for rainwater, supplementing urban water needs and substantially reducing the overall cost of urban water use.

The construction cost of rain gardens mainly comes from engineering materials and labor costs. Among them, the engineering materials mainly include artificial soil, filler, gravel, drainpipe and so on. At present, the cost of rain garden in China is 400–1,000 CNY/m^2.

3.1.3.6 Application Examples

Rain garden technology was specifically implemented in the parking lot of the Binshui park of Future Science and Technology City, Beijing [19]. The rain garden covers an area of 58 m^2, and the actual picture and structure are depicted in Figs. 3.19 and 3.20, respectively. The rain garden primarily comprises a water storage layer, a vegetation planting layer, and a drainage layer. Among them, the water storage layer is 15 cm, the vegetation planting layer is 30 cm, the sand transition layer is 5 cm and the drainage layer is 50 cm. Perforated PVC pipes, with a diameter of 200 mm, are arranged in the drainage layer for efficient rainwater drainage after seepage. This rain garden functions by purifying rainwater runoff through soil layer filtration, adsorption and absorption by plant roots, and microbial systems. The purified rainwater either infiltrates into the soil, replenishing groundwater, or is discharged into municipal pipes.

Fig. 3.19 Actual picture of rain garden in parking lot of the Binshui park of Future Science and Technology City, Beijing

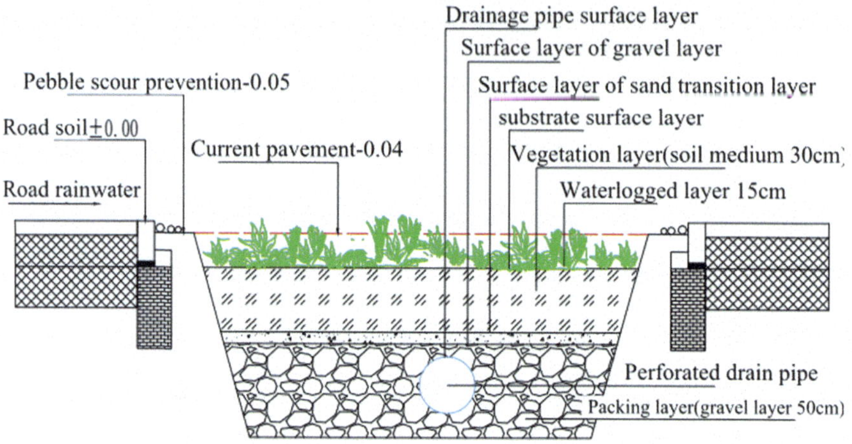

Fig. 3.20 Structure of rain garden in parking lot of the Binshui park of Future Science and Technology City, Beijing

During 2016 to 2017, the inflow and outflow of rain garden during the main rainfall in flood season were monitored on the spot. The findings revealed that rain gardens effectively reduced rainfall runoff by 15% to 85%, and significantly decreased pollutant emissions by 35% to 95%. Additionally, they showcased an enhanced ecological landscape effect. The rainfall runoff reduction process of rain garden on August 2, 2017 was monitored. The rainfall on that day was 22.5 mm, the total runoff control rate was 73.9%, the peak flow reduction rate was 49.0%, and the peak delay time was 25 min, which met the design requirements of the demonstration area.

3.1.4 Permeable Pavement

3.1.4.1 Technical Overview and Principles

Permeable pavement represents an innovative urban pavement technique. It utilizes a layer with a large pore structure or drainage infiltration facilities, enabling rainwater surface runoff to infiltrate in situ through infiltration facilities. This method aims to enhance rainwater infiltration and reduce surface runoff. Additionally, special pavement materials are employed to facilitate the retention, adsorption, absorption, and degradation of pollutants during the rainwater infiltration process. This approach effectively reduces the presence of pollutants in the runoff.

Permeable pavement can be divided into permeable brick pavement, permeable cement concrete pavement and permeable asphalt concrete pavement according to different surface layer materials. Grass-embedded brick, cobblestone in the garden pavement, gravel also belongs to permeable pavement. Three typical permeable pavement profiles are as follows:

(1) Permeable brick pavement, in which small permeable pavement bricks are placed on the street pavement, as shown in Fig. 3.21;
(2) Permeable cement concrete pavement, constructed with concrete with a porosity of 15% to 25%, and also commonly constructed with sandless concrete, as shown in Fig. 3.22;
(3) Permeable asphalt concrete pavement, commonly known as asphalt concrete. Artificially selected with a certain grading composition of mineral materials,

Fig. 3.21 Permeable brick pavement

Fig. 3.22 Permeable cement concrete pavement

Fig. 3.23 Permeable asphalt concrete pavement

crushed gravel, stone chips or sand, mineral powder, etc., and a certain proportion of road asphalt materials, mixed under strictly controlled conditions into the mixture, as shown in Fig. 3.23.

3.1.4.2 Structural Characteristics and Design Parameters

i. Structure

Permeable pavement shall consist of four layers: surface layer, base layer, subbase and bedding layer, as shown in Fig. 3.24.

ii. Structural Design Requirements

(i) *Design Requirements for Permeable Surface Layer*

(1) When the permeable brick is used as the permeable surface layer, the permeability coefficient should exceed 1×10^{-4} m/s. Porosity must be greater than 20%, while the compressive strength should be higher than 35 MPa, and the flexural strength should exceed 3.2 MPa.
(2) When the permeable cement concrete is used as the permeable surface layer, the thickness of cement concrete should adhere to professional road design standards. High-strength grade slag silicate cement is recommended. The thickness of cement concrete in parking lots should be 100 to 150 mm, and the thickness of cement concrete in road should be 150 to 300 mm. The porosity should be 15% to 21%.

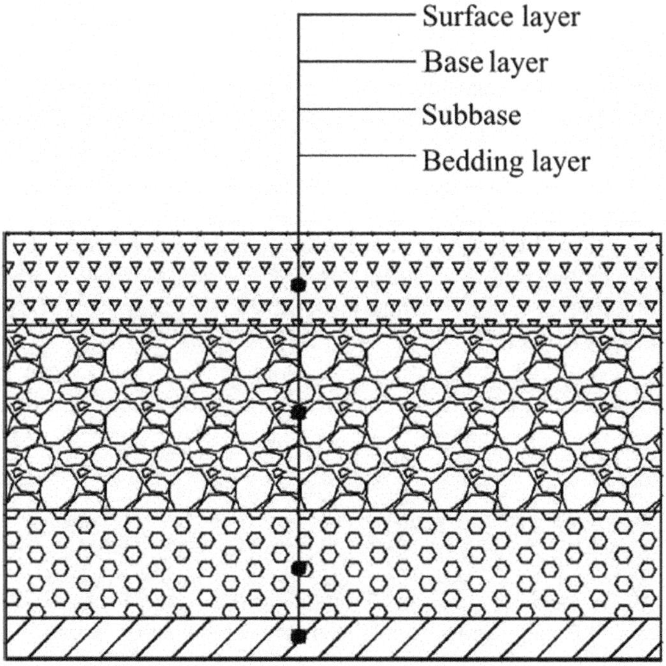

Fig. 3.24 Schematic diagram of basic structure of permeable pavement [20]

(3) When the permeable asphalt concrete is used as the permeable surface layer, the thickness of asphalt concrete in parking lots should be 50 to100 mm, and the thickness of asphalt concrete in road should be 100–150 mm. The porosity should be greater than 16%.

(ii) *Design Requirements for Permeable Base Layer*

(1) When permeable cement concrete or permeable asphalt concrete is used as permeable surface layer in travel lanes with light traffic flow, the degree of compaction of base layer soil is usually 94% or more, and seepage control facilities are provided, and drainage facilities are required in the permeable bedding layer.
(2) When permeable bricks, permeable cement concrete, etc. are used for parking lots and sidewalks, the degree of compaction of base layer soil is usually around 90%.

(iii) *Design Requirements for Permeable Bedding Layer*

The thickness of permeable bedding layer should not be less than 150 mm, and the porosity should be at least 30%. During the design process, the thickness of permeable bedding layer should be determined according to the design standards, the storage water requirements and the drainage time of stored rainwater specified in relevant codes. A thicker permeable bedding layer allows for a greater amount of rainwater storage.

For instance, considering a requirement to store 100 mm of net rainfall in the permeable pavement, with a permeable bedding layer porosity of 35%, a subsoil permeability coefficient of 1×10^{-6} m/s, and a desired rainwater emptying time of 36 h, the minimum thickness of permeable bedding layer to meet the design objective requirement is calculated as: $Dp = 0.10/0.35 = 0.286$ m = 286 mm; On the other hand, the maximum thickness of permeable bedding layer to fulfill the rainwater emptying time requirement is calculated as: $Dp = (1 \times 10^{-6} \times 36 \times 3600)/0.35 = 0.370$ m = 370 mm. Therefore, the final design thickness of permeable bedding layer is determined as 286 mm.

Permeable bedding layer should be used continuous grading gravel and single grading gravel and other permeable materials. The particle size of single grading gravel bedding layer should be 5–10 mm, and the particle size of continuous grading gravel bedding layer should be 5–40 mm.

(vi) *Design requirements for permeable bedding layer with non-infiltration of base layer soil*

(1) Set a 150 mm sand filter layer, sand should meet the corresponding requirements.
(2) A permeable geotextile layer shall be set between sand filter layer and permeable bedding layer, geotextile should meet the corresponding requirements.
(3) The perforated pipe is set up horizontally along the road under the sand filter layer.
(4) The diameter of perforated pipe should be 100–150 mm, and the drilled holes should meet the corresponding requirements.
(5) The transverse slope of the perforated pipe along the permeable road surface shall be more than 1%.
(6) The perimeter of the perforated pipe is wrapped in a gravel trench of the same material as the permeable bedding layer, with a trapezoidal cross-section, 0.5 m wide at the top of the trapezoid, 0.2 m wide at the bottom and 0.2 m high. A 5 cm gravel layer was placed above and below the perforated pipe.
(7) Perforated pipes can be connected to inspection wells for discharge or infiltration wells for infiltration;
(8) An impermeable layer should be set between the sand filter layer and the base layer soil. The impermeable layer materials should comply with road professional design standards. The impermeable layer should encompass the gravel tank, but it does not obstruct the passage of rainwater into the gravel tank.
(9) If the soil has poor or no infiltration capacity and the permeable bedding layer is unable to infiltrate rainwater effectively, it becomes necessary to discharge the rainwater. To maintain the quality of discharged rainwater, a sand filter layer is required for filtration. The filtered rainwater is then directed to drainage facilities, as illustrated in Fig. 3.25. The drained rainwater can be connected to inspection wells for discharge or to infiltration wells for infiltration. The structural design is outlined in Fig. 3.26.

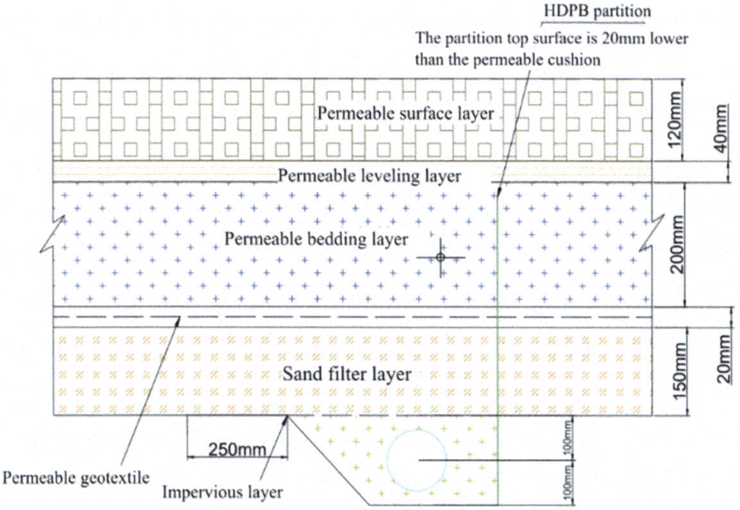

Fig. 3.25 Structure of sand filter layer and subsurface drainage layer under permeable bedding layer

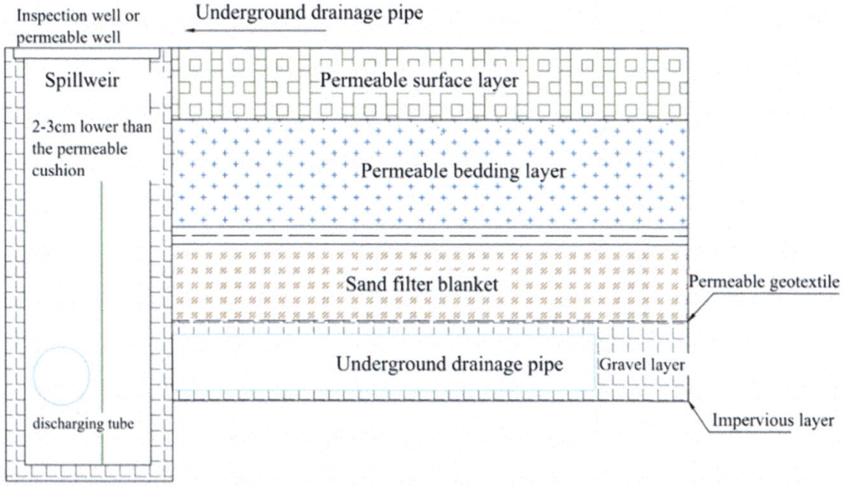

Fig. 3.26 Permeable pavement underground drainpipe connected to infiltration wells or inspection wells

(10) The trapezoidal cross-section of the gravel tank is used to make the rainwater in the sand filter layer have a larger contact area to enter the gravel tank and be discharged through the perforated pipe. Permeability coefficient of permeable leveling layer should be greater than 5×10^{-4} m/s, and its thickness should be 20–50 mm. The

leveling layer should be made of coarse sand, fine stone and permeable concrete with fine stone.

iii. Overflow Design Requirements

Although permeable pavement has good control of total runoff and external flood peaks, overflow facilities need to be installed on permeable pavement in the following two cases.

(1) After long time use of permeable pavement, the permeable surface layer is clogged severely and its permeability is greatly reduced;
(2) Under extreme conditions such as continuous rainfall, the water storage capacity of permeable pavement is saturated.
(3) Permeable pavement overflow facilities are laid out in the form shown in Fig. 3.27; its relationship with the overflow of curb, infiltration wells, bioretention, grassed swales and other low-impact development facilities is shown in Fig. 3.28.

The overflow flow rate of permeable pavement, that is, the difference between the runoff volume flowing through the permeable pavement and the effective rainwater

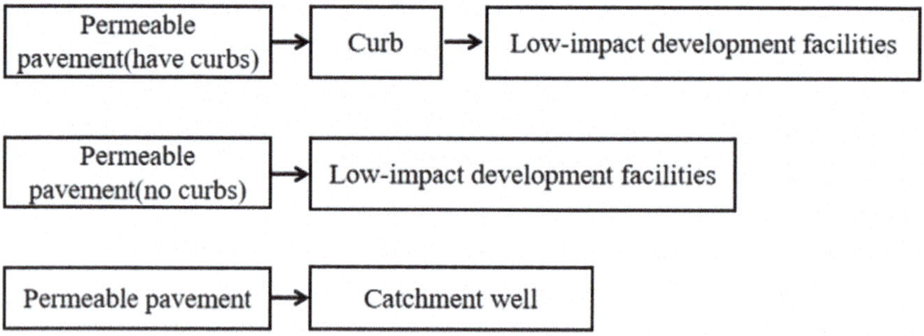

Fig. 3.27 Layout form of overflow facilities on permeable pavements

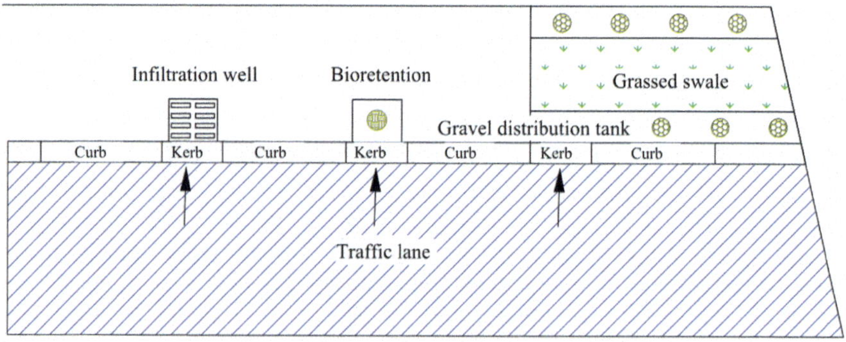

Fig. 3.28 Schematic diagram of overflow mode of permeable roadway

storage capacity of the permeable pavement. Where the calculation of the effective rainwater storage capacity shall be calculated according to the following formula (3.7):

$$V_p = 0.5 \times \left(D_p \times \theta_p + D_c \times \theta_c\right) \times L_{pmax} \times W_p \quad (3.7)$$

where, V_p-effective rainwater storage capacity of permeable pavement, m^3;

D_p, Dc-thickness of permeable bedding layer and permeable surface layer, respectively, m;

θ_p, θ_c-the porosity of permeable bedding layer and permeable surface layer, respectively;

L_{pmax}-partition length of permeable pavement, m;

W_p-width of permeable pavement, m.

If rainwater is directed into the permeable bedding layer, its effective storage volume should be deducted from the total rainwater volume. When permeable pavement is equipped with underground drainage facilities, the positive drainage impact of the sand filter layer in these facilities will not be considered in calculations due to the slow drainage rate. Additionally, the permeable leveling layer, having a limited thickness and porosity, will not be factored into the calculations for its positive rainwater storage impact.

Considering the following reasons, the effective rainwater storage capacity in the rainy season adopts 0.5 times the maximum storage capacity, (1) The actual storage capacity of permeable pavement is reduced due to clogging after a long period of time; (2) There is a slope in the road, and the actual storage thickness is reduced due to the slope; (3) Considering the influence of the rainfall in the previous period, the actual storage capacity of the road will be reduced as well.

iv. Construction Process of Permeable Pavement

Initially, the roadbed excavation must reach the designated depth, and the original soil layer should be thoroughly compacted. The compaction coefficient for loam and clay roadbeds must exceed 90%, ensuring the base of the roadbed is level. The longitudinal and transverse slopes of the subgrade and the edge line should align with the design specifications. Construction of the permeable pavement shall adhere to the process outlined in Fig. 3.29.

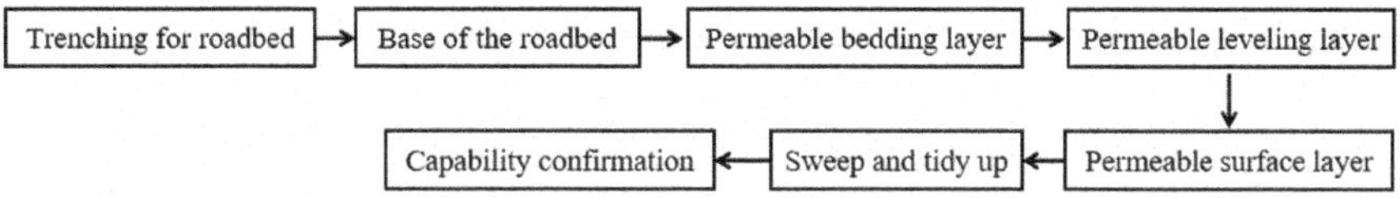

Fig. 3.29 Construction processes of permeable pavement

3.1.4.3 Scope of Application and Advantages and Disadvantages

Permeable bricks, in comparison to regular bricks, have a higher porosity and excellent water permeability. This allows rainwater to swiftly infiltrate the ground, effectively slowing down runoff and reducing water volume. During light to moderate rainfall, the pavement generates minimal runoff, significantly enhancing the reduction of rainfall runoff.

Permeable brick pavement is mostly used for sidewalks; permeable cement concrete pavement is mostly used for light-load sidewalks, motorized vehicles or convenient roads in parks, residential, etc.; permeable asphalt concrete pavement is mostly used for urban expressways, highways and other motorized vehicles needing to have a certain load capacity.

Based on the above analysis, permeable pavement has the following three advantages:

(1) It can increase rainwater infiltration and reduce rainwater runoff;
(2) The water quality of rainfall runoff has a certain purification effect, and can protect the safety of groundwater resources;
(3) It has the role of temperature regulation, humidity regulation, dust reduction, noise reduction, the use of grass bricks type of permeable pavement can also increase the area of urban green space, landscaping.

Permeable pavement also suffers from the following three disadvantages:

(1) It is easy to be clogged in the course of use;
(2) In the subsequent management and maintenance, it needs to be cleaned regularly and requires high maintenance;
(3) Higher price compared with traditional pavement.

d. Operation and Maintenance Requirements

In order to ensure that permeable pavement has the ability to infiltrate rainfall runoff and reduce pollutants, permeable pavement should be maintained with good permeability over time.

Clogging of porous permeable pavement materials poses a significant challenge for pavements. Typically, after prolonged use, the clogging occurs within the top 2.0 cm of the paved surface. Regular inspection and cleaning are essential to address this common issue, often requiring methods such as drenching, negative pressure suction, and high-pressure water flow flushing. Prompt repairs are necessary if the surface layer of permeable pavement is damaged. Additionally, it's advisable to refrain from using chemicals like snowmelt and pesticides on the pavement surface to prevent groundwater contamination. Common maintenance requirements of permeable pavement are shown in Table 3.9.

Table 3.9 Common maintenance requirements of permeable pavement

Maintenance content	Maintenance priorities and objectives	Maintenance cycle
Road sweeping	Removal of debris from roadways	Regular sweeping in accordance with sanitation requirements
Permeable surface finishing	First use high-energy vacuum cleaner to clean, then use high-pressure cleaner to clean; blow off with compressed air; according to the manufacturer's requirements to use professional equipment to clean up	Determined based on permeable pavement inspection results; once every 3 to 7 years or so, depending on pavement sanitation
Replacement of permeable surface tiles	Replacement of permeable surface bricks	Depending on the sanitary condition of the pavement, usually after 5 to 15 years of use; after the permeable surface tiles are broken
Replacement of permeable surface layer, permeable leveling layer, permeable bedding layer, sand filter layer	Replacement of permeable surface layer, permeable leveling layer, permeable bedding layer, sand filter layer	Usually after 10 to 25 years of use

3.1.4.4 Technology Effectiveness and Costs

Permeable pavement systems offer significant advantages over traditional impermeable pavements by enhancing rainwater infiltration rates, reducing surface runoff, mitigating peak flow rates, and delaying the onset of peak flow. A study conducted by Fassman et al. [21] demonstrated that permeable asphalt pavements can retain 70% more rainwater runoff than their impermeable counterparts, yielding runoff coefficients as low as 0.29 to 0.67. In Eastern North Carolina, Collins et al. [22] observed that both permeable interlocking pavement and permeable grids could store 6 mm of rainfall without runoff, equating to approximately 30% of moderate rainfall. Another study by Zhao Fei et al. [23] revealed that the installation of rainwater collection pipes within the structural layer of permeable pavement increased its rainwater runoff reduction rate to 40% to 90%, it is about 10% higher than that without collection systems. Additionally, permeable pavement can absorb the surface runoff produced by impermeable pavement in the same area. However, the reduction capacity of permeable pavement on rainwater runoff significantly diminished when the permeable pavement area fell below 1/3 of the total area.

Permeable pavement exhibits remarkable efficacy in purifying various pollutants present in stormwater runoff, in addition to its capacity to manage runoff and regulate flood flow. A study by Jin Jianrong et al. [24] investigated the pollutant removal efficiency of different permeable pavement structures, including slit permeable bricks with cement-stabilized gravel as base, slit permeable bricks with gravel as base, and

Table 3.10 Common investment cost of permeable pavement

Permeable pavement facilities	Permeable brick	Permeable cement concrete	Permeable Asphalt Concrete
Investment (CNY/m^2)	50~100	200~400	100~200

Note The investment price of permeable pavement is the increased cost over traditional practices

permeable concrete. The results indicated significant removal rates for TP, SS, COD, Cr, Mn, Cu, Zn, Pb, and petroleum in all three types of facilities. However, a lower removal rate was observed for TN, and all three types of facilities exhibited a notable NO_3^--N release phenomenon. Notably, slit permeable bricks with cement-stabilized gravel as base, slit permeable bricks with gravel as base exhibited better removal of NH + 4-N compared to permeable concrete pavement. In a study conducted by Zhao Xianyong et al. [25], different permeable pavement structures demonstrated varying pollutant reduction effects. Permeable turf bricks exhibited the highest reduction effect on COD and TN, while permeable bricks showed exceptional efficacy in TP reduction, achieving an impressive reduction rate of 94.6%. Interestingly, the purification efficiency of pollutants on permeable pavement improved with lower rainfall intensity. Brown et al. [26] found notable differences in pollutant concentrations among various permeable pavement types. Porous asphalt (PA) outflow had significantly higher TN, NO_3-N, and NH + 4-N concentrations compared to permeable interlocking concrete (PICP) and permeable concrete (PC). The elevated TN concentrations in PA outflow were attributed to its higher pH, exceeding the optimal range for nitrifying bacteria. Additionally, the orthophosphate concentration in PA effluent was lower than that in the other two permeable pavement options, potentially due to rapid orthophosphate precipitation with metal cations in PA. Furthermore, permeable bricks demonstrated superior heavy metal removal compared to permeable grids. However, the concentrations of heavy metals in water treated by both permeable pavements met the required standards.

Permeable pavement costs more than traditional surface pavement, but reduces the expense of laying traditional drainage pipelines and control systems. The investment costs for various types of permeable pavement are shown in Table 3.10.

3.1.4.5 Application Examples

Permeable pavement technology has been successfully implemented in the Future Science and Technology City, Beijing, covering a total area of approximately 600,000 m^2. Various permeable pavement solutions, including permeable bricks in central enterprise plots, park roads, walkways, squares, and permeable concrete pavement, have been utilized. This technology significantly contributes to rainwater infiltration within the park, enhancing local water resources, managing rainfall runoff, and mitigating runoff pollution. According to the monitoring results in 2015 and 2016, compared with impermeable pavement, permeable pavement can reduce runoff by

30%–70% and pollutant discharge by 40%–60%, and it has good ecological and environmental effects.

3.1.5 Infiltration Facilities (Infiltration Wells (Pipes), Infiltration Depressions, Infiltration Trenches)

3.1.5.1 Technical Overview and Principles

Infiltration facilities are facilities that infiltrate rainwater into the underground soil and include infiltration wells (pipes), infiltration depressions, and infiltration trenches.

3.1.5.2 Structural Characteristics and Design Parameters

i. Structure

(1) Infiltration wells (pipes) are utilized to facilitate rainwater infiltration, employing either standalone infiltration wells or the combination of infiltration wells and pipes. The fundamental structure comprises infiltration inspection (catchment) wells, infiltration pipes, gravel aquifers surrounding the infiltration well (pipes), water intake facilities, overflow facilities, and more, as illustrated in Fig. 3.30.

When the total runoff and flood peak flow to be controlled are small, independent infiltration wells can be used to realize rainwater infiltration. Conversely, a combination of infiltration wells and infiltration pipes can be employed to achieve effective rainwater infiltration.

Infiltration wells (pipes) necessitate inspection wells to facilitate maintenance and ensure their proper operation. Additionally, collecting wells are essential as water intake facilities. In cases where both inspection wells and collecting wells are needed, catchment inspection wells can be set up.

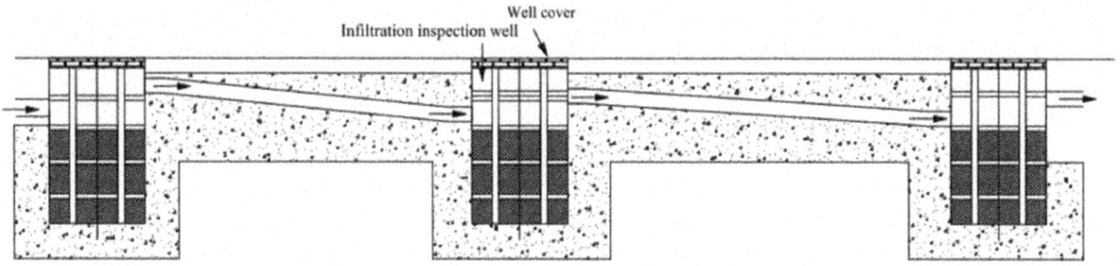

Fig. 3.30 Structure schematic diagram of infiltration well (pipes)

(2) The basic structure of the infiltration depression includes: planting area, water intake facilities, water storage area, planting soil layer, geotextile layer, aquifer, and overflow facilities, etc., as shown in Fig. 3.31.

The maximum water storage depth of the infiltration depression should be set at 10–30 cm, and it is advisable to set a slope larger than 2: 1 (h: v) around the water storage area. The thickness of the soil layer in the infiltration depression should be 20–40 cm, and the permeability coefficient should be between 5.0×10^{-5} and 1.0×10^{-4} m/s.

(3) The basic structure of the infiltration trench includes: pea gravel filtration layer, aquifer, sand filtration layer, and overflow facilities, as detailed in Fig. 3.32.

The thickness of pea gravel filtration layer should be 10–20 cm, the grading should be 5–10 mm, and it should be composed of washed pea gravel. At the bottom of the infiltration trench, a sand filtration layer should be installed with thickness of 15–20 cm. It's important to note that the infiltration trench is different from the infiltration depression. Unlike the infiltration depression, it doesn't use planting soil layer for rainwater filtration; instead, it employs pea gravel layer for filtering rainwater. The water quality in its aquifer is relatively worse than that in the infiltration depression,

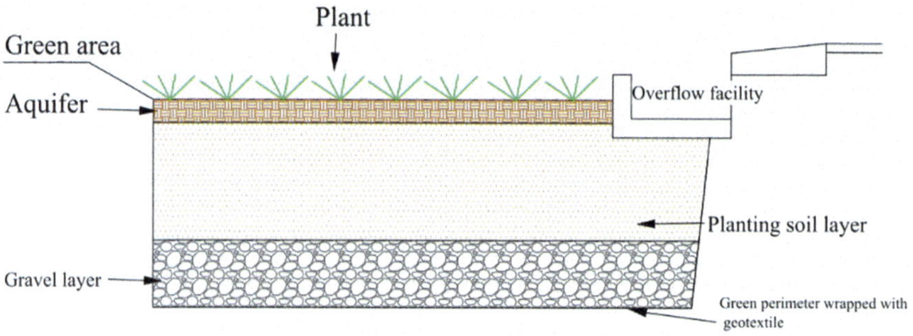

Fig. 3.31 Structure schematic diagram of infiltration depression

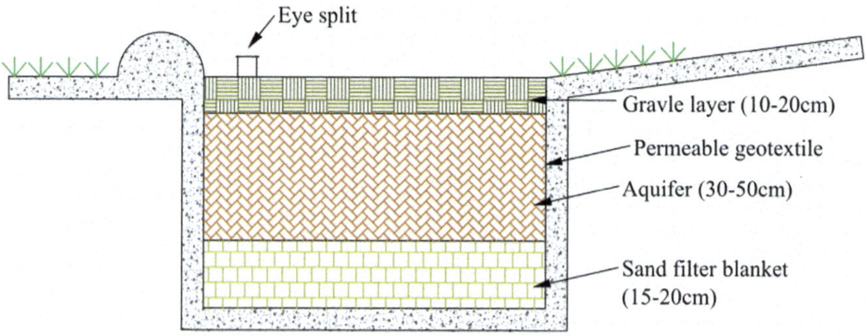

Fig. 3.32 Structure schematic diagram of infiltration trench

so a sand filter layer should be set under the aquifer. The infiltration trench should be equipped with overflow facilities, which can be implemented using inlets or overflow weirs. It's advisable to install an observation hole every 50 m along the length of the infiltration trench.

ii. Structural Design Requirements

(i) *Design Requirements for Infiltration Wells (Pipes)*

The bottom of the infiltration facility should be at least 0.6 m above the water table or impermeable layer, and the soil beneath the infiltration facility should have a permeability coefficient between 4×10^{-6} and 1×10^{-3} m/s. Ideally, the infiltration facility should be constructed on terrain with a slope of less than 15%. When infiltration facilities are installed between the roof of underground buildings and the overburden layer, the superstructure of the roof can be used to build infiltration depressions. The roof of underground buildings is usually covered with a certain thickness of soil for greening. In order to make the greening plants grow normally, it is necessary to arrange infiltration pipes or water delivery plates on the roof of buildings. This arrangement allows excess rainwater collected in the infiltration depression to be diverted away, ensuring that rainwater continues to seep down effectively.

(1) Infiltration wells (pipes) should adhere to the following requirements: The bottom of the infiltration well must feature a sand chamber with a depth greater than 0.2 m. The upper section of the chamber should have a seepage area, with gravel placed outside this region. The outer layer of gravel should be wrapped with geotextile. Catchment infiltration wells must be equipped with interceptor baskets. The appropriate borehole diameter in infiltration area is 15–20 mm, spaced at intervals of 10–15 cm. At least three layers of holes should be arranged within the infiltration area.
(2) The setting of infiltration wells (pipes) should adhere to the following requirements: Infiltration pipes should be made of perforated PVC or HDPE pipes, sandless concrete or drainpipes and other permeable materials. The diameter of the infiltration pipe should not be less than 150 mm, and the slope of the infiltration pipe laying should be between 0.5% and 2%. The infiltration layer of infiltration pipe should consist of gravel with a minimum thickness of 300 mm. The gravel layer should be wrapped with geotextile. The spacing of infiltration inspection wells should not be more than 150 times of the diameter of infiltration pipes. The elevation of outlet pipe of infiltration inspection well should be higher than that of inlet pipe, but not higher than that of outlet pipe of upstream adjacent well. Infiltration pipes should not be installed beneath driving surfaces, and the overburden depth should not be less than 0.7 m if installed under driving surfaces. The design of the water intake for infiltration wells (pipes) should be decentralized and multi-point, ensuring the effective utilization of the infiltration capacity and sedimentation chamber. This approach avoids situations where the sedimentation chamber of the intake well needs cleaning while downstream chambers remain unused.

(3) Volume design of infiltration wells (pipes): Infiltration wells (pipes) are used to control total rainwater runoff and peak flow. The effective storage volume must not be less than the rainwater retention infiltration volume V_R and rainwater retention volume V_D required by the infiltration wells (pipes) for control purposes. The effective storage volume of the infiltration wells (pipes) is equivalent to its storage volume at the average water storage elevation. The average water storage elevation is calculated as the average between the elevation of the outlet pipe of the influent infiltration well and the elevation of the overflow port. This calculation should encompass the volume of the infiltration well, the volume of the infiltration trench, and the volume of the gravel layer multiplied by the porosity.

(ii) *Design Requirements for Infiltration Depressions and Infiltration Trenches*

The design of infiltration depressions and infiltration trenches consists mainly of the design of planting soil layers, aquifers and planting plants.

(1) Design of planting soil layers

The thickness of planting soil layer in infiltration depressions and trenches is typically 20–40 cm. These depressions and trenches are designed to facilitate rapid entry of rainwater into the aquifer. Therefore, the permeability coefficient of the planting soil layer should be higher than that of the natural soil. However, to maintain the quality of rainwater entering the aquifer, a maximum permeability coefficient can also be specified accordingly.

(2) Design of aquifers

Gravel or water storage module should be used in aquifer of infiltration depressions and infiltration trenches, the thickness of the gravel layer should be 30–50 cm; water-permeable geotextile should be set up between the aquifer and the planting soil layer; the water storage volume of the aquifer and the water storage area should be greater than the maximum of the rainwater retention infiltration volume V_R and the rainwater retention volume V_D.

(3) Design of planting plants

The thickness of the planting soil layer in infiltration depressions and trenches determines that trees cannot be planted. For infiltration depressions and trenches with a rainwater emptying time of 24 h, flood-resistant plants should be selected for planting. Flood-resistant herbaceous plants or shrubs are suitable choices for these areas.

iii. Overflow Design Requirements

The infiltration wells (pipes) must have overflow facilities to guarantee its effective storage volume. Depending on the overflow facilities installed in the infiltration wells (pipes), two distinct rainwater infiltration systems, off-line and on-line, are created. Among them, the overflow pipe is installed in the upstream infiltration well to form an off-line rainwater infiltration system, while the overflow pipe is placed in the end infiltration well to form an on-line rainwater infiltration and drainage system, The specific differences between the two systems are outlined in Table 3.11.

Table 3.11 Comparison of in-line rainwater infiltration and drainage systems and off-line rainwater infiltration systems

Systems	Advantages	Disadvantages
In-line rainwater infiltration and drainage systems	Replacement of drainage systems	Poor rainwater infiltration and pollutant removal
Off-line rainwater infiltration Systems	Requires a separate drainage system	Highly efficient rainwater infiltration and pollutant removal

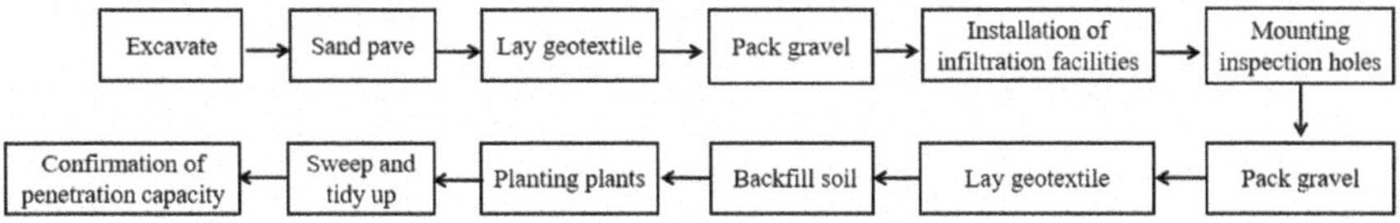

Fig. 3.33 Construction process diagram for infiltration facilities

iv. Construction Process for Infiltration Facilities

Infiltration wells (pipes), infiltration depressions and infiltration trenches and other facilities should be constructed to ensure the accuracy of the construction and installation, there should be reliable measures to protect the finished product for the complete set of products, and the construction site should be clean to prevent sediment and other sinks from infiltrating into the infiltration facilities, which will affect the infiltration capacity and normal use of the facilities. Infiltration facilities should be constructed in accordance with the procedures in Fig. 3.33.

3.1.5.3 Scope of Application and Advantages and Disadvantages

Infiltration facilities are primarily utilized in the green areas surrounding buildings and communities, roads, and parking lots. When applying infiltration wells in areas where rainfall runoff is heavily polluted, and the bottom of the facility is less than 1.0 m away from the seasonally highest groundwater level or rock layer and less than 3.0 m away from the building foundation (horizontal distance), necessary precautions must be taken to prevent secondary disasters. These facilities should not be constructed in hazardous locations prone to collapses and landslides, and they should avoid geological areas with special soils such as self-weighted collapsible loess, expansive soils, and high salt content.

Infiltration facilities shall not affect the foundations of other structures, roads, pipelines, etc. The distance between infiltration facilities and structures shall meet the requirements of Table 3.12. When the distance between infiltration facilities and the foundation of roads and pipelines cannot meet the requirements, an impermeable layer shall be used to separate the aquifer from the foundation.

Table 3.12 Distance requirements of infiltration facilities and structures

Structure	Building foundations	water intake well	septic
Minimum distance requirement (m)	3.0	15.0	30.0

Table 3.13 Maintenance requirements of infiltration wells (pipes)

Maintenance content	Maintenance Priorities and Objectives	Maintenance Cycle
Inspection well	Clear siltation from grit chamber; cleaning of infiltration holes in infiltration zones	Twice a year; determine based on inspection results
Overflow facility	Clear siltation from overflow facilities	Twice a year; determine based on inspection results
Infiltration pipe, gravel and geotextile around infiltration well	Replacement of infiltration pipes, gravel and geotextiles around infiltration wells	Inspection results show poor rainwater infiltration and drainage, usually after 5 to 10 years of use

Generally speaking, infiltration facilities occupy a small area and have low construction and maintenance costs, but their water quality and quantity control functions are limited.

3.1.5.4 Operation and Maintenance Requirements

Infiltration depressions shall be maintained in accordance with planting requirements after completion, and pretreatment of infiltration facilities shall be maintained in accordance with the maintenance requirements for that facility. Maintenance requirements of infiltration wells (pipes), infiltration depressions, and infiltration trenches are shown in Tables 3.13, 3.14 and 3.15.

3.1.5.5 Technology Effectiveness and Costs

Infiltration facilities play an important role in reducing rainfall runoff volume and increasing runoff infiltration. At the same time, infiltration facilities are highly efficient in removing pollutants such as SS, COD and heavy metals from rainfall runoff.

The annual maintenance cost of infiltration facilities is approximately 5% to 20% of their construction cost. Table 3.16 summarizes the investment construction costs for each type of infiltration facilities.

Table 3.14 Maintenance requirements of Infiltration depressions

Maintenance content	Maintenance priorities and objectives	Maintenance cycle
Planting maintenance	Replanting, fertilizing, and removing weeds to ensure that plantings grow	Periodically as required by plants; determine based on inspection results
Plant pruning	Landscape needs	Three times a year; determine based on inspection results
Dredge	Clear siltation from water storage areas; clear siltation from water intake and overflow facilities	Twice times a year; determine based on inspection results
Water storage area	Repair of slope failures in water storage areas; replacement of top 5 cm of soil; grading of water storage areas	Once a year; determine based on inspection results
Aquifers and geotextiles	Replacement of aquifer gravel and geotextiles	Inspection results show poor rainwater infiltration, usually after 5 to 10 years of use

Table 3.15 Maintenance requirements of Infiltration trenches

Maintenance content	Maintenance priorities and objectives	Maintenance cycle
Dredge	Clear siltation from pea gravel filter layers; clear siltation from overflow facilities	Twice times a year; determine based on inspection results
Pea gravel filter layers	Replacement of pea gravel layers	Once two years; determine based on inspection results
Aquifers, geotextiles and sand filter layers	Replacement of aquifers, geotextiles and sand filter layers	Inspection results show poor rainwater infiltration, usually after 5 to 10 years of use

Table 3.16 Investment Construction Costs for Various Types of Infiltration Facilities

Infiltration facility	Infiltration well (pipe)	Infiltration depression	Infiltration trench
Investment (CNY/m^2)	–	200–800	200–800

3.1.5.6 Application Examples

During the construction of urban rainfall runoff pollution control facilities in Fengxi New Town, Xixian New District, Shaanxi Province, an artificial rainwater rapid infiltration well was built at Xianyang Vocational and Technical College, as depicted in Fig. 3.34. The well has a depth of 2.75 m, an outer diameter of 3 m, a volume of 13.49 m^3, and an infiltration rate of 136.5 m/d.

Fig. 3.34 Artificial rainwater rapid infiltration well

3.2 Storage Technologies

3.2.1 Rainwater Tank

3.2.1.1 Technical Overview and Principles

Rainwater tanks, also known as rain barrels, are simple collection and utilization facilities, whether above ground or underground. They can be constructed from various materials such as plastic, fiberglass, or metal. Rainwater tanks have a small storage capacity, typically ranging from tens to hundreds of liters. They function as independent water storage systems and are not connected to indoor plumbing. They are primarily used for outdoor activities such as watering flowers and cleaning.

3.2.1.2 Structural Characteristics and Design Parameters

i. Structure

Rainwater tanks generally consist of water inlets, filter screens, tank bodies, outlets and overflow components.

ii. Design Requirements

Rainwater tanks are commonly utilized in conjunction with rainfall runoff control strategies, including grassed swales, infiltration trenches, and detention ponds. They are commercially available as proven products for purchase or can be customized and constructed based on user requirements.

3.2.1.3 Scope of Application and Advantages and Disadvantages

Rainwater tanks are a suitable option for collecting and utilizing rainwater from the roof of a single building. Ground-type rainwater tanks can be utilized for areas with a small amount of rainwater runoff. Underground rainwater tanks can be utilized in situations where there are specific demands for the building's exterior and an ample supply of rainwater, alongside the availability of suitable surroundings for tank placement.

Rainwater tanks as rainwater storage facilities, generally located in the low-impact development of the front end of the rainwater system, should be located close to the perimeter of the building from which the rainwater is to be collected, while maintaining a visually appealing overall landscape without compromising the building's appearance.

Rainwater tanks generally have the following advantages: They facilitate the collection, storage, and reuse of rainwater from roofs, which can reduce the consumption of tap water for external drainage, green irrigation, and other purposes. Most rainwater tanks are molded products, making them easy to construct and install. They have low maintenance requirements, especially when equipped with grills and other dirt interception facilities, which further reduces the need for maintenance.

Rainwater tanks also have the following drawbacks: The rainwater collected in the tank should be utilized within two consecutive rainfall intervals to maximize its storage capacity, minimize external drainage, and prevent the deterioration and odor of rainwater.

3.2.1.4 Operation and Maintenance Requirements

(1) When there is a blockage or accumulation in the water inlet, resulting in poor water flow, it is necessary to promptly clean up garbage and sediment;
(2) Timely remove sediment from the rainwater tank;
(3) In northern regions, before the arrival of winter, the water in the rainwater tank and its connecting pipelines should be emptied to prevent damage from freezing;
(4) When warning signs for preventing misconnection, misuse, or accidental consumption are damaged or missing, they should be repaired and improved in a timely manner.

3.2.1.5 Technology Effectiveness and Costs

Rainwater tanks enhance the utilization of rainwater resources. Rainwater has a high content of dissolved oxygen, so rainwater irrigation is good for plant growth. Moreover, rainwater is acidic, so rainwater irrigation can improve the quality of the soil to a certain extent.

The cost of rainwater tanks depends on factors such as the chosen material, volume, and accessories. Domestic rainwater tanks typically range in cost from 2,000 to 4,000 CNY.

3.2.1.6 Application Examples

Wuxi has a community roof area of 6,000 m^2, with 80% of the roof dedicated to rainwater collection for community landscape water recharge, green watering, and vehicle washing purposes. The rainwater reuse system includes several components such as rainwater wells, abandoned wells, rainwater tanks, storage tanks, inclined plate sedimentation tanks, and clear water tanks. Specifically, there are 5 rainwater tanks available, each with a capacity of 15 m^3 for storing rainwater. The maximum daily water consumption of the miscellaneous water network system is 44.9 m^3, which is allocated as follows: 20.0 m^3 for landscape water replenishment, 21.0 m^3 for watering green areas, and 3.9 m^3 for vehicle washing.

3.2.2 Reservoir

3.2.2.1 Technical Overview and Principles

Reservoirs are rainwater harvesting and utilization facilities that have the function of storing rainwater and also serve to reduce peak flows, and mainly include landscape water body within green buildings and community, reinforced concrete reservoirs, brick and stone masonry reservoirs, and plastic water storage module assembled reservoirs. In urban areas where land use is limited, underground closed reservoirs are commonly employed.

3.2.2.2 Structural Characteristics and Design Parameters

i. Structure

Reservoirs are water storage facilities built with artificial materials and have impermeability properties. Reservoirs can be constructed either above ground or underground, depending on the terrain and soil conditions. Therefore, they can be categorized as either open or closed systems. Reservoirs typically consist of pools and associated supporting infrastructure.

The main supporting facility include: (1) The water diversion ditch (canal), which is an integral part of the reservoir. Depending on terrain conditions, it can generally be lined to connect water source with the reservoir. Ideally, the water diversion ditch should be connected to the natural rainwater rushing ditch on the slope, and its length should be sufficient to ensure an adequate amount of water diversion. (2) The sedimentation tank plays a crucial role in precipitating sediment particles larger than the specified size, thereby ensuring that the sediment content meets water quality requirements and matches the sand-carrying capacity of downstream channels. It allows the sediment from slope surface runoff to deposit within the settlement tank, while also facilitating the removal of sediment, preventing excessive accumulation in water storage tanks, which reduces water storage capacity and increases the amount of dredging. (3) Trash-rack should be installed at the inlet of the sedimentation tank or the filtration tank to intercept debris such as dry branches, leaves, and weeds carried by the runoff. (4) Inlet pipes (channels) are mainly used to transport rainwater to the reservoirs. (5) Guardrails are primarily used to prevent human and livestock entry, protect human and livestock safety, and ensure engineering safety. The overall height should be greater than 0.7 m, with a bottom solid wall height of 0.2 m.

ii. Design Requirements

(1) The reservoir should be constructed on a stable foundation with sufficient load-bearing capacity. The settlement of the foundation should not have a detrimental effect on the integrity of the reservoir when it is filled with water.
(2) When constructing a reservoir in areas with collapsible loess, the technical requirements stated in the "*Code for Building Construction in Collapsible Loess Regions*" (GB 50,025) should be followed.
(3) The strength of the reservoir structure should be greater than the maximum water pressure. After the reservoir is filled with water, it should not experience engineering accidents such as cracking or collapse.
(4) Reservoirs used for different purposes should have different waterproofing grades.
(5) Priority should be given to collecting rooftop rainwater in the reservoir, and it is not suitable to collect rainwater from heavily polluted underlaying surfaces such as motorized roadways. Before entering the reservoir, the rainwater should undergo silt and sand separation or coarse filtration. If the reservoir has sedimentation or filtration treatment functions and the water quality of the effluent meets the requirements, a clear water tank may not be necessary. When using rainwater treated in a reclaimed water clear water tank, the tank should have sufficient capacity to accommodate the rainwater.
(6) The reservoir should be equipped with an overflow system, overflow outlets including overflow risers and spillways. The drainage capacity should be determined based on the drainage capacity of downstream rainwater drains or the excessive rainwater runoff drainage system. Both the overflow pipe and vent pipe should have measures in place for pest control.
(7) The reservoir composed of plastic modules and siliceous sand blocks should meet the requirements of ground and soil bearing capacity for its structural strength. The plastic modules should satisfy the anti-floating requirement, and the outer layer should be wrapped with impermeable geomembrane or materials with equivalent performance. The internal structure of the reservoir should facilitate the removal of deposited silt and sediment, and when equipped with

filtering function, it should allow for the removal of filtered sediments. The siliceous sand block reservoir should have a reinforced concrete bottom slab.

3.2.2.3 Scope of Application and Advantages and Disadvantages

Reservoirs are suitable for buildings and communities with a demand for rainwater reuse and urban green spaces. Depending on the specific rainwater reuse purposes, corresponding rainwater purification facilities should be built. Reservoirs are not suitable for areas without a demand for rainwater reuse or areas with severe rainfall runoff pollution. Generally, reservoirs are located at the end of low-impact development rainwater systems, after the rainfall runoff pollution control facilities and before the rainwater overflow discharge outlets, in order to fully utilize their capacity for storing runoff rainwater.

In general, reservoirs have the following advantages: space saving, easy access to rainwater pipes and channels, avoidance of direct sunlight, prevention of mosquito breeding, and large water storage capacity. Rainwater can be reused for irrigation, road flushing, and vehicle washing. At the same time, they also have some disadvantages, mainly including high construction costs and high maintenance and management requirements.

3.2.2.4 Operation and Maintenance Requirements

(1) When water and soil erosion occur at the water inlet or overflow port due to scouring, it is necessary to promptly install gravel buffers or take other anti-erosion measures;
(2) When the water inlet or overflow port is blocked or silted, resulting in poor water flow, it is necessary to promptly clean up garbage and sediment;
(3) When the sediment in the sedimentation tank exceeds the designed dredging height, it should be promptly dredged;
(4) The relevant equipment such as pumps and valves should be regularly inspected to ensure their proper functioning;
(5) When warning signs for preventing misconnection, misuse, or accidental consumption, guardrails and other safety precautions and early warning systems are damaged or missing, they should be repaired and improved in a timely manner.

3.2.2.5 Technology Effectiveness and Costs

The main function of a reservoir is to collect rainwater for storage and reuse. The amount of rainwater collected is closely related to the reservoir's capacity; the larger the volume of the reservoir, the more water can be stored.

reservoirs have relatively high costs, and the expenses mainly depend on their scale. The costs primarily include excavation and repair and masonry expenses during the construction phase, as well as equipment power consumption and maintenance costs during operation.

3.2.2.6 Application Examples

The case study takes place in a public building situated in Hefei City, Anhui Province. The project adheres to the two-star green building design labeling requirements and complies with the "*Assessment Standard for Green Building*" (GB/T 50,378). As per the standard, the project should achieve a total annual runoff control rate of 55%. The project comprises four high-rise buildings, each with a two-level basement. The project site area is 30225 m^2, the roof area is 9043 m^2, the green area of the site is about 6051 m^2, and the building footprint is 8859 m^2. The road and hard pavement cover an area of 15,315 m^2, with the road area accounting for roughly 4434 m^2 and the hard pavement area totaling 10,881 m^2. Permeable bricks are used for the hard pavement, enhancing rainwater seepage rates and minimizing surface runoff. The reservoir has a volume of 160 m^3 and caters to the green irrigation area, road washing area, and garage washing area, which measure 6051 m^2, 4434 m^2, and 12,390 m^2, respectively.

3.2.3 Wet Pond

3.2.3.1 Technical Overview and Principles

Wet ponds, also known as wet detention basins or wet detention ponds, are landscape water bodies that utilize rainwater as the main source of water supply, and possess both rainwater storage and purification functions. Wet ponds are sometimes designed as multifunctional detention water bodies in combination with green spaces and development spaces, typically serving normal landscape, recreational and leisure functions, while also contributing to rainwater detention and purification during rainfall events, thereby achieving synergistic treatment of runoff volume and pollutants.

3.2.3.2 Structural Characteristics and Design Parameters

i. Structure

Wet ponds typically consist of water inlets, forebays, main ponds, overflow outlets, slope protection, revetments, and maintenance pathways, as shown in Fig. 3.35.

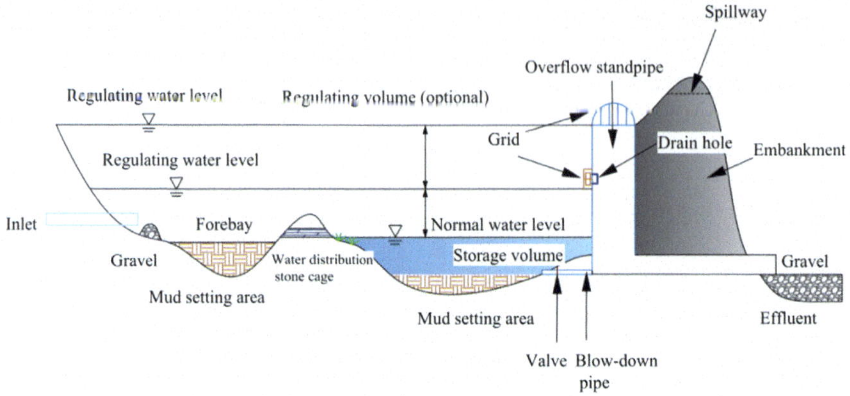

Fig. 3.35 Schematic diagram of typical configuration of a wet pond [27]

ii. Design Requirements

(1) Water inlets and overflow outlets should be equipped with energy dissipating facilities such as gravel and energy dissipating steps to prevent flow scour and erosion.
(2) The forebay serves as the primary pretreatment facility for the wet pond, facilitating the sedimentation of coarse pollutants in the runoff. The bottom of the pond is typically constructed with concrete or rock structures to facilitate sediment removal. Clearing channels and protective structures should be installed in the forebay, with ecological soft revetments being the preferred form. The slope ratio (vertical: horizontal) of the forebay is generally 1:2 to 1:8. The volume of the sedimentation zone in the forebay should be determined based on the clearing cycle and the pollutant load from the collected rainfall runoff.
(3) Generally, the main pond consists of two components: the permanent volume and the storage volume located below the permanent water level. The water depth of the permanent volume typically ranges from 0.8 to 2.5 m. The storage volume is determined based on the "unit area control volume" specified in the regional planning. The wet pond, which is equipped with peak flow reduction function, incorporates a regulating volume that must be emptied within 24 to 48 h. It is advisable to establish an aquatic planting area (rainwater wetland) between the main pond and the forebay. The revetments of the main pond should be an ecological soft revetments, and the slope of the side slope (vertical: horizontal) should not be greater than 1:6.
(4) Overflow outlets encompass both overflow risers and spillways, and the drainage capacity should be determined by considering the drainage capacity of downstream storm drains or overflow rainwater runoff discharge systems.
(5) Safety protection measures and precautions, such as guardrails and warning signs, should be installed in wet ponds.

3.2.3.3 Scope of Application and Advantages and Disadvantages

Wet ponds are suitable for sites with spatial conditions such as buildings, residential areas, urban green spaces, and squares.

Overall, wet ponds can effectively reduce the total runoff volume, runoff pollution, and peak flow in larger areas, making them an important component of urban rainfall runoff pollution control. However, wet ponds also have some drawbacks, such as strict site requirements and higher construction and maintenance costs.

3.2.3.4 Operation and Maintenance Requirements

(1) When water and soil erosion occur at the water inlet or overflow port due to scouring, it is necessary to promptly install gravel buffers or take other anti-erosion measures;
(2) When the inlet or overflow port is blocked or silted, resulting in poor water flow, it is necessary to promptly clean up garbage and sediment;
(3) When the sediment accumulation in the forebay/pre-treatment tank exceeds 50%, it should be promptly dredged;
(4) When warning signs for preventing misconnection, misuse, or accidental consumption, guardrails and other safety precautions and early warning systems are damaged or missing, they should be repaired and improved in a timely manner;
(5) When slope protection collapse occurs, reinforcement should be done promptly;
(6) The relevant equipment such as pumps and valves should be regularly inspected to ensure their proper functioning;
(7) Harvesting, replanting, pruning of plants, and removal of weeds should be carried out in a timely manner;

3.2.3.5 Technology Effectiveness and Costs

The design structure, pollutant concentration in runoff, site soil background values, and types and combinations of planted vegetation all influence the pollutant removal effectiveness of wet ponds in rainfall runoff. Generally, wet ponds can achieve removal rates of around 85% for suspended solids (SS), 25% for total nitrogen (TN), and 40% for total phosphorus (TP).

The construction cost of wet ponds mainly includes geotechnical engineering and plant planting, and the annual operation and maintenance cost is generally 2% to 10% of the construction cost.

Fig. 3.36 Artificial wet ponds in phase two of central green corridor of phase two in Fengxi New Town in Xixian New Area

3.2.3.6 Application Examples

In order to effectively regulate urban rainwater runoff and control runoff pollution, a large artificial wet pond has been constructed in the central green corridor of phase two in Fengxi New Town in Xixian New Area, as shown in Fig. 3.36.

The total area of the wet pond is 123,000 m^2, with a total volume of 110,700 m^3. The natural soil infiltration rate is 15.12 mm/h, the average water depth is 100 mm under normal water level, and the storage depth is 800 mm. Overflow is adopted as the discharge method for the outflow of the wet pond.

The monitoring results of a 21-h rainfall event show that the wet pond has achieved good effectiveness in regulating runoff volume and purifying runoff pollutants. The total runoff control rate is 82.43%, with a peak delay time of 25 min and a peak reduction rate of 49.56%. The removal rates for SS, TN, TP, COD, and NH_3-N are 52.23%, 58.23%, 50.23%, 52.36%, and 59.22% respectively.

3.2.4 Rainwater Wetland

3.2.4.1 Technical Overview and Principles

Rainwater wetlands are generally located in depressed areas with trees or bushes on landscaped green spaces, and covered with bark or ground cover plants. They supplement groundwater by retaining and permeating rainwater and reduce the peak runoff of stormwater on the surface. They can also reduce runoff pollution through adsorption, degradation, ion exchange, and volatilization, making them an efficient runoff pollution control facility.

Rainwater wetlands can be classified into two forms: surface rainwater wetlands and small-scale subsurface rainwater wetlands. Surface rainwater wetlands refer to natural depressions, ponds, and landscape water bodies near the rainwater pipeline. They are modified to meet the goals of controlling rainwater runoff pollution and peak flow, while also considering landscape design and easy maintenance. On the other hand, small-scale subsurface rainwater wetlands have similar structures and functions to subsurface flow wetlands used in wastewater treatment.

3.2.4.2 Structural Characteristics and Design Parameters

i. Structure

Surface rainwater wetlands should include an inlet and pretreatment tank, deep channels, shallow areas, outlet tank, and outlet and overflow facilities. The total area of surface rainwater wetlands should not be less than 1.0% of the catchment area, nor should it exceed 15 hm^2. The basic structure of surface rainwater wetlands is shown in Fig. 3.37.

Small-scale subsurface rainwater wetlands should include water distribution facilities, filling layer, water storage and plant areas, and overflow facilities. Small-scale subsurface rainwater wetlands can use surface distribution or underground perforated pipe distribution, and their basic structure is shown in Fig. 3.38.

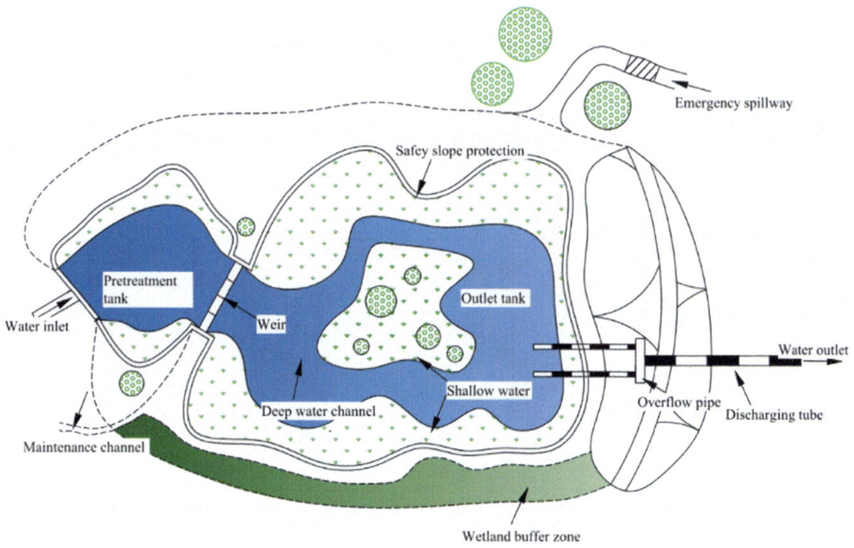

Fig. 3.37 Schematic diagram of surface rainwater wetland structure [28]

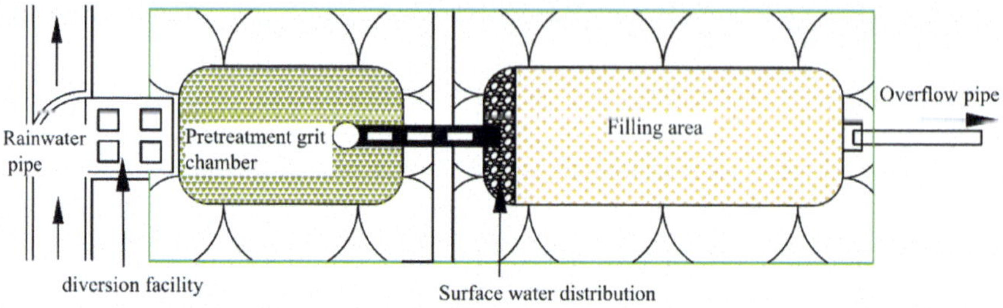

Fig. 3.38 Schematic diagram of small-scale subsurface rainwater wetland structure

ii. Design Requirements

Surface rainwater wetlands are suitable to be set up as online systems, while small-scale subsurface rainwater wetlands are suitable to be set up as offline systems due to their small scale and limited rainwater retention capacity. Figure 3.39 illustrates the combination of rainwater wetlands. This technology is not suitable for treating rainwater with surface water quality classified as Class V or higher. To prevent clogging of the rainwater wetland due to the relatively large amount of sediment, it is recommended to install sand settling facilities for rainwater runoff with Class III and IV water quality.

When designing a rainwater wetland, it is essential to calculate the water balance based on factors such as catchment area, evapotranspiration, infiltration, and the amount of rainwater retained within the wetland. This calculation is necessary to ensure that the wetland remains sufficiently hydrated, even during a 30-day drought period. The study findings indicate that if the surface rainwater wetland's total area is less than 1.0% of the catchment area, there will be a significant decrease in pollutant removal efficiency due to the occurrence of drying out. Therefore, it is imperative to establish a minimum area requirement for rainwater wetlands. In cases where the surface flow wetland area is limited, the use of fill materials such as gravel can enhance pollutant removal and retention capacity.

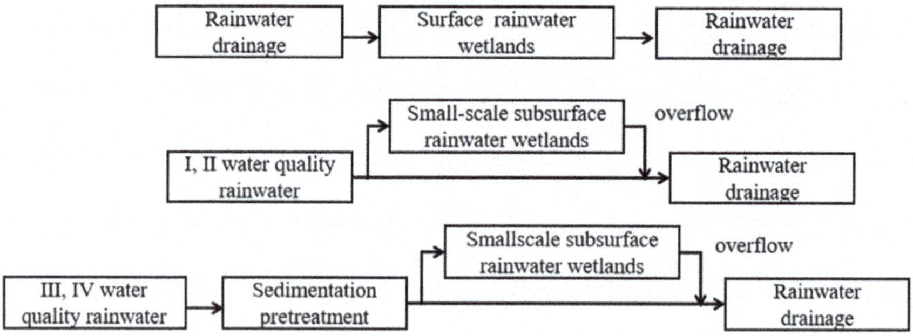

Fig. 3.39 Rainwater wetland combination forms

The water storage depth in small-scale subsurface rainwater wetlands should be between 15 and 30 cm, and the slope of the storage area should have a minimum ratio of 2:1 (H:V). If the slope of the small-scale subsurface rainwater wetland is less than 2% and the permeability coefficient of the bottom soil is greater than 1.0 $\times$ 10^{-7} m/s and higher than the groundwater level, an impermeable layer should be installed. Due to the small scale of the subsurface rainwater wetland and the use of gravel fill layer, poor water quality may easily clog the distribution pipe and fill layer. Therefore, it is advisable to employ sand pre-treatment facilities such as sand settling tanks or wells. Small-scale subsurface rainwater wetlands should be equipped with overflow facilities, which may comprise of overflow pipes or wells. The elevation of the overflow port must be equal to the maximum water storage elevation.

(1) Design requirements for deep water channel of the surface rainwater wetlands are as follows:

   1) The deep water channel capacity and water depth requirements are necessary for maintaining the landscape requirements of the rainwater wetland and for facilitating the water required for water quality treatment.
   2) To ensure an extended hydraulic retention time when setting up deep water channels, the following measures should be taken: lengthening the channel; creating meandering and winding water flow paths to both increase the channel length and reduce flow velocity; employing the use of gravel or cobblestone to construct check dams, similar to grassed swales design, to further prolong the hydraulic retention time.

(2) surface rainwater wetlands should be designed with normal water level, retention water level, and overflow water level. The design requirements for the normal water level are as follows:

   1) In the rainwater wetland, over 35% of the area should have a water depth of less than 15 cm, and over 65% of the area should have a water depth of less than 50 cm, to meet the requirements for landscaping and the growth of deep-water plants.
   2) If the normal water level in the rainwater wetland is too shallow, it may lead to easy drying of the rainwater wetland, which is unfavorable for plant growth and pollutant removal.
   3) If the normal water level in the rainwater wetland is too deep, it will also affect the growth of aquatic plants, as well as the landscape effect and rainwater retention capacity of the wetland. The elevation of the surface rainwater wetland at the edge should be at least 30 cm higher than the overflow port.
   4) When the water depth at the edge of the surface rainwater wetland exceeds 1.2 m, a two-level platform should be designed for slope protection. The platform should meet the following conditions: The lower platform, with a width greater than 1.0 m, should be located 0.5 m below the normal water

level. The upper platform, with a width greater than 1.0 m, should be located 0.5 to 0.8 m above the normal water level.

(3) Design requirements for the outflow tank of surface rainwater wetlands are as follows:

Setting an outflow tank in surface rainwater wetlands is to prevent sediment from being washed up, further improve the removal efficiency of pollutants, and meet the requirements for pollutant load removal. The water depth of the normal water level in the outflow tank should be between 0.8 m and 1.2 m. The capacity of the outflow tank at normal water level should not be less than 5% of the total capacity of the wetland at the normal water level.

(4) Design requirements for flood prevention standards in surface rainwater wetlands:

1) Compliance with relevant flood prevention and drainage laws, regulations, and plans in the project location.
2) The flood prevention standard should adopt a 10-year 24-h design rainstorm.

(5) Design requirements for water distribution in small-scale subsurface rainwater wetlands:

Small-scale subsurface rainwater wetlands can be equipped with surface water distribution or perforated pipe water distribution, as shown in Fig. 3.40 and Fig. 3.41.

The design requirements for subsurface perforated pipe water distribution facilities are as follows:

1) The diameter of the perforated pipe should be 150–300 mm.
2) The slope of the perforated pipe should be 1%-2%.

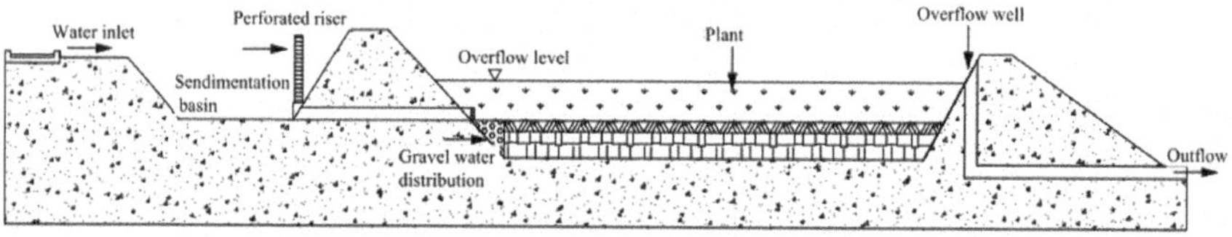

Fig. 3.40 Small-scale subsurface rainwater wetlands with surface water distribution

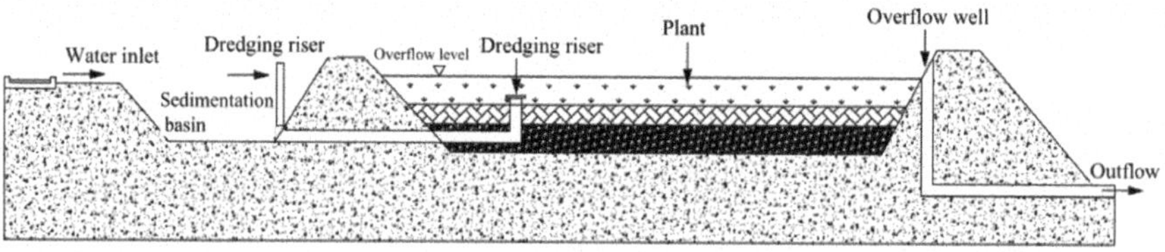

Fig. 3.41 Small-scale subsurface rainwater wetlands with perforated pipe water distribution

3) The perforated pipe should be surrounded by gravel, and the gravel should be wrapped in permeable geotextile.
4) A perforated pipe should be installed every 5 m in the width direction.
5) Each perforated pipe should be equipped with a silt flushing riser.

(6) Design requirements for the fill layer of small-scale subsurface rainwater wetlands (see Fig. 3.42):

1) Planting soil layer: Configured according to the growth requirements of the plants, typically with a thickness of 5–10 cm.
2) Pea gravel layer: 5 cm thick.
3) Gravel layer: 40–100 cm thick. Porous materials such as zeolite or anthracite can be added to the gravel layer to enhance pollutant removal efficiency. The particle size distribution of the gravel layer is shown in Table 3.17.

(7) Selection of water plants for rainwater wetlands

Water plants should be planted around the deepwater zone, shallow water zone, slope protection, and outlet tank in the rainwater wetlands. The selection of water plants should meet the following requirements:

1) The wetland should be divided into zones based on water level and depth. Water plants should be selected and configured according to the normal water level

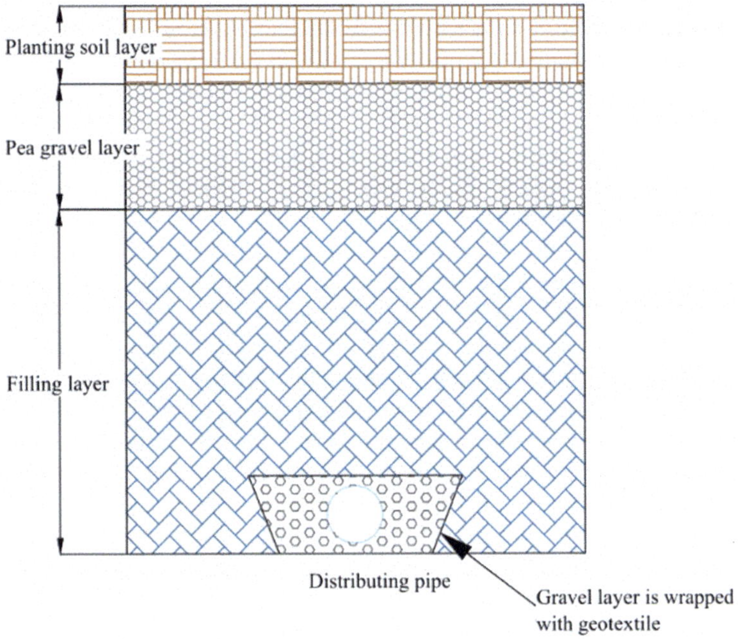

Fig. 3.42 Fill layer for small-scale subsurface rainwater wetlands

Table 3.17 Particle size distribution of the gravel layer

Sieve hole(mm)	passing rate(%)
9.5	100
4.75	50~70
3.35	10~30
2.36	0~10

and depth of each zone. One type of water plant should be planted in each zone. The total number of water plant species in the wetland should be 5–7, and for small-scale subsurface rainwater wetlands, it is recommended to plant at least 3 different types of water plants.

2) Local water plants should be used as they are easy to cultivate and maintain. Examples of local water plants include Lythrum salicaria, Acorus calamus, Sagittaria sagittifolia, Cyperus involucratus, Typha orientalis, Phragmites communis, etc.
3) The selected water plants should meet the requirements of landscape design and have well-developed root systems.

3.2.4.3 Scope of Application and Advantages and Disadvantages

Surface rainwater wetlands utilize their larger retention capacity to achieve the goals of rainwater runoff pollution control and peak flow control. Due to the typically larger surface area and volume of surface flow wetlands, they can retain a significant amount of rainwater with a certain depth of storage. As a result, they are usually constructed in relatively sparsely populated areas such as parks, highways, and industrial zones. In regions with high groundwater levels or poor soil permeability, such as coastal areas, it is advisable to use small-scale subsurface rainwater wetlands for runoff pollution control. These wetlands are relatively easy to install and can also achieve moderate control of peak flows.

Overall, rainwater wetlands possess several advantages, including effective pollutant reduction, a certain degree of runoff and peak flow control, and the provision of natural habitats for wildlife. Additionally, there are some drawbacks. Rainwater wetlands only operate reliably for prolonged durations when stringent water balance requirements are met. Furthermore, wetland vegetation maintenance requires frequent attention, especially in terms of pruning, harvesting, and replanting. Compared to other engineering measures for controlling runoff pollution, rainwater wetlands require larger land areas. If wetland maintenance is inadequate, it may lead to significant proliferation of invasive plant species. Failure to promptly collect and treat fallen branches and leaves following wetland vegetation wilting can cause secondary pollution. Finally, the construction and maintenance costs of rainwater wetlands are relatively high.

Table 3.18 Rainwater wetlands maintenance requirements

Maintenance content	Maintenance priorities and objectives	Maintenance cycle
Maintenance of plantings	Replanting, removing weeds and ensuring that plantings grow	Regular maintenance according to plant requirements.; determine based on inspection results
Pruning of plants	Landscape requirements	Three times a year;determine based on inspection results
Dredge	Clean up the sedimentation in the water storage area; clean up the sedimentation in the inlet and overflow facilities	Twice a year;determine based on inspection results
Water storage area	Repair the collapsed slopes in the water storage area; level the water storage area	Once a year;determine based on inspection results

3.2.4.4 Operation and Maintenance Requirements

The operation and maintenance requirements for rainwater wetlands include the following seven main points. Among them, the maintenance priorities and objectives and maintenance cycles are shown in Table 3.18.

(1) When water and soil erosion occur at the water inlet or overflow port due to scouring, it is necessary to promptly install gravel buffers or take other anti-erosion measures;
(2) When the inlet or overflow port is blocked or silted, resulting in poor water flow, it is necessary to promptly clean up garbage and sediment;
(3) When the sediment in the pre-treatment tank is more than 50%, it should be promptly dredged;
(4) When warning signs for preventing misconnection, misuse, or accidental consumption, guardrails and other safety precautions and early warning systems are damaged or missing, they should be repaired and improved in a timely manner;
(5) When slope protection collapse occurs, reinforcement should be done promptly;
(6) The relevant equipment such as pumps and valves should be regularly inspected to ensure their proper functioning;
(7) Harvesting, replanting, pruning of plants and removal of weeds should be carried out in a timely manner.

3.2.4.5 Technology Effectiveness and Costs

Rainwater wetlands have shown good effectiveness in reducing SS, COD, NH_4^+-N, TN, and TP in rainfall runoff, with removal rates reaching 87.4%, 87.6%, 86.2%, 67.8%, and 63.8% respectively.

Table 3.19 Cost of construction of different types of rainwater wetlands

Rainwater wetlands	Investment (CNY/m^2)
Surface rainwater wetlands	500–1,000
Small-scale subsurface rainwater wetlands	300~600

The construction cost of rainwater wetlands mainly depends on factors such as their structure, configuration, site selection, and location. Generally, the investment costs of different types of rainwater wetlands are shown in Table 3.19, and the annual operation and maintenance costs are approximately 2% to 5% of the construction cost.

3.2.4.6 Application Examples

Surface rainwater wetlands have been applied in New weisha Wetland of Fengxi New Town, with a total area of 33,030 m^2. The natural soil infiltration rate is 21.24 mm/h, and the groundwater depth ranges from 10.5 m to 13.1 m. The evaporation rate is 1289 mm, and the hydraulic retention time is 51.6 h. Through monitoring a rainfall event on March 12, 2017, it was found that the surface rainwater wetlands achieved a total runoff control rate of 59.89%, with a peak delay time of 19 min and a peak reduction rate of 41.23%. The removal rates for SS, COD, NH_3-N, TN, and TP were 41.23%, 37.23%, 45.26%, 52.36%, and 48.56%, respectively.

3.3 Transmission Control Technologies

Transmission control technology is one of the main technologies used in the source reduction of rainfall runoff pollution. Mainly through the construction of rainwater ecological transmission channels, so that rainfall runoff in the transmission process to achieve runoff infiltration and pollutant reduction. Commonly used transmission control techniques include grassed swales and seepage pipes/channels.

3.3.1 Grassed Swale

3.3.1.1 Technical Overview and Principles

A grassed swale, also known as a vegetated swale, is a type of surface ditch that is planted with vegetation, is shallow in depth, and has a gentle slope. As a facility for collecting, conveying and discharging rainwater runoff, it can be used to connect other single-use rainwater utilization facilities and rainwater pipeline systems, to restore the rainwater runoff to its natural state as much as possible, increase natural infiltration of rainwater, and delay the time it takes for runoff to be conveyed. During the transfer process, it can also play an initial purification role on the rainwater runoff through the actions of plants, soil, and microorganisms.

3.3.1.2 Structural Characteristics and Design Parameters

i. Structure

According to the different processes of surface runoff transportation and retention in grassed swales, grassed swales can be divided into drainage grassed swales and infiltration grassed swales. Drainage grassed swales mainly utilize their extended flow accumulation time and retention capacity, as green drainage facilities to control peak flow discharge and also facilitate rainwater infiltration. Infiltration grassed swales utilize the infiltration capacity of the soil to achieve rainwater infiltration. The structural forms of the two are shown in Fig. 3.43 and Fig. 3.44, and they can be combined to design infiltration and drainage grassed swales.

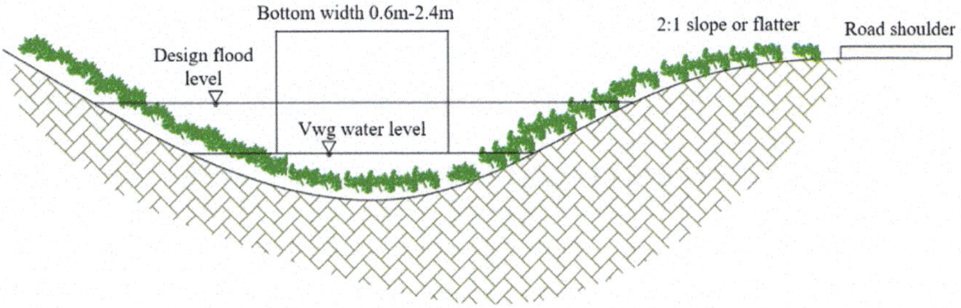

Fig. 3.43 The cross-section of drainage grassed swales [28]

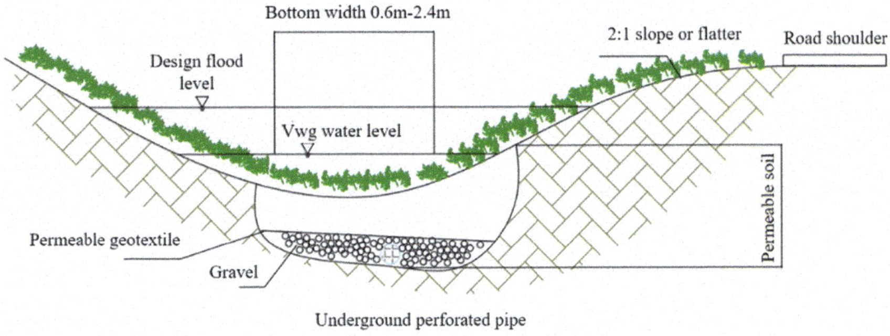

Fig. 3.44 The cross-section of infiltration grassed swales [28]

ii. Structural Design

When designing, one can choose one of the forms shown in Fig. 3.43 and Fig. 3.44 based on the actual characteristics, or a combination of the two. For example, in a design project where the soil has good infiltration capacity and the terrain slope is around 1% to 3%, and the grassed swale has a certain width, it is possible to design an infiltration grassed swale that can achieve both rainwater infiltration and drainage.

(i) *Soil Design for Grassed Swales*

The maximum depth of grassed swales is generally between 0.20 to 0.40 m. For drainage grassed swales, the depth of the groundwater table and impermeable layer should be greater than 0.60 m to ensure that the groundwater is not polluted and will not infiltrate into grassed swales. As the primary function of drainage grassed swales is to drain water, there are no special requirements for the soil permeability coefficient. Infiltration grassed swales, on the other hand, require the installation of an underground water storage layer and the successful infiltration of rainwater, so the depth of the groundwater table and impermeable layer should be greater than 1.20 m. The soil permeability coefficient should be within the range of 4×10^{-6} to 1×10^{-4} m per second. This ensures that groundwater is not polluted and that there is adequate space for rainwater drainage.

(ii) *Pre-Treatment Facilities for Grassed Swales*

As mentioned earlier, for Class I and Class II water quality, the rainwater runoff has good water quality and can directly enter the grassed swale through road curbs, pipes, or overland flow. However, for Class III and Class IV water quality, the rainwater runoff needs to be treated before entering the grassed swale. Pre-treatment facilities for grassed swales can include sedimentation facilities, rain gardens, sand filtration equipment or facilities. Drainage grassed swales are recommended to be set up in an online configuration, while infiltration grassed swales are recommended to be set up in an offline configuration. The combination form shown in Fig. 3.45 is recommended for grassed swales.

Class III and Class IV rainwater contain a large amount of sediment and oil-based pollutants. If not treated, these pollutants will spread out in the grassed swale, causing

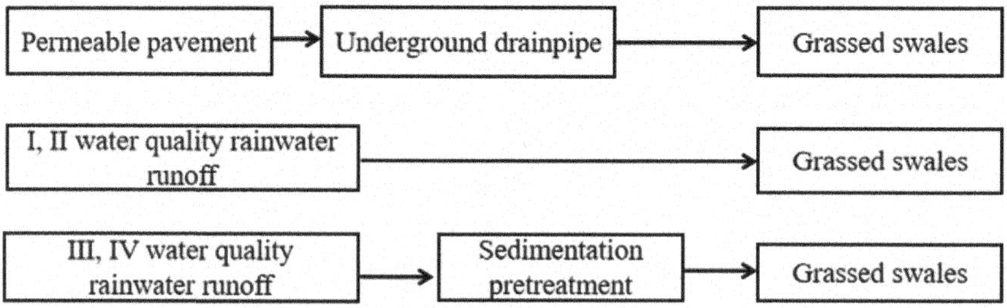

Fig. 3.45 Combination form of grassed swales

widespread pollution that is difficult to clean up. The use of sedimentation facilities, rain gardens, sand filtration equipment or facilities can concentrate pollutants for easy removal. When the following situations arise, it is recommended to use water distribution facilities, with specific technical requirements as follows:

(1) Rainwater runoff enters the grassed swale through pipes.
(2) The water level in the grassed swale should exceed 15 cm when rainwater enters.
(3) When the grassed swale crosses a roads, pipes are used for connection.

(iii) *Slope Design of Grassed Swales*

Drainage grassed swales should have a slope of 1% to 5%, with a minimum slope requirement for easy drainage and a maximum slope requirement to avoid excessive flow velocity and erosion of the grassed swale. Infiltration grassed swales should have a slope of less than 2%, with a slope requirement for rainwater storage and enhanced rainwater infiltration. The catchment area of the grassed swale should not exceed 2 hm^2. Permeable pavement can be used in combination with grassed swales, the structural form shown in Figs. 3.46 and 3.47.

(iv) *Water Inlet Design of Grassed Swales*

It is required that the inflow of the grassed swale should flow smoothly and evenly, without causing erosion to the grassed swale soil. When rainwater enters the grassed

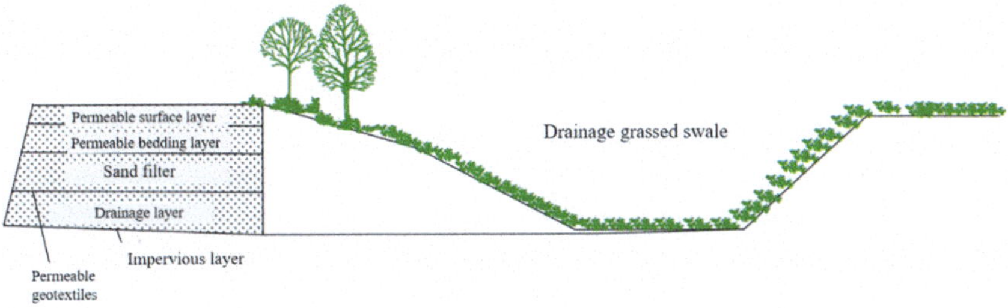

Fig. 3.46 Combination of permeable pavement and drainage grassed swales

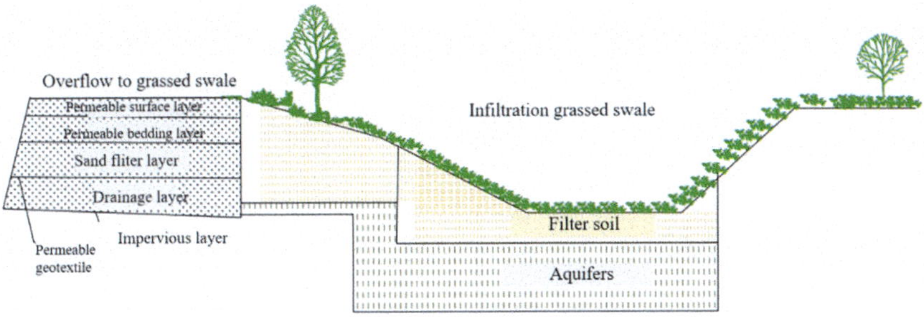

Fig. 3.47 Combination of permeable pavement and infiltration grassed swales

swale through a pipe or drop structure, distribution facilities are needed to ensure that the inflow is even and does not erode the grassed swale.

(v) *Cross-Section Design of Grassed Swales*

The cross-section design calculation method of the grassed swale is shown in Eq. (3.8). Drainage grassed swales should meet the drainage requirements of a two-year rainfall event.

$$Q = \frac{A_g \cdot r^{\frac{2}{3}} \cdot S_g^{\frac{1}{2}}}{n} \tag{3.8}$$

where, Q-design flow rate, m^3/s;

A_g-cross-sectional area, m^2, parabolic section: $A_g = 2/3 \cdot d \cdot B^2$;trapezoidal section: $A_g = B \cdot d + z \cdot d^2$, d is maximum depth of the section, typically ranging from 20 to 40 cm; B is bottom width of trapezoidal section, m; z is slope of trapezoidal section(H:V);

r-hydraulic radius, m.

$$\text{Parabolic section}: r = \frac{2 \cdot d \cdot B^2}{3 \cdot B^2 + 8 \cdot d^2} \tag{3.9}$$

$$\text{Trapezoidal section}: r = \frac{\mathrm{B} + z \cdot d}{z + \sqrt{1 + z^2}} \tag{3.10}$$

S_g-slope gradient of grassed swales,%;

n-manning coefficient, values are listed in Table 3.20.

Infiltration grassed swales should also meet the requirements of rainwater draining within 24 h and total control of rainwater runoff volume.

According to the time of rainwater emptying, the maximum depth of the cross section of infiltration grassed swales is determined.

$$\mathrm{d} = \mathrm{k_g} \cdot \mathrm{t_g} \tag{3.11}$$

Table 3.20 Values of Manning's coefficient n in grassed swales

Forms of grassed swales	Plant species	Minimum	Maximum	Average
A:Linear grassed swales	1 low-growing plants (<10 cm), are weeds	0.022	0.033	0.027
	2 low-growing plants (<10 cm), abundant weeds	0.026	0.033	0.030
B:curvilinear grassed swales	1 high-growing plants (>10 cm), are weeds	0.026	0.040	0.032
	2 high-growing plants (>10 cm), bundant weeds	0.030	0.045	0.035

where, k_g-permeability coefficient of surface soil, m/s;;

t_g-rainwater emptying time, s, <86,400.

Sections are sized based on meeting total runoff control requirements.

$$V_R = A_{ave} \cdot L_g \tag{3.12}$$

$$A_{ave} = \alpha \cdot A_g \tag{3.13}$$

where, V_R—the rainwater retention and infiltration control capacity of grassed swales, m^3;

A_{ave}—the maximum average storage cross-sectional area, m^2;

α—the cross-sectional storage coefficient;

L_g—length of grassed swales, m.

The length of grassed swales should not normally be less than 30 m. If a grassed swale is too short, its pollutant-removal capacity will be greatly affected. The values of cross-sectional storage coefficients for infiltration grassed swales are shown in Table 3.21.

Based on the above design, the following requirements should also be met:

(1) The cross-sectional dimensions of the drainage grassed swale should be calculated using the Manning formula, considering both parabolic section and trapezoidal section designs. The Manning coefficients can be found in Table 3.20. The maximum depth of the section should be between 20–40 cm.
(2) When determining the maximum water depth for the infiltration grassed swale, if the maximum depth (d) falls between 20–40 cm, the soil should be prepared to ensure that the soil permeability coefficient meets the requirement of 24-h drainage time for rainwater.
(3) To allow for greater water storage capacity in the infiltration grassed swale, it is designed with a certain slope. Hence, the maximum average water storage cross-sectional area is not equal to the maximum cross-sectional area.
(4) The bottom width of the infiltration grassed swale should range between 0.5–2.5 m, and the ratio of the bottom width to the maximum depth should be less than 12.

Table 3.21 Values of cross-sectional storage coefficients for infiltration grassed swales

Slope of grassed swales	α (setting embankments)	α (Without setting embankments)
<0.5%	0.90	0.70
[0.5%,1.0%)	0.85	0.65
[1.0%,1.5%)	0.80	0.60
[1.5%,2.0%]	0.75	0.55

(5) In cases where the width of the green area cannot meet the requirements of the infiltration grassed swale, a water storage layer can be utilized. The water storage layer should adhere to the following requirements:

1) The water storage layer can be underground, and the thickness of the upper soil layer should be between 20–40 cm. The permeability coefficient should be between 5×10^{-5} to 1×10^{-4} m/s to allow rainwater to quickly enter the water storage layer.
2) The porosity of the water storage layer should be no less than 30%. It can be constructed using graded gravel layer, perforated pipes, or pre-made water storage modules.
3) The thickness of the water storage layer should be between 30–50 cm.
4) A permeable geotextile should be placed between the water storage layer and the soil.
5) The water storage layer should not have any impact on structures, roads, pipelines, and other foundations.
6) Grassed swales with water storage layers can accumulate large amounts of water and may impact various types of infrastructure. In such cases, a waterproofing layer can be used to separate the water storage layer from the foundations of structures.

(vi) *Flow Velocity Design for Grassed Swales*

The maximum flow velocity in grassed swales under the design storm of once every two years should not exceed the requirements in Table 3.22.

When the grassed swale cannot meet the maximum design flow velocity requirements, a embankment can be used to reduce the flow velocity and alleviate erosion caused by rainwater on the grassed swale. Typically, large stones are used for the embankment, but it is important to ensure that it does not affect the drainage requirements, is not washed away by water, and is aesthetically pleasing.

When the slope of the drainage grassed swale is greater than 3% and the slope of the infiltration grassed swale is greater than 1%, a rainwater embankment should be set.

Table 3.22 Maximum flow rate requirements for grassed swales

Soil behaviour	Vegetation height(cm)	Maximum Flow Rate Requirements(m/s)	
		Vegetation in average condition	Vegetation in good condition
Chalky soil, sandy soil, loamy soil	5–15	0.6	0.9
	15–30	0.7	1.0
Powdery clay, sandy clay	5–15	0.9	1.2
	15–30	1.1	1.3
Clay	1–15	1.2	1.5
	15–30	1.4	1.7

The rainwater embankment should meet the following requirements: use 150–200 mm stones with good grading and cleanliness; the height of the embankment top should be lower than the top of the grassed swale by 10 cm; the width of the embankment should be set to ensure that it will not be washed away by rainwater; the spacing of the embankment should be calculated according to the formula (3.14).

$$L_{gmax} = \frac{d}{1.5 \cdot S_g} \tag{3.14}$$

where, L_{gmax}-the spacing of the embankment,m;

(vii) *Plant Design for Grassed Swales*

The drainage grassed swale is primarily designed for drainage and should be planted with grass and lichens, among other shorter plants. The infiltration grassed swale, on the other hand, is primarily designed for rainwater retention and infiltration, and should be planted with flowering herbs, grass, and other similar plants. The plants selected for grassed swales should have a minimum tolerance time of 24 h for water-logging and large trees and shrubs should not be planted in the grassed swale.

iii. Overflow Design

(1) The overflow facilities for grassed swales should be calculated and determined based on their infiltration capacity, retention capacity, and drainage capacity. The overflow facilities can be set at the end or in the middle of the grassed swale.
(2) To ensure the water storage capacity of the grassed swale, the overflow port should be set at a height of 3–5 cm below the maximum water storage level.
(3) The placement of overflow port should not be too dense and should generally not be less than 30 m apart.

iv. Construction Process of Grassed Swales

The cross-sectional shape and slope of the grassed swale have a significant impact on its operational effectiveness and must strictly adhere to the design and construction requirements. It is particularly important to avoid cross-sectional inconsistencies in flow capacity and prevent situations of water surges or drops. When planting vegetation, it is necessary to first plant on slopes and embankments before planting on the trench bottom. During the rainy season, erosion control measures should be implemented to prevent water erosion. The specific construction procedures are shown in Fig. 3.48.

3.3.1.3 Scope of Application and Advantages and Disadvantages

The primary function of a grassed swale is to control rainfall runoff pollution and peak flow. It can retain or infiltrate rainfall and also serve as a drainage facility.

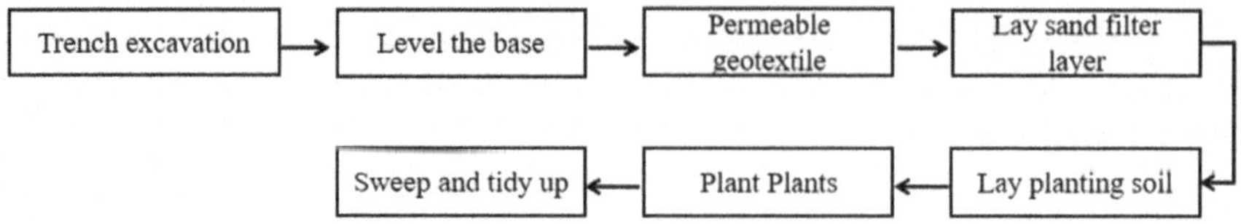

Fig. 3.48 Grassed swale construction procedures

Grassed swales can be applied to various land types. Drainage grassed swales can be used when the soil has poor permeability, while infiltration grassed swales can be used when the soil has good infiltration capacity.

Grassed swales can accumulate a certain amount of water and have a certain flow velocity. Therefore, they should not be constructed in high-risk areas prone to collapse or landslides. In areas where self-weight collapsible loess is susceptible to water immersion, rapid soil structure damage and significant additional sinking can occur under certain pressure. Similarly, in areas with high saline soil content, salt crystallization can occur with increased soil moisture. Grassed swales should also not be constructed in these areas.

In general, grassed swales have the following advantages: they can effectively reduce the total volume of rainfall runoff, achieve peak flow regulation, and use surface vegetation to filter, adsorb, absorb, and biodegrade pollutants in the runoff. Additionally, grassed swales can increase the greening rate of urban areas and residential areas, to some extent, play a role in air purification and noise reduction. Furthermore, grassed swales have simple design and construction, relatively low cost, and require minimal maintenance. When properly designed, they can also have a certain aesthetic appeal.

However, grassed swales also have the following disadvantages: they require a large land area and are not suitable for areas with excessive rainfall, steep slopes, or poor drainage capacity.

3.3.1.4 Operation and Maintenance Requirements

After the construction of grassed swales, maintenance work should be carried out according to the requirements of the planted vegetation. The operational and maintenance requirements for grassed swales are outlined in Table 3.23.

3.3.1.5 Technology Effectiveness and Costs

The control rate of grassed swales for total rainfall runoff volume is about 56%-90%, and the removal rate of SS can reach 36%-85%. At the same time, they can effectively reduce suspended solid particles and organic pollutants, remove some metal ions such as Pb, Zn, Cu, Al, and oil substances. Grassed swales also have beautification and landscape functions, soften space boundaries, reduce the cold and hard feeling of the

Table 3.23 Operational and maintenance requirements for grassed swales

Maintenance content	Focus and objectives of maintenance	Maintenance cycle
Maintenance of planted vegetation	Reseeding, fertilizing, weed removal, ensuring plant growth	Regular maintenance in accordance with plant requirements; determined based on inspection results
Pruning of planted vegetation	Landscape requirements, ensuring acceptable Manning's coefficient	Three times a year; determined based on inspection results
Dredge	Cleaning up overflow facilities and water distribution facilities clogged with debris; clearing sediment accumulation at the bottom of grassy channels	Twice a year; determined based on inspection results; within 24 h after rainfall
Cross-sectional shape and slope	Repair the collapsed section, maintaining the cross-sectional shape; level the bottom of the grassy channel, maintaining the channel slope; restore the setting of embankments	Once a year; determined based on inspection results; within 24 h after rainfall
Water storage layer (if provided)	In accordance with the requirements for infiltration facilities in the specification	When the inspection results show poor rainwater infiltration and failure to meet the design requirements for drainage time; typically after 5–10 years of use

ground boundary, and improve spatial sense. In addition, grassed swales can replace traditional rainwater pipes, enhance ecological functions, and protect biodiversity.

The construction cost of grassed swales is low. It is estimated that the construction cost of a grassed swale is only 50 CNY/m^2, and the maintenance cost is about 1 CNY/m^2.

3.3.1.6 Application Examples

Grassed swales have been applied in the Future Science and Technology City in Beijing [19]. In the construction of Road No. 2 in the northern area, the length of the road's rainwater ecological grassed swales is 15.82 km, while in the southern area, the length of the road's rainwater ecological grassed swales is 14.60 km. The on-site effects are shown in Fig. 3.49, and the structural diagram is shown in Fig. 3.50.

In 2017, the rainfall runoff reduction and pollutant reduction effects of ecological grassed swales are shown in Table 3.24. Monitoring results indicate that ecological grassed swales can reduce the total runoff volume by 22% to 69% and delay peak flow by 40 to 510 min. They can also reduce pollutant emissions by 43% to 93% while providing a favorable ecological landscape.

Fig. 3.49 Real scene of ecological grassed swales in Future Science and Technology City, Beijing

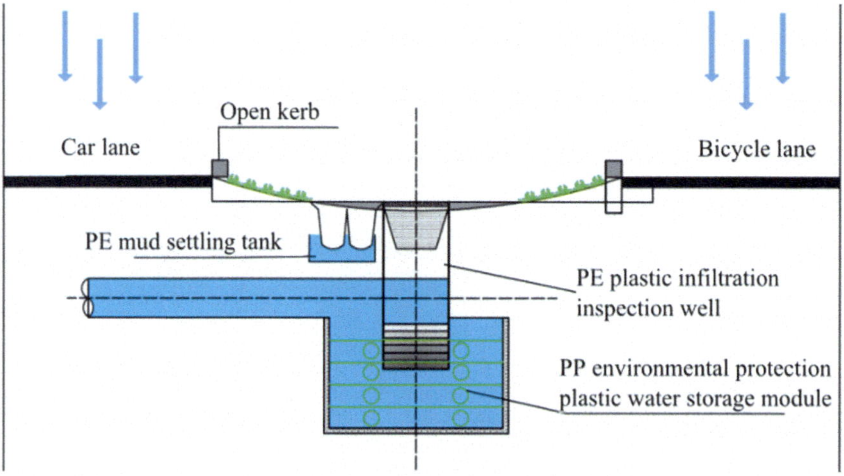

Fig. 3.50 Structure diagram of ecological grassed swales in Future Science and Technology City, Beijing

Table 3.24 Runoff reduction effect of ecological grassed swales

Date	Rainfall volume (mm)	Drainage lag time (min)	Peak reduction rate (%)	Runoff coefficient	Total reduction rate (%)
2017.6.21	10.5	40	31.0	0.544	69.0
2017.6.23	11.5	35	50.6	0.589	49.4
2017.8.11	31.1	60	37.2	0.890	62.8
2017.8.5	42.7	10	78.1	0.324	21.9
2017.8.2	48.8	510	48.8	0.277	51.2
2017.6.22	91.45	320	63.0	0.729	37.0

3.3.2 Seepage Pipe/Channel

3.3.2.1 Technical Overview and Principles

Seepage pipes/channels refer to rainwater pipes/channels with permeability, which can be composed of perforated plastic pipes, non-sand concrete pipes/channels, and gravel (broken) stones and other materials.

3.3.2.2 Structural Characteristics and Design Parameters

i. Structure

Seepage pipes/channels are a type of rainwater conveyance facility that is typically used in combination with low-impact development facilities such as grassed swales, rain gardens, and constructed wetlands. Generally, seepage pipes are buried below the soil layer, with a layer of permeable geotextile wrapping around the outside of the pipeline and a layer of gravel filling the outermost space. The structure of seepage channels is similar to seepage pipes, but they are constructed above the ground, and the channel typically made of non-sand concrete, with a layer of permeable geotextile at the bottom and a gravel layer beneath the geotextile.

The typical construction of seepage pipes/channels is shown in Fig. 3.51, which includes horizontal collecting pipes, collecting wells, and inspection wells.

(i) *Horizontal Collecting Pipes*

Collecting pipes are typically perforated concrete pipes. When the water volume is small, clay pipes or cast iron pipes can also be used. There are also options for using dry-stacked stone blocks with gaps and assembled concrete culverts. The internal diameter of reinforced concrete collecting pipes is generally between 600 and 1,000 mm. The specific pipe diameter required should be determined based on calculations according to the actual conditions.

There are two types of inlet holes on the pipe wall: round holes and slit holes. The diameter of the round holes is 20–30 mm; the length of the slit holes is 200 mm, and the width is 60–100 mm. The hole size should be larger on the inside and smaller

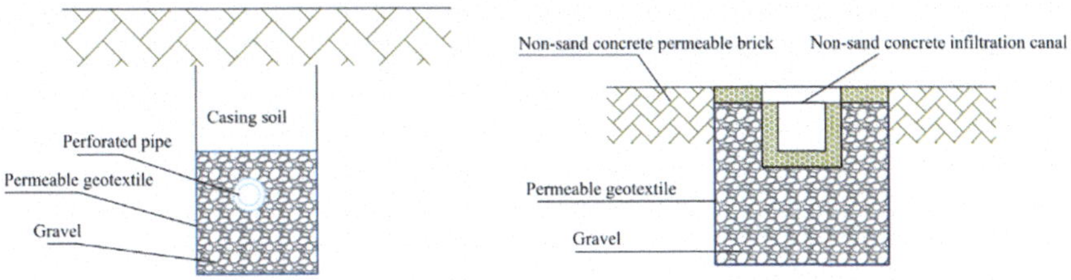

Fig. 3.51 Typical configuration of a seepage pipe/channel [27]

on the outside, and they should be staggered at 1/2 to 2/3 of the circumference of the upper part of the pipe channel. The net distance between the holes should be considered based on the strength requirements of the collecting pipes. While meeting the strength requirements, the porosity should be as large as possible to increase the infiltration amount, but it should not exceed 15% at maximum. When using slit holes, attention should be paid to satisfy the strength and reinforcement requirements of the reinforced concrete pipes. In order to prevent blockage of the inlet holes on the collecting pipes, an artificial antifilter layer should be set up outside the collecting pipes. The antifilter layer should be comprised of three layers, namely the inner layer, middle layer, and outer layer, with filter media grain sizes of 10–35 mm, 5–10 mm, and 1–5 mm, respectively. The antifilter layer must be between 0.2–0.3 m thick. The filter media size must be slightly larger than the size of the inlet holes; otherwise, blockages may occur.

(ii) *Collecting Wells*

The diameter and depth of the collecting well should be determined based on the diameter and burial depth of the collecting pipe. In principle, the diameter of the collecting well should be 1.5 to 2.0 times the diameter of the collecting pipe. The minimum well diameter should not be less than 1.5 m. The bottom of the well should be 0.3 to 0.5 m lower than the bottom of the collecting pipe. The construction material for the well structure can be either concrete or mortar brickwork.

(iii) *Inspection Wells*

To facilitate inspection and cleaning, inspection wells should be installed at regular intervals of 50 to 100 m along the straight sections of the collecting pipes, as well as at the ends, corners, and where there are changes in cross-section. For inspection wells located below the flood level, the well cover should be sealed and secured with bolts to prevent floodwater from flowing in and causing sediment accumulation in the seepage channel.

ii. Design Requirements

(1) Seepage pipes/channels should be equipped with pre-treatment facilities such as grassed swales and sediment (sand) tanks.
(2) The opening ratio of seepage pipes/channels should be controlled between 1 and 3%, and the porosity of non-sand concrete pipes should be greater than 20%.
(3) The laying gradient of the seepage pipes/channels should meet the drainage requirements.
(4) Gravel or other porous materials should be filled around the seepage pipes/channels, with a permeable geotextile layer outside the gravel layer. The overlapping width of the geotextile should not be less than 200 mm.
(5) When the seepage pipes/channels are installed under a roadway, the depth of soil cover should not be less than 700 mm.

3.3.2.3 Scope of Application and Advantages and Disadvantages

Seepage pipes/channels are suitable for areas with lower flow rates in buildings, residential areas, and public green spaces. They are not suitable for areas with high groundwater levels, severe runoff pollution, and the risk of structural collapse where rainwater infiltration is not appropriate (such as when rainwater pipes are located under roadways for motor vehicles).

In general, seepage pipes/channels have low space requirements but they come with higher construction costs, are prone to blockage, and harder to maintain.

3.3.2.4 Operation and Maintenance Requirements

(1) When water and soil erosion occur at the water inlet due to scouring, it is necessary to promptly install gravel buffers or take other anti-erosion measures
(2) If the capacity for detention or overflow is inadequate due to sediment accumulation within the facility, the sediment should be cleared in a timely manner.
(3) If the emptying time of the seepage well's storage space for rainwater exceeds 36 h, the filling material should be promptly replaced.

3.3.2.5 Technology Effectiveness and Costs

Seepage pipes/channels can achieve a removal rate of pollutants (measured in SS) of 35% to 70%, and their construction and maintenance costs are generally higher.

3.3.2.6 Application Examples

In the process of controlling rainfall runoff pollution, Fengxi New Town has constructed seepage pipe/channel facilities for the transportation of rainwater. The entire seepage channel has a length of 6,000 m and a volume of 0.09 m^3/m, as shown in Fig. 3.52.

3.4 Application of Combined Technologies

In the engineering construction of urban rainfall runoff pollution control in our country, due to the significant differences in available space of development zones with different intensities and the differences in land area occupied by different source reduction measures, there is often a situation where the urban space is narrow and

Fig. 3.52 Actual scene of seepage pipe/channel in Fengxi New Town

many plots cannot be used for the construction of source reduction measures. Therefore, it is necessary to optimize the combination of various source reduction technologies in combination with the area size, pattern distribution, soil texture and micro-terrain conditions of the source reduction reconstruction space in plots with different development intensities.

Building on the foundation of a large number of individual technologies described in the previous text, based on the characteristics of the underlying surface, a combination of source reduction technologies is developed to form a combined process, providing inspiring ideas and case studies for reducing pollution at the source of rainfall runoff.

3.4.1 Combined Application of Source Technologies for Rainfall Runoff Pollution Control in Urban Road

Urban roads, which represent a high percentage of the built-up area of a city, are usually hardened. They can be divided into pedestrian walkways, traffic lanes, and road peripheries based on their functions. Due to their poor permeability and complex pollutant composition, they quickly generate runoff during rainfall, leading to urban non-point source pollution.

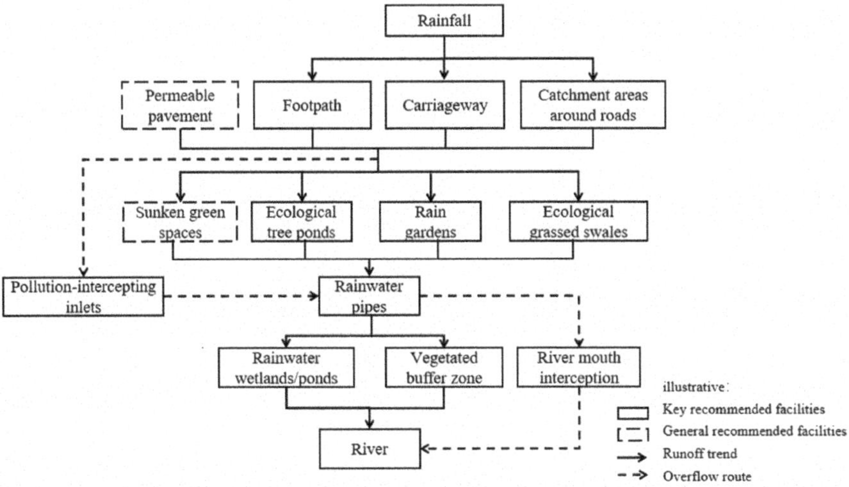

Fig. 3.53 Combined process flow diagram of source reduction technology for runoff pollution control in road

Pollutants in road rainwater runoff mainly come from tire wear, antifreeze usage, vehicle leaks, pesticide and fertilizer application, and discarded waste. The pollutants primarily include organic or inorganic compounds, nitrogen, phosphorus, metals, oils, etc. With the rapid growth of urban road construction and traffic flow, the total amount of pollutants on road surfaces inevitably shows a significant increase. As road runoff or snowmelt enters water bodies, it poses a threat to urban water environments.

Common individual technologies used in road construction include inlets, rainwater inspection wells, infiltration pipes, grassed swales, and rain gardens. The usual combination pattern is between inlets, infiltration pipes, grassed swales, and rain gardens, and the process flow is shown in Fig. 3.53.

In the demonstration project of rainfall runoff pollution control in Wanguo Road, Jiaxing, source reduction technologies such as permeable pavement, sunken green spaces, ecological tree ponds, rain gardens, and ecological grassed swales are efficiently combined. The combined source reduction technologies are then effectively integrated with non-point source pollutant process control technologies such as pollution-intercepting inlets and rainwater pipes. The rainwater from the outlet pipes enters rainwater wetlands/ponds, vegetated buffer zone, etc. for further purification and treatment, and ultimately enters water bodies.

Monitoring results show that the implementation of this combined technology reduces the total volume of rainfall runoff discharged from the two monitoring points on Wanguo Road by approximately 50%. The reduction rate of suspended solids (SS) ranges from 68.8% to 85.9%, achieving a SS reduction rate in rainfall runoff of not less than 40% and a total volume reduction rate of rainfall runoff of not less than 30%.

3.4.2 Combined Application of Source Technologies for Rainfall Runoff Pollution Control in Buildings and Communities

Buildings and communities have the largest proportion in urban land areas, resulting in a high load of rainfall runoff pollution. They are a key component of source control for runoff pollution and significantly impact the effectiveness of reducing both the volume of runoff and pollutant load in the entire watershed.

Common individual technologies for controlling rainfall runoff pollution in buildings and communities include permeable pavements, sunken green spaces, green roofs, rain gardens, grassed swales, rainwater disconnect devices, rainwater collection and purification systems, and rainwater collection networks. Due to the close relationship between buildings and communities and human life, certain landscape requirements need to be met during construction. A commonly used combination technique is to connect green roofs with rainwater disconnect pipes, where the rainwater from the disconnected pipes enters buffering tanks for energy dissipation, and then enter rain gardens or sunken green spaces. When the ground technologies reaches a saturated water-holding state, the excess rainwater then enters the rainwater pipes through the rainwater overflows set up in rain gardens or sunken green spaces, through which it is smoothly drained away. The process flow is illustrated in Fig. 3.54.

In the construction and community runoff pollution control project in Jiaxing City, several technologies are employed. Firstly, the green roofs, rainwater collection devices, and rainwater purification devices are utilized to effectively collect and purify the rooftop rainfall. Secondly, source reduction technologies such as sunken

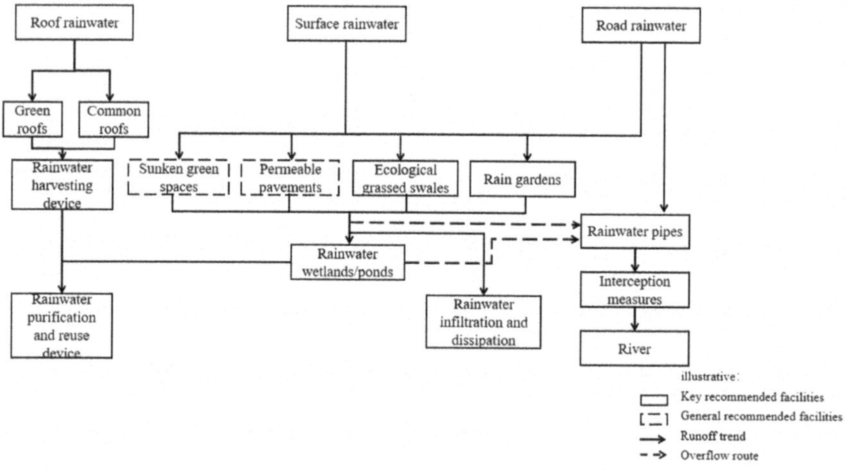

Fig. 3.54 Combined process flow diagram of source reduction technology for runoff pollution control in building and community

green spaces, permeable pavements, grassed swales, and rain gardens are used for initial purification of surface runoff. In the event of heavy rainfall, the road runoff water is first treated by source reduction technologies before overflow, which enters the rainwater pond/wetland for collection and treatment. The road rainwater within the buildings and communities is directed into the rainwater wetland/pond through rainwater pipes for collection and treatment, ultimately flowing into the river.

By using a combination of these technologies, the total reduction rates of rainwater runoff outflow and pollutants from the six monitoring points of the Shihe District were between 12.1% to 100% and 69.6% to 100%, respectively, effectively controlling the rainfall runoff pollution in the district.

3.4.3 Combined Application of Source Technologies for Rainfall Runoff Pollution Control in Centralized Green Spaces

Urban centralized green spaces are an important component of cities and play a significant role in regulating the urban environment and creating a good urban landscape. Due to their usually large size, centralized green spaces have greater potential for retention, storage, and utilization of rainwater, which should be fully utilized to control, mitigate and purify runoff pollution from rainfall.

As places with higher density of population movement, the construction of urban centralized green spaces should also consider their landscape effects. Common individual technologies used include rain gardens, cobblestone ditches, green roofs, ecological corridors, sunken green spaces, grassed swales, inlets, rainwater wells, and rainwater pipes. The combination of these technologies is similar to that of building communities, with typically final runoff rainwater entering the park landscape water bodies. The process flow is shown in Fig. 3.55.

In the rainwater runoff pollution control project of ecological greenway network in Jiaxing City, the use of sunken green spaces, permeable pavements, rain gardens, ecological grassed swales, and ecological tree ponds is employed to collect and treat rainwater from centralized green spaces. When there is heavy rainfall, a portion of the overflow rainwater enters the rainwater pipe network, while another portion goes into rainwater wetlands/ponds for further purification, and ultimately flows into rivers and lakes.

By implementing this combination of technologies, the rainwater runoff pollution control project in Jiaxing City's ecological greenway network achieves 100% control of rainwater runoff when rainfall is $\leq$20 mm. The removal rates for SS, COD, NH_3-N, TN, and TP range from 65.4% to 89.4%, 55.2% to 83.6%, 68.5% to 94.0%, 32.5% to 87.1%, and 60.3% to 94.8% respectively. It creates excellent landscape effects within the project implementation area and effectively mitigates the pollution caused by regional rainwater runoff.

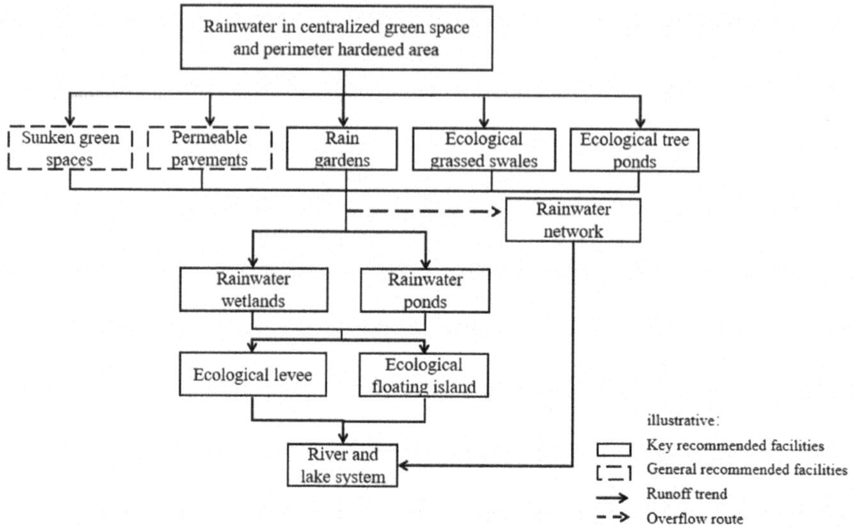

Fig. 3.55 Combined process flow diagram of source reduction technology for runoff pollution control in centralized green space and perimeter hardened areas

3.5 Development and Selection of Typical Equipment and Material

Developed countries such as Europe and the United States, which started earlier in the process of urban development, have made greater progress in the research and development of rainfall runoff pollution control equipment and materials for low-impact development. In China, research and application of rainwater runoff pollution control in urban areas started relatively late but has made significant progress in recent years. This section summarizes the equipment and materials used in the source reduction of rainfall runoff pollution, and provides lessons that can be learned for related engineering design and scientific research.

3.5.1 Equipment and Materials for Source Reduction of Rainwater Runoff Pollution

Source reduction is a key component of urban rainwater runoff pollution control, and the development of equipment and materials for this purpose is a focus of research both domestically and internationally. The equipment and materials developed abroad primarily include five categories: permeable pavements, green roofs, infiltration drainage wells/canals/pipes, rainwater inlet waste intercepting filter devices, and grass-planting parking lots. Among them, the equipment and materials for permeable pavements include seam permeable bricks, climate bricks, permeable concrete and permeable asphalt. The equipment and materials for grass-planting parking lots mainly consist of concrete grass-planting brick. The equipment and materials for green roofs mainly refer to integrated waterproofing and planting systems for roofs. The equipment and materials for infiltration drainage wells/canals/pipes include infiltration wells, infiltration trenches, and high-strength film-coated spiral corrugated steel pipes. The equipment and materials for rainwater inlet waste intercepting filter devices include rainwater inlet interception devices, horizontal gratings, and inspection wells. The foreign equipment and materials for source reduction and their performance characteristics are summarized in Table 3.25.

Domestic research and development of equipment and materials are similar to those developed abroad. Among them, the equipment and materials for permeable pavements include concrete permeable bricks, ecological sand-based permeable bricks, structural joint permeable bricks, recycled permeable bricks, ceramic silica sand permeable bricks, self-cleaning permeable bricks, water-retaining permeable

Table 3.25 Foreign reduction and control equipment and materials and their performance characteristics

Serial number	Form	Equipment and materials	Performance characteristics
1	Permeable pavements	Seam permeable bricks	Usually, the gap between square tiles is 5–10 mm; the compressive strength grade of permeable tiles without parking sidewalk is not less than C40; the compressive strength grade of parking sidewalk is not less than C50; and it meets the resistance to freezing and thawing cycles for 50 times
2		Climate bricks	Brick surface porous, can absorb 30% of precipitation, can to a certain extent alleviate the pressure of heavy rainfall on urban drainage facilities, can regulate the regional microclimate through transpiration, but also to alleviate soil salinization
3		Permeable concrete	Generally contains 15–25% penetrating pores, has good water permeability and air permeability, but its strength is weaker than that of ordinary concrete
4		Permeable asphalt	Adopting skeleton-void structure, the void ratio is around 15%–25%, which has many advantages such as drainage, noise reduction and anti-skid performance

(continued)

Table 3.25 (continued)

Serial number	Form	Equipment and materials	Performance characteristics
5	Grass-planting parking lots	Concrete grass-planting brick	A permeable pavement product in Germany, using hole-type concrete bricks, brick holes filled with humus mixed soil, in the holes can grow some weeds, is a large number of applications and popularization
6	Green roofs	Integrated waterproofing and planting systems for roofs	German Green Roof Derivative Equipment, an integrated system for waterproof and root-proof planting, featuring uniform root watering, light load, and even plant growth
7	Infiltration drainage wells/canals/pipes	Infiltration wells	A UK underground rainwater management facility for infiltrating small amounts of rainwater, filled with stone or plastic grids. Requires pre-treatment and needs to be drained within 12–72 h
8		Infiltration trenches	A small subsurface infiltration facility in the U.S. used to collect runoff and infiltrate it, typically filled with stone or plastic grids that work well with linear projects such as highways
9		High-strength film-coated spiral corrugated steel pipes	A Korean transfer pipe with a long life and a mesh-type external protection network of impact-resistant materials to prevent coating damage during transportation and construction of the pipe
10	Rainwater inlet waste intercepting filter devices	Rainwater inlet interception devices	A UK rainwater interceptor, the core material is metal, fibers, fabrics, geotextiles, in the shape of hanging baskets, cartridges, "socks", nets, etc.
11		Horizontal gratings	Germany a fully automatic operation for filtering mixed sewage or overflow rainwater cleaning equipment, can retain floating material and coarse material, with fully automatic cleaning grid, head loss is small, large discharge capacity, in the strong corrosive sewage still maintain a long service life and operational efficiency and so on
12		Inspection wells	An Australian interception and purification plant where water is treated through three stages of cascade filtration and can be discharged or collected for reuse for non-potable purposes

bricks, permeable cement concrete, permeable asphalt concrete, permeable plastics, and permeable adhesive stones. The equipment and materials for grass-planting parking lots include permeable multifunctional concrete grass-planting bricks and grass grids. The equipment and materials for green roofs include lightweight materials for rooftop greening construction base, containerized roof greening, irregular roof planting bags, and lightweight new roof greening. The equipment and materials for infiltration drainage wells/canals/pipes include infiltration drainage boards, infiltration pipes, infiltration channels, and linear drainage ditches. The equipment and materials for rainwater inlet waste intercepting filter devices include plastic water collection and permeation inspection wells, three-dimensional turbine raindrop generators, rainwater inlet automatic waste interception devices, rainwater collection pre-treatment devices, and interceptor basket sedimentation devices. The domestic equipment and materials for source reduction and their performance characteristics can be found in Table 3.26.

Table 3.26 Domestic reduction and control equipment and materials and their performance characteristics

Serial number	Form	Equipment and materials	Performance characteristics
1	Permeable pavements	Concrete permeable bricks	Low production cost, simple and easy to handle production process
2		Ecological sand-based permeable bricks	By "destroying the surface tension of water" principle of water permeability, it effectively solves the problem that traditional water permeable materials are easily blocked by dust through the pore space
3		Structural joint permeable bricks	Stable structure, good water permeability, low attenuation rate
4		Recycled permeable bricks	High water permeability, wear-resistant and durable
5		Ceramic silica sand permeable bricks	Water permeability, air permeability, anti-slip, good resistance to freezing and thawing, noise reduction and many other features
6		Self-cleaning permeable bricks	It has strong anti-fouling and self-cleaning, air and water purification functions, which effectively solves the problem that the pores of permeable tiles are easily blocked and pollute the urban soil

(continued)

Table 3.26 (continued)

Serial number	Form	Equipment and materials	Performance characteristics
7		Water-retaining permeable bricks	By increasing the thickness of the brick body, a water storage cavity is set up on the side of the brick body, taking into account both water permeability and water storage
8		Permeable cement concrete	High strength and good water permeability
9		Permeable asphalt concrete	Porosity greater than 15%, higher strength than permeable cement concrete, but higher cost
10		Permeable plastics	Adopting environmentally friendly resin super adhesive and EPDM rubber as the main material, environmentally friendly, non-toxic, odorless
11		Permeable adhesive stones	With honeycomb continuous pore space, porosity not less than 20%, water permeability coefficient not less than 1.5 mm/s
12	Grass-planting parking lots	Permeable multifunctional concrete grass-planting bricks	In addition to the pores, the brick body itself also has a certain degree of water permeability, grass planting survival rate is high
13		Grass grids	More than 95% grass planting area, unique flat insertion lap, labor-saving, fast, adjustable expansion joints
14	Green roofs	Lightweight materials for rooftop greening construction base	Low cost, lightweight and can store and retain water
15		Containerized roof greening	The container is made of PP material for longer life
16		Irregular roof planting bags	Simple construction process, short cycle, small impact on the surrounding environment, suitable for shaped and slope not more than 60°of the roof
17		Lightweight new roof greening	The most important feature is no soil, no watering, permanent growth
18	Infiltration drainage wells/canals/pipes	Infiltration drainage boards	Mostly used in basement drainage, at the same time both seepage, drainage and water storage functions

(continued)

Table 3.26 (continued)

Serial number	Form	Equipment and materials	Performance characteristics
19		Infiltration pipes	It can be made of a combination of perforated plastic pipes, sandless concrete pipes and gravel (crushed) stones, etc., with a small footprint and good water permeability
20		Infiltration channels	Flexible layout, easy installation, high volume ratio
21		Linear drainage ditches	Adopting U-shaped section, divided into flat counting type, slit type two types of cover forms
22	Rainwater inlet waste intercepting filter devices	Plastic water collection and permeation inspection wells	Infiltration, retention and purification of rainwater to improve drainage return period
23		Three-dimensional turbine raindrop generators	Reduce the degree of rainwater pollution, the use of three-layer three-dimensional form of drainage structure, greater water discharge, maintenance is simple, easy to inspect
24		Rainwater inlet automatic waste interception devices	Automatic adjustment of rainwater runoff and waste accumulation overflow in order to ensure smooth drainage of rainwater outlets
25		Rainwater collection pre-treatment devices	Filters and retains sediment and waste during rainwater harvesting
26		Interceptor basket sedimentation devices	The shell material is PE, built-in stainless steel basket and filter, can effectively intercept larger pollutants in the rainwater collection process

3.5.2 Rainwater Storage and Control Equipment

Currently, there are various types of rainwater storage and control equipment both domestically and internationally. Typical examples of foreign equipment include the Q-Bic rainwater tank, the Grundfos rainwater storage tank, buoyancy weir, and the floating outlet device for detention ponds. Typical examples of domestic equipment include rainwater tanks, plastic water tanks, PP water retention modules, silica sand honeycomb water storage purification modules, glass-fibre reinforced plastic rainwater storage devices, arch-shaped storage devices, High-precision flow limiting valve, and delayed regulating pools. The types of equipment and their performance characteristics for rainwater storage and control, both domestically and internationally, are shown in Table 3.27.

Table 3.27 Domestic and foreign rainwater storage and control equipment and its performance characteristics

Serial number	Area	Equipment	Performance characteristics
1	Foreign	The Q-Bic rainwater tank	A German rainwater storage device with a honeycomb design with uniform top and side loading properties, ensuring a long service life of 50 years
2		The Grundfos rainwater storage tank	Combines mixing, cleaning and aeration functions; easy installation, maintenance and low operating costs
3		Buoyancy weir	A German regulating device mainly used for overflow and diversion of pipe networks and rivers, as well as for flooding of reservoirs and dams
4		The floating outlet device for detention ponds	A U.S. regulation device for use in detention ponds to regulate water quality and quantity in detention ponds
5	Domestic	Rainwater tanks	Made of PE material, rainwater can be collected and utilized locally
6		Plastic water tanks	With sterilization, antibacterial, durable, wear-resistant, acid-resistant, alkaline-resistant, etc., life expectancy of more than 10 years
7		PP water retention modules	Easy to assemble with a water storage void ratio of not less than 95% and a compressive strength of more than 6 t
8		Silica sand honeycomb water storage purification modules	The permeable impermeable sand laid at the bottom has the function of permeability and impermeability, realizing the ion exchange between the water body and the bottom layer, and playing a role in preserving the water quality
9		Glass-fibre reinforced plastic rainwater storage devices	High bearing pressure, can be made into rainwater storage tanks (pools) or regulating tanks
10		Arch-shaped storage devices	Provide a complete runoff pollution control system including filtration, collection, storage and infiltration
11		High-precision flow limiting valve	A flow control device based on vortex technology, no power requirement, self-starting, no moving parts, no maintenance, large overflow aperture, no risk of clogging, easy installation, space saving
12		Delayed regulating pools	In order to ensure the original regulating capacity of the regulating pool, increase the delayed regulating volume

References

1. DeNardo, J. C., Jarrett, A. R., Manbeck, H. B., et al. (2005). Stormwater mitigation and surface temperature reduction by green roofs. *Transactions of the ASAE, 48*(4), 1491–1496.
2. VanWoert, N. D., Rowe, D. B., Andresen, J. A., et al. (2005). Green roof stormwater retention: Effects of roof surface, slope, and media depth. *Journal of Environmental Quality, 34*(3), 1036–1044.
3. Carter, T. L., & Rasmussen, T. C. (2006). Hydrologic behavior of vegetated roofs. *Journal of the American Water Resources Association, 42*(5), 1261–1274.
4. Volder, A., & Dvorak, B. (2014). Event size, substrate water content and vegetation affect storm water retention efficiency of an un-irrigated extensive green roof system in Central Texas. *Sustainable Cities and Society, 10*, 59–64.
5. Teemusk, A., & Mander, Ü. (2007). Rainwater runoff quantity and quality performance from a green roof: The effects of short-term events. *Ecological Engineering, 30*(3), 271–277.
6. Gregoire, B. G., & Clausen, J. C. (2011). Effect of a modular extensive green roof on stormwater runoff and water quality. *Ecological engineering, 37*(6), 963–969.
7. Jarrett, A. R., Hunt, W. F., & Berghage, R. D. (2006). Annual and individual-storm green roof stormwater response models. *ASAE Annual Meeting. American Society of Agricultural and Biological Engineers*, 1.
8. Wang, S., He, Q., Zhang, J., et al. (2012). The concentrations distribution and composition of nitrogen and phosphor in stormwater runoff from green roofs. *Acta Ecologica Sinica, 32*(12), 3691–3700. (in Chinese).
9. Wang, S., He, Q., Sun, X., et al. (2012). Analysis on the effect of runoff retention and pollution load control for two vegetation roofs during rainfall events. *Journal of Chongqing University, 35*(05), 137–142. (in Chinese).
10. Berndtsson, J. C., Bengtsson, L., & Jinno, K. (2009). Runoff water quality from intensive and extensive vegetated roofs. *Ecological Engineering, 35*(3), 369–380.
11. Berndtsson, J. C., Emilsson, T., & Bengtsson, L. (2006). The influence of extensive vegetated roofs on runoff water quality. *Science of the Total Environment, 355*(1–3), 48–63.
12. Major national water pollution control and treatment project. Project name: Study and demonstration of stormwater system for green building and low impact development of residential area. Project Number: 2010ZX07320–001. (in Chinese)
13. Cong, X., Ni, G., Hui, S., et al. (2006). Simulative analysis on storm flood in typical urban region of Beijing based on SWMM. *Water Resources and Hydropower Engineering, 04*, 64–67. (in Chinese).
14. Tian, S. (2013). Effect analysis on rainwater conservation benefit of concave herbaceous field in Jinan City. *Advanced Materials Research. Trans Tech Publications Ltd, 726*, 3685–3689.
15. Qu, C. (2017). *Study on Rainwater Infiltration Efficiency and SWMM Model Simulation of Sunken Green Space and Permeable Pavement-A Case Study of Chang'an Campus in Northwest University*. Northwest University. (in Chinese)
16. Song, Z. (2017). Formula and Experimental Demonstration of Sunken Lawn Runoff Reduction. *Green Science and Technology, 14*, 119–121. (in Chinese).
17. Yang, L., Zhang, L., Li, Y., et al. (2015). Water-related ecosystem services provided by urban green space: A case study in Yixing City (China). *Landscape and Urban Planning, 136*, 40–51.
18. Fan, Q. (2006). *Study on the effect of runoff storage and pollution reduction of urban green space*. East China Normal University. (in Chinese).
19. Major national water pollution control and treatment project. Project name: Research and demonstration of urban stormwater runoff management and runoff pollution control technology. Project Number: 2013ZX07304–001. (in Chinese)
20. Guo, Y. (2020). *Optimal Design of Low Impact Development Based on SWMM Model and Study on the Effect of Rain and Flood Control*. North China Electric Power University. (in Chinese)
21. Fassman, E. A., & Blackbourn, S. (2010). Urban runoff mitigation by a permeable pavement system over impermeable soils. *Journal of Hydrologic Engineering, 15*(6), 475–485.

22. Collins, K. A., Hunt, W. F., & Hathaway, J. M. (2008). Hydrologic comparison of four types of permeable pavement and standard asphalt in eastern North Carolina. *Journal of Hydrologic Engineering, 13*(12), 1146–1157.
23. Zhao, F., Zhang, S., Chen, J., et al. (2011). Study on rainwater infiltration collection and runoff reduction technology of permeable pavement. *Water Supply and Drainage, 47*(S1), 254–258. (in Chinese).
24. Jin, J., Li, T., & Shi, Z. (2017). Performance of Applying Scale Permeable Pavements for Control of Runoff Pollution in an Area with High Groundwater Level. *Environmental Science, 38*(06), 2379–2384. (in Chinese).
25. Zhao, X., Cheng, F., Zhang, X., et al. (2012). Removal efficiency of pollutants in rainwater runoff by permeable pavements of different structures. *Journal of Tianjin College of Urban Construction, 18*(04), 280–285. (in Chinese).
26. Brown, R. A., & Borst, M. (2015). Nutrient infiltrate concentrations from three permeable pavement types. *Journal of Environmental Management, 164*, 74–85.
27. Ministry of Housing and Urban-Rural Development, People's Republic of China. Technical Guide for Sponge City Construction—Construction of Stormwater systems for low Impact Development (Trial), 2014. (in Chinese)
28. SZDB/Z 49–2011, Technical Specification for Rainwater Utilization Engineering. (in Chinese)

Chapter 4
Urban Rainfall Runoff Pollution Process Control Technology

Urban rainfall runoff pollution process control refers to the use of physical, chemical, biological, and management techniques to reduce the volume and pollution load of rainwater runoff after it enters the collection system (including combined sewer networks, separate rainwater pipes, and storage facilities), in order to achieve the goal of runoff pollution control. This process involves the control techniques used during the convergence, collection, and discharge (overflow) of rainwater runoff, including front-end control techniques for drainage networks, rainwater inlet technologies, diversion techniques, design and maintenance of drainage pipes, pollution control and purification strategies for overflow and initial rainwater pollution, as well as key purification technologies.

4.1 Front-End Control Techniques and Strategies for Drainage Networks

Front-end control of drainage networks refers to reducing the total volume, peak flow, and pollution load of runoff entering combined sewer networks and separate rainwater pipes from both water quality and quantity aspects. For combined sewer networks, it effectively reduces the frequency and volume of overflows and the pollution load associated with overflows, thereby reducing the size of overflow treatment structures. For separate rainwater pipes, it reduces the total volume of rainfall runoff and the volume of initial rainwater to be treated, thus reducing the direct discharge of pollutants into rivers.

Common front-end control techniques and strategies for drainage networks include urban sanitation management, soil and water erosion control, rainwater inlet management, and daily rainwater cleaning.

D. Sun and F. Qi, *Urban Rainfall Runoff Pollution Control Technology and Application*,
https://doi.org/10.1007/978-981-96-4174-1_4

4.1.1 Urban Sanitation Management

The majority of garbage that clogs rainwater outlets consists of various solid wastes, mainly including domestic waste, sediment, and fallen leaves.

Regarding domestic waste, it is necessary to strengthen supervision and inspection throughout the entire process of cleaning, collection, transportation, and disposal, to ensure timely removal of domestic waste. Reasonable arrangements should be made for collection and transportation routes, avoiding densely populated areas as much as possible, strictly preventing the scattering, airborne, and open-air dumping of domestic waste, and ensuring regular, fixed-point, and sealed collection and transportation. Harmless treatment of domestic waste should be carried out promptly. In addition, the hygiene of rainwater outlets in restaurant gathering areas is a prominent issue. There is not only accumulated domestic waste but also food residues, which not only cause blockage of the rainwater outlets but also generate runoff pollution such as COD, NH_3-N, TN, TP and new pollutants. It is one of the areas that requires strict control.

The accumulation of sediment and fallen leaves is also a common problem at rainwater outlets, and in severe cases, it can cause these outlets to lose their normal drainage function. Some of this sediment and fallen leaves enter the rainwater outlets/wells with the rainwater, while others are swept into the rainwater outlets/wells by sanitation workers when cleaning the road surface. Rainwater grates cannot completely prevent their entry into the wells, and the sediment and fallen leaves that accumulate in the wells are also difficult to be washed away by rainwater. Over time, this leads to the blockage of rainwater wells. Compared to domestic waste, the cleaning difficulty and blockage severity are greater for accumulated sediment and fallen leaves. They easily cause the blockage of pipelines, leading to the paralysis of large-scale drainage systems. The accumulated water near the blocked rainwater outlets poses serious traffic hazards and greatly inconveniences citizens' travel.

In recent years, with the rapid development of automated and intelligent mechanical equipment, a mechanized cleaning and joint operation mode has been adopted for road flushing, cleaning, and sweeping. Among them, main roads and branch roads with suitable conditions for mechanical sweeping are mainly cleaned mechanically, supplemented by manual sweeping, which can greatly achieve front-end runoff pollution control of drainage networks.

Therefore, enhancing the level of sanitation cleaning and transportation operations, improving the rate of mechanized cleaning, cleaning efficiency and quality, and strengthening urban cleanliness are important aspects of front-end control technology and strategies for drainage networks.

4.1.2 Soil Erosion Control

Green belts are usually set up in the middle or around the roads to prevent soil erosion caused by rainwater. During rainy days, the soil in the green belts can be easily washed away by the rain and carried into the rainwater inlets, resulting in pipe sedimentation and blockage. Therefore, regular inspections of the soil and plants in the green belts are important measures to prevent soil erosion and control runoff pollution at the front end of the drainage network.

4.1.3 Rainwater Inlets Management

Typically, rainwater inlet management includes preventing sewage mixing, strengthening cleaning, and routine maintenance of the inlets.

During construction and street-facing shops, there is often a problem of directly discharging wastewater into the rainwater inlets, leading to sedimentation and blockage of the inlets and pipes, thus increasing overflow pollution. Construction wastewater contains substances that are prone to coagulation and settling, such as soil, sand, and cement slurry. According to the "*Water Quality Standards for Sewage Discharge to Urban Sewer Systems*" GB/T 31962–2015, construction wastewater should be pretreated on-site (sedimentation or clarification) before being discharged into the urban sewage collection system after meeting the standards. However, construction companies often discharge untreated wastewater into the rainwater inlets in violation of the regulations, leading to direct accumulation and blockage of the inlets. Moreover, due to changes in hydraulic conditions, sediments such as silt and cement slurry can easily settle in the drainage pipes. The solidification of cement slurry makes it difficult to clean after sedimentation, gradually causing pipe blockage and affecting the entire drainage system. When houses along the road are converted into street-facing shops, there is also a problem of unauthorized discharge of untreated sewage into the rainwater inlets, especially in the catering industry or small shops.

Urban rainwater inlets are widely distributed and close to buildings, often serving as access points for scattered drainage. To prevent blockage of rainwater inlets, inspections and management should be strengthened, and intelligent operations should be implemented to effectively prevent and control sewage from entering the rainwater inlets [1]. Additionally, enhancing public awareness, education, supervision, and assessment of individuals and organizations, and increasing social and industry attention to runoff pollution control in the front end of the drainage network can help control pollution from heavy rainfall to some extent [2].

4.1.4 Rainwater Inlets Cleaning

Usually, Rainwater inlets and sedimentation tanks are used to intercept domestic waste, fallen leaves, and sediments. Over time, these rainwater inlets and sedimentation tanks will accumulate a certain amount of debris. If they are not cleaned in a timely manner, it can easily lead to blockages in the rainwater inlets. Once there is heavy rainfall, the water cannot be drained quickly and properly, resulting in surface water accumulation. This can cause problems ranging from mild road flooding to serious urban waterlogging, leading to significant damage to the town's economy and safety. Therefore, regular cleaning of rainwater inlets is necessary to ensure their normal drainage function.

4.2 Rainwater Inlets Technology

4.2.1 Technical Overview and Principles

Rainwater inlets are located at the population points of the drainage network and are used to collect rainwater runoff. Traditional rainwater inlets consist of rainwater inlets (or curb inlets) and catchment tanks. They are used to capture the debris, sediment, and pollutants brought by rainwater runoff from the road surface. They serve as the first line of defense in preventing rainwater runoff pollutants from entering the drainage system and urban rivers.

In recent years, in order to enhance the pollutant interception capacity of rainwater inlets, pollutant-intercepting rainwater inlets have been widely developed and applied.

4.2.2 Rainwater Inlets Structural Characteristics and Design Requirements

4.2.2.1 Rainwater Inlets Structural Characteristics

The structure of traditional rainwater inlets consists of an inlet, a well tube, and a connecting pipe, as shown in Fig. 4.1. Accumulation of pollutants is common in traditional Rainwater inlets, especially in ordinary flat rainwater inlets. Pollutants can accumulate in the rainwater inlets and enter the water body or sediment in the pipeline during rainfall flushing. Therefore, strengthening the cleaning management and maintenance of rainwater inlets, and timely removal of pollutants from the rainwater inlets, can effectively prevent pipe blockages, corrosion, and other issues.

In order to enhance the treatment of rainfall runoff pollution, improvements have been made to traditional rainwater inlets. Sediment baskets have been added and

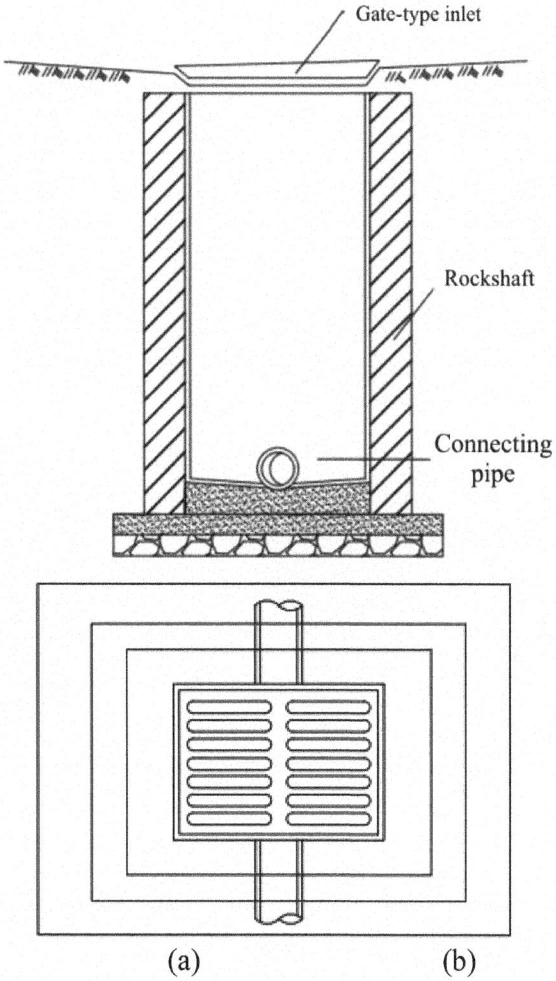

Fig. 4.1 Sectional drawing and vertical view of traditional flat grate rainwater inlet **a** sectional drawing; **b** vertical view

sedimentation areas have been set up to remove debris, sand, and other sediment pollutants. This has resulted in the development of environmentally friendly rainwater inlets with pollution interception capabilities, as shown in Fig. 4.2. In order to prevent pollutants accumulated at the bottom of the rainwater inlet from being suspended and entering the pipeline under the impact of water flow, the sedimentation area should have a depth of at least 0.4 m, as shown in Fig. 4.3. Additionally, for catchment areas with suspended pollutants such as oil, a rainwater inlet as shown in Fig. 4.4 can be used. By setting up floating barriers, oil and other suspended pollutants can be trapped. Filtration can also be achieved through the use of filter screens or permeable concrete walls to treat the rainwater.

Fig. 4.2 Different forms of environmentally friendly rainwater intercepting baskets [3]

4.2.2.2 Design Requirements for Rainwater Inlets

The form, quantity, and arrangement of rainwater inlets should be determined based on the runoff volume generated by the catchment area, the drainage capacity of the rainwater inlets, and the road conditions. The location of rainwater inlets should ensure the rapid and effective collection of surface runoff. Other specific design requirements are as follows:

(1) Bricks for Masonry: Varieties, specifications, appearance, strength, quality should comply with the current national standard "*Sintered Ordinary Brick*" GB/T 5101 provisions, and meet the design requirements. Generally, the strength of bricks used for masonry should not be lower than MU10, and they should have factory product quality certificates and test reports. Upon arrival at the construction site, samples should be sent for retesting and must meet the qualified standards.
(2) Cement: Generally, 32.5 grade ordinary Portland cement and slag Portland cement are commonly used. Cement should have product qualification certificates and factory inspection reports. Upon arrival at the construction site,

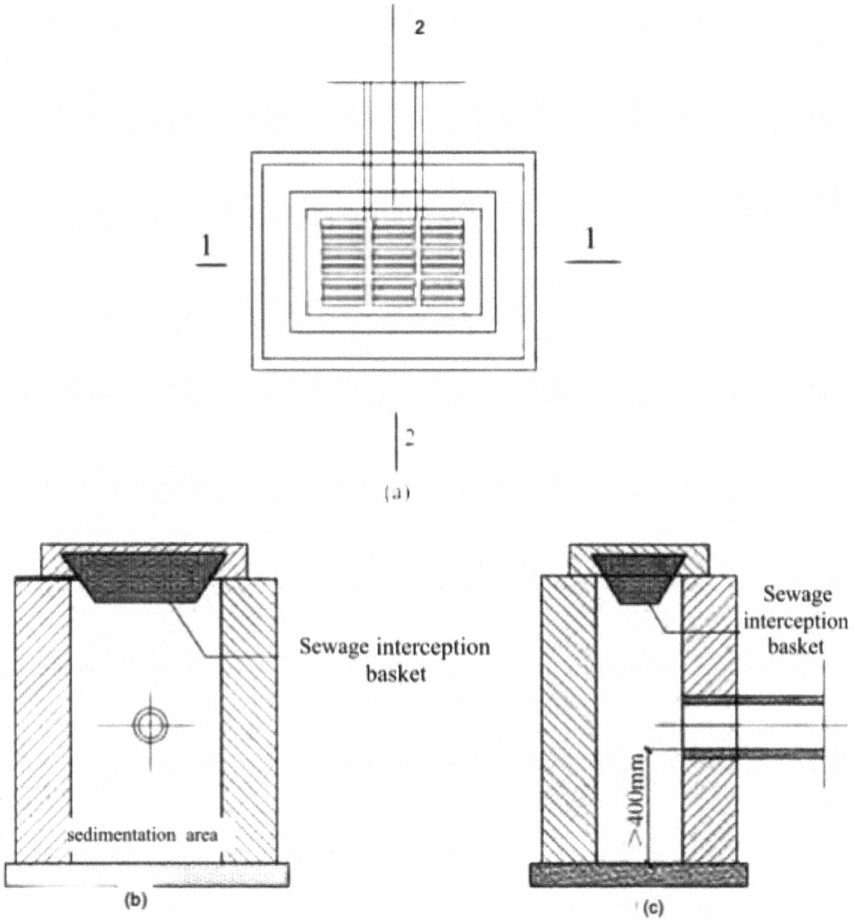

Fig. 4.3 Environmental protection rainwater outlet sedimentation area Plan Fig.; **b** 1-1 sectional drawing; **c** 2-2 sectional drawing [3]

samples should be retested for strength, stability, and other necessary performance indicators. The quality of cement must comply with the current national standards, such as "*General Portland Cement*" GB 175. If there are doubts about the quality of cement or if the cement has been stored for over 3 months, it must be retested before use, and the results of the retest should be followed. Different types of cement should not be mixed.

(3) Concrete Pouring and Water: It is recommended to use drinking water. When using other water sources, the water quality should meet the requirements of the current national standard "*Water for Concrete*" JGJ 63.

(4) Sand: It is advisable to use medium-coarse sand with hard texture, good gradation, and cleanliness. The mud content in the sand should not exceed 3%, and its quality should meet the requirements of the current national standard

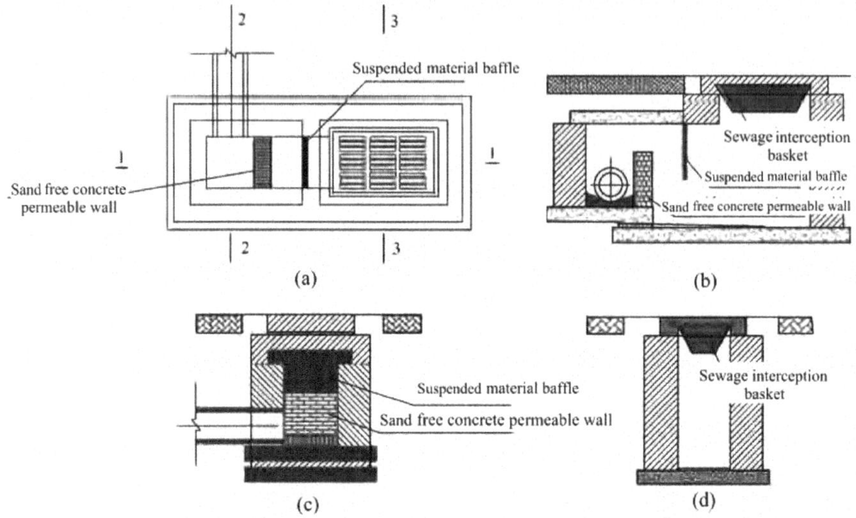

Fig. 4.4 Rainwater inlet with filtration and oil separation functions **a** Plan Fig.; **b** 1-1 sectional drawing; **c** 2-2 sectional drawing; **d** 3-3 sectional drawing

"*Standard for Quality and Inspection Methods of Sand and Stone for Ordinary Concrete*" JGJ 52. Upon arrival at the construction site, samples should be taken for retesting and must meet the qualified standards.

(5) In order to ensure the smoothness between the rainwater inlet and the road surface, the rainwater inlets well ring and cover should be installed before the asphalt surface is applied, according to the design elevation.

(6) The elevation of the top surface of the road rainwater inlets should be 30 mm lower than the elevation of the adjacent road surface and should be flush with the nearby road surface.

4.2.3 Rainwater Inlets Operation and Maintenance Requirements

4.2.3.1 Ensure that the Rainwater Outlets Are not Blocked During Construction

During the construction of rainwater inlets, blockages may occur due to negligence of workers or accidental pouring of cement mortar into the rainwater outlets. Therefore, the management unit needs to strengthen the supervision and acceptance of construction units.

4.2.3.2 Prevent Sewage from Entering the Rainwater Inlets

The discharge of construction wastewater is a key focus of inspection. Construction wastewater often contains substances such as soil, sand, and cement slurry that are prone to coagulation and settlement. Therefore, preventing sewage from entering the rainwater outlets is of great significance for effectively controlling the pollution load of rainfall runoff.

4.2.3.3 Prevent Garbage from Entering the Rainwater Inlets

Rainwater outlets are set below the ground level and have holes with a certain area. While effectively collecting rainwater, they are also susceptible to debris entering. It is common for road cleaning personnel to sweep ash, soil, leaves, and other debris into the rainwater inlets, which can cause blockages. Relevant drainage management departments should establish management systems, coordinate with environmental sanitation departments, carry out publicity, education, supervision, and assessment to effectively reduce blockages caused by human activities.

4.2.4 Scope and Pros and Cons of Rainwater Inlets

4.2.4.1 Scope of Application of Rainwater Inlets

Rainwater inlets are commonly used to collect rainwater from roads, ground surfaces, and green spaces. They are applied in combined sewer systems, separate rainwater drainage systems, and collection facilities.

4.2.4.2 Advantages and Disadvantages of Rainwater Inlets

Advantages: By setting up sediment baskets in rainwater inlets, they can capture debris, sediments, and pollutants, preventing solid pollutants from clogging the pipes. They can achieve a certain level of pollutant reduction before the pollutants enter the collection system.

Disadvantages: The main issues that rainwater inlets face are blockages and compressive strength. Since rainwater outlets are constructed on the sides of roads or within green spaces, the received rainfall runoff contains complex water quality and high concentrations of SS. Prolonged sedimentation will cause blockages in rainwater outlets. Additionally, road rainwater inlets have high requirements for their materials and compressive strength. If ordinary materials are used, the safety requirements may not be met.

4.2.5 Application Examples of Rainwater Inlets

Figure 4.5 depicts a schematic diagram of a new environmentally friendly rainwater inlets, including inlet pipes, water spreader covers, pollution interception devices, filtration and drainage pipes, and outlet pipes.

The surface runoff flows into the inlet pipes and falls onto the top cover of the water spreader. Under the force of gravity, it scatters onto the inclined plate connected to the top of the cover and enters the pollution interception device through the sides of the rainwater outlet. The surface runoff first passes through the pollution interception device to remove larger particles and then enters the filtration and drainage pipes, quartz sand layer, gravel layer, and surrounding soil. When the water level in the rainwater well reaches the height of the outlet pipe, the accumulated water in the rainwater outlet flows out through the outlet pipe. This technology has been successfully applied in the auxiliary road of Beijing Airport Expressway and residential areas under the jurisdiction of Zuo jiazhuang Street, Chaoyang District, Beijing. It has been in operation for four rainy seasons, successfully intercepting at least 90% of the garbage entering the rainwater outlets and rainwater pipes, saving over 90% of pipeline maintenance costs.

This technology effectively alleviates the problem of urban drainage pipeline blockage, improves the operation and maintenance efficiency of the drainage network, saves maintenance costs, and helps to alleviate urban flooding and reduce pollution from rainfall runoff. It has become one of the 36 advanced technologies for sponge cities in the first batch of 2016 issued by the Ministry of Housing and Urban–Rural Development.

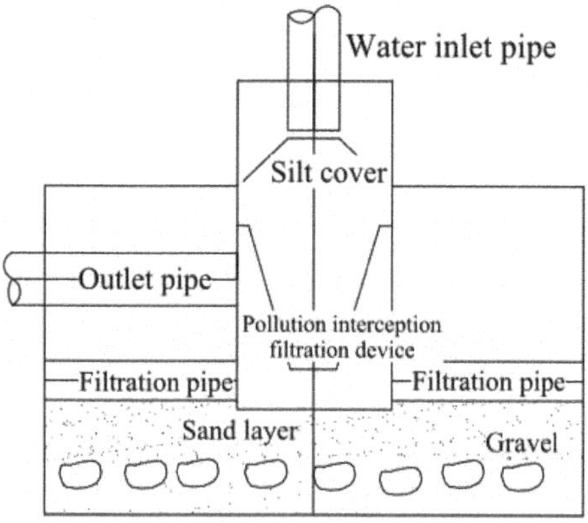

Fig. 4.5 Schematic diagram of a new environmentally friendly rainwater inlet

4.3 Diversion Technology

4.3.1 Overview and Principles of the Technology

Due to the high concentration of pollutants in initial rainwater, it is not convenient for resource utilization. Therefore, the initial 25 mm (10–20 min) of rainfall runoff is diverted through diversion facilities. After storage and centralized treatment, it can be discharged into rivers or reused for landscape water, domestic washing, and groundwater replenishment. The remaining cleaner rainwater can be directly utilized after simple treatment, achieving differentiated treatment for different qualities of rainfall runoff.

4.3.2 Structural Characteristics and Design Requirements of Diversion Facilities

4.3.2.1 Structural Characteristics of Diversion Facilities

Diversion facilities consist of rainwater inlet pipes, overflow wells, initial rainwater pipes, float ball valves, rainwater outlet pipes, rainwater sensors with time relays, and liquid level gauges, as shown in Fig. 4.6.

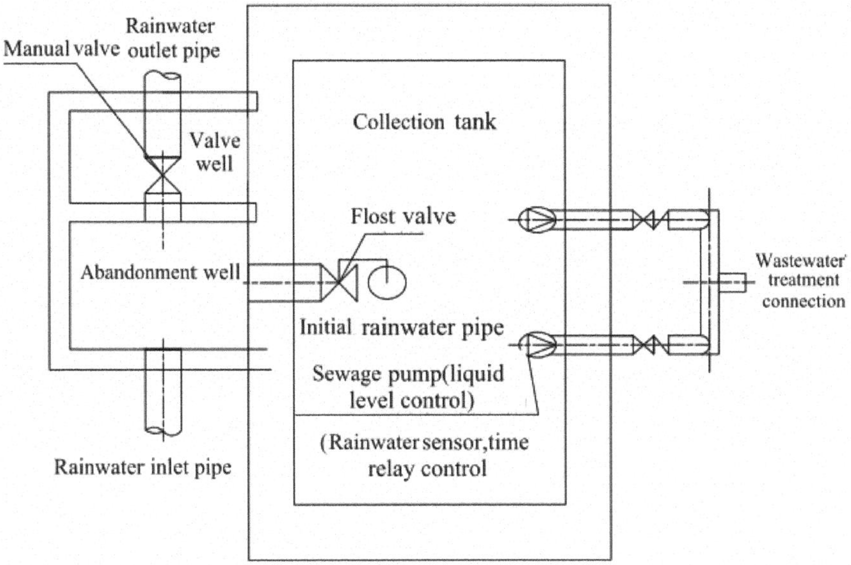

Fig. 4.6 Process schematic diagram of diversion facilities [4]

When rainwater enters the overflow well, it first enters the collection pool through the initial rainwater pipe. When the water level reaches a certain height, the float ball valve automatically closes, stopping the flow of initial rainwater, and the rainwater flows out and is discharged into the surrounding environment through the rainwater outlet pipe.

During rainfall, the rainwater sensor cuts off the circuit of the sewage pump, preventing it from starting. When the rain stops, the rainwater sensor controls the circuit to be disconnected after a certain period of time with the help of a time relay. The circuit of the sewage pump is reconnected, and under the control of the liquid level gauge, the sewage pump automatically sends the water in the collection pool to the wastewater treatment system for treatment. After the water level drops to the lower limit, the sewage pump automatically stops and waits for the next rainfall.

4.3.2.2 Design Requirements for Diversion Facilities

(1) The initial rainwater diversion facilities should be set up in a decentralized manner and can include initial runoff pools, permeable overflow wells, or finished overflow devices.
(2) The volume of initial rainwater to be diverted should be determined based on the rainfall amount or duration, usually ranging from 2 to 5 mm or 10–20 min.
(3) The diversion capacity of the initial rainwater diversion facilities should be determined based on the design volume or quality of the diverted rainwater, clearly separating the initial rainwater.
(4) The design of the water inlet and outlet pipes, as well as the size of the overflow well and collection pool of the initial rainwater diversion facilities, should be reasonable.
(5) A rainwater standpipe should be installed at the water inlet of the initial rainwater diversion facilities, and a inspection hole should be placed at the top for easy cleaning.

4.3.3 Operation and Maintenance Requirements for Diversion Facilities

(1) In case of blockage or sediment accumulation at the water inlet or outlet, garbage and sediment should be promptly cleaned.
(2) If sediment accumulation causes insufficient volume in the overflow well, timely desilting should be carried out.

4.3.4 Range of Application and Pros and Cons of Diversion Facilities

4.3.4.1 Range of Application of Diversion Facilities

Diversion facilities are suitable for handling initial rainfall runoff of 2–5 mm or 10–20 min. They are also suitable for the front-end of rainwater collection facilities such as rainwater pipes from roofs and concentrated inlets (rainwater inlets) for runoff.

4.3.4.2 Advantages and Disadvantages of Diversion Facilities

Advantages: Automatic control of initial rainwater collection and pumping can be achieved without manual operation. It does not involve the opening and closing of electric valves, resulting in low failure rate. It can also be operated remotely.

Disadvantages: Due to the certain difference in liquid level between the collection pool inlet pipe and the rainwater pipe, some late-stage rainwater entering the collection pool increases the load of wastewater treatment.

4.3.5 Application Examples

Diversion technology has been applied in the "Kaid · Visual field" residential project in Gongshu District, Hangzhou City [5]. In this project, rainwater is collected and enters the initial discharge device, and the initial discharge rainwater is discharged into the municipal sewage pipe nearby and finally enters the urban sewage treatment plant for treatment and discharge. This not only prevents pollution from pollutants such as COD, heavy metals, and volatile phenols in the initial rainwater but also creates conditions for further treatment and utilization.

The rainwater collected after the initial discharge is collected in a storage tank, and the rainwater in the storage tank is pumped to an aeration biofilter by a pump. Before entering the filter tank, coagulant is added using a coagulation and dosing device to remove SS. Since the quality of rainwater after the initial discharge is relatively stable and the SS content is low, the coagulated floc enters the filter tank for direct filtration, and then it goes through disinfection and enters the clear water tank. The treated rainwater is mainly used for replenishing water in the community's landscape lake and for sprinkling for greening. The volume of the storage tank is determined based on the daily water consumption for greening and landscape lake replenishment in the community and the storage period. The water consumption for sprinkling is about 16 m^3 per day, and the water for landscape lake replenishment is about 4 m^3 per day.

4.4 Design and Maintenance of Drainage Pipelines

Given the vast expanse and climate variations in our country, the distribution of annual rainfall runoff is uneven, with the majority of rainfall occurring during the summer season. This often leads to the rapid accumulation of surface runoff if not promptly collected and treated. The direct entry of such runoff pollutants into surface water bodies can pose significant risks to urban water security and water quality.

In China, urban drainage systems employ both combined sewer systems and separate rainwater drainage systems to collect and transport rainfall runoff and its pollutants. The combined sewer system is a pipe system that combines domestic wastewater, industrial wastewater, and rainwater, consisting of structures such as rainwater inlets, rainwater pipes, diversion wells, detention tanks, and outfalls. During dry weather, the diversion pipes transport domestic and industrial wastewater to sewage treatment plants. During rainy weather, as the rainfall increases, the diversion pipes transport a mixture of domestic wastewater, industrial wastewater, and rainwater to the sewage treatment plants. On the other hand, the separate rainwater drainage system comprises structures such as rainwater inlets, rainwater pipes, inspection wells, and outfalls. Its main purpose is to promptly collect and remove surface runoff generated from heavy rainfall to prevent urban flooding and ensure the safety of lives, properties, and normal order of life and production.

Within urban drainage systems, pipelines are a crucial component. Carrying out proper design of drainage pipelines, determining appropriate diversion ratios, and gradually implementing rain and sewage separation transformation are important means to achieve effective control of combined sewer systems. Additionally, the repair and maintenance of drainage pipelines are critical to ensuring the normal operation of both combined sewer systems and separate rainwater drainage systems.

4.4.1 Design of Drainage Pipelines

4.4.1.1 Determination of Interception Ratio

The interception ratio (n_0) refers to the ratio of the diverted rainfall volume in combined sewer systems to the dry weather sewage volume. The size of n_0 directly affects the quantity of Combined Sewer Overflows (CSOs). A larger n_0 value corresponds to a larger diverted rainfall volume and smaller CSOs, leading to lower environmental pollution levels.

As early as 1915, Engberding proposed that the diverted rainfall volume should be ten times the dry weather sewage volume for controlling CSOs pollution. However, in foreign countries, the diversion ratio is generally set between 2 and 5 [6]. The national standard "*Technical code for urban drainage engineering planning*" GB 50318-2017 recommends a diversion ratio between 2 to 5 for combined sewer systems. However, in practice, for cost savings, a lower diversion ratio of 0.5–1.0 is commonly adopted, resulting in significant overflow volumes in combined sewer systems during rainfall and consequently causing severe pollution to the receiving water bodies.

The operating water level of the pipelines should not exceed their design filling level, with a maximum filling level of not higher than 0.9. The operating water level of pumping stations ideally should not exceed the top of the inflow pipe. Combined sewer systems without diversion pipes should have them installed, and their diversion ratios should comply with the requirements of the current national standard "*Code for Outdoor Drainage Design*" GB 50014. For combined sewer systems with existing diversion pipes, the original design diversion ratios should be restored. Rainwater pipelines should not be used as combined or sewage pipelines.

4.4.1.2 Rainwater and Sewage Diversion

Rainwater and sewage diversion, also known as separate sewerage, is the transformation of a combined sewerage system into a separate sewerage system, separating rainwater and sewage within the drainage system to fundamentally eliminate the occurrence of CSOs. For many years, most of the urban areas in China with combined sewerage systems have developed their drainage plans based on this concept.

However, there are also certain drawbacks and issues with separate sewerage transformations. In the early stages of urban development, rainwater often contains high levels of pollutants. If the separated rainwater is not effectively treated before being discharged into receiving water bodies, it can pose a threat. Additionally, the transformation process is subject to various constraints, such as road networks and underground spaces, which can result in significant workloads, high costs, long durations, or even difficulties in implementation.

In addition, due to reasons such as inadequate regulations, ineffective management, and insufficient enforcement, there are serious issues of illegal connections and mixed sewage and rainwater discharges in China's urban separate sewerage systems. The drainage management department in Shenzhen conducted random inspections on approximately 25% of the rainwater pipelines in the earliest developed districts of Luohu and Shangbu and found over 260 concentrated points of sewage discharge [7]. This indicates that the rainwater and sewage drainage systems in these two districts are almost entirely mixed. It can be seen that establishing an ideal separate sewerage system or converting a combined sewerage system into a fully separate system also faces challenges.

Therefore, the selection or transformation of drainage systems requires in-depth research, meticulous comparative evaluations, and careful decision-making based on the national conditions and specific circumstances of each urban area. For newly developed urban areas, separate sewerage systems can be considered, along with the formulation of strategies for controlling rainwater runoff pollution. In older urban areas with relatively established combined sewerage systems, limited space, and lower rainfall, alternative approaches can be taken, such as implementing effective CSOs control facilities, diversion measures, pollution interception, and repairing and improving aging pipelines within the existing combined sewerage system [8].

4.4.2 *Maintenance of Drainage Pipelines*

After the design and construction of drainage pipelines, it is important to regularly inspect the pipelines for blockages and damages and plan for their maintenance. This is one of the key elements in ensuring the long-term stable operation of drainage pipelines and effectively controlling the pollution of urban rainfall runoff. The maintenance of drainage pipelines mainly includes controlling leakage and infiltrations, as well as cleaning and repairs. Additionally, for addressing the limited capacity of drainage pipelines during rainy days, the addition of detention and retention facilities can be considered as part of the solution.

4.4.2.1 Control of Leakage and Infiltrations in Pipelines

Damaged pipelines can allow sewage to infiltrate into the ground, polluting groundwater. Conversely, when the groundwater level is high, groundwater can infiltrate into the pipelines, increasing the amount of external water entering the pipes and causing increased overflow during the rainy season. Excessive infiltration of external water can increase the collection, conveyance, and treatment costs of sewage, affect the operation and efficiency of urban wastewater treatment plants, and reduce the safety of drainage systems.

To prevent leakage and infiltrations in pipelines, the following measures can be taken:

(1) Use high-quality pipeline materials to improve the lifespan of pipelines and reduce the rate of damage.
(2) Strengthen the supervision of pipeline construction to improve engineering quality and prevent leakage and infiltrations at pipe joints.
(3) Conduct pipeline monitoring and maintenance to promptly detect pipeline damages, carry out necessary repairs, and address leakage and infiltration issues.

4.4.2.2 Pipeline Cleaning Techniques

i. Unblock and maintain

The sedimentation of pollutants in pipelines during the dry season is an important source of pollution for combined sewer overflow and direct discharge of rainwater into rivers in separate sewer systems. If the sediment in the drainage pipelines is not removed in a timely manner, it will be washed into the receiving water body by runoff during rainfall. The occurrence of foul odors in urban water bodies during rainy days is closely related to the flushing of sediment in drainage pipelines. Additionally, the sedimented pollutants in the pipelines occupy the volume of the pipes, affect the hydraulic function and conveyance capacity of the drainage pipelines.

Regularly flushing the pipelines, collecting and treating the deposited pollutants on-site or transporting them to wastewater treatment plants, can reduce the pollutant discharge from combined sewer overflow during rainy days and the pollution entering rivers from separate rainwater pipelines.

Unblocking and maintaining drainage pipelines can effectively remove the accumulated sludge, improve the hydraulic function of the pipes, and reduce the amount of pollutants discharged into water bodies. Common methods include rod pushing, rotating rod, winch, jet truck, and hydraulic flushing. The characteristics of various unblocking and maintenance methods for drainage pipelines are as follows:

(1) Rod pushing: This method involves manually pushing tools such as bamboo strips, steel bars, and hooks into the pipelines to remove blockages. It has the advantages of simple equipment and low cost and is mainly used for small pipes.
(2) Rod Rotating: This method involves rotating the unblocking rod to clear blockages inside the pipelines. Currently, most rotating rod unblocking machines use spring-loaded rods and are mainly used for community or indoor drainage pipes.
(3) Winch: This method uses a winch to tow tools for unblocking the pipelines. Winches can be divided into manual winches and motorized winches. Depending on the pipe cross-sectional size, blockage type and volume, and sludge moisture content, different trenchers or other winch unblocking equipment can be selected.
(4) Water jetting truck for pipeline unclogging: It is a specialized vehicle that combines high-pressure pump, high-pressure hose, winch, nozzle, water tank, and other water jetting devices. It is also called a high-pressure flushing truck. There are various types of water jetting trucks, including light-duty water jetting trucks, medium-duty water jetting trucks, and combination flushing trucks.
(5) Hydraulic flushing: This method involves increasing the water level difference between upstream and downstream and increasing the flow velocity to unblock the pipelines.

ii. Inspection well and rainwater inlet maintenance

The sludge from the drainage pipeline clearing is cleaned and transported through the inspection wells and rainwater inlets. Therefore, regular maintenance and desilting of the remaining sludge in the inspection wells and rainwater inlets are necessary. Commonly used equipment includes mud suction trucks, mud grabbing trucks, and combination flushing trucks. The characteristics of various equipment are as follows:

(1) Mud suction truck: There are mainly vacuum mud suction trucks and fan-type mud suction trucks. Vacuum mud suction trucks are suitable for drainage pipelines with sewage and rainwater, using water as a medium to carry away the sludge. In the suction pipe, the mud and water are in a solid water column state. Fan-type mud suction trucks use the kinetic energy of air to suck up the sludge and debris using air as a medium, and the suction height is not affected by atmospheric pressure. They are suitable for drainage pipelines with less or no water.
(2) Mud grabbing truck: Equipped with large and small grab buckets, each grab bucket is composed of 4 claws. When opened, it resembles a lotus flower shape and can be freely replaced according to the excavation site. It can grab and

clean various pollutants (sludge, sand, stones, bricks, debris, etc.) in deep and shallow wells without leaving dead spots. This equipment has the advantages of unlimited depth, simple operation, high efficiency, and low cost.

(3) Combination flushing truck: A flushing truck with both water jetting and vacuum mud suction functions is called a combination flushing truck. The advantage of a combination flushing truck is high flushing efficiency, but it has a higher cost.

4.4.2.3 Pipeline Repair Technology

Pipeline repair can be divided into major repairs and minor repairs, which should be based on the local economic conditions. When damage is found in the pipeline system, it should be repaired promptly to prevent the damage from expanding and causing accidents. The main engineering and management measures include timely repair of damaged pipelines, reducing pipeline roughness, increasing flow capacity, reducing occurrences of overloading and backflow, and reducing the deposition of pollutants.

For defects and misconnections in drainage pipelines and inspection wells, various techniques such as excavation or non-excavation should be used to repair and treat the problem of rainwater and sewage mixing, restoring the separation of rainwater and sewage.

Traditional excavation repair techniques have disadvantages such as long construction time, traffic disruption, impact on surrounding pipelines and structures/buildings safety, and high maintenance costs. On the other hand, non-excavation repair techniques generally do not affect the safety of surrounding pipelines and structures/buildings, ensure smooth traffic, save infrastructure investment, and achieve energy-saving and emission reduction goals. Non-excavation repair of drainage pipelines mainly includes local non-excavation repair technology and overall non-excavation repair technology.

i. Excavation repair technology

Excavation repair involves using excavators or manual excavation to remove the old pipe, reinstalling a new pipe, and then backfilling the trench [9, 10]. It is a pipeline repair method. After excavating the damaged pipe section, it is repaired in place according to the original diameter, slope, and elevation, ensuring that the water-carrying capacity of the pipeline remains unchanged after repair. Prior to patching, both sides of the pipeline need to be sealed to ensure that the pipeline is dry and free of water accumulation.

Applicable scope: When the pipeline repair index (RI) > 7 or the pipeline experiences a grade 4 damage, excavation and buried pipe repair method can be adopted. It is suitable for pipelines with severely damaged structural sections, where the entire pipeline section is excavated and renewed.

Excavation repair technology has the advantages of low cost and long service life for newly laid pipes. The disadvantages of this technology are that it occupies at least one road during construction, causing significant traffic impact. In addition, it requires breaking and repairing the existing road, which often affects the traffic on surrounding streets and causes inconvenience to business operations.

Excavation repair generally has two methods: sloping excavation and support excavation.

i) *Sloping excavation*

Generally applicable to open spaces with good soil conditions, no surrounding buildings, and no underground pipelines. When the slope height exceeds 5 m, it should be excavated in stages. The advantages of this method are low cost and fast construction; the disadvantages are large volume of excavation and backfill soil, and prone to collapse in rainy weather.

ii) *Support excavation*

It can be divided into panel support and channel steel support. Panel support is suitable for geological conditions, and when the trench depth is less than 3 m, while channel steel support is generally used for excavating pipelines with a burial depth less than 3 m. The advantages of this method are that channel steel has good durability and can be reused, and it is convenient to construct with a short construction period; the disadvantages are that it cannot block groundwater and small particles. If the groundwater level is high, proper water isolation and dewatering operations are required.

ii. Local non-excavation repair techniques

i) *Stainless steel sleeve method*

After the stainless steel sleeve, which is wrapped with waterproofing material, expands, a sealed pipe lining is formed between the original pipeline and the stainless steel sleeve, blocking the leakage point. This method is mainly used for repairing local defects such as pipe disconnection and leakage. It has the advantages of good water-stopping effect, stable quality, cost-saving, and fast repair. However, it is not suitable for repairing structural defects such as pipe rupture, severe misalignment of interfaces, and severe linear deformation of pipelines. This method is applicable to pipe diameters ranging from 150 to 1350 mm. Common waterproofing materials for stainless steel sleeves include water-stop rubber and foam rubber, as shown in Fig. 4.7.

ii) *Point-wise in-situ solidification method*

The point-wise in-situ solidification method involves fixing fiber materials soaked in room-temperature curing resin at the damaged site, injecting compressed air to tightly compress the fiber materials against the inner wall of the pipeline, and forming a new pipe lining through solidification. This method is mainly used for repairing defects such as pipe disconnection, leakage, and rupture, but it is not suitable for repairing structural defects such as pipe rupture, severe misalignment of interfaces, and severe linear deformation of pipelines. This method is applicable to pipe diameters ranging from 50 to 1500 mm. The main repair materials used are glass fiber and various

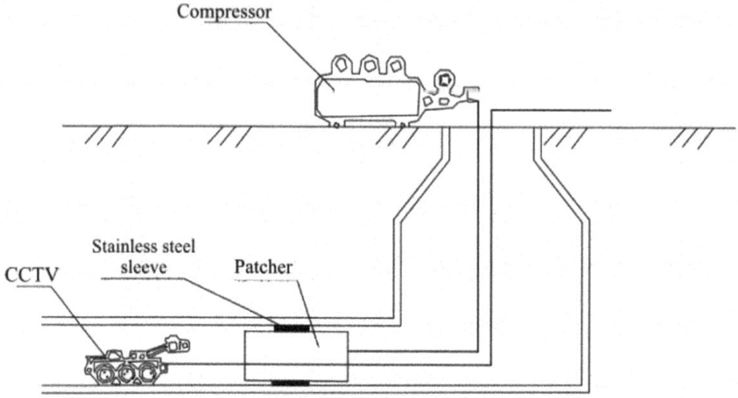

Fig. 4.7 Schematic diagram of stainless steel sleeve valve repair construction

types of epoxy and polyurethane resins. Common curing processes include natural curing, heat curing, and UV curing, which can be chosen according to specific needs. Please refer to Fig. 4.8 for a schematic diagram of the point-wise in-situ solidification method.

iii) *Stainless Steel Double Expansion Ring Repair Method*

The stainless steel double expansion ring repair method involves using a circular rubber sealing tape and stainless steel expansion rings. The rubber sealing tape is installed at the pipe interface or the damaged area, and then fixed in place with 2–3 stainless steel expansion rings to achieve a watertight seal. This repair method is suitable for fixing defects such as pipe deformation, misalignment, disconnection, and leakage, with interface misalignment less than 3 cm. However, this repair method requires the pipe's foundation structure to be basically stable, with no significant

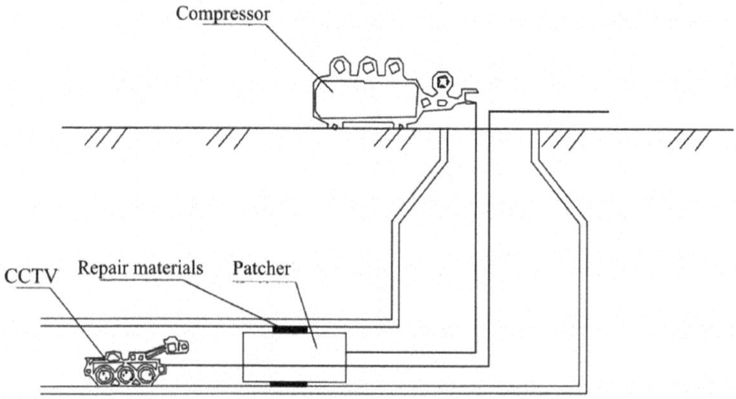

Fig. 4.8 Schematic diagram of repair construction using spot in-situ solidification method

changes in pipe alignment and the pipe walls to be solid and not corroded. It serves as a temporary sealing measure for pipe interfaces and does not provide structural strength. Additionally, it is limited by the rubber's weak corrosion resistance and aging resistance, resulting in a relatively short service life after repair. This method is suitable for pipes with a diameter larger than 800 mm and extra-large drainage pipes. The repair materials used are stainless steel double expansion pressure bars and specially made water-stop rubber, which offer advantages such as simple and fast construction and good sealing effects. The schematic diagram of the stainless steel double expansion ring repair method is shown in Fig. 4.9.

iv) *Chemical grouting method*

This method involves injecting various chemical grouts into the external bedding soil and soil voids around the damaged section of the pipeline using specific equipment. The chemical grouts quickly solidify to achieve waterproofing, leakage prevention, soil stabilization, and void filling. It is suitable for repairing identified leakage points and damages inside various types of pipelines. The main repair material used in this method is specialized chemical grout. By utilizing the fluidity and rapid solidification of the grout, it seals and strengthens the external part of the pipeline, simultaneously reinforcing the surrounding soil during leakage repair. It has the advantage of quick repair. The schematic diagram of the chemical grouting method is shown in Fig. 4.10.

iii. Overall Non-excavation Repair Technology

i) *Hot water in-situ curing method*

This method involves inserting a hose impregnated with thermosetting resin into the existing pipeline using water pressure inversion. After heating and curing, a new lining is formed inside the pipeline for the repair of various structural defects. It is suitable for drainage pipelines of different geometric shapes and can repair pipe diameters ranging from 100 to 270 mm. It has the advantages of short construction

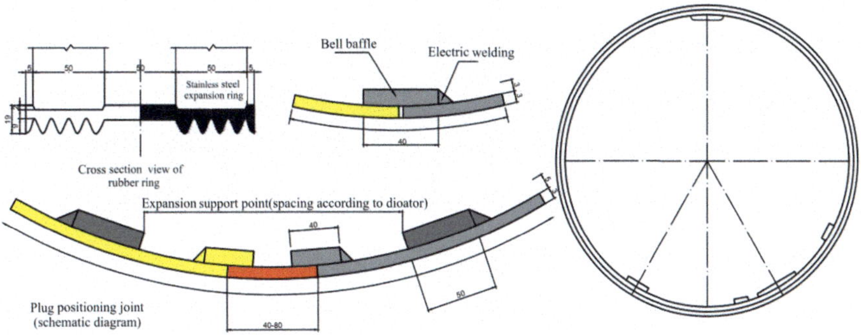

Fig. 4.9 Schematic diagram of repair construction using stainless steel double expansion ring repair method

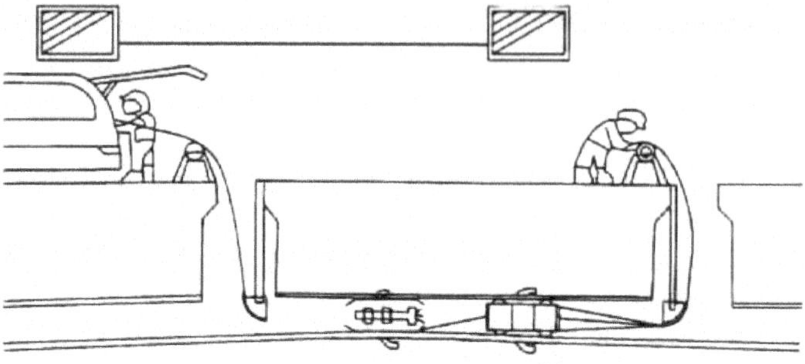

Fig. 4.10 Schematic diagram of chemical grouting method for repair construction Trenchless rehabilitation technology

time, small footprint, long service life, strong overall integrity after repair, smooth surface, and minimal impact on the surrounding environment.

Before proceeding with the repair, it is necessary to ensure that the pipeline to be repaired meets the repair conditions for the hot water in-situ curing method. For sections with severe deformations, collapses, and other defects, local excavation repair can be combined with the hot water in-situ curing repair process for repair construction. The schematic diagram of the hot water in-situ curing method for repair is shown in Fig. 4.11.

ii) *UV light in-situ curing method*

This method involves inserting a hose impregnated with photosensitive resin into the existing pipeline and curing it through UV light irradiation, forming a new lining inside the pipeline. It can be used for the repair of various structural defects and

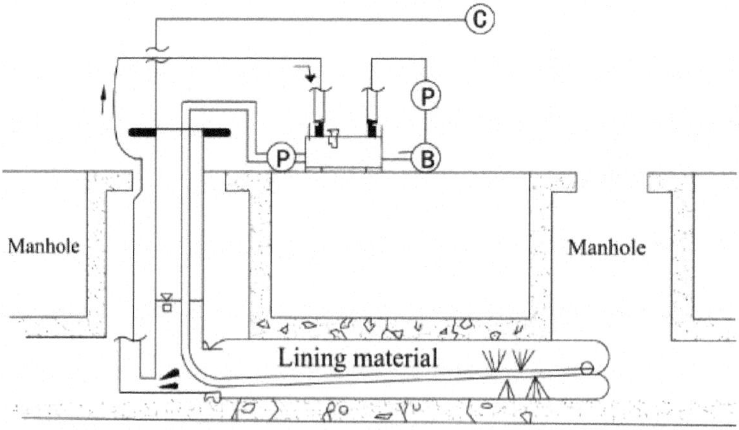

Fig. 4.11 Schematic diagram of chemical grouting method for repair construction [11]

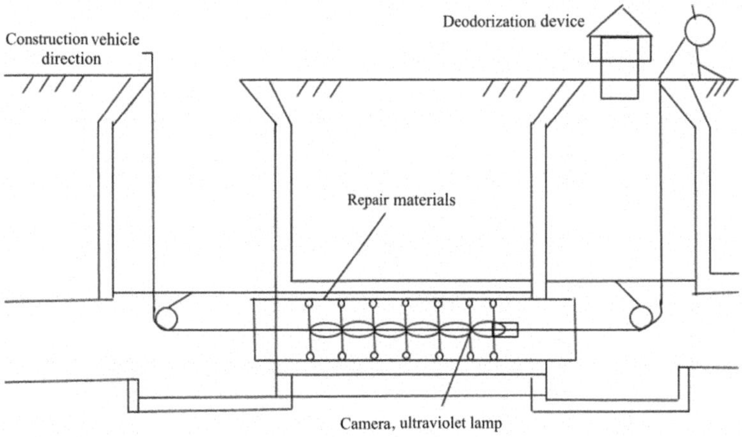

Fig. 4.12 Schematic diagram of chemical grouting method for repair construction

can repair pipe diameters ranging from 150 to 1600 mm. It is suitable for drainage pipelines of different geometric shapes. This method has the advantages of short construction time, small footprint, long service life, strong overall integrity after repair, smooth surface after repair, and minimal impact on the surrounding environment. It can seal existing holes, cracks, and gaps, isolate infiltration, and prevent upward seepage. It is widely used in the repair of structural defects in drainage pipelines. Compared to the hot water in-situ curing method, the UV light in-situ curing method has a faster curing speed and higher pipe strength after repair. The schematic diagram of the UV light in-situ curing method for repair is shown in Fig. 4.12.

iii) *Helical winding method*

This method uses mechanical winding to form a new lining inside the existing pipeline using strip-shaped materials. It can be used for the repair of various structural defects and can be operated with water in the pipeline. There are mainly two processes: expansion method and fixed-caliber method. It is suitable for drainage pipelines of different geometric shapes and can repair pipe diameters ranging from 450 to 3000 mm. This method has the advantages of short construction time, small footprint, long service life, strong overall integrity after repair, smooth surface after repair, and minimal impact on the surrounding environment. It can seal existing holes, cracks, and gaps, isolate infiltration, and prevent seepage. It is widely used in the repair of structural defects in drainage pipelines.

iv) *Pipe liner method*

This method involves splicing PVC sheet materials inside the existing pipeline to form a new pipe and filling the gaps between the new pipe and the original pipe. It is used for the repair of defects such as rupture, dislocation, and leakage. The shape of the pipeline is not restricted, and it is suitable for pipe diameters ranging

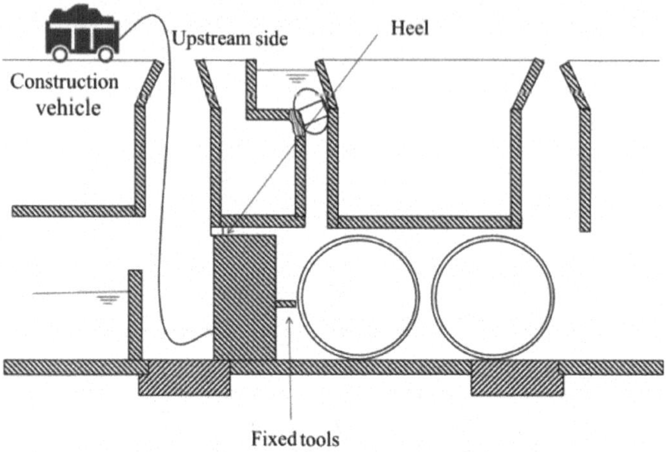

Fig. 4.13 Schematic diagram of segment lining method repair construction

from 800 to 3000 mm. After repair, the liner pipe and the original pipe form a high-strength composite pipe, with a strength equal to or higher than that of a new pipe. However, there is a certain loss in the hydraulic section of the drainage pipeline after repair. The repair process is illustrated in Fig. 4.13. This method has a small construction footprint and can be done in curves or in local sections. There is no limitation on the length of a single construction and construction can be temporarily interrupted. However, this method is not suitable for repairing structural defects such as severe misalignment of pipes, fractures or breakages in pipe foundations, and severe deformation of pipes.

4.4.2.4 Storage and Regulation Techniques

In order to effectively reduce and regulate the peak flow of rainwater runoff, the combined flow of peak flow polluted water and the diversion of initial rainwater can be stored until the flow decreases, and then discharged. These stored waters are transported through pipelines to wastewater treatment plants for treatment, aiming to reduce peak flow and control rainfall runoff pollution. Common techniques for rainwater runoff storage and regulation include storage tanks and deep tunnels.

i. Storage Tank

By appropriately setting up storage tanks in the drainage system, the initial rain-water mixed with pollutants can be intercepted, reducing the overflow of combined sewer during heavy rainfall. During the initial stage of rainfall, the rainwater mixed with pollutants directly enters the wastewater treatment plant. If the capacity of the drainage system is exceeded, the excess rainwater mixed with pollutants overflows

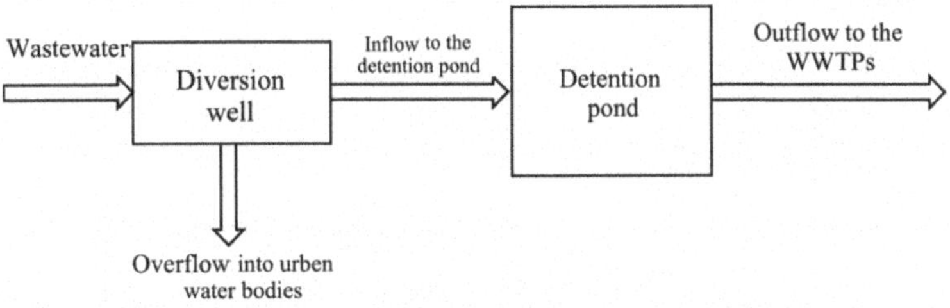

Fig. 4.14 Overflow detention pond process flowchart

into the storage tank. After the rainfall ends and the water capacity of the pipeline network returns to normal, the stored rainwater mixed with pollutants is sent to the wastewater treatment plant for treatment or discharged into the receiving water body after on-site treatment. Additionally, the storage tank also plays a role in purifying the wastewater. The process flow and changes in the storage flow rate of the storage tank are illustrated in Figs. 4.14 and 4.15, respectively. Photos of rainwater storage tanks can be seen in Figs. 4.16 and 4.17.

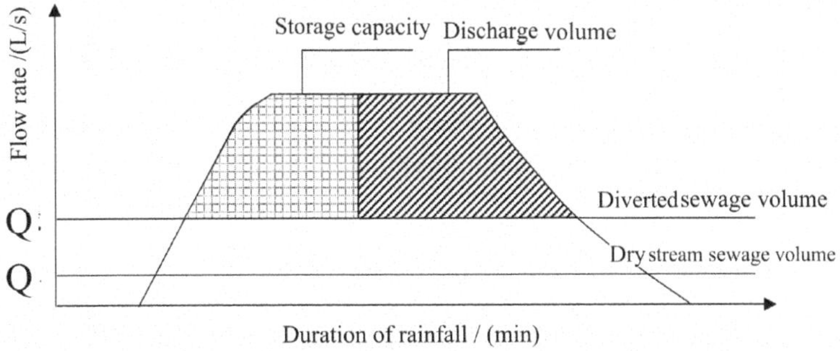

Fig. 4.15 Illustration of the flow rate of the retention tank in the combined sewer system [12]

Fig. 4.16 Rainwater overflow storage tanks in Germany

Fig. 4.17 Rainwater storage tank with separation partition

i) *Principles of storage tank setup*

(1) The outlet of the storage tank should be connected to the wastewater pipe network. If the downstream wastewater treatment system capacity is not sufficient to meet the requirements for emptying the storage tank, on-site treatment facilities should be installed.
(2) The location of the storage tank should be determined based on considerations such as the drainage system, pipe network, downstream water level of overflow pipes, and surrounding environment. In areas with suitable conditions, mathematical models can be used to optimize the design plan. The burial depth of the storage tank should be determined based on the depth of the upstream and downstream drainage pipes, taking into account factors such as land availability, project investment, construction difficulty, and operational energy consumption.
(3) Pollution interception and storage tanks can be set up in conjunction with the construction of underground utility tunnels.

ii) *Flushing methods for storage tanks*

Devices for flushing and cleaning the sediment at the bottom of the storage tank should be installed. The flushing method for storage tanks should be selected based on the project characteristics and the design of the tank, considering factors such as safety, environmental protection, energy efficiency, and ease of operation. Hydraulic self-cleaning and equipment flushing methods are recommended. For storage tanks located in pump rooms, it is preferable to prioritize flushing methods that require low equipment maintenance, have simple controls, and are hydraulically driven. A comparison of various flushing methods can be found in Table 4.1.

Table 4.1 Table of various advantages and disadvantages

Sequence number	Flushing method	Advantages	Disadvantages
1	Manual cleaning	No mechanical equipment required, no need for maintenance	High risk, and heavy labor intensity
2	Mobile cleaning equipment	Cost-effective investment, easy maintenance	Suitable only for open-bottom detention tanks, cleaning equipment (sweepers, forklifts, etc.) requires manual operation
3	Intelligent sprayers	Automatic cleaning with aeration process to reduce odors, suitable for most tank types, requires the construction of a cleaning water reservoir and related equipment	Higher operating costs, equipment located at the bottom of the tank, susceptible to pollution and wear
4	Submersible agitators	Agitation driven by water flow, self-cleaning, cost-effective investment	Poor cleaning efficiency, equipment located at the bottom of the tank, prone to entanglement, pollution, and wear
5	Hydraulic bucket flush	No need for electricity or mechanical drive, simple control	It requires a pressurized external water source for bucket flushing. High operating costs, limited bucket capacity, and limited flushing range
6	Continuous trench self-cleaning	No need for electricity or mechanical drive, no external water supply required	It relies on sewage flow for self-cleaning, difficult to achieve thorough cleaning, and may lead to secondary deposits. The structure of continuous trenches increases the depth of pump station construction
7	Gate-type self-cleaning	No need for electricity or mechanical drive, no external water supply required, simple control system. Long flushing distance for a single flushing wave, flexible adjustment, can be controlled manually or electronically. Low operating costs, high efficiency	Higher initial investment

iii) *Operation and maintenance of storage tanks*

(1) Inspection and maintenance

The inspection and maintenance cycle for storage tanks is generally 1 to 12 months. The focus should be on removing pollutants and debris, and attention should be paid to the leakage of the storage tank.

(2) Safety measures

1) Strictly adhere to the principle of "ventilation first, testing second, and operation last". It is strictly forbidden for personnel to enter the storage tank for work without ventilation and testing.
2) Protective barriers, grates, covers, and warning signs should be installed at the entrances and exits of the storage tanks. When visibility is poor, warning lights should be used.
3) In prominent areas outside the storage tanks, caution zones, caution lines, and caution signs should be set up. The installation should comply with relevant national regulations.
4) Personnel should wear isolated protective masks and, if necessary, be equipped with a lifeline. Workers should wear anti-static work clothes and work shoes, use appropriate explosion-proof low-voltage lighting fixtures, tools that do not generate sparks, and be equipped with combustible gas alarms, etc.
5) In the event of an accident, supervisors should promptly report to the relevant authorities, and rescue personnel should ensure their own protection and be equipped with necessary respiratory equipment and rescue tools. Blindly rescuing and causing the accident to escalate is strictly prohibited.

iv) *Application case of storage tanks*

(1) The Xiamen Xinglin Bay Jiutian Lake Comprehensive Remediation Project: The Xiamen Xinglin Bay Jiutian Lake area is a combined sewerage system. The integrated remediation project is located beneath the green belt on the south side of the Xiamen Xinglin Bay waterfront, west of Jiutian Lake Bridge. The size of the storage tank is $L \times B = 198.8 \text{ m} \times 28 \text{ m}$, with a depth of 5 m and an effective water depth of 4.5 m, providing an effective volume of approximately 26,000 m^3. Initially, rainwater flows through horizontal grates in the diversion well, filtering out suspended and floating solids. Specifically, there are two self-cleaning horizontal grates with dimensions of $5950 \text{ mm} \times 1283 \text{ mm}$. Additionally, eight intelligent jet sprayers are installed for aeration flushing of the storage tank.

On sunny days, when the flow of sewage in the pipes is less than the maximum treatment capacity of the sewage treatment plant, the sewage is directly sent to the treatment plant for processing. During the early stages of rainfall, when the combined flow of rainwater and sewage is still less than the maximum treatment capacity of the plant, the mixed sewage is also directly sent to the sewage treatment plant for processing. However, when the combined flow of rainwater and sewage exceeds the maximum treatment capacity of the plant, a portion of the mixed sewage (rainwater and sewage) is treated at the sewage treatment plant, while the remaining mixed sewage passes through self-cleaning horizontal grates and enters the storage tank, where suspended and floating solids are intercepted. When the storage tanks are full, the electric gates close to stop the inflow of water into the storage tanks. To prevent odor, the intelligent jet sprayers intermittently aerate the storage tank. If the rainfall continues and the water level in the buffer tank rises to the emergency discharge level, the later-stage rainwater is discharged directly into natural water bodies through overflow weirs.

When the flow rate in the buffer tank is less than the maximum treatment capacity of the sewage treatment plant, the submersible sewage pump starts to transfer the rainwater from the storage tank to the sewage treatment plant for processing. At the same time, the intelligent jet sprayers activate their stirring function to facilitate the transportation of sediment in the sewage to the treatment plant. When the water level in the storage tank drops to the bottom of the tank, the intelligent jet sprayers begin to flush the tank bottom. The flushed sewage is then pumped to the sewage treatment plant by the submersible sewage pump for processing.

(2) Combined Sewer System Improvement Project in the old urban area of a certain city in Anhui Province: In the old urban area of a certain city in Anhui Province, due to the adoption of combined sewer systems in some areas and separate sewer systems in other areas, there is a serious problem of misconnection and disorderly connection of pipelines. Additionally, the aging of sewage interception pipes and high infiltration of groundwater lead to overflow and water pollution, as well as backflow issues. The purpose of this project is to control the overflow water quality, prevent water pollution and backflow, and reduce the load on the sewage treatment plant. As a newly constructed support facility for sewage interception pipes, the project is mainly used for storage and treatment of CSOs and rainwater, aiming to achieve on-site treatment and meet the emission standards, thus relieving the load on the sewage treatment plant. The project provides a suitable water source for ecological treatment facilities.

The design of the storage tank for this project is a combined sewer overflow storage tank, which is divided into sedimentation, flow-through, and overflow zones. The sedimentation zone is the most frequently used, followed by the flow-through zone, and the overflow zone has the least frequency of use. The partitioned design improves the efficiency of each zone since the inflow frequency and concentration vary in different zones. This leads to different operation and maintenance cycles and costs, which helps maintain the equipment condition and efficiency in long-term operation, while reducing the overall operation and maintenance costs.

The inflow section of the storage tank is equipped with a flat grille, hydraulic grit separator, gantry automatic flushing system, and float control flow valve. The collection area of the storage tank is 128 ha., the storage volume is 11,500

m^3 (with volumes of 4,500, 4,500, and 2,500 m^3 for each zone, respectively), the effective water depth is 3.1 m, the overflow flow rate is 1,500 L/s, and the burial depth of the storage tank is 7.5 m. The project has an estimated investment of approximately 70 million yuan.

The storage tank is completely buried, and the ground is used for municipal parking lots and leisure squares. A diversion well is set at the front end of the storage tank. During non-rainy periods and the initial stage of rainfall, incoming water flows into the sewage pipeline through a flotation control valve. The overflow water, after passing through a coarse grille and controlled by flotation control facilities, steadily enters the storage tank. Under normal circumstances, incoming water first enters the collection tank through a distribution channel. When the collection tank is full, the water flows into the through tank, and when the through tank is full, it enters the comprehensive tank. The supporting gantry automatic flushing system flushes the bottom of the storage tank. The concentration of incoming water is highest in the collection tank. After the rain ends, all the water is pumped into the sewage network. All the water except the upper layer in the through tank is also sent to the sewage network. The comprehensive tank only needs to send the lower layer of water and the flushed sewage into the sewage network. Hydraulic particle separators are installed in the through tank and the comprehensive tank to perform in-depth treatment on the upper layer water in the through tank, upper layer water in the comprehensive tank, and continuous overflow water, and the treated overflow is discharged. Flat fine grilles are installed at the overflow outlets to perform simple treatment on excessive overflow water before it is discharged. All overflow water from the tanks can enter the ecological treatment facilities on the riverbank for further treatment before entering the receiving water body. When the storage tank is full and extreme rainfall occurs, with the incoming water exceeding the continuous overflow capacity, the emergency overflow channel is opened, and the water is directly discharged from the diversion facilities into the receiving water body to ensure flood control and facility safety. Ventilation equipment is installed in the storage tank to prevent the accumulation of harmful gases. Inspection holes and necessary access channels are installed in the storage tank for easy daily maintenance of the facilities. After the equipment is operational, it is monitored through cameras or manual inspections to ensure normal operation.

When the storage tank is full and extreme rainfall occurs, with the incoming water exceeding the continuous overflow capacity, the emergency overflow channel is opened, and the water is directly discharged from the diversion facilities into the receiving water body to ensure flood control and facility safety. Ventilation equipment is installed in the storage tank to prevent the accumulation of harmful gases. Inspection holes and necessary access channels are installed in the storage tank for easy daily maintenance of the facilities. After the equipment is operational, it is monitored through cameras or manual inspections to ensure normal operation.

The detention basin is fully underground, with the surface being a municipal parking lot and a leisure plaza. At the front end of the detention basin, there is a diversion well. During non-rainy periods and at the beginning of rainfall, incoming water is sent into the sewer pipes through a float-controlled flow control valve. The overflow water, after passing through a coarse grid, is controlled by float-controlled facilities and is directed steadily into the detention basin. Under normal circumstances, incoming water first enters the collection pool through a distribution channel. When the collection pool is full, it flows into the through-pool, and when the through-pool is full, it enters the comprehensive pool. A complementary gantry-style automatic flushing system flushes the bottom of the detention basin. The water in the collection pool has the highest concentration and is pumped into the sewage pipe network after the rainfall has ended. All water in the through-pool, except for the upper layer, is sent to the sewage pipe network. The comprehensive pool only needs to send the lower layer of water and flushing wastewater into the sewage pipe network. Hydraulic particle separators are installed in the through-pool and the comprehensive pool to perform deep treatment on the upper layer water in the through-pool, upper layer water in the comprehensive pool, and continuous overflow water. After treatment, the overflow is discharged. Flat fine grids are installed at overflow points to perform simple treatment on excessive overflow water before discharge. All overflow water from the ponds can enter the riverbank ecological treatment facility, undergo further treatment, and then enter the receiving water body. If the detention basin becomes full and an extreme rainfall event occurs, with the inflow greatly exceeding the continuous overflow rate, the emergency overflow channel is opened to discharge water directly from the diversion facility into the receiving water body, ensuring flood control and facility safety. Ventilation equipment is installed in the detention basin to prevent the accumulation of harmful gases. Inspection holes and necessary inspection pathways are provided in the detention basin for ease of daily facility maintenance. After the equipment is operational, it is monitored through cameras or manual inspections to ensure proper functioning.

Before the project implementation, the overflow frequency at the riverbank discharge outlet exceeded hundreds of times annually. After project implementation, the overflow frequency reduced to approximately 5–10 times per year, and emergency overflows decreased to 1–3 times annually. The effluent COD and SS concentrations decreased by 60–80%.

ii. deep-level tunnel

To address the prominent issues of urban flooding, such as local flooding in already developed urban areas and combined sewer overflow pollution, cities both domestically and internationally are actively undertaking the transformation and improvement of their existing drainage systems. However, the comprehensive upgrade and renovation of drainage systems face enormous challenges due to various factors, including spatial constraints, traffic considerations, construction timelines, relocation of residents or businesses, and financial constraints. This is especially challenging in older or central urban areas, where many rainwater management measures are difficult to implement and may not yield quick results. Achieving a significant and comprehensive increase in flood control standards and meeting the requirements for combined sewer overflow pollution control in a short period of time is also very difficult [8, 13].

In recent years, tunnels have gained significant attention as an effective engineering solution. Some cities, such as Beijing, Shenyang, and Guangzhou, have initiated discussions, research, and preliminary planning to address urban flooding issues through tunnel construction. Meanwhile, cities like Shanghai, Hefei, and Jiaxing are planning to construct tunnels to reduce CSOs pollution.

Tunnels can rapidly, flexibly, and efficiently alleviate localized urban flooding and the problem of combined sewer overflow pollution. Rainwater control tunnels, often built at deep underground levels (referred to as "deep tunnels"), avoid the various factors affecting the urban surface or shallow underground spaces and conflicts with other infrastructure. Additionally, modern and efficient construction techniques like tunnel boring machines (TBM) provide strong support for the application of deep tunnels. However, due to the scale and high cost of such projects, it's important to first consider their suitability and applicability under specific conditions.

Generally speaking, in conditions where there are numerous and densely distributed overflow outlets with a significant overflow volume, or where there are many and severe flooding points, or when traditional surface and underground discharge and storage facilities lack space or are difficult to implement rapidly, deep tunnel systems can be considered a viable and effective solution. In some developed countries, cities have addressed the problems of urban flooding and combined sewer overflow pollution by constructing tunnels to capture, store, and divert large volumes of rainwater runoff that their existing drainage systems cannot handle. Tunnel projects designed to control flooding and combined sewer overflow pollution have seen numerous successful applications. Some cities began building such tunnels as early as the 1970s and 1980s, such as the Chicago Deep Tunnel project initiated in 1975 and the Osaka Deep Tunnel project in 1985 [14, 15]. Additionally, some cities are currently constructing or planning to construct similar tunnel systems, including the Indiana Storage Tunnel currently under construction and the London Storage Tunnel planned for future development.

i) *Classification of Deep Tunnel Systems*

Tunnels designed for the control of rainfall-induced flooding and the mitigation of overflow pollution are collectively referred to as deep tunnel systems. Before urban planners and decision-makers embark on the planning and construction of rainwater control tunnels, they must first define the conditions, objectives, and suitable tunnel types for their specific needs. In reality, deep tunnel systems come in various types based on specific conditions, functions, and operational methods.

Based on their functions and control objectives, deep tunnel systems can be categorized into three main types: flood control, pollution control, and multi-purpose systems. Different types of deep tunnel systems have distinct design methods, scales, interconnections, and downstream outlets.

The deep tunnels designed for flood control purposes can be classified into flood storage tunnels and flood diversion tunnels based on the site, rainfall runoff discharge, and operational conditions.

Flood Storage Tunnels: These tunnels primarily collect and store rainfall runoff exceeding the capacity of existing drainage pipes or pump stations. They can be positioned vertically along the path of rainwater runoff. By intercepting upstream flash floods or river flooding, they reduce the flood risk downstream. A prominent example is the flood storage tunnel in Osaka, Japan [15].

Flood Diversion Tunnels: Flood diversion tunnels are mainly responsible for diverting and accommodating upstream floodwaters or flows exceeding the capacity of river channels. Typically, these tunnels discharge into rivers or other receiving bodies of water downstream. They are often located along the main arterial roads within flood-prone areas. Notable examples include the West Island Tunnel in Hong Kong and the Outer Peripheral Flood Diversion Tunnel in Tokyo [16, 17].

In addition, factors such as urban expansion leading to increased peak flows, encroachment on urban river channels, limitations in channel cross-sections, and vertical constraints often contribute to the occurrence of urban flooding. The generation of urban flooding is also closely related to inadequate flood discharge capacity in the river channels and the downstream water level rising. In such situations, tunnels are often constructed parallel to the river channels or located directly beneath the channels to address the issue of insufficient drainage capacity in the channels. For example, the Wolli Creek Flood Diversion Tunnel [18], in domestic cities like Beijing and Guangzhou, has considered placing certain tunnels beneath the river channels to alleviate flooding issues.

A deep-level tunnel designed for the purpose of pollution control is commonly referred to as a storage tunnel or CSOs storage tunnel. These are predominantly used in older urban areas with combined sewer systems, and in some cases, they extend into newer urban areas. Their primary function is to collect combined sewer overflow wastewater that exceeds the capacity of the interceptor sewer pipes, and in a few instances, they may also collect rainwater runoff from separated sewer systems. In the United States, the CSOs storage tunnel in South Boston [19], for instance, either treats the collected wastewater on-site or conveys it to a sewage treatment plant for further processing before discharge. These types of tunnels are typically positioned along the overflow points of the combined sewer system, running parallel to interceptor sewers, rivers, or coastlines. They effectively connect multiple overflow points, functioning somewhat like a larger interceptor sewer and a detention basin. Due to their location downstream in the drainage system, they are primarily utilized for storing and treating combined sewer overflow wastewater that exceeds the interceptor sewer capacity and generally cannot address upstream drainage area flooding issues.

There is another type of multi-purpose tunnel, which, through thoughtful design and operational adjustments, can serve various functions such as flood control, pollution control, and transportation. For instance, in combined sewer systems, apart from managing overflow pollution, there's also a need to address urban flooding. Therefore, it's crucial to comprehensively consider the location and scale of the tunnel

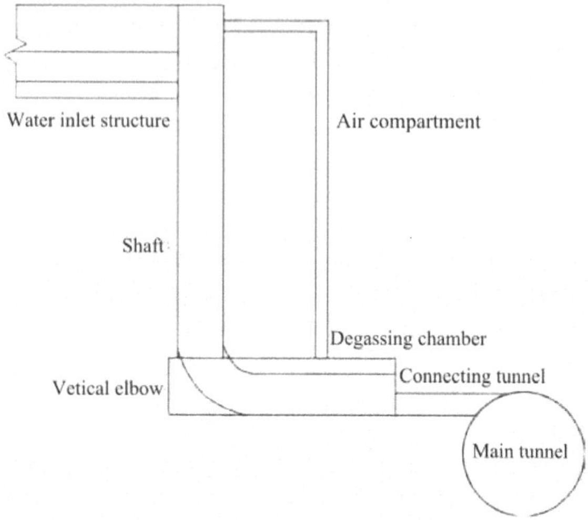

Fig. 4.18 Schematic diagram of tunnel connecting facility construction

and ensure a rational connection between existing pipeline systems, overflow points, and flood-prone areas, with the aim of minimizing both urban flooding and pollution. Typical examples of such multi-functional tunnels include the "Deep Tunnel" in Chicago, USA, and the "SMART Tunnel" in Kuala Lumpur, Malaysia [20].

ii) *Tunnel Construction and Operation*

The construction of tunnels can vary depending on the type of tunnel, but it typically includes components such as the main tunnel, connecting facilities, ventilation system, exit facilities, and control center. Connecting facilities generally consist of inlet structures, vertical shafts, vertical bends, and tunnel connections [21]. Exit facilities usually include end drainage pump stations, pollution control treatment facilities, bottom sediment flushing, and discharge auxiliary facilities. Common tunnel connecting facility structures are shown in Fig. 4.18.

The design of the main tunnel is the most crucial aspect in tunnel design, and its scale is typically determined through simulation analysis. The layout of the tunnel and its proper integration with the existing drainage system directly impact the tunnel's functionality and investment efficiency.

The connecting facilities are auxiliary structures that link the existing pipeline system, ground facilities, overflow outlets, collection points, and the main tunnel. Among these, vertical shafts can also store a certain amount of rainwater. Their diameter can be designed reasonably based on factors such as overflow volume, accumulation volume, inflow rate, tunnel operation mode, etc. The stored water in the vertical shafts is drained into the main tunnel through vertical bends and connections.

The ventilation system is often integrated with connecting facilities to introduce fresh air into the tunnel system or remove odorous gases generated during treatment. It is a necessary requirement for the safe operation and maintenance of tunnel systems.

The scale of an end-of-line drainage pump station should be designed based on its functionality, operational mode, efficiency requirements, and other factors. Among these factors, the choice of design flow rate will impact the overall project size and construction cost. For flood prevention or drainage tunnels, the pump station should be designed to meet the drainage requirements. In the case of storage tunnels used for pollution control, it is essential to ensure that the designed storage volume within the tunnel can be transferred to the wastewater treatment plant within a reasonable time frame. The pump station flow rate should be designed based on the capacity of the wastewater treatment plant to prevent excessive settling of pollutants due to prolonged retention time, which could pose difficulties for the subsequent operation and maintenance of the system.

Based on the site, wastewater treatment plant, and regional economic conditions, pollution control tunnels can choose to treat rainwater on-site or transport it to a distant wastewater treatment plant for processing. The former requires specialized design, while the latter needs to consider the characteristics of stored rainwater in terms of water quality and quantity, as well as the alignment and adjustments of the existing wastewater treatment plant's capacity and processes.

The flushing and removal of tunnel sediment is also an important issue, directly affecting the normal operation and efficiency of the tunnel.

The control center conducts 24 h monitoring of all connection points and pumping stations in the tunnel system. Operators track, monitor, and report all real-time data, promptly assess the system's operational status, and make necessary adjustments as needed.

iii) *Typical Applications of Deep Tunnel Systems*

(1) Pollution control deep tunnel: The coastal drainage system in the northern suburbs of Sydney serves approximately 416 km^2 of communities in the northwest of the city. In its early stages, domestic sewage was directly discharged into the ocean. However, after several upgrades, domestic sewage is now transported to the Northern Sewage Treatment Plant for primary treatment. Due to sewage pipe damage and improper connections with rainwater pipes, a significant amount of rainwater infiltrated the sewage system, leading to overflow discharges that severely polluted the water environment of Sydney Harbor. As a result, a large storage tunnel system has been constructed in the suburban areas of Sydney, consisting of main tunnels, overflow points, on-site treatment facilities, ventilation systems, control centers, and other components, as shown in Fig. 4.19. The section from Lane Cove to the sewage treatment plant serves as the main tunnel, while the segments from Tarban Creek Park to Scotts Creek are branch tunnels. This tunnel system connects major overflow points and sewage treatment plants, including Lane Cove, Scotts Creek, Tarban Creek Park, Gladesville Bay, Shelly Beach, among others. Simultaneously, it involves the upgrading of existing drainage pumping stations and sewage treatment plants and the construction of new pumping stations, thereby improving the overall system's operational efficiency. The operation modes of this tunnel can be

divided into four categories: standby, rainy day operation, tunnel maintenance, and bypass of sewage treatment plants.

Backup mode, achieved through proper design and operation, ensures that overflow water first fills the tunnel, and during most non-rainfall periods, the tunnel remains in a backup mode, i.e., it remains empty. The rainy-day operational mode is characterized by the fact that after overflow, the overflow volume continues to enter the tunnel and replace the previously stored water. The previously stored volume is discharged at the end of the overflow location or transferred to the sewage treatment plant for treatment, and the pump station continues to operate until the tunnel returns to an empty state. The scale of the end drainage pump station is determined based on the size of the tunnel and the design storage capacity of rainwater and sewage. For local rainfall events of 70–80%, the pump station can empty the tunnel within 2.5–6 h. Depending on the size and timing of rainfall, the main tunnel may go through the "filling-emptying-overflow" process multiple times during a single rainfall event. Figure 4.20 provides a schematic representation of the system. Tunnel maintenance mode involves regular inspection and maintenance of underground facilities in the tunnel, daily maintenance of surface facilities at overflow outlets, and timely flushing and transfer of tunnel sediments. In case of equipment failure or regular maintenance of the sewage treatment plant, to prevent sewage from directly discharging into receiving water bodies, sewage is diverted to the storage tunnel, forming a bypass mode for the sewage treatment plant.

In addition to this, cities in the United States such as Atlanta, Boston, Portland, Indianapolis, and others have also constructed or planned to construct storage tunnels to control CSOs pollution and reduce the impact of rainfall runoff pollution loads on receiving water bodies.

(2) Deep flood control tunnel: Tunnels, as a large-scale flood control measure, whether in direct discharge mode or controlled storage and discharge mode, can significantly improve urban flood control and anti-flood standards. They quickly prove effective in the prevention and control of waterlogging in local areas.

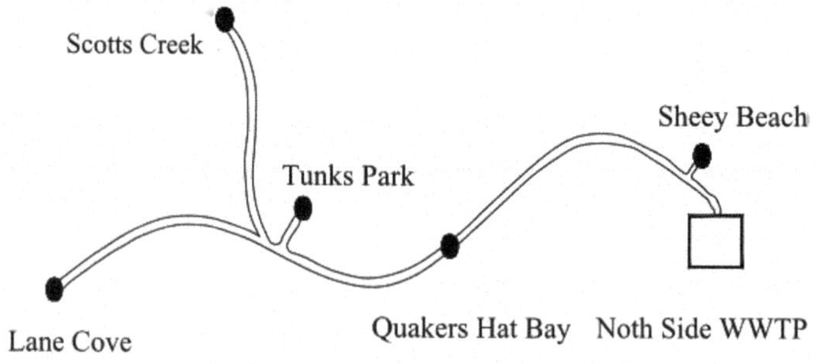

Fig. 4.19 Sydney northern suburbs tunnel system schematic [22]

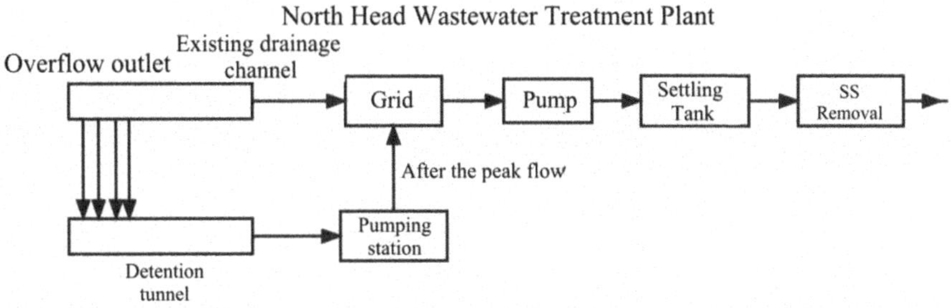

Fig. 4.20 Tunnel system rainy day operation mode [22]

1) Rainwater Discharge Tunnel on Hong Kong Island's Western Side: The geographical characteristics of Hong Kong include many mountains and limited flat areas, with mountain torrents being one of the main reasons for urban flooding. In the low-lying areas of the northern part of Hong Kong Island, factors such as climate conditions, rapid urban development, and aging drainage pipelines make it highly susceptible to flooding disasters. In order to reduce the serious impact and economic losses caused by urban flooding, the Hong Kong Drainage Services Department initiated the "Hong Kong Island Northern Rainwater Discharge Overall Plan Study" in its early stages. This study rigorously examined multiple options, including "expansion and improvement of traditional rainwater systems," "flood detention schemes," and "rainwater interception tunnels." After comprehensive consideration of factors such as land, environment, transportation, underground space, and investment, the solution of tunnel interception was adopted, leading to the construction of the Rainwater Discharge Tunnel on Hong Kong Island's Western Side, located halfway up the mountain.

   The project involved the construction of multiple inlet points, vertical shafts, and connecting tunnels in the mid-levels of Hong Kong Island to intercept rainwater runoff from the mid-levels catchment area. The collected rainwater is then discharged into the sea near Cyberport through the main tunnel. This has significantly reduced the risk of urban flooding in downstream areas. Unlike some other tunnels that require pumps to facilitate drainage due to their greater depth, the Hong Kong Island Western Rainwater Discharge Tunnel utilizes its own gravity drainage capacity, resulting in energy savings. These types of tunnels, which widen or create new upstream drainage channels and improve drainage standards, are similar to the construction of larger rainwater mains and are considered a typical form of direct discharge tunnel. In addition to the Hong Kong Island Western Rainwater Discharge Tunnel, Hong Kong has also constructed the Lai Chi Kok Rainwater Discharge Tunnel and the Tsuen Wan Rainwater Discharge Tunnel in the West Kowloon and Tsuen Wan areas, respectively, to intercept mountain torrents.
2) Waller Creek Flood Control Tunnel in the United States: The Waller Creek watershed in the United States consists of two sub-watersheds: one is an

upstream area with higher terrain in downtown Austin, with an area of approximately 13.1 km^2, and the other is a downstream area along the Twelfth Street with lower terrain, covering approximately 1.6 km^2. Both sub-watersheds are located in the center of Austin. Over the past few decades, the Waller Creek watershed has experienced severe flooding disasters multiple times.

The Waller Creek Flood Control Tunnel is located in the downstream business district of downtown Austin, following the course of Waller Creek and passing through First to Twelfth Streets. It consists of four components: entrance facilities, side weir facilities, the tunnel main body, and exit facilities. The entrance facilities are situated at the downstream end of the upper sub-watershed within Waterloo Park, where they receive and divert approximately 85% of the total floodwater from the upstream segment of Waller Creek. This water is then conveyed to the exit facilities after coarse filtration. There are two sets of side weir facilities located at Fourth and Eighth Streets, designed to capture floodwaters with elevated water levels in the downstream section, accounting for approximately 15% of the total floodwater. The exit facilities are directly connected to the lake without the presence of water treatment facilities. Once completed, the Waller Creek Flood Control Tunnel has the capacity to divert the once-in-a-century floods in the watershed to Lady Bird Lake.

The primary purpose of this tunnel is to control rain-season flooding, with the direct aim of reducing river peak flow, and the indirect goal of preventing urban flooding. Water quality control is limited to the retention, filtration, and sedimentation of certain particle pollutants. The operation of the tunnel system is illustrated in Fig. 4.21. During rainfall, rainwater is slowly discharged into the lake through the tunnel, with larger particles being retained at the entrance, while smaller particles are filtered out or settle within the tunnel and are removed through regular maintenance. On sunny days, lake water is pumped back into the Waller River through the tunnel to maintain the ecological stability of the river and alleviate water scarcity issues in certain river sections.

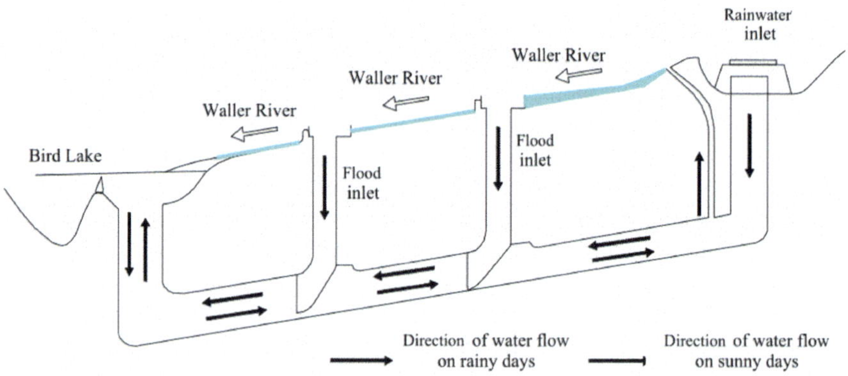

Fig. 4.21 Schematic cross-section of the Austin Waller river flood discharge tunnel [22]

(3) Multi-Function Tunnel: These types of tunnels are designed to address multiple urban issues such as flooding, rainwater runoff pollution, traffic congestion, etc. Through sensible design and operational scheduling, they achieve multi-objective control, thereby saving investment, conserving space, and maximizing the comprehensive benefits of the tunnel.

1) Kuala Lumpur Smart Tunnel: The Kuala Lumpur city center often experiences flooding disasters along the Klang River, leading to inundation of surrounding urban areas and traffic congestion. Since the construction of the "Intelligent Tunnel", local flooding and traffic issues have been effectively alleviated. For instance, in March 2009, a heavy rainfall event resulted in severe urban flooding in Kuala Lumpur's Petronas Twin Towers and the surrounding areas. However, the "Intelligent Tunnel" diverted 700,000 m^3 of floodwater from the catchment area, preventing flooding disasters in its service area.

The "Intelligent Tunnel" consists of a three-layer structure, comprising a permanent drainage layer at the lowest level and a double-deck highway tunnel. The bottom tunnel has a length of 9.7 km. The "Intelligent Tunnel" diverts upstream floodwaters to a bypass tunnel for temporary storage before releasing them into downstream suburban reservoirs, thus mitigating the issue of river backflow and severe flooding in critical areas.

The highway tunnel within the "Intelligent Tunnel" has a total length of 4.0 km, connecting the southern gateway to the city center. This significantly reduces travel time between these two locations and alleviates traffic congestion during peak hours. Ventilation systems or emergency exits are installed at every 1 km interval within the tunnel, ensuring proper airflow and the safety of personnel during sudden heavy rainfall events.

The "Intelligent Tunnel" has three operating modes, as shown in Fig. 4.22. In clear weather or light rain, it operates in the first mode, with both levels of the highway tunnel open for regular traffic. During normal rainfall, it switches to the second mode, where the lower level of the highway tunnel is closed off and used as a drainage channel, while the upper level remains open for traffic. When encountering heavy rainfall, it enters the third mode, where the highway tunnel is completely closed off. Automatic control gates are used to allow heavy rainwater to pass through, and the tunnel is reopened for traffic after the floodwaters recede.

2) Chicago's "Deep Tunnel" and Large Storage Reservoirs: Chicago and its surrounding areas have long been plagued by drainage issues, with combined sewer overflows leading to pollution of Lake Michigan and severe urban flooding. To address these challenges, the city of Chicago constructed deep tunnels and large storage reservoirs both below ground and on the surface. Due to the substantial investment required, the project was divided into two phases: Phase I and Phase II. In Phase I, the primary focus was on constructing four main tunnels and their associated facilities. These tunnels were designed with appropriately sized shafts, allowing excess flow to bypass the tunnels and divert CSOs rainwater, ultimately transporting it to wastewater treatment plants for processing. By the time Phase I was completed, the tunnel system had stored and treated

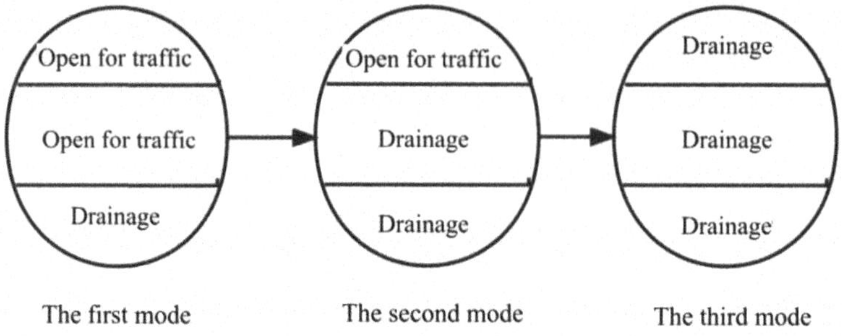

Fig. 4.22 The schematic diagram of the operating modes of the "Intelligent Tunnel" in Kuala Lumpur [22]

approximately 8.7 million cubic meters of CSOs rainwater. Given the limited capacity of the tunnels, Phase II was initiated to create more storage space. This phase involved the construction of large storage reservoirs, auxiliary tunnel branches, and supporting facilities. The primary goal of Phase II was to reduce urban flooding while also addressing CSOs pollution control. The "Deep Tunnel" intercepted rainwater and sewage, diverting it to the large surface storage reservoirs. After the river levels subsided, the collected rainwater and sewage were transported to wastewater treatment plants. Currently, the operational components of the project have effectively mitigated the risk of urban flooding and CSOs pollution in the city.

4.5 Rainfall Runoff Pollution Process Control Strategies and Ideas

4.5.1 Combined Sewer Overflow Pollution and Remediation Strategies

4.5.1.1 Definition and Hazards

Combined Sewer Systems, during rainy or snowmelt conditions, overflow directly when the flow exceeds the design capacity of urban drainage systems or wastewater treatment plants. This phenomenon is known as CSOs. CSOs comprise pollution from various sources, including urban domestic sewage, some industrial wastewater, rainfall runoff, and dry weather pipe sediment. They are characterized by high levels of pollutants and a complex mix of contaminants, including suspended solids, organic matter, nutrients, pathogens, heavy metals, emerging contaminants, and more.

The flow of CSOs varies significantly depending on the magnitude of rainfall events. Factors such as climate, rainfall intensity, characteristics of the drainage area (land use types, vegetation cover, street area, and paving methods) all have a significant impact on the concentration of pollutants in CSOs.

In heavy rainfall conditions, the initial stages of CSOs are characterized by peak water flow due to the rapid accumulation of surface runoff, which converges into the sewer system. During this phase, there is a peak in water volume. Additionally, the initial heavy rainfall also washes accumulated pollutants from the surface and within the sewer system, leading to a peak in pollutant concentration. As the volume of runoff from rainfall increases, the CSOs become diluted, and pollutant concentrations decrease to average levels. The degree of pollution caused by CSOs can vary depending on the hydrological and hydraulic conditions of the receiving water body. When the receiving water body has a higher flow velocity, it possesses greater dilution and self-purification capabilities, which mitigate the impact of CSOs pollution. However, in areas where the receiving water body has lower flow velocities and smaller volumes, the pollution caused by CSOs can be relatively severe. Discharging untreated urban CSOs directly into water bodies can result in flooding disasters and pollution of the receiving water bodies.

i. Flood Disaster

During the rainy season, the intensity of rainfall and the scale of urban rainwater and wastewater collection and treatment systems determine the magnitude of overflow. When the rainfall intensity is high and the overflow volume is substantial, the overflow directly enters the receiving water body, causing a rise in the water surface and even resulting in the phenomenon of water backing up into the urban low-lying areas. This, in turn, leads to the rainwater and wastewater collection system in the urban area operating at full capacity, causing severe waterlogging and impeding the drainage of rainwater, resulting in urban flooding.

ii. Pollution of Receiving Water Bodies

The discharge of a large amount of suspended solids in CSOs significantly reduces the visual quality of the receiving water body, affecting the sensory perception of urban aquatic landscapes. The discharge of a substantial amount of nitrogen and phosphorus nutrients in CSOs triggers eutrophication in urban slow-flowing water bodies, leading to abnormal algae proliferation, decreased dissolved oxygen levels, and a brown-green coloration of the water. This disrupts the water body's aesthetics and affects the growth of aquatic life. CSOs contain a high concentration of bacteria and viruses, and their direct discharge can reduce the biological safety of the receiving aquatic landscape, posing a threat to aquatic ecosystems and the health of nearby residents.

4.5.1.2 Advancements in Research on CSOs Pollution Control Abroad

Developed countries abroad initiated research and practices related to CSOs pollution control quite early. In response to CSOs pollution issues, the United States developed CSOs control strategies, offering operational guidelines for coordinating and implementing CSOs control plans among various stakeholders. Germany addressed CSOs through engineering measures such as rainwater retention basins. London, in the United Kingdom, constructed a deep drainage tunnel to store CSOs wastewater. Japan released the "*Guidelines and Provisional Guidelines for Overflow Countermeasures in Combined Sewer Systems*", but did not mandate a transition from combined sewer systems to separate systems; instead, it proposed CSOs control measures for combined sewer systems.

The CSOs Control Strategy issued by the United States in 1989 includes the following components:

(1) Ensuring no overflows occur during dry weather.
(2) Ensuring that all CSOs discharge points during rainy weather meet the technical and water quality-based discharge limits established by the Clean Water Act (CWA).
(3) Minimizing impacts on water quality, aquatic life, and human health.

In 1991, the U.S. Environmental Protection Agency began developing CSOs control policies to provide detailed guidance on these control strategies in order to expedite compliance with CWA requirements. The official release of these control policies occurred in 1994. These control policies outline the responsibilities of CSOs discharge permittees, permit writers, the National Pollutant Discharge Elimination System (NPDES) permitting authority, and state water quality standard agencies. They also provide operational guidelines for coordinating and implementing CSOs control plans among various stakeholders [23].

In the late 1980s, Germany gradually included urban rainwater pollution control as one of the three major objectives of water pollution control. Germany attached great importance to the combination of source pollution control and CSOs pollution control [24]. While constructing a large number of rainwater retention basins and detention tanks to intercept and treat CSOs, Germany also implemented decentralized ecological measures at the source to reduce rainfall runoff and purify rainwater. These measures primarily included infiltration basins, underground infiltration channels, permeable pavements, as well as various "dry" and "wet" ponds or small reservoirs. Through these ecological measures, rainwater is stored, retained, purified, or infiltrated, extending the discharge time of rainwater. The primary objectives are to reduce peak flows, control runoff, purify rainfall runoff, and replenish groundwater [25].

The United Kingdom is also one of the countries that began addressing CSOs relatively early. The main drainage system in London, which is still in use today, was constructed around 1860 and is a combined sewer system. However, due to a significant increase in population from the original 2 million people to the current 8 million people, urban development and increased surface impermeability have led to frequent CSOs and substantial pollution loads in the combined sewer system. To address this issue, London initiated a CSOs control program in 2007. The program proposed a pollution control technology solution that retains combined sewers within the urban area and constructs deep drainage tunnels to store and control overflow sewage. The alternative option of implementing a regional rainwater separation scheme was rejected due to its high cost (over £14 billion) and long construction period (over 25 years). The proposed solution is expected to be completed by 2023, with a total project investment of approximately £4.2 billion. Upon completion of the project, it is anticipated that the frequency of CSOs overflows in London will decrease from approximately 50 times per year to 4.6 times per year. The average annual overflow sewage volume will be reduced from the current 3950–235 m^3, significantly alleviating pollution in the Thames River [26].

In Japan, there are 192 cities (which account for 8.7% of all Japanese cities) covered by combined sewer systems, with a total coverage area of 2,270 km^2 (which accounts for 20% of all Japanese cities). These combined sewer systems serve approximately 30% of the population. As early as 1982, Japan released the "*Guidelines for Countermeasures and Provisional Guidelines for Combined Sewer Overflow*" which did not mandate the conversion of combined sewer systems into separate systems. Instead, it proposed overflow control measures for combined sewer systems, with control standards equivalent to the pollution levels of separate sewer systems. In the Tokyo Metropolitan area, combined sewer systems account for 82% of the drainage systems. Various measures have been adopted to control CSOs, including source reduction of rainwater, the construction of storage conduits and tanks, rapid filtration treatment, and modifications to overflow outlets. These measures have proven to be effective in reducing CSOs. Furthermore, in the Tokyo Metropolitan area, wastewater treatment plants in combined sewer systems have undergone enhancements specifically to address the large fluctuations in flow during rainy seasons. Specifically, the first-stage treatment section of wastewater treatment plants has incorporated rapid filtration technology, which allows the treatment capacity on rainy days to reach 3 to 4 times that of dry days. Under this system, wastewater with design flow rates equivalent to dry weather conditions enters the biological treatment tanks for biochemical processing, while excess wastewater undergoes first-stage enhanced treatment before discharge, with different discharge standards for the two streams of treated wastewater [26].

4.5.1.3 Progress in Research on CSOs Pollution Control in Our Country

Research on CSOs control in our country started relatively late and has gone through several stages of development, including "construction of combined sewerage systems," "comprehensive transformation of urban drainage systems," "sponge city construction," "treatment of black and malodorous water bodies," and "improving the quality and efficiency of urban sewage treatment." The issue of CSOs has received increasing attention and has played a crucial role in each of these stages.

With the rapid development of urbanization in our country, the gradual increase in residents' awareness of ecological environmental protection, and the requirements of our country's environmental protection planning, many cities have invested a substantial amount of funds in controlling CSOs in existing combined sewerage systems to improve the water environment quality and reduce the pollution load discharged into receiving waters.

In Beijing, downstream of the existing combined sewerage system, overflow wells and diversion pipelines were constructed to intercept overflow sewage that used to be directly discharged into watercourses, diverting it to sewage treatment plants for processing. Additionally, in some urban areas of Beijing, there has been an effort to separate rainwater from sewage in the combined sewer system. However, these measures only addressed the direct discharge of dry weather sewage from combined sewer systems, and the problem of CSOs pollution during rainy days remained severe. In 2011, Beijing formulated a plan to renovate the combined sewerage system in the central urban area. This plan involved refurbishing aging and damaged pipelines, increasing the capacity of diversion pipelines, constructing combined sewer retention tanks, and enhancing the rainy-day treatment capacity of sewage treatment plants. These measures improved the combined sewerage system's ability to collect, transport, and treat rainy-day CSOs sewage, reducing the pollution discharged into receiving waters.

Since 1988, Shanghai has invested heavily in a series of engineering projects to address CSOs pollution. These projects include Phase I of combined sewage treatment, the comprehensive environmental improvement of Suzhou River Phases I, II, and III, and the construction of the Suzhou River Combined Flow Regulation and Storage Reservoir. These initiatives involve intercepting, regulating, and treating sewage before discharging it into the river, thereby reducing CSOs pollution entering the Suzhou River. In recent years, Shanghai has also planned to construct storage reservoirs and large underground storage tunnels in the urban areas with combined sewer systems to increase storage capacity, further reducing the pollution load entering the receiving water bodies during rainy days [27].

Kunming City has implemented pollution interception projects in existing combined sewer overflow areas, achieving partial regional separation of rainwater and sewage; carried out watershed division and constructed combined flow regulation reservoirs; and conducted enhanced treatment during the rainy season at urban wastewater treatment plants to improve their capacity to treat combined sewage during heavy rainfall events, reducing CSOs pollution.

Cities like Nanjing, Suzhou, Shenyang, and Wuxi have mainly adopted measures to retrofit existing combined sewer systems for rainwater and sewage separation to control CSOs pollution. Chongqing and Dongguan have implemented projects in recent years to intercept and treat dry-weather sewage from combined sewer systems.

Overall, China has made significant efforts in controlling CSOs pollution during rainy weather, but further strengthening is still required.

4.5.1.4 CSOs Pollution Control Approach

The current approach in China for controlling CSOs pollution involves various measures: Implementing green initiatives at the source to achieve peak flow reduction, peak flow diversion, and reduction of pollution loads during rainfall events.

Retrofitting rainwater outlets, intercepting sewer networks, modifying overflow points, and constructing rainwater and sewage storage facilities to achieve in-process pollution control. Combining multiple measures, such as on-site treatment at the outlet or retrofitting sewage treatment facilities at wastewater treatment plants, to effectively control CSOs pollution.

In urban construction, residential areas, roads, squares, and green spaces, the use of permeable pavement, green roofs, eco-grass ditches, sunken green spaces, and rain gardens, among other green low-impact development technologies, is adopted. These techniques allow for the infiltration, retention, and recharge of rainfall runoff, reducing the volume of rainwater entering the combined sewer system, thus decreasing overflow occurrences.

By renovating traditional rainwater outlets, including intercepting, diverting, or bypassing, it is possible to purify rainwater pollution, achieve separate collection and treatment of different types of pollutants, and reduce the pollution load from CSOs.

During the rainy season, rainfall runoff enters the combined sewer system with a high volume and fast flow rate, which washes away pollutants that have accumulated in the sewer system during dry weather. As a result, the concentration of pollutants in rainwater during rain events is much higher than the pollution levels in surface runoff. Therefore, regular cleaning and maintenance of the pipes is also one of the effective measures to control pollution from CSOs.

Constructing storage facilities to store combined sewer system wastewater that exceeds the capacity of sewage treatment plants during rainy days; during dry weather, when the sewage treatment plant's capacity returns to normal, the stored wastewater from the storage facilities is then sent to the sewage treatment plant for processing. This helps reduce overflow pollution loads and significantly alleviates river pollution. When rainfall is excessive and the storage tank capacity is limited, making overflow unavoidable, cyclonic separators and rapid filtration devices can be installed after the storage tank or diversion well to quickly treat the overflow rainwater on-site, thus achieving the goal of purifying the overflow wastewater.

During the rainy season, rainwater and wastewater storage systems, including storage tanks and drainage networks, have a certain storage capacity. Sewage treatment plants should also have additional treatment capacity during the rainy season. By fully utilizing the storage capacity and treatment capacity of both systems and dynamically controlling them, it is possible to effectively reduce the discharge of overflow pollution loads from the entire drainage system during the rainy season.

4.5.2 Separate System for Early Rainwater Pollution and Purification Strategies

4.5.2.1 Definition and Hazards

Currently, there is no unified and precise definition of early rainwater. Various researchers both domestically and internationally have proposed corresponding quantitative definitions. Generally, early rainwater refers to the rainfall at the initial stage of a precipitation event and is associated with the initial flushing effect. It can be quantitatively defined based on rainfall volume, where the rainwater within a certain range of initial rainfall is considered early rainwater. Alternatively, it can be defined based on the duration of rainfall, where the rainwater before a certain time point during a precipitation event is regarded as early rainwater. Different intensities of rainfall exhibit significant variations in runoff volume and flushing characteristics within the same time frame. Moreover, the types and concentrations of pollutants carried by early rainwater are also markedly influenced by factors such as land surface characteristics, soil background conditions, and the level of drainage facility completeness.

Due to the early rainfall, a significant amount of pollutants, such as acidic gases, vehicle exhaust, industrial emissions, and other contaminating substances present in the air, dissolve into the rainwater. After falling onto the underlying surfaces, including rooftops, roads, and others, and due to the scouring effect of rainwater on these surfaces and drainage pipes, the initial rainwater contains a high concentration of pollutants. This results in a high level of contamination in the early rainwater, even exceeding the pollution levels found in typical urban domestic wastewater. If this initial rainwater is collected through separate rainwater drainage systems and discharged directly into receiving bodies of water without treatment, it can cause severe pollution to these water bodies.

The water quality of initial rainfall varies significantly in different regions. Apart from factors like regional rainfall characteristics and underlying surface conditions, the incomplete construction of drainage networks in some areas, insufficient separation of rainwater and wastewater, issues such as misconnections, mixing, pipeline sedimentation, and groundwater infiltration can also lead to further deterioration of the water quality in initial rainfall. Additionally, regional population density, level of economic development, and industrial structure can indirectly affect the water quality of initial rainfall. Initial rainfall typically contains organic compounds, nitrogen compounds, heavy metals, and some pathogens, all of which pose threats to the water quality and aquatic ecosystem of receiving water bodies.

The water quality and quantity of initial rainfall are both characterized by instability, primarily due to the uncertainty of rainfall, which leads to significant fluctuations in its generation. In recent years, rapid urbanization in China has resulted in significant changes in urban land surfaces, leading to substantial variations in the water quality of initial rainfall.

4.5.2.2 Research Progress in Urban Initial Rainwater Pollution Control at Home and Abroad

Countries around the world have proposed a series of new concepts and methodologies for urban initial rainwater management, taking into account their respective characteristics. The United States began researching initial rainwater pollution issues as early as the 1970s, introducing the concept of BMPs. In recent years, the United States has developed low-impact development technologies that employ decentralized small-scale measures for source control of rainwater runoff. To address the issue of initial rainwater runoff pollution, the United States primarily focuses on controlling the quantity of water to achieve water quality control for initial rainwater. This is done by constructing a comprehensive rainwater management index system and water quality control volume standards, with water quality as the main indicator. Moreover, many rainwater management guidance documents, such as the "District of Columbia rainwater Management Manual" (January 2020 revised edition), are based on this system. These documents provide detailed regulations for various aspects, including calculations for rainwater retention volume and rainwater treatment volume at the source, definitions and treatment methods for initial rainwater during the rainwater collection process, and post-runoff rainwater treatment measures.

Australia has introduced water-sensitive urban design, and the United Kingdom has proposed sustainable urban drainage systems. In Australia, the guidelines for rainwater collection and reuse emphasize the importance of both the water quality and quantity of initial rainwater. For example, the pollution load index in rainfall runoff focuses on water quality control, while the analysis of runoff volume frequency emphasizes water quantity control. However, these guidelines lack specific provisions regarding the definition of initial rainwater and the characteristics of water quality pollution.

The collection and treatment of initial rainwater in our country are still in the early stages. The "*Outdoor Drainage Design Standard*" GB 50014.2021 includes relevant provisions for controlling initial rainwater in terms of drainage systems, drainage channels, comprehensive rainwater utilization, rainwater outlets, rainwater detention tanks, rainwater infiltration facilities, etc. Regarding the definition of initial rainwater, this standard specifies that when used for runoff pollution control in separate drainage systems, the storage volume of rainwater detention tanks can be calculated based on rainfall and may range from 4 to 8 mm. In addition, based on different factors such as detention purposes and drainage systems, three different methods for calculating the effective volume of rainwater detention tanks are proposed for runoff pollution control in combined drainage systems, runoff pollution control in separate drainage systems, and reducing peak flow in drainage pipes. The first two methods are primarily for water quality control, while the third method mainly focuses on water quantity control but also considers reducing the total pollutant load. Existing detention-related standards, such as the "*Technical Specification for Urban Rainwater Detention Engineering*" GB 51174.2017 and the "*Technical Regulations for*

Urban Runoff Pollution Control and Detention Tanks" CECS 416-2015, provide corresponding volume calculation formulas for rainwater detention tanks. In both cases, the storage volume for separate drainage systems is specified as 4–8 mm of rainfall, consistent with the provisions of GB 50014.2021.

The "sponge city" concept proposed in our country has been applied to rainwater management and runoff pollution control. It metaphorically likens the city to a sponge with the functions of absorbing, retaining, and releasing rainwater. By combining natural and artificial measures, it maximizes the accumulation, infiltration, attenuation, and purification of rainwater in urban areas, thus promoting the utilization of rainwater resources and ecological environmental protection. Sponge cities, through enhanced urban planning, construction, and management, fully utilize ecosystems such as buildings, roads, green spaces, and water bodies to accumulate, infiltrate, attenuate, and purify rainwater, effectively controlling rainfall runoff. This achieves a city development model based on natural accumulation, natural infiltration, and natural purification. The "Technical Guidelines for Sponge City Construction—Low Impact Development Rainwater System Construction (Trial)" require, in terms of initial rainwater control, the use of specific methods or devices to exclude initial rainfall runoff that may have a significant initial scouring effect or higher pollutant concentration. This is done to reduce the difficulty of subsequent rainwater treatment. The guidelines also list common methods and forms of initial rainwater exclusion.

In addition, several local standards in China also address initial rainwater management. Beijing has its local standard, "*Design Standard for Sponge City Construction*" (DB11/T 1743), which provides specific requirements for initial rainwater runoff facilities in terms of reducing rainfall runoff pollution, rainwater retention and discharge, water quality maintenance, and more. It also summarizes the applicability of various sponge city technologies. Shenzhen has its local standard, "*Sponge City Design Atlas*" (DB4403/T 24), which proposes the use of infiltration technology, storage technology, regulation technology, transportation technology, and pollution interception and purification technology to achieve the main functions of infiltration, retention, storage, purification, utilization, and discharge. Regarding pollution interception and purification, this standard provides detailed specifications for initial rainwater runoff discharge volume, the direction of initial rainwater runoff discharge interception, requirements for discharge devices and their settings, grass ditch design, initial rainwater runoff pool design, infiltration pit design, green roof design, and more. Yunnan Province, in its "*Technical Specifications for Initial Rainwater Interception in Plateau Lakes and Urban Rivers*" (DB53/T 950), offers clear calculation methods for the control of initial rainwater runoff pollution and lists various initial rainwater interception technologies. It also specifies the range of initial rainwater runoff pollution concentrations for different urban functional areas.

From the perspective of the focus and implementation methods of rainwater management, both domestic and international regulations or standards for initial rainwater control currently involve the control of rainwater quality. In China, rainwater quality control is mainly achieved by specifying water quality indicators or controlling the quantity of water. Currently, it is primarily implemented through engineering measures, which lack specificity, especially in terms of focusing on initial rainwater control. In contrast, in foreign countries (such as the United States and Australia), initial rainwater control begins with a macroscopic hydrological approach, achieving the control objectives of both water quality and quantity simultaneously. Therefore, domestic regulations or standards should incorporate relevant provisions related to macroscopic hydrological control and non-engineering measures. Due to significant regional differences in China, these provisions should be tailored to local conditions. It may be considered to categorize different types of areas and establish regional standards accordingly. Encouraging local authorities to develop their own standards based on local conditions is also recommended.

From the perspective of objectives and systems of rainwater runoff control, the goals of rainwater runoff control mainly include rainwater utilization, ecological environmental protection, and flood and drought prevention. The United States' rainwater management system integrates these three aspects comprehensively and exhibits a strong systemic approach. However, domestically, there has not yet been established or introduced a rainwater management system, technical specifications, or standards that effectively consider all three aspects. Among the current domestic specifications and standards for rainwater runoff control that have been issued and implemented, there are varying emphases, and specific provisions for initial rainwater runoff control also differ to some extent. This leads to challenges in practical implementation, as the lack of a strong systemic approach makes it difficult to apply these standards effectively. This situation is related to the fact that the responsibilities for regulating these three aspects in China are divided among different authorities, namely urban construction, ecological environment, and water resources, resulting in a somewhat fragmented regulatory landscape.

4.5.2.3 The Train of Thought on Initial Rainwater Pollution Control

Currently, foreign countries place great emphasis on the issue of initial rainwater pollution. Collecting, treating, and utilizing initial rainwater not only helps reduce the pressure on urban water supply and drainage networks but also effectively improves the aquatic ecological environment. This is of significant importance for the sustainable development of urban areas. After years of exploration, the United States and Europe have established a comprehensive approach to rainwater collection, treatment, and management.

Handling initial rainwater in China is still in the exploratory and experimental stage. Summarizing domestic and foreign experiences, the governance of initial rainwater should be approached from the following four aspects:

i. Source Reduction and Treatment of Initial Rainwater:

By reducing the runoff coefficient of the rainwater catchment area, promoting rainwater infiltration into the ground, and reducing surface runoff during heavy rains.

By implementing decentralized source treatment facilities to treat rainwater before it enters the drainage system.

ii. Storage and Regulation of Initial Rainwater:

By constructing rainwater storage reservoirs to collect initial rainwater, which can be treated in wastewater treatment plants after the rainy season, reducing pollution from direct discharge of initial rainwater into rivers.

iii. Management of Initial Rainwater Drainage Networks:

Regularly clearing rainwater pipelines and repairing any damages to ensure the continuous and effective operation of the network, preventing issues like blockages, overflows, and rainwater contamination.

iv. End-of-Pipe Treatment of Initial Rainwater:

Constructing on-site treatment facilities for initial rainwater, enhancing pollutant removal, and reducing the pollution load entering rivers. Upgrading and retrofitting urban wastewater treatment plants with separate drainage systems to enhance initial rainwater treatment during the rainy season. Implementing coordinated operation between rain and wastewater pipelines and treatment plants to effectively treat initial rainwater. Implementing purification treatments for initial rainwater through urban water bodies' artificial wetlands, river/lake buffer zones, and near-natural wetlands.

4.5.3 Key Technologies and Application Cases

Setting up purification treatment facilities at CSOs points and rainwater discharge points in separate systems is an essential step in purifying runoff during the rainwater collection process. These technologies aim to reduce pollution entering rivers by effectively managing and treating overflow and initial rainwater in both combined and separate drainage systems. Controlling overflow pollution and purifying initial rainwater commonly involve key technologies such as: Integrated Technology for Overflow Pollution Control in Intercepted Combined Sewer Systems. Identification and Retrofitting Technology for rainwater Network Mixing in Separate Drainage Systems. Integrated Pollution Reduction Technology for the Source-Flow-Sink System of Combined Sewer Networks. Dual-Peak Control Technology for Water Quality and Quantity in Intercepted Channels. Diagnosis and Retrofitting Technology for rainwater Pollution Mixing in Separate Drainage Systems. Retrofit Technology for Efficient Pollution Interception Systems in Drainage Pump Stations and Collection Pools. Enhanced Coagulation and Cyclone Separation Technology.

4.5.3.1 Integrated Technology for Overflow Pollution Control in Intercepted Combined Sewer Systems

i. Basic Principles

By optimizing the rainwater operation mode and detention tank settings of the combined sewer system using a hydraulic model, the volume of rainwater overflow sewage on rainy days is reduced. An improved SWMM model with coupled bottom sediment flushing is used to rationally determine the interception ratio of the intercepted combined sewer system, in order to maximize pollutant removal efficiency and the environmental benefits of drainage system investment. This is achieved by combining on-site treatment of overflow rainwater to reduce the pollution load from rainwater overflow on rainy days.

ii. Process Flow

This integrated technology includes: In-line storage technology optimized based on hydraulic modeling; detention tank design method based on hydraulic modeling; interception ratio optimization method based on the improved SWMM model and unit investment environmental benefit; and on-site treatment technology for overflow sewage. The process flow of this integrated technology is shown in Fig. 4.23.

i) *Line In-System Storage Technology Optimized Through Hydraulic Modeling and Control*

This technique involves establishing the hydraulic relationship between the generalized Hefei Xinghua drainage system and the rainy-day gravity outflow into the

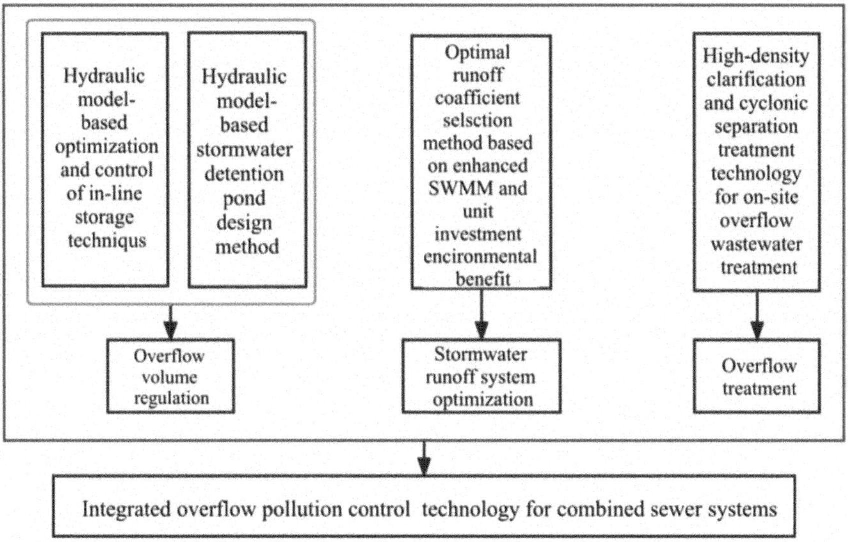

Fig. 4.23 Process flow diagram of overflow pollution control integrated technology for combined sewer systems

Nangang River based on equivalent connecting pipes. It also involves creating a hydraulic model of the pre-renovated Xinghua drainage system to simulate the impact of different rainfall characteristics and renovation methods on the additional pipeline storage capacity of the Xinghua drainage system. A comprehensive system construction plan is formulated to fully utilize the additional pipeline storage capacity of the Xinghua drainage system and an optimized operational plan for the renovated drainage system. Scenario analysis is conducted based on six months of actual operation data from the project and the local rainfall characteristics, as well as the probability of drainage system capacity. The results indicate that the newly optimized and controlled system can reduce overflow volume by 30% compared to the older Xinghua drainage system.

ii) *Design Method for Detention Reservoirs Based on Hydraulic Modeling*

Using a hydraulic model of the drainage network, this study investigated the design parameters and operational methods of detention reservoirs. Long-term continuous simulations were conducted based on the rainfall processes of representative years to analyze the operational efficiency of detention reservoirs with different design and operational parameters under existing interception ratios. The study assessed the impact of detention reservoir volume on the annual reduction rate of runoff, annual frequency of overflow events, and annual frequency of reservoir filling. It also analyzed the efficiency of detention reservoir volume utilization. A reliable method for designing detention reservoirs based on the local drainage network conditions, receiving water body environmental capacity, and rainfall characteristics was established.

iii) *Optimization Method for Interception Ratios Based on Improved SWMM and Environmental Benefit Per Unit Investment*

An improved SWMM model coupled with bottom sediment transport was established, significantly enhancing the accuracy of water quality simulations in the post-rainfall period of combined sewer systems. The relative error in predicting total COD concentration was reduced from 22.70 to 15.15%. Some simulation results are shown in Fig. 4.24.

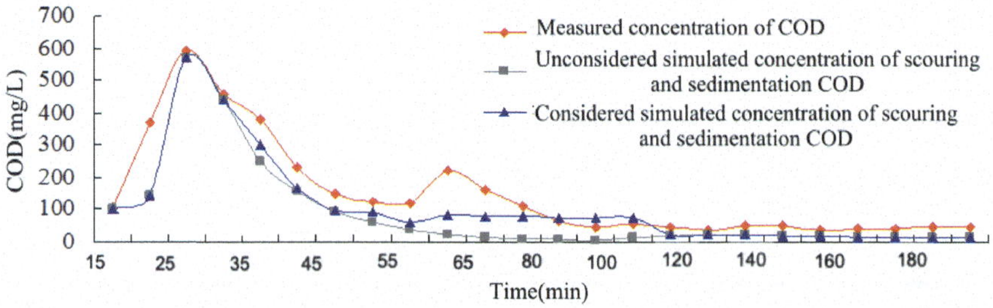

Fig. 4.24 Simulation results of COD (Chemical Oxygen Demand) using the improved SWMM model coupled with bottom sediment transport

An optimization method for interception ratios in combined sewer systems was developed based on the improved SWMM model and environmental benefit per unit investment. Using the improved SWMM model, it predicted the runoff efficiency and pollutant removal rates for specific pollutants throughout the year for different interception ratios in combined sewer systems. Based on the engineering investment required for interceptor pipes and wastewater treatment facilities under different interception ratios, it was determined that an interception ratio of 40 achieved the maximum pollutant removal rate per unit investment in the old city area of Chaohu City.

iv) *On-site Treatment Technology for Overflow Sewage*

This technology utilizes cyclone separation and high-density clarification on-site treatment equipment and has constructed an on-site treatment project for overflow rainwater and sewage with a processing capacity of 500 m^3/h. Among them, the high-density clarification equipment achieved an average removal rate of 80% for SS and 60% for COD, while the cyclone separation facility achieved an average removal rate of 20% for SS and 13% for COD.

iii. Application Examples

In Hefei, a combined rainwater and sewage control and storage project was constructed in the Xinghua drainage system. While ensuring drainage safety, measures such as the transformation of the drainage system, expansion of storage capacity in the pipe network, and optimization of pump station control were implemented. These measures achieved the goals of reducing rainfall overflow by 30% annually and reducing overflow pollutant loads by more than 30%.

In the upstream of the Nanfei River in Hefei, a rapid purification project for overflow sewage with a capacity of 500m^3/h was constructed to treat rainwater and sewage exceeding the drainage capacity. The demonstration project achieved a 30% reduction in pollution loads during rainy days.

4.5.3.2 Rainwater Diversion Drainage System Rainwater Network Mixture Identification and Retrofit Technology

i. Basic Principles

By analyzing the causes of rainwater network mixing, the sources of rainwater network mixing and the volume of mixed water are determined. Based on this, the basic plan for retrofitting the rainwater network mixing is established. For the areas intended for diversion retrofitting, targeted on-site investigations of the mixing are carried out to determine the location and volume of mixing points. The comprehensive decision support platform for retrofitting is used to demonstrate the feasibility of diversion retrofitting and optimize the technical scheme for retrofitting the mixing.

ii. Process Flow

This technology mainly includes: rainwater network mixing pollution cause analysis technology, on-site investigation technology for rainwater network mixing sources, and optimization technology for rainwater network mixing retrofit.

i) *Rainwater Network Mixing Pollution Cause Analysis Technology*

Comprehensive use of the water balance analysis of the combined sewage and rainwater network and the monitoring of water quality characteristic factors to determine the sources of rainwater network mixing and the proportion of mixed water. Firstly, the sewage generation, sewage collection, and discharge under different operating conditions of the dry weather rainwater network in the study area are analyzed for water balance, determining the sources of mixed water (sewage, groundwater, river water, etc.). Secondly, water quality characteristic factor indicators for different types of mixing are monitored, and through pollution mass balance analysis of inflow and outflow, the causes of rainwater network mixing and the volume of mixing water (different types of domestic sewage and industrial wastewater from different industries, etc.) are clarified. Through the analysis of the causes of rainwater network mixing pollution, the basic scheme for mixing retrofit is preliminarily determined (with diversion retrofit as the main focus or end-of-pipe interception as the main focus).

ii) *Rainwater Network Mixing Source Field Investigation Technology*

Investigate the network water volume and water quality characteristic factors of the survey area unit; conduct on-site investigations using a multifunctional rainwater network detection device to determine the location and volume of the main mixing sources; for systems with significant river backflow and network leakage, investigate points of river backflow and points of pipeline damage.

iii) *Rainwater Network Mixing Retrofit Optimization Technology*

Utilizing a decision support system for rainwater network mixing retrofit, the retrofit plans for the rainwater network mixing are optimized. The rainwater network mixing retrofit decision support system integrates hydraulic network models, GIS systems for the network, and real-time network operation monitoring systems. Among these, the network GIS system can visually display information such as the locations of rainwater and sewage mixing points, mixing volumes, depth of buried pipes, and the direction of rainwater and sewage pipelines. The real-time network operation monitoring system continuously monitors the water levels in the rainwater and sewage network dynamically. The hydraulic network model, in conjunction with the network GIS system and the real-time network operation monitoring system, assesses the hydraulic efficiency of the sewage network before and after mixing retrofit. This evaluation determines the feasibility of diversion retrofit and the necessity of locally retrofitting sewage network segments.

iii. Application Example

This technological achievement was applied in the rainwater network mixing retrofit project of the Caohajing Drainage System in Shanghai. The Caohajing Drainage System in Shanghai covers an area of 3.74 km^2 and was constructed in 1986. It was designed as a combined sewer system, but it had issues with rainwater and sewage mixing. To address this problem, the rainwater network mixing pollution analysis technology and on-site investigation technology were employed to determine the types of mixing (including domestic sewage, industrial wastewater, groundwater, and river water) and the mixing volumes in the area. Over 200 mixing points and their respective volumes, more than 40 river water backflow points, and damaged sections of rainwater pipes were identified. By combining the rainwater network mixing retrofit decision support system, the feasibility of retrofitting the mixing points was assessed. Rainwater and sewage separation retrofit was implemented at three major high-flow domestic and industrial mixing points, accounting for approximately 35% of the total mixing volume. For over 200 dispersed sources with lower flow rates, end-of-pipe municipal pumping stations were implemented for dry weather flow diversion (with a diversion capacity of 21,600 m^3/day). Additionally, measures such as sealing and repairing were carried out at over 40 points where river water was backflowing into the rainwater network. After the application of this technology and the implementation of the engineering projects, the discharge of dry weather flow into the river was substantially eliminated in the service area, directly improving the water quality of the central urban backbone rivers in the surrounding area.

4.5.3.3 Integrated Pollution Reduction Technology for Combined Sewer Systems: Source-Flow-Confluence

i. Basic Principles

This technology is an integrated approach composed of various individual techniques, primarily aimed at reducing urban combined sewage pollution. It systematically applies targeted techniques from the source, during transport, to the overflow and confluence of combined sewage, taking into account the runoff process of combined sewage. Through technology integration, it addresses issues such as load matching and functional coupling in the purification chain from the source to the confluence in the combined sewer system, as well as the transport behavior of pollutants within the system and their response to operational parameters of the sewer network. This improves the pollution interception efficiency of the combined sewer network. Through systematic integration and summarization, comprehensive solutions for overflow pollution in urban combined sewer systems are proposed from the perspectives of planning, management, and engineering.

ii. Process Flow

This technology consists of three main components: source reduction, process control, and end-of-pipe treatment.

i) *Source Reduction*

The primary sources of overflow pollution include:

(1) Domestic sewage discharged from residential areas.
(2) Industrial wastewater from industrial zones that meets discharge standards into urban sewers.
(3) Runoff pollution generated during rainfall, including solid waste and household garbage disposed of improperly.
(4) Surface runoff from roads, yards, and other areas during rainfall.

Source reduction aims to minimize the generation of overflow pollutants by reducing inputs from different pollution sources. This can be achieved through various measures, including: Effective control of domestic sewage, such as reducing water usage or implementing water recycling. Reducing the open dumping of solid waste and household garbage or ensuring proper disposal. Decreasing impervious surfaces or enhancing permeable surfaces. Enhancing rainfall runoff detention and storage.

ii) *Process Control*

Building upon source reduction, for pollutants that cannot be reduced at the source (such as the volume of domestic sewage or industrial wastewater), or in locations where source reduction is not feasible, techniques like detention and quality-based diversion are employed. These techniques obstruct the transport of pollutants or extend their migration pathways, thus minimizing the volume of pollutant migration and dispersion. Quality-based diversion involves setting different diversion factors for regions with varying pollution levels to maximize the diversion of pollutants. For some overflow pollutants that are difficult to physically intercept, it may be necessary to treat and purify them in-situ. In overflow pollutants, COD, N, and P content is usually relatively high and exhibits characteristics such as solubility in water, easy migration, and diverse forms. Special control measures are required for their deep treatment and purification. These measures typically fall into three categories: reduction of sediment within pipelines, integrated detention and treatment purification systems, and physicochemical combination rapid in-situ treatment measures. Through the combined physical, chemical, and biological processes in these technologies, the goal is to maximize the removal of overflow pollutants like nitrogen and phosphorus, which are challenging to reduce, from the combined sewer network system.

iii) *End-of-Pipe Treatment*

This component comprises two main aspects: The deep purification system at the end of the sewer network, which includes the construction of biological, ecological, and physicochemical purification measures at overflow points. This further controls the pollutants discharged into receiving water bodies. After maximizing the removal of overflow pollutants, a reevaluation and restoration of the entire receiving water body system are necessary to achieve the healthy and sustainable development of the combined sewer network area. Key activities include the restoration of the aquatic ecology within the receiving water body system, making it a primary producer in the new ecosystem, a habitat builder for important organisms, a driver of nutrient uptake and transformation, and a promoter of suspended particle settling. Rebuilding the basic structure of the ecosystem, including primary producers, consumers, and decomposers, is essential to establish a food web with recycling functions. Building upon this basic structure, ecological engineering measures are employed to restore and enhance biodiversity, stabilize the system, and ultimately improve the self-repair and self-purification abilities of the receiving water body system, transitioning it from a damaged state to a healthy and stable state.

iii. Application Examples

The technological achievements have been successfully applied in the transformation projects of combined sewer systems in the old urban area of Zhenjiang City. Typical cases include the "Zhenjiang City Ancient Canal Middle Section Comprehensive Improvement Project" and the "Zhenjiang City Old Town Yizheng Road Residential Area Combined Sewer Network Transformation Project." Among them, the Ancient Canal transformation project covers a design watershed area of 7.8 km^2 with nearly 38 km of drainage pipelines. The Yizheng Road residential area involves the transformation of 1,800 m of drainage pipelines and the modification of 15 septic tanks. Using the technological achievements, comprehensive pollution reduction systems were constructed for the Ancient Canal along the river (specifically including demonstration projects for flow-controlled combined sewer drainage network overflow pollution control and in-situ rapid treatment of sewage pump station overflow pollution) and a staggered diversion system for residential area sewage. The construction of these technology application projects can annually reduce COD entering the river by 320 tons and ammonia nitrogen by nearly 22 tons, delivering significant social and environmental benefits.

4.5.3.4 Technology for Water Quality and Quantity Control in Pollution Interception Channels with Dual Peak Shifting

i. Basic Principles

Utilizing the large capacity of pollution interception channels and the time difference between peak concentrations and peak flows, optimization of operation and control is carried out to achieve efficient interception of urban rainfall runoff, peak shifting, peak reduction, and efficient interception of rainfall runoff pollution.

ii. Process Flow

The technological process flow is shown in Fig. 4.25.

Selecting a typical pollutant as the control indicator. When stricter control requirements for nutrient indicators are needed, TN can be used as the control indicator; when organic pollution is severe, SS and COD can be used as control indicators.

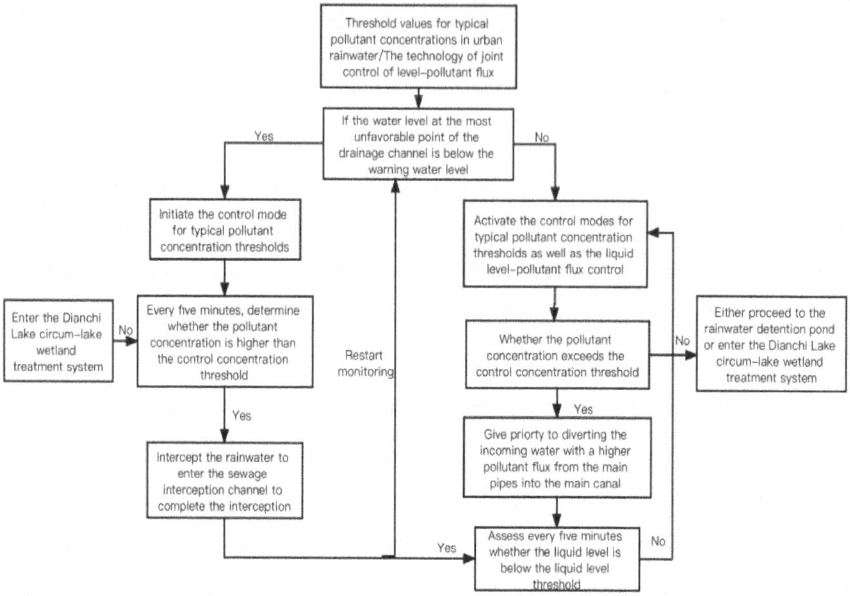

Fig. 4.25 Process flow diagram of dual peak shifting and storage control technology for water quality and quantity in pollution interception channels

Through model simulation and optimization, set the concentration threshold for typical pollutants (SS = 120 mg/L or TN = 5 mg/L). At the point where the rainwater main pipe connects to the pollution interception channel, install a swirl-type flow control/diversion switching well (hereinafter referred to as the "switching well"). Inside the switching well, install flow and typical pollutant concentration online monitoring instruments (the monitoring duration interval should not exceed 5 min), as well as electric valves and sedimentation facilities. By controlling the opening and closing of the electric valves in the switching well, the pollution interception channel can be used to control the diversion and discharge of urban rainfall runoff. Depending on different control modes and data from online monitoring instruments, the electric valves in the switching well are operated. The sedimentation facilities installed in the switching well can reduce the pollution of receiving water bodies by rainwater discharge.

Install a liquid level sensor at the most critical point of the pollution interception channel (where overflow is most likely to occur) and set a warning water level (4.5 m, which is the channel's top height). When the water level at the most critical point of the pollution interception channel falls below the preset warning water level, enter the typical pollutant concentration threshold control mode. When the concentration of a certain typical pollutant in the rainwater main pipe exceeds the control concentration threshold, the electric valve opens, and the rainwater flows through the switching well into the rainwater channel of the pollution interception channel, achieving flow interception. When the concentration of a certain typical pollutant in the rainwater main pipe is below the control concentration threshold, the electric valve is closed, the water level in the switching well rises, and the rainwater overflows into the constructed wetland after undergoing swirl sedimentation treatment in the switching well, then discharged into the receiving water body.

When the water level at the most critical point of the pollution interception channel exceeds the preset warning water level, the typical pollutant concentration threshold control mode is executed simultaneously with the liquid level-pollutant flux control mode (i.e., the typical pollutant concentration threshold/liquid level-pollutant flux joint control mode). Priority is given to diverting rainwater from the main pipe with the highest pollutant flux by closing the electric valve in the switching well connected to the rainwater main pipe with the lowest pollutant flux, thereby diverting its flow. After one control operation, if the water level at the most critical point of the pollution interception channel remains above the preset warning water level, it is necessary to continue with the typical pollutant concentration threshold/liquid level-pollutant flux joint control mode, recalculate the pollutant flux of the rainwater main pipes connected to the switching wells with electric gates open, and continue to divert rainwater from the main pipe with the lowest pollutant flux. If the water level at the most critical point of the pollution interception channel drops below the preset warning water level, the typical pollutant concentration threshold/liquid level-pollutant flux joint control mode ends and switches to the typical pollutant concentration threshold control mode.

iii. Application Example

This technology has been applied in the pollution interception channel around Dianchi Lake. The fourth control room of the pollution interception channel is located on the east bank of Dianchi Lake, serving the Baoxiang River area in the eastern bank provincial investment section, with an area of approximately 20.025 km^2, including 8.012 km^2 of urban areas. To fully utilize the pollution control and purification efficiency of the pollution interception system around the lake, conductivity probes and control cabinets were installed, and a self-control system program was set up to monitor conductivity in real-time. This enables the implementation of an efficient rainwater interception method based on water quality changes, allowing for rapid identification and efficient interception of rainwater. This technology provides technical support for pollution control in the Dianchi Lake Basin and the improvement of water environmental quality. The arrangement of on-site data collection instruments is shown in Fig. 4.26.

4.5.3.5 Diagnosis and Retrofitting Technology for Rainwater and Sewage Mixed Drainage Systems

i. Basic Principles

Rainwater and sewage mixed drainage systems are like a black-box system, and accurately determining the sources of mixed pollution and their respective quantities

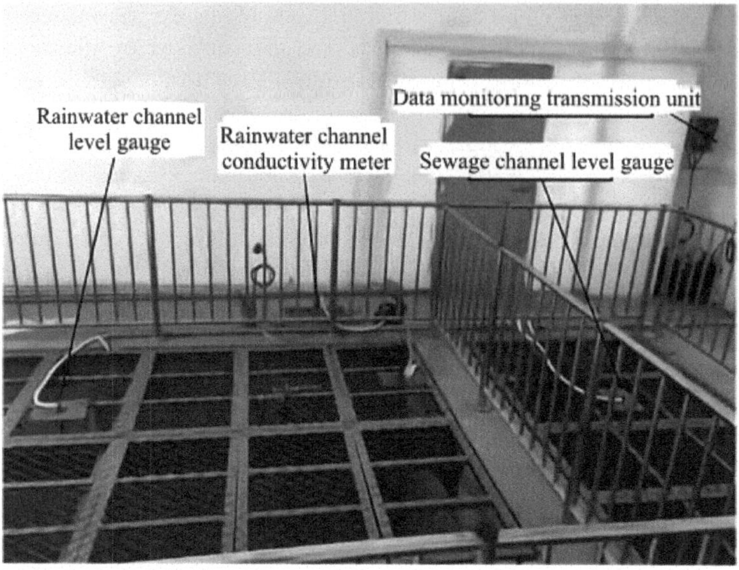

Fig. 4.26 Layout of data collection instruments in the fourth control room of the Dianchi Lake pollution interception channel

is crucial for the success of the retrofitting process. By analyzing node flow rates and water quality characteristic factors, a node-based chemical mass balance can be established, allowing for the reverse analysis of mixed pollution types. Based on this, utilizing Lagrangian tracking of point source water flow movement, dynamic waveform inversion of node outflow rates can be performed, thereby tracing back to the origins of mixed pollution. This provides quantitative predictive information for on-site investigations and retrofits, enabling a quantitative assessment of the results of the mixed pollution investigation.

This technology is an integrated system, with its core technologies including mixed pollution source diagnosis based on water quality characteristic factors, mixed pollution source inversion based on Lagrangian water flow path tracking, economic analysis mathematical models for mixed pollution retrofitting, and a mixed pollution retrofitting decision support system that integrates mixed pollution diagnosis with network modeling.

ii. Process Flow

The process flow includes the following steps: Mixed pollution quantitative prediction and tracing—On-site investigation of mixed pollution—Evaluation of mixed pollution investigation results—Economic analysis of mixed pollution retrofitting technology—Mixed pollution retrofitting decision information platform.

(1) Collect basic information about the study area, including information on pollution sources and total discharge/emission of rainwater and sewage network,

conduct regional water balance analysis to determine the overall sources of mixed pollution (sewage, groundwater, external inflow, etc.).

(2) Divide the drainage system into grid nodes, carry out monitoring of node flow rates and water quality characteristic factors, analyze the types of mixing and the proportions of mixed water in pipeline network nodes, and identify key mixing areas.
(3) Utilize Lagrangian water flow tracking and reverse optimization algorithms to reverse trace and locate the sources of mixed pollution and water flow in various pipeline segments of the network.
(4) Conduct on-site investigations of mixing points, compare the results of on-site investigations with the quantitative predictions and trace-back results from steps 1–3, and assess the reliability of on-site investigation results.
(5) Based on the mixed pollution diagnosis and on-site investigation results, conduct a technical and economic analysis of mixed pollution retrofitting options to determine the optimized retrofitting plan.
(6) Establish a decision information platform for network retrofitting based on mixed pollution diagnosis, network GIS system, and network water quantity and quality models to evaluate network retrofitting options.

iii. Application Examples

This technology was applied in the drainage systems of the Shipyard Street and Lujiazui Street in Pudong New Area, Shanghai (serving an area of 1.58 and 3.08 km^2, respectively). The project was designed as a diversion drainage system, with underground rainwater pump stations. To ensure the safe operation of the rainwater pump stations, when the water level in the rainwater pipe network reaches 1.0 m or above during dry weather, the rainwater pumps are activated to discharge rainwater forcefully, mixing it with sewage and infiltrated groundwater before directly entering the Huangpu River.

In the application area of this project, the quantitative prediction and tracing method for mixing were used to monitor the water quantity and water quality characteristics of grid nodes in the rainwater main pipe system. This helped identify key areas where groundwater infiltration and domestic sewage mixing occurred. Through tracing and location techniques, the main residential areas where domestic sewage was predominantly mixed within these key areas were determined, including areas upstream of Lujiazui such as Laosan Village and Laosan Wucun. In response to this, rain and sewage diversion retrofitting was implemented in the key mixing source areas within residential areas.

Currently, the first and second-phase retrofitting projects within the scope of the project application have been completed, achieving a reduction of over 20% in dry weather mixed pollution discharge.

4.5.3.6 High-Efficiency Pollution Interception System Retrofit Technology for Drainage Pump Stations

i. Basic Principles

This technology enhances the pollutant retention capacity of drainage pump stations by adding controllable sewage interception tanks, reducing the risk of rainwater overflow pollution. It can be applied to the design of new drainage pump stations and the upgrade of existing ones.

ii. Process Flow

By using walls equipped with self-control gates to separate sewage collection tanks from rainwater collection tanks, controllable sewage interception tanks and drainage pump station collection tanks are added to the drainage pump station. These tanks can be connected under specific conditions to allow the transfer of wastewater from sewage collection tanks and early-stage rainwater. Self-control gates automatically separate the drainage pump station collection tank based on the operating conditions and status of the drainage system. Various measures can be used for their implementation, including but not limited to vertically opening and closing electric or pneumatic gates, or gates that open and close with vertical axial rotation, or gates with horizontal axial rotation. The bottom elevation of the self-control gate on the partition wall is set higher than or equal to the water stop level of the rainwater pump, and the outlet of the sewage interception pump is connected to the external sewage pipeline network to transport dry weather sewage and initial rainwater to the sewage treatment plant. The outlet of the rainwater pump is connected to the rainwater pipe for discharge into the receiving water body.

iii. Application Examples

This technology was applied in the construction project of the Drainage System at Health Road Pump Station in Chaohu City. By incorporating controllable sewage interception tanks into the design of the drainage pump station's dry weather and early-stage rainwater pollution interception system, the effective pollutant load retention capacity of the drainage pump station was improved, reducing the risk of rainwater overflow pollution. The efficiency of intercepting suspended impurities during dry weather increased by over 20%, indirectly reducing the pollution load discharged into the river during rainy days.

References

1. Gao, T., Yu, Y., & Zuo, G. (2006). On administration and maintenance of urban road gully. *Shanxi Architecture, 32*(03), 184–185. (in Chinese)
2. Yan, Q. (2007). Management and maintenance of urban stormwater outlets. *Tianjin Construction Science and Technology, S1*, 277–278. (in Chinese)

3. Pan, G. (2008). *Study on pollution load and control measures for different drainage system*. Beijing University of Civil Engineering and Architecture. (in Chinese)
4. Weng, H. (2021). Construction of initial rainwater collection and control system in industrial enterprises. *Modern Industrial Economy and Informationization, 11*(10), 93–95. (in Chinese).
5. Liu, G. (2009). Brief introduction of the rain water recovery and reuse system of the Kaide Shijie residential project. *Zhejiang Construction, 26*(03), 72–73. (in Chinese).
6. Zhou, J. (2005). *Research on disinfection and black odor removal technology for stormwater pump station direct discharge channels*. Tongji University. (in Chinese)
7. Feng, G., Liu, S., & Hu, Y. (2002). Exploration of the implementation of diversion system for drainage in Shenzhen city. *China Water & Wastewater, 18*(10), 24–26. (in Chinese)
8. Che, W., & Tang, L. (2012). Study on the renovation of China's urban combined sewage systems and strategies for controlling overflow pollution. *Water & Wastewater Engineering, 48*(03), 1–5. (in Chinese).
9. Li, W. (2020). Analysis of the advantages and disadvantages of non-excavation repair and excavation repair. *China Construction Metal Structure, 20*(12), 132–133. (in Chinese)
10. Du, X. (2018). The project of combination of trenchless technology and trench technology to repair municipal drainage pipeline. *China Resources Comprehensive Utilization, 36*(06), 171–173+176. (in Chinese)
11. https://www.sohu.com/a/290655157_99921552.
12. Xu, Y. (2010). Influencing factors and pollution control measures on combined sewer overflows. *Shanxi Architecture, 36*(34), 170–172. (in Chinese)
13. Tang, L., Che, W., Zhao, Y., et al. (2012). Study on the systematic decision-making of combined sewer overflow pollution control. *Water and Wastewater Engineering, 48*(07), 28–34. (in Chinese).
14. Scalise, C., & Fitzpatrick, K. (2012). Deep tunnel design and construction. In *Structures Congress* (pp. 1485–1495). Chicago.
15. Yamamoto, S. (2002). Mechanism behind Manole cover ejection phenomenon and its prevention measures. *Global Solutions for Urban Drainage, 2002*, 1–16.
16. Drainage services department design construction of Hong Kong west drainage tunnel (2013)
17. Bobylev, N. (2007). Vulnerability analysis of critical underground infrastructure. In *Managing critical infrastructure risks* (pp. 445–469). Dordrecht: Springer.
18. Reis, B.K., & Espey, W.H. (2008) Waller creek tunnel project. In *World Environmental and Water Resources Congress* (pp. 1–8). Austin, Texas.
19. MWRA (2009) CSO control in south Boston.
20. Koo, D. H., Jung, J. K., & Lee W. (2013). Sustainability applications for storm drainage systems minimizing adverse impacts of global climate change. *Better Pipeline Infrastructure for a Better Life 2012*, 36–47.
21. Water Environment Federation. (2011). *Prevention and control of sewer system overflows* (p. 2011). Alexandria: WEF Press.
22. Tang, L. (2013). *Research on technology and strategy of combined sewer system retrofits and overflow pollution control*. Beijing University of Civil Engineering and Architecture. (in Chinese)
23. Yang, X., Che, W., Li, J., et al. (2008). Control and management of combined sewer overflow pollution at home and abroad. *China Water & Wastewater, 24*(16), 7–11. (in Chinese)
24. Che, w., Wu, L., Jun, Q., Chen, H.P, et al. (2005). Strategic thinking on drainage systems in urban planning and construction. *Journal of Kunming University of Science and Technology, 30*(3), 72–76 (in Chinese).
25. Brombach, H. (1989). Combined-sewer-overflow control in West Germany—History, practice and experience. *Design of Urban Runoff Quality Controls*, 359–374.
26. Wang, J., Hu, Y., Zhang, C., et al. (2018). Some perspectives on combined sewer system and its overflow control in China. *Environmental Protection, 46*(17), 14.19. (in Chinese)
27. Xu, G., Chen, C., & Zhang, H. Y. (2006). Study on the influence of controlling pollution of the overflow from the initial rainwater storage tanks to Suzhou creek. *Advances in Water Science, 17*(05), 705–708. (in Chinese)

Chapter 5
Backend Treatment Technology for Urban Rainfall Runoff Pollution

The backend control technology for urban rainfall runoff pollution refers to the pollution control technology adopted before rainfall runoff flows out of the pipeline system or directly into the receiving water body without passing through the pipeline. The pollutants removed include nutrients such as nitrogen and phosphorus, organic pollutants, microorganisms, etc., which are the final measures for controlling urban rainfall runoff pollution. The common backend governance technologies can be roughly divided into two categories. One is physical and chemical treatment technology, and its pollution control concept is to implement on-site treatment of rainwater and sewage at the rainwater outlet. This type of technology has advantages such as fast and efficient, small footprint, short construction period, and convenient maintenance and management; The second type is bioecological treatment technology, whose pollution control concept is to use near natural biological and ecological purification principles for rainwater, sewage, and directly discharged rainwater before entering the receiving water body, using the combined action of microorganisms, substrates, plants, etc. for treatment. This type of technology has advantages such as low treatment cost, long service life, and convenient operation and maintenance.

5.1 Physical and Chemical Technology

The physical and chemical technologies for the back-end treatment of urban rainfall runoff pollution mainly include methods such as grid, cyclone separation, coagulation sedimentation, filtration, and disinfection.

D. Sun and F. Qi, *Urban Rainfall Runoff Pollution Control Technology and Application*,
https://doi.org/10.1007/978-981-96-4174-1_5

5.1.1 Grille Technology and Devices

5.1.1.1 Principles of Grille Technology

Grille is an economical and effective method for removing floating and solid materials, usually composed of parallel steel bars, wire mesh, grating bars, or perforated plates. The principle of removing solid substances is to directly filter out solid substances with larger spacing than the grid, or to intercept substances with smaller volume by using solid substances that have already accumulated on the grid. Grille is usually divided into two types: coarse grille and fine grille. The spacing between the bars of coarse grille is usually 4–8 cm, and the spacing between the bars of fine grille is usually 0.3–1.3 cm.

5.1.1.2 Grille Device Design Parameters

The load of the grid in the control of overflow pollution in the combined flow pipeline network is usually 3.70×10^{-9}–8.23×10^{-8} m^3/L, and the corresponding ratio of peak flow to hourly flow can be 2:1–20:1. The design parameters for different types of grilles can refer to Tables 5.1 and 5.2 [1].

After the water flows through the grid, there will be a certain head loss, which can be estimated by the following equation:

$$h_L = (1/0.7) \times \left[\left(V^2 - v^2\right)\right]/2g \tag{5.1}$$

where, h_L—Head loss, m;

Table 5.1 Design parameters for fixed grille

Hydraulic load (L/m/s)	20.7–37.3
Grid inclination angle (°, angle from vertical direction)	35
Slot distance (μ M)	250–1600

Table 5.2 Design parameters for rotating grille

Parameter	Rotating grille
Grid spacing (μm)	74–167 (recommendation 105)
Rotation speed range (r/min)	30–65 (recommendation 55)
Peripheral velocity (cm/s)	4.3–4.9
Water flow density ($m^3/h/m^2$, submerged grid)	170–366
Hydraulic efficiency (%, inflow)	75–90
Pressure (kg/cm^2)	3.5

0.7—Empirical emission coefficient of turbulence loss;

V—Flow velocity through the opening of the grating, m/s;

v—The flow velocity of the upstream channel, m/s;

g—Gravitational acceleration, m/s^2.

5.1.1.3 Scope of Application and Advantages and Disadvantages of Grilles

i. Applicable scope

Grilles are suitable for intercepting solid substances with a diameter larger than the opening of the grille.

ii. Advantages and disadvantages

Advantages: Easy to clean slag, simple structure, and convenient maintenance.

Disadvantages: The effect of removing fine particles by the grid is relatively poor; Particles and larger impurities are prone to entanglement with the rake, shortening the service life and lifespan of the facility.

5.1.1.4 Operation and Maintenance Requirements

When the grid is blocked due to the initial rainwater erosion effect, it should be cleaned in a timely manner to ensure that its head loss is minimized; Due to the intermittent discharge characteristics of overflow rainwater and initial rainwater, the sprinkler system of the grid should work regularly to prevent solid substances from adhering to the grid and increasing head loss.

5.1.1.5 Technical Processing Effect

The removal rate of SS by the grid is between 25 and 90%, which is closely related to its design size.

5.1.1.6 Application Examples

The initial rainwater in the MaShaXi diversion drainage area of Fuzhou City contributes more than 40% to the water pollution of MaShaXi. The nearby sewage treatment plant did not consider treating the initial rainwater, and reconstruction would require a large amount of funds. In order to reduce the initial rainwater pollution load, a regulating and storage tank consisting of two chain plate grilles was built

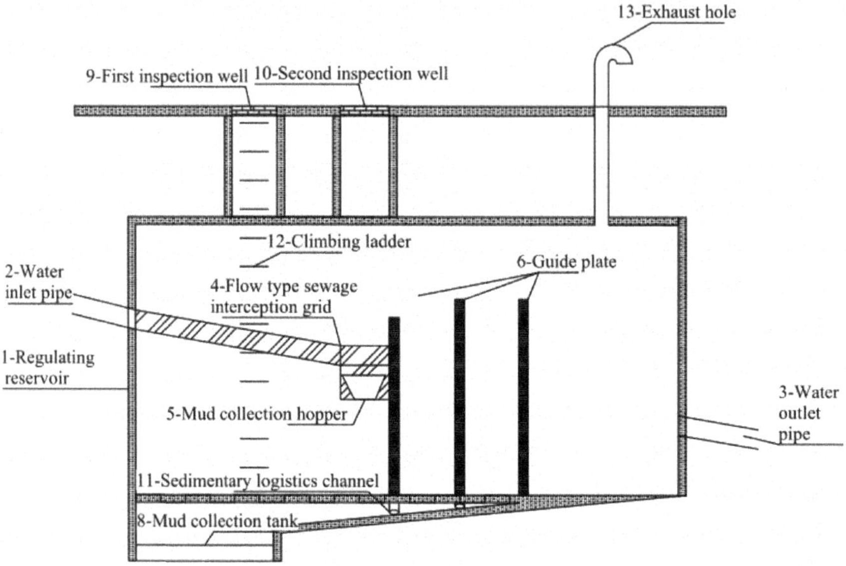

Fig. 5.1 Schematic diagram of rainwater runoff control device containing flow type sewage interception grid network [3]

at the Banyan Tree Park downstream of the river. The initial rainwater was treated and used as a water source for river replenishment, with a maximum replenishment amount of up to 5,000 m^3 each time [2].

The rainwater from a certain city's road was directly discharged into the receiving water body, and there was also a small amount of mixed sewage mixed in, which caused significant pollution to the receiving water body. At present, the use of filter type grid to intercept pollution can effectively reduce the pollution load of rainfall runoff, as shown in Figs. 5.1 and 5.2.

Rainwater and sewage are collected and stored in the space below the overflow type interception grid. Floating debris and large particle sediment in the rainwater and sewage are intercepted, and the intercepted sediment slides into the mud collection bucket under the action of water flow erosion. It is regularly lifted to the surface for disposal through the second inspection well. Rainwater and sewage flow through each compartment unit in sequence, and particles in the rainwater and sewage settle at the bottom of the regulating tank through sedimentation. Sediments at the bottom of the regulating tank flow towards the inlet sludge collection tank along the slope of the tank bottom. Sediments in the sludge collection tank are regularly lifted to the ground for disposal through the first inspection well.

The use of a drop over flow sewage interception grid to treat road rainwater resulted in SS and COD removal rates of 65.4%–89.4% and 55.2%–83.6%, respectively, greatly reducing the inflow load of SS and COD into the river.

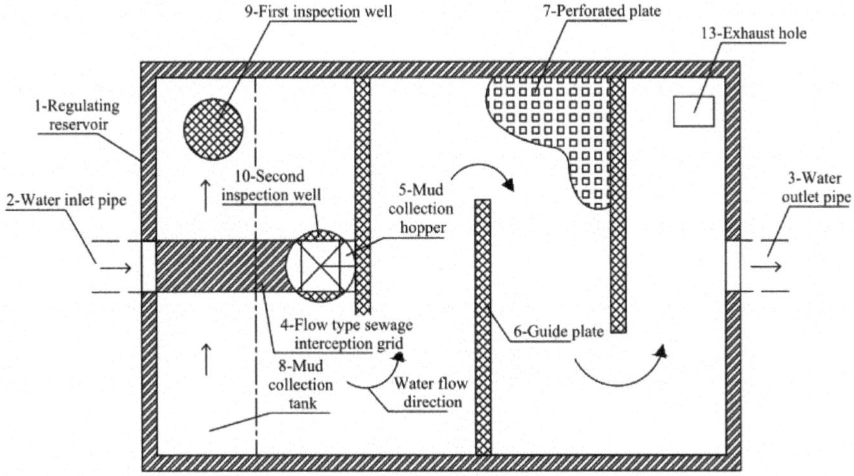

Fig. 5.2 Top view of rainwater runoff control device containing flow type sewage interception grid network [3]

5.1.2 Cyclone Separation Technology and Equipment

5.1.2.1 Principle of Cyclone Separation Technology

The basic principle of cyclone separation technology is that under the guidance of weir plates, rainwater and sewage enter the cyclone separator in a tangential direction, undergo initial interception and energy dissipation by inclined or straight weirs, then settle along the wellbore cyclone, and pass through the grid area for secondary interception and energy dissipation. They flip under the weir and flow out of the tank with the outlet pipe. The two energy dissipation actions accelerate the sedimentation of sand particles, thereby improving the sand removal effect. The energy dissipation grid can also effectively prevent the secondary suspension of settled particles. The three-dimensional structure of the simple cyclone separation device and the cross-section of the energy dissipation grid are shown in Figs. 5.3 and 5.4.

5.1.2.2 Structural Characteristics and Design Parameters of the Device

i. Structural characteristics

The design of the cyclone separation device is a cylindrical pool body, which is easy to install in the rainwater outlet or rainwater well and can withstand the pressure of the surrounding soil; Ensure that the inlet and outlet pipes are on the same horizontal line, in order to facilitate the construction and renovation of existing pipe top flush inlet and outlet pipes; A semi full flow weir is installed in the cylinder to form a

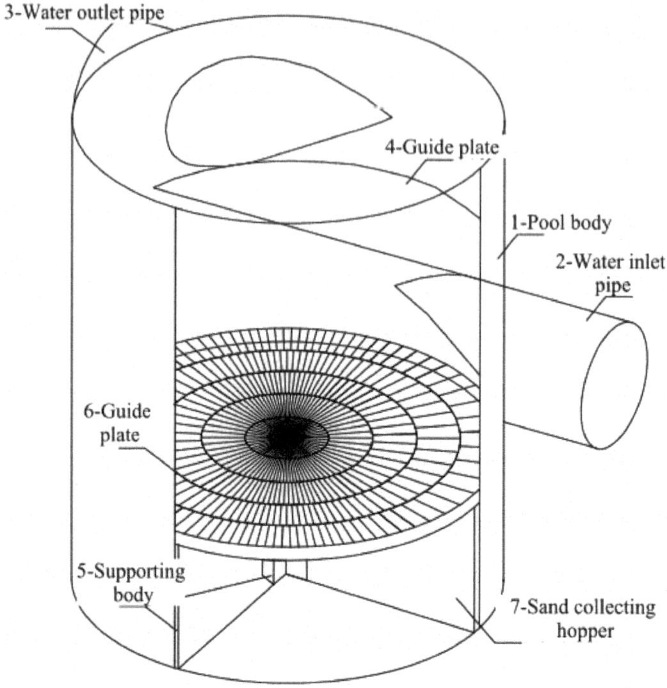

Fig. 5.3 Schematic diagram of the three-dimensional structure of a simple cyclone separation device [4]

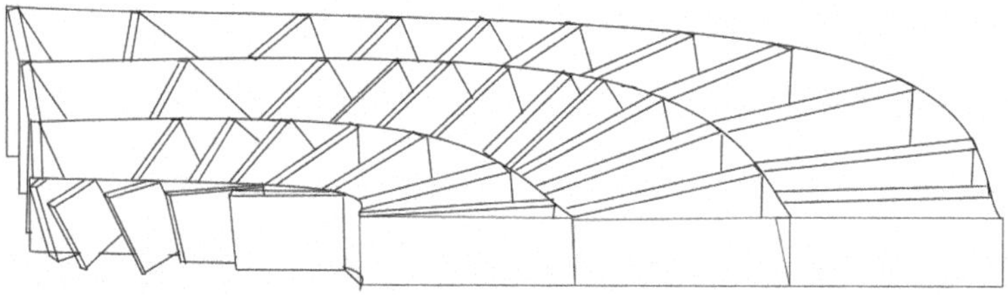

Fig. 5.4 Cross section view of energy dissipation grid [4]

barrier when the water flow is small, avoiding short flow and increasing the retention time of the water flow to achieve effective sand removal. When the flow is large, it is convenient for the water flow to pass through the weir to form short flow and effectively discharge floodwater; Set up energy dissipation grilles to settle more sand particles and alleviate the problem of already settled sand particles being carried out again by the water flow of the outlet pipe.

Simplified pool bodies should use inclined or straight weirs instead of diversion cylinder weirs, with the outlet pipe directly connected to the pool type cylinder, and the difficult to construct cone shaped surface in the settlement area should be removed.

When designing the pond, it should be ensured that the overflow area of the pond in front of the weir is not less than the area of the inlet pipe, and the overflow area from the weir to the upper part of the grid is not less than the area of the inlet pipe to prevent the water flow rate from increasing. Based on this standard, the size range of the inclined weir or straight weir and the distance from the weir to the upper part of the grid can be defined.

During construction, a sand lifting bucket is installed at the bottom, with a lightweight integrated design similar to the grid, which facilitates regular cleaning of the bottom sand and maintenance of the pool body. The use of a simpler structural design reduces the high construction cost of using a cyclone separation device for rainwater pipelines.

ii. Design parameters

The material used for the inner chamber of a cyclone separator is usually cement, and the inner wall may not be strictly constructed as a circular shape. Prefabricated steel structures can be considered for stronger and easier maintenance; Flat plates can also be used, but for drainage purposes, the minimum slope from the wall to the center is 2%. A maintenance hole with a diameter of 60–80 cm should be installed on the cyclone column to facilitate access for maintenance in case of blockage; A side door should be installed next to it to regularly remove floating debris, and the size of the door depends on the size of the water inlet and the size of the floating debris; For safety and aesthetics, a top cover can be installed for the cyclone separator.

When the separation particle size is coarse and the processing water volume is large, a large diameter cyclone separator is generally selected; On the contrary, when the separation particle size is small, choose a small diameter cyclone separator. After determining the diameter (D) of the cyclone separator, other structural parameters can be determined according to the following general requirements, including: inlet pipe diameter (0.15–0.25) D, overflow pipe diameter (0.20–0.30) D, overflow pipe insertion depth (0.50–0.80) D, bottom flow pipe diameter (0.07–0.10) D, and cylindrical section length (0.70–2.00) D. Generally speaking, for the treatment of fine-grained influent, a cone angle of 10° to 15° is used; Coarse grained inlet water can use a cone angle of 20° to 45°.

The inlet of the cyclone separator should be set in the middle of the cyclone separator as much as possible to accelerate the sedimentation of solid substances. The water inlet should be directly connected to the inner chamber of the cyclone separator to ensure that the water flow can enter from the tangent direction. The size of the overflow port can be adjusted according to the size of the gate on the drainage pipeline.

It is recommended to have an automatic cleaning device for the cyclone separator. The cleaning water usually uses the effluent after cyclone separation, and can also be treated with cleaner water sources and receiving water bodies with better water quality through cyclone separation. When using rainwater as the water source, it is necessary to set up a reservoir near the cyclone separator to store rainwater for cleaning purposes; If receiving water is used as the water source, channels and pumps need to be installed.

Under peak flow conditions during the dry season, it should be possible to ensure that rainwater and sewage directly enter the receiving water body through the cyclone separator, and the corresponding diameter of the bottom flow pipe discharge outlet should not be less than 20 mm, preferably between 25–30 mm. For cyclone separators used online, the maximum allowable flow rate should not be less than 4 times the design flow rate; For cyclone separators used offline, the maximum allowable flow rate is generally equal to the design flow rate.

5.1.2.3 Scope of Technical Application and Advantages and Disadvantages

i. Applicable scope

The cyclone separation device can be used for combined and separated pipelines, inspection wells, rainwater outlets, and overflow outlets to achieve the separation of suspended solids and sand particles in rainwater runoff. It can be installed in the rainwater inlet or rainwater well of municipal pipeline branches, or in the interception well at the main pipe.

ii. Advantages and disadvantages

Advantages: The cyclone separation device has a high sand removal rate, which can effectively separate solid and liquid from rainwater and wastewater, and remove SS and COD from rainwater and wastewater; At the same time, it also has advantages such as simple structure, small footprint, high processing capacity per unit volume, strong impact load resistance, easy design, installation and regulation, and low cost.

Disadvantages: The main issue is severe wear and tear on the equipment, especially at the inlet and outlet of the underflow pipe. To avoid equipment wear and prolong its service life, high-strength wear-resistant materials should be used at the inlet and outlet of the bottom flow pipe.

5.1.2.4 Operation and Maintenance Requirements

After each rainfall, the cyclone separation device should be checked. Use automatic flushing or manual cleaning to remove suspended and large particulate matter accumulated in the swirl pool.

5.1.2.5 Technical Processing Effect

When the cyclone separation device controls overflow pollution, its lower limit of particle size is 100–200 μm (corresponding to a settling speed of 3.6 m/h), the SS removal rate can exceed 80%; When controlling rainwater runoff pollution, the lower limit of particle size is 50 μm. The SS removal rate is about 60%; But when the particle size is less than 50 μ When m is reached, the removal rate will be greatly reduced.

5.1.2.6 Application Examples

The cyclone separation technology has been demonstrated and applied in the rapid treatment project of bamboo forest CSOs in Changzhou. The Zhulin area was originally a combined drainage area, with service areas mainly consisting of residential areas, public buildings, and commercial areas, covering an area of 1.0 km^2. After the transformation of rainwater and sewage diversion, most of the sewage is connected to the sewage pipeline, and the rainwater enters the Guanhe River through the original drainage system. A interception pump station is set up at the end of the rainwater for sewage that is not completely separated from rainwater. On sunny days, the sewage is lifted into the nearby sewage pipe network through the interception pump station. The current problem with the drainage system is that the capacity of the interception pump station is small, the interception ratio is small on rainy days, and a large amount of rainwater overflows into the Guanhe River, polluting the water body.

To this end, an initial rainwater interception pipe is established for rainwater and sewage that exceeds the capacity of the interception pump station on rainy days, and it is treated through a "self-cleaning overflow grille + cyclone separation" process before being discharged into the Guanhe River, thereby reducing pollution to the Guanhe River. The project was completed and put into use in May 2018, and can intercept rainfall of no less than 5.0 mm. Compared with before the construction of the project, it reduces COD by 88.4 t/a, accounting for 29.1% of the annual runoff pollution COD load.

5.1.3 Coagulation and Sedimentation Technology and Equipment

5.1.3.1 Principles of Coagulation and Sedimentation Technology

The coagulation precipitation method is achieved by adding coagulants such as aluminum salts, iron salts, polyaluminum, polyiron, and polyacrylamide to aggregate and settle suspended and colloidal substances in runoff rainwater under the action of coagulants, effectively removing suspended and colloidal substances in runoff rainwater.

5.1.3.2 Structural Characteristics and Design Parameters of the Device

The coagulation and sedimentation device includes mixing facilities, flocculation facilities, and sedimentation tanks.

i. Design requirements for mixed facilities

Polymer coagulants have advantages such as good flocculation effect, decolorization ability, and simple operation, and can be given priority consideration when selecting coagulants.

The closer the distance between the mixed facility and the subsequent processing structure, the better. Direct connection should be used as much as possible. When using pipeline connection, the flow rate can be taken as 0.8–1.0 m/s, and the residence time inside the pipeline should not exceed 2 min.

The mixing process requires speed and intensity. Usually completed within 10 to 30 s, with a maximum of 2 min; The stirring intensity is calculated according to the speed gradient, and the G value is generally within 700–1000 s^{-1}.

ii. Design requirements for flocculation facilities

The small flocs in the mixed water cannot settle naturally yet. The purpose of flocculation facilities is to increase the chances of particle contact and collision, gradually forming larger flocs for subsequent sedimentation. The flocculation time should be 8–20 min, and the depth of the mechanical flocculation tank is generally 3–4 m. The flocculation tank is generally not less than 2 groups.

There are generally 3–4 levels of mixers installed in the pool, and each level can be separated by partition walls or perforated walls to avoid short flow. The linear speed range at the center of the slurry plate of the mixer is generally 0.2–0.5 m/s. The total area of the slurry plate on each mixer should be 10–20% of the cross-sectional area of the water flow in the flocculation tank, and should not exceed 25%, to prevent the water from synchronously rotating with the slurry plate and weakening the flocculation effect. The length of the paddle board shall not exceed 75% of the impeller diameter, and the ratio of the paddle board width to length shall be 1:10–1:15. The paddle board width is generally 0.1–0.3 m. The top of the upper paddle plate of the vertical axis agitator should be located about 0.3 m below the water surface of the reaction tank, and the bottom of the lower paddle plate should be located 0.3–0.5 m away from the bottom of the tank. The distance between the outer edge of the paddle plate and the side wall of the tank should not exceed 0.25 m. All mechanical equipment such as mixing shafts and impellers should take anti-corrosion measures; The bearings and shaft frames should be placed outside the pool to prevent mud and sand from entering, resulting in severe wear of the bearings and breakage of the shaft rods.

iii. Design requirements for sedimentation tanks

Considering the discontinuity of rainfall, natural sedimentation is recommended for rainwater sedimentation. The sedimentation tank can be designed as a static sedimentation tank, which can be shared with the rainwater storage tank. The sedimentation tank should adopt reinforced concrete structure or brick and stone structure, and small-scale sedimentation tanks can also use organic materials such as plastic or fiberglass. Rainwater sedimentation tanks should be built underground to meet the elevation requirements of urban land scarcity and rainwater collection.

The rainwater sedimentation tank can be designed in the same way as traditional sewage sedimentation tanks and independent sedimentation tanks. Traditional sedimentation tanks also include sedimentation methods such as swirl and vertical flow, and the maximum design flow rate should not exceed 0.5 m/s, and the residence time of the highest flow rate should not be less than 30 s; When using an independent sedimentation tank, the design method of the sedimentation tank in GB50014 can be based on the characteristics of rainwater quality, flow rate, etc. The surface hydraulic load of the sedimentation tank can be 15–25 $m^3/(m^2 \cdot h)$, not exceeding 30 $m^3/(m^2 \cdot h)$. A longitudinal slope and mud area of no less than 0.01 should be set up in the sedimentation tank, and sediment can be discharged into the initial runoff tank through the sewage pipe for removal. The diameter of the sewage pipe should not be less than 100 mm.

5.1.3.3 Scope of Technical Application and Advantages and Disadvantages

i. Applicable scope

The coagulation and sedimentation technology is suitable for the treatment of initial rainwater and combined overflow rainwater and sewage under the impact of rainy season. While removing suspended solids, it can also partially remove organic matter, TP, and TN.

ii. Advantages and disadvantages

Advantages: Coagulation and sedimentation are carried out by adding chemicals and carriers, which have the advantages of fast start-up speed, strong impact resistance, high removal efficiency, mature and stable treatment methods, simple operation, and low power consumption.

Disadvantage: The dosage of medication must be determined through experiments, and excessive dosage of medication itself can have an impact on the water body.

5.1.3.4 Operation and Maintenance Requirements

When the alum flowers at the end of the coagulation tank are in good condition, the turbidity of the water is low, and the effluent of the sedimentation tank carries alum flowers, the hydraulic surface load of the sedimentation tank should be reduced; When the alum particles at the end of the coagulation tank are small and the water is turbid, and the turbidity of the effluent from the sedimentation tank increases, the dosage should be increased in a timely manner; When the alum flowers at the end of the coagulation tank are large and loose, and the effluent of the sedimentation tank is exceptionally clear, but the effluent carries a large amount of alum flowers, the dosage should be reduced. When the water temperature drops, inorganic polymer coagulants and other coagulants that are less affected by water temperature should be used, or the method of adding coagulants can be used.

If the uneven weir plate causes short flow in the sedimentation tank, the weir plate should be leveled; If the short flow in the sedimentation tank is caused by temperature changes, effective rectification measures should be taken at the inlet of the sedimentation tank; When the coagulation strength is insufficient, strengthen operation scheduling and try to ensure sufficient flow velocity in the mixing zone; If the alkalinity of the incoming water is insufficient, lime should be added in a timely manner to supplement the alkalinity; When the mud accumulation at the end of the coagulation tank blocks some of the holes on the inlet perforated wall, or the mud accumulation in the sedimentation tank reduces the effective tank capacity, the tank should be stopped for mud cleaning.

Regular water quality inspection and cleaning of rainwater sedimentation tanks are required. When encountering rainstorm, it is necessary to strengthen the monitoring and cleaning of sedimentation tank; In the absence of rain for a long time, the water level should be maintained at a low level. If the water quality becomes black and foul smelling, an appropriate amount of alum should be added to the water tank to ensure clear water quality; Clean up any suspended solids promptly.

5.1.3.5 Technical Processing Effect

In general, the removal rate of suspended solids and TP by coagulation sedimentation technology is greater than 85%, the removal rate of COD is 40–70%, the removal rate of BOD5 is 30–50%, and the removal rate of TN is around 20%.

5.1.3.6 Application Examples

The mixed connection renovation project of the rainwater pipe network in the Caohejing drainage system in Shanghai adopts the technology of regulating and storing concrete, and the process flow is shown in Fig. 5.5 [5]. This technology mainly involves setting up a mixing system at the front end of the inlet pipe of the regulating tank. In the mixing system, the initial rainwater and coagulant quickly mix before entering the regulating tank. At this point, the internal structure of the storage tank has been modified to effectively allow the mixed initial rainwater to react and settle in it, and after experiencing the optimal settling time to achieve the maximum removal effect of pollutants, it is discharged into the water body. Under actual rainfall conditions in 2009, the volume of the regulating tank in the Shanghai Caohejing drainage system was 5959 m^3, and its annual coagulation and regulation interception capacity was 24%.

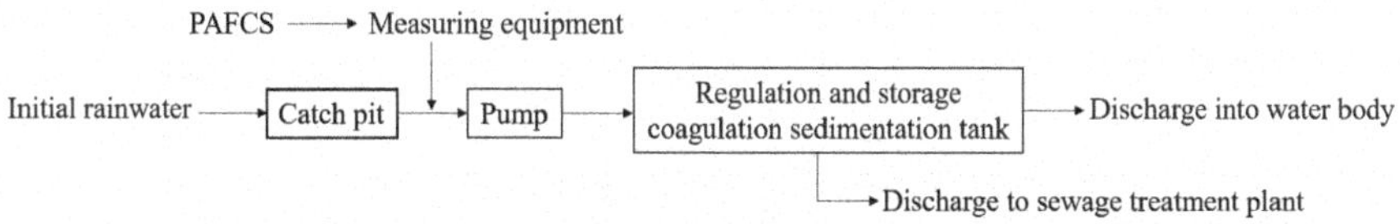

Fig. 5.5 Process flow of regulation, storage, and coagulation linkage technology

5.1.4 Filtering System

5.1.4.1 Principle of Filtration System

A filtration system uses quartz sand, gravel, or fibers as media to filter and remove pollutants from rainwater runoff. While reducing runoff pollution, it can also retain runoff and delay peak runoff.

5.1.4.2 System Structural Characteristics and Design Parameters

i. System structural characteristics

The filtration system is shown in Fig. 5.6, and the physical image is shown in Fig. 5.7. It mainly consists of a pre sedimentation tank and a filter bed. The main function of the pre sedimentation tank is to remove precipitable solid particle pollutants carried by rainfall runoff; The filter bed mainly includes the inlet pipe channel, drainage tank, filter medium (filter material layer), filter medium support layer (cushion layer), and water distribution (drainage) system. The inlet and outlet water of the system is transmitted through the inlet pipe channel and drainage tank, and the diameter of the medium filter material is usually less than 2 mm.

ii. Parameter design

The maximum water depth of the filtration system is calculated using the following formula [6]:

$$H = 0.75\mathrm{V}/\mathrm{S}_{\mathrm{forebav}} + S_{\mathrm{sand\ filter}}) \tag{5.2}$$

where, V—the required runoff to be treated under design rainfall conditions, m^3;

$S_{\text{Front tank}}$—Area of front sedimentation tank, m^2;

$S_{\text{Sand filter}}$—The area of the sand filter, m^2.

The minimum area A_f of the sand filter is calculated using the following formula:

$$A_f = V \times d_F/k/t/(h_A + d_F) \tag{5.3}$$

where, d_F—thickness of the filter bed, m;

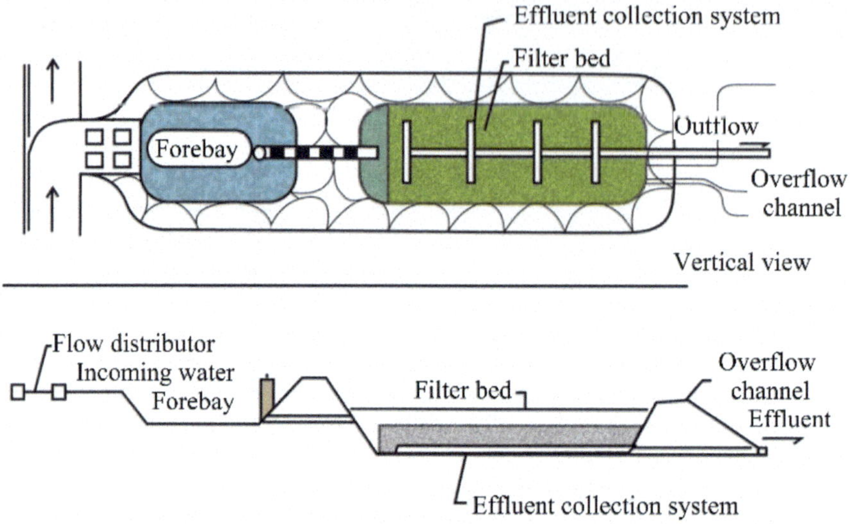

Fig. 5.6 Design schematic diagram of surface sand filtration system [1]

Fig. 5.7 Physical photos of surface sand filtration system [1]

k—Permeability coefficient of filter bed, m/d, recommended value of 1 m/d;

t—The recommended time for runoff to pass through the filter bed is 40 h or 1.66 days;

$h_A + d_F$—total head loss, m.

The depth of the pre sedimentation tank should be 1–2 m, the slope gradient is generally 1:4, and the internal flow velocity is 1.2 m/s; The thickness of the filter bed is generally between 46–60 cm, and the number of filter beds is not less than 2. When the number of filter beds is less than 5, it is advisable to use a single row arrangement, otherwise a double row arrangement can be used; When the area of a single filter bed is greater than 50 m^2, a central drainage channel can be set up in the pipe gallery; The area of a single filter bed is generally not greater than 100 m^2, and the aspect ratio is mostly between 1.25:1 and 1.5:1; When the area of a single filter bed is less than 30 m^2, 1:1 can be used. When using rotary surface flushing, 1:1, 2:1, and 3:1 can be used.

For a single-layer quartz sand filter bed, the design filtration rate is generally 8–10 m/h. When low runoff or overflow SS is required, the design filtration rate for a single-layer sand filter is 4–6 m/h, and for a double-layer coal sand filter, it is 6–8 m/h; The designed flow rate of the fiber filter layer can be increased to 20–27 m/L; The water depth above the filter layer is generally 1.5–2.0 m, and the ultra-high of the filter tank is generally 0.3 m.

The design working cycle of the filter bed is generally 12–24 h, and the head loss before flushing is generally 2.0–2.5 m; The washing intensity of single-layer filter material filtration is generally 12–15 L/(s·m^2), and the washing time is 7–5 min; The filtration and washing intensity of the double-layer filter material is 12–16 L/(s·m^2), and the washing time is 8–6 min.

5.1.4.3 Scope of Application and Advantages and Disadvantages

i. Applicable scope

The filtration system is suitable for treating combined overflow rainwater and wastewater, as well as initial rainwater purification with high pollutant concentrations in the diversion system. It is suitable for placement in densely populated areas such as residential and commercial areas, and should not be chosen in areas with high sediment loads.

ii. Advantages and disadvantages

Advantages: The filtration system has the advantages of high pollution load, fast and efficient treatment, small footprint, and no additional increase in drainage system pressure. It has a high removal rate for suspended particles, heavy metals, nitrogen, pathogens, and other pollutants; Compared with other runoff control measures, the effluent quality is better; Suitable for areas with high population, building density, and impermeable proportions; The filtration system can be designed according to local conditions and is flexible, making it easy to use; For larger catchment areas, multiple units can be combined to increase the capacity of rainfall runoff treatment.

Disadvantages: High construction and maintenance costs, and strict selection of filter material particle size; The backwash strength of the filter material is high, which requires high mechanical strength of the filter material; Moderate control of backwash intensity is required to avoid loss of filter material.

5.1.4.4 Operation and Maintenance Requirements

After installing new filter media in the filter, it should be soaked in a solution with a chlorine content of 0.3 mg/L or above for 24 h. After passing the inspection of the filtered water, it should be rinsed twice or more before it can be put into use. When the filter is not used for a long time, the water level in the tank should be kept above the drainage tank to prevent the filter material from drying out. A sampling inspection of the filter layer with a total area of 20% should be conducted annually to ensure that the mud content is less than 3%, and the annual loss rate of filter material should not be too high.

After rainfall, it is necessary to check the filtration system in a timely manner, clean the packing regularly, and remove sediment. When the accumulation height of sediment exceeds 15 cm or the head loss reaches 1.5–2.5 m, backwashing should be carried out; Before flushing the filter, the vent valve on the washing water pipeline must be opened, and the filter can only be flushed after the residual gas is released. When flushing, the drainage tank and drainage pipeline should be unobstructed and there should be no water blockage. The flushing intensity should be 8–17 L/(s·m^2), and the air pressure for air water flushing should be determined based on its flushing effect. Overpressure is strictly prohibited from causing sand running; After adjusting the pressure, it must be operated at a constant pressure, and the expansion rate of the filter material during flushing should be 40–50%.

5.1.4.5 Treatment Effect of Filtration System

The removal rates of SS, COD, NH3-N, TN, and TP in rainwater and sewage by filtration systems are usually between 80 to 85%, 45% to 55%, 70% to 80%, 20% to 35%, and 40% to 55%, respectively.

5.1.4.6 Application Examples

The filtration system has been applied at the Shangcun Pump Station and Jinling Middle Road Rainwater Pump Station in Zhonglou District, Changzhou City.

i. Application of Shangcun Pump Station in Zhonglou District, Changzhou City

The project adopts a suspended fast filter with a diameter of 2.4 m, and the filter materials are ethylene vinyl acetate (EVA) foam plastic and polyester fiber. The filtration speed is above 20 m/h, and the rapid processing capacity is 102.7 m^3/h. It can achieve automatic backwashing. On sunny days, sewage is intercepted by the pump station and enters the municipal sewage pipeline network. In rainy days, excess rainwater and sewage are combined and stored in the regulating tank, greatly reducing overflow in rainy days. The excess capacity is treated by the suspension rapid filtration device and discharged into the receiving water body. The reduction rate of SS load in six monitoring sessions has reached over 30%.

ii. Application at Jinling Middle Road Rainwater Pump Station in Changzhou City

This project collects rainwater runoff from the road of Jinling Middle Road in front of the pump station, with a catchment area of 16,000 square meters. The collected rainwater runoff is discharged into the nearby Beitang River through a rainwater pump. Due to the high pollution load of rainwater in the early stages of the road, it has a significant impact on the water quality of Beitang River. Therefore, it is necessary to treat the rainwater on the road.

The treatment project adopts a combination of cyclone separation and sand filtration system. The initial rainwater is first collected into the initial rainwater tank, then pumped into the cyclone separator for the separation of large suspended particles, and then enters the sand filtration tank. After sand filtration treatment, the effluent SS is less than 10 mg/L. The removal rates of COD, NH_3-N, TN, and TP are above 50%, 70%–90%, 15%–40%, and 60%–80%, respectively.

5.1.5 Disinfection Technology and Equipment

5.1.5.1 Principles of Disinfection Technology

Disinfection refers to various technologies that use disinfectants or other disinfection methods to inactivate the vast majority of pathogens in rainfall runoff, so that the microbial content in the runoff meets the requirements. Common disinfection techniques mainly include ultraviolet disinfection, chlorine disinfection, ozone disinfection, and peracetic acid disinfection.

5.1.5.2 Equipment Structural Characteristics and Design Parameters

When designing disinfection systems for rainfall runoff pollution, consideration must be given to dealing with pollution loads and flow fluctuations. Overflow wastewater, due to its intermittent discharge, short duration, and unstable flow, usually has high concentrations of suspended solids and bacteria in the initial rainwater. As rainfall continues, the concentration of pollutants gradually decreases. The concentration of pollutants in rainfall runoff is influenced by the characteristics of the watershed, drainage network, previous dry season conditions, and rainfall intensity. Therefore, the design of disinfection systems should be able to meet the local pollution load characteristics, while also being able to cope with high levels of pollutants during initial flushing.

i. UV disinfection

In order to control the reasonable water flow pattern and fully exert the irradiation effect, the design of the ultraviolet irradiation channel should meet the following requirements: the water flow in the irradiation channel should be evenly distributed, and the length of the channel before and after the lamp tube should not be less than 1 m; The water depth should meet the submergence requirements of the lamp tube; The number of ultraviolet radiation channels should not be less than 2. When using 1 channel, it is advisable to set up an overtaking channel; As the main disinfection process, the effective dose of ultraviolet radiation should not be less than 40 mJ/cm^2; UV water disinfection equipment should use tubular disinfection equipment.

ii. Chlorine disinfection

The amount of chlorine added to treated rainwater should be determined based on experimental data or similar operating experience. In the absence of experimental data, the amount of chlorine added to treated rainwater can be 6–15 mg/L, and the amount of chlorine added to recycled water should be determined according to hygiene indicators and residual chlorine; Chlorine dioxide should be prepared on-site using chemical methods and added. The storage tanks, equipment, and pipes in the disinfection system should have good sealing and corrosion resistance. In buildings where disinfection system equipment is installed, all building components and walls that may come into contact with raw materials or reaction products should be treated with anti-corrosion measures.

iii. Ozone disinfection

The amount of ozone disinfection depends on the quality of runoff rainwater, and the exact value should be determined through experiments or experience; The disinfection dosage is generally 5–15 mg O_3/L water, with a disinfection contact time of 6–15 min, or a residence time of 10–15 min after contact; For early rainwater or overflow sewage with severe pollution, the ozone concentration and action time can be increased.

If COD degradation is required, considering that the half-life of ozone is only 20 min, and significant degradation of COD requires a longer contact time, 3–6 diffusion chambers can be set up, with each diffusion chamber having a contact time of 8–15 min. The ozone diffusion device is similar to the aeration system of the aeration tank, using a microporous aerator; The designed water depth is generally greater than 5 m, and the pool top is covered with a tail gas recovery device to avoid ozone leakage.

iv. Peroxyacetic acid disinfection

Pure peroxyacetic acid has extremely unstable properties and can only be stored when it forms a stable system with hydrogen peroxide and acetic acid; The storage capacity should not be too large, especially attention should be paid to using plastic containers for storage, and the use of metal containers such as iron or aluminum is strictly prohibited; The decomposition rate of peracetic acid is influenced by temperature, concentration, and purity. Low concentration ($<20\%$) peracetic acid disinfectants have stable properties and must be stored in a cool and dark place at low temperature, with ventilation measures taken.

To determine the safe and economical dosage concentration during use, it is necessary to ensure continuous disinfection effect while minimizing the contribution of decomposition products to COD.

5.1.5.3 Scope of Technical Application and Advantages and Disadvantages

i. Applicable scope

After the rainwater is treated by physical, chemical and other methods, the bacterial content in the water decreases, but the absolute value may still be high, and there is a possibility of pathogenic bacteria. According to the purpose of rainwater reuse, if there are requirements for bacteriological indicators, disinfection treatment should be carried out before reuse. However, when rainwater is reused in water bodies that do not come into direct contact with humans, disinfection can be used as a backup measure.

ii. Advantages and disadvantages

The advantages and disadvantages of common disinfection methods are shown in Table 5.3.

Table 5.3 Advantages and disadvantages of common disinfection methods [8]

Disinfection methods	Advantage	Shortcoming
Ultraviolet radiation	Can kill viruses and spores, without the need for chemical agents, without residual disinfectants, easy to operate, not affected by pH, extremely short contact time, and small footprint	Due to the significant impact of SS, there is no sustained bactericidal effect in the water after disinfection. Each lamp tube has limited water treatment capacity and requires regular cleaning and replacement, which is costly
Chlorine	Mature and reliable technology, simple operation process, easy access to chemicals, with subsequent disinfection effect, without the need for large equipment	Difficult to control dosage, greatly affected by SS, residual chlorine has toxic effects on receiving water bodies
Ozone	The disinfection effect is better than chlorine, with no trihalomethane by-products, increasing dissolved oxygen in water, not changing the pH of water, removing iron and manganese ions, controlling odor, and short contact time	Large investment, generation of NO_x and HNO_3 is corrosive, gaseous, unstable, and prone to decomposition. It is not suitable for transportation and needs to be produced and used on site
Peracetic acid	Good disinfection effect, no disinfection by-products, unaffected by pH, and short contact time	Poor stability, complex operation, and greatly affected by temperature

5.1.5.4 Operation and Maintenance Requirements

The maintenance of urban rainwater runoff sewage disinfection facilities is similar to the maintenance of intermittent operation facilities. After each rainfall, the facilities need to be inspected and chemical agents added again.

Regularly inspect all copper pipes and fittings, and replace any corroded pipes; Inspect pipelines and containers for initial signs of leakage (moisture or metal discoloration).

Taking liquid chlorine disinfection as an example, the evaporator needs to be checked for sludge accumulation every year or after using 200 tons of chlorine. The pipes and connections of the evaporator need to be checked every 6 months; Replace the chlorine gas filter every 6 months; Chlorine pressure reducing valves should be cleaned with isopropanol or trichloroethylene, and spring valves should be replaced every 25 months; Clean the injector every 6 months; Perform regular maintenance on the booster pump. In order to avoid harm to human health, chlorine disinfection equipment should be equipped with sufficient ventilation, safety measures, eye wash pipes and showers, emergency respiratory protection, emergency kits, telephone and information for emergency plan staff, and safety operation training for staff.

5.1.5.5 Application Rxamples

The rainwater collection and treatment system in a large overpass area of the Beijing section of the Beijing Taiwan Expressway adopts a rainwater treatment process of "three-stage sedimentation + swirling sedimentation + fully automatic filtration + ultraviolet disinfection" (see Fig. 5.8) [9]. Among them, the ultraviolet disinfection system mainly radiates multi band ultraviolet rays into water, used for the removal of microorganisms such as algae and bacteria, as well as the degradation of some organic pollutants. It not only kills bacteria but also reduces COD in water. The treatment method is simple and the effect is significant. In Fig. 5.8, the COD, BOD5, turbidity, TP, and chromaticity of the clear water tank were reduced by 52.94%, 54.83%, 70.28%, 53.85%, and 75.00% compared to the reservoir, respectively.

5.2 Biotechnology and Ecological Technology

The back-end biological and ecological treatment technologies for urban rainfall runoff pollution mainly include waterfront buffering technology, artificial wetland technology, and pond wetland combination technology.

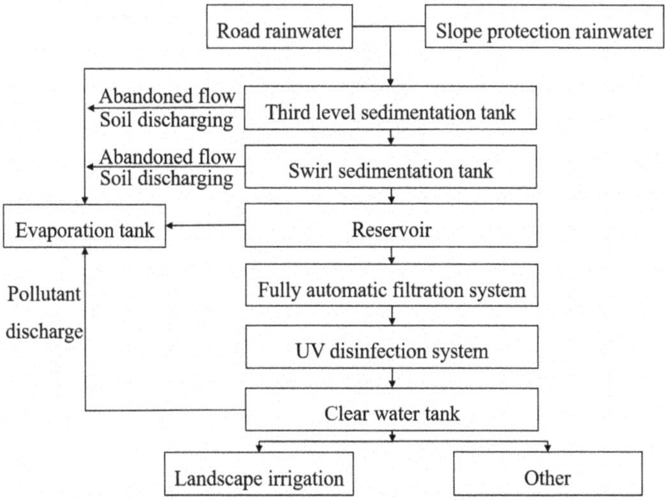

Fig. 5.8 Rainwater collection and reuse process in a large overpass area of the beijing section of the Beijing Taiwan Expressway

5.2.1 Waterfront Buffering Technology

5.2.1.1 Principles of Waterfront Buffering Technology

Waterfront buffer zone, also known as forest buffer zone, refers to a vegetation area or buffer zone built between aquatic and terrestrial environments that has the function of intercepting and purifying nutrient pollutants in runoff. It is generally built along the banks of rivers or around lakes and wetlands.

Waterfront buffer zones are usually distributed in a strip pattern along the water body from high to low, and the specific shape depends on the local geographical conditions and runoff migration pathways. When used as a pollution control measure for rainfall runoff, some pollutants in the runoff are removed through infiltration, adsorption, filtration, and other processes in the waterfront buffer zone before the rainfall runoff enters the receiving water body, reducing the pollution of rainfall runoff on the receiving water body.

The purification mechanism of the waterfront buffer zone on rainfall runoff pollution is shown in Fig. 5.9, mainly including the interception of particulate matter, nitrification denitrification of vegetation and microorganisms for nitrogen removal, sedimentation and fixation of phosphorus, and biological adsorption and oxidation of organic pollutants.

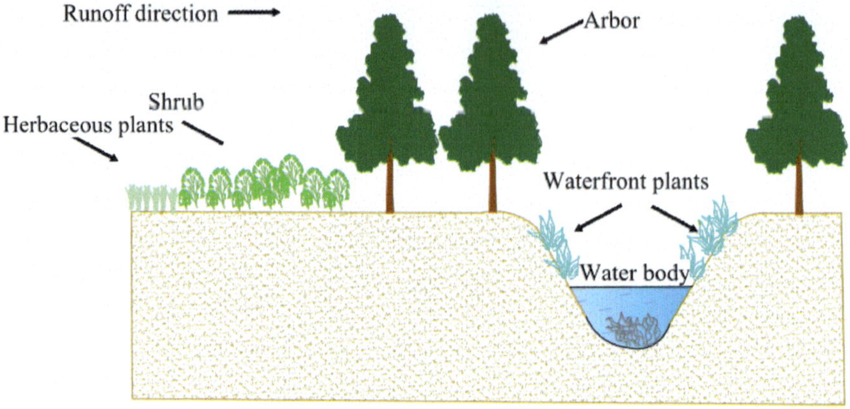

Fig. 5.9 Schematic diagram of the mechanism of waterfront buffer zone

5.2.1.2 Structural Characteristics and Design Parameters

i. Technical structure

The waterfront buffer zone has three main characteristics: a narrow and elongated shape along the water body; Structurally, it serves as a link between highland vegetation and water bodies, while functionally, it is a typical open system for material and energy exchange; Compared to the surrounding area, the vegetation buffer zone of the waterfront buffer zone has an abnormally high richness of plant species.

The waterfront buffer zone also has four-dimensional structural characteristics, which can be divided vertically into upstream and downstream, horizontally into riverbed and flood plain, vertically with surface runoff, underground seepage, and groundwater structures, as well as seasonal changes and community succession of the buffer zone in the time dimension. From a horizontal perspective, the riparian vegetation buffer zone can be divided into three parts: nearshore waters, waterfront areas, and nearshore land areas.

There are three common types of waterfront buffer zones: ① Slope contour buffer zones: roughly equivalent to China's contour plant hedgerows, applied between crops and forests on gentle slope farmland, but with emphasis on controlling runoff pollution in design; ② Buffer zones around water bodies: generally set up along rivers, lakes, and reservoirs, emphasizing the protection function of water quality; ③ Wind erosion buffer zone: equivalent to China's wind and sand resistant forest belts.

ii. Plant configuration

The vegetation in waterfront buffer zones usually includes forests, shrubs, grasslands, mixed vegetation, and swamp wetlands. Buffer zones of different vegetation types can exhibit different functional characteristics. Guo et al. [10] studied the waterfront buffer zones of four different vegetation patterns in the Wenyu River in Beijing, namely grassland zone, farmland zone, grassland farmland mixed zone, and artificial grassland zone. Research has shown that the distribution of soil TN, TP, available nitrogen (AN), available phosphorus (AP), and organic matter (OM) in the waterfront buffer zone shows a gradually decreasing trend from highlands to rivers, while in grassland areas, different trends are observed; Grassland with lower slopes has higher nutrient and organic matter content. At the same time, the content of TN, TP, AN, and AP near the riverbank is higher on the downhill slope of the grassland belt, which has poor nutrient retention effect. Li et al. [11] studied the waterfront vegetation buffer zone of four different vegetation types in the Taihu Lake Basin, namely the arbor herbaceous buffer zone composed of *Populustomentosa* and ryegrass (*Loliumperenne*), the Joe-shrub-grass waterfront buffer zone composed of *Populustomentosa*, *Broussonetiapapyrifera* and *Phragmites australis*, a shrub-grass buffer zone consisting of shrub-like trees, *Amorphafruticosa* and herbaceous *Humulus scandens* and I*mperatacylindrica*, as well as herbaceous buffers composed of perennial reeds. Research has shown that under certain conditions, the waterfront buffer zone composed of shrubs and grasses has a higher removal rate of pollutants; The effect of the waterfront buffer zone composed of herbaceous reeds is lower than that of other vegetation types, showing seasonal differences, with summer being higher than winter; The buffer zone composed of Populus tomentosa and ryegrass has a higher decontamination effect only in winter compared to other types of vegetation zones; The efficiency of the buffer zone composed of trees, shrubs and grasses is only higher than that of the buffer zone composed of trees and grasses. In addition, many research conclusions at home and abroad have also confirmed that forest land has a better pollutant removal effect compared to grassland [12–15], because the organic carbon storage in forest land is larger, making its denitrification effect stronger compared to grassland.

iii. Design requirements

Scientifically designing waterfront buffer zones is the foundation for their better performance. In the design, four elements should be considered: site selection, scale, vegetation type configuration, and management and maintenance.

When laying out the waterfront buffer zone, it is advisable to choose a place with sufficient sunlight, so that the ground can dry thoroughly during the interval between two rainfall periods; The site selection is generally at the downhill position of the slope, arranged perpendicular to the flow direction of runoff; For long slopes, multiple buffer zones can be set along the contour line to reduce the energy of the water flow.

The slope of vegetation buffer zones is generally 2–6%, and it is better to have a slope below 5%; The length of runoff in the catchment area should be less than 45 m, not more than 90 m, and the width should not be less than 2 m.

The plant configuration in the waterfront buffer zone can use a combination of trees, shrubs, and herbaceous plants, and its overall effect is relatively ideal; Attention should be paid to the use of local plant varieties, and the introduction of foreign plant varieties should be very cautious to ensure the stability of the ecosystem.

When the groundwater level is high and close to the surface, water level spreaders or sand dams should be installed to ensure that the contact time with runoff is not less than 5 min; The flow velocity of runoff entering the buffer zone should preferably be less than 0.45 m/s and not exceed 1.5 m/s.

5.2.1.3 Scope of Application and Advantages and Disadvantages

i. Applicable scope

The waterfront buffer zone is mainly suitable for areas with high potential generation of urban surface runoff and severe soil erosion in the watershed.

ii. Advantages and disadvantages

Advantages: The waterfront buffer zone has good ecological and environmental benefits, and has a good removal effect on pollutants in rainfall runoff; It can also provide habitats for birds, insects, etc., maintain soil and water, reduce erosion, increase urban green space area, and enrich biodiversity.

Disadvantages: The construction of waterfront buffer zones requires a large amount of trees, shrubs, and grass, resulting in high construction costs; The prevention and control of large-scale pests and diseases, as well as daily maintenance such as vegetation pruning, also require high financial support.

5.2.1.4 Operation and Maintenance Requirements

The waterfront buffer zone requires a certain degree of management and maintenance to ensure its function of controlling runoff pollution. Maintenance measures include: trimming the grass appropriately to promote nutrient absorption and system operation; Thinning the trees to enhance sunlight exposure in the buffer zone and promote the growth of understory shrubs and herbaceous plants.

5.2.1.5 Technical Processing Effects

Overall, the removal rates of SS, NH_3-N, NO_3^--N, TN, and TP by waterfront buffering technology are 40%–90%, 50%–80%, 10%–30%, 10%–50%, and 30%–60%, respectively. The investment cost of waterfront buffer zones mainly includes vegetation cost, construction cost, and daily maintenance cost, and the corresponding costs vary in different regions, construction areas, and construction requirements. The usual cost is between CNY100-1000 /m^2, and its maintenance cost ranges from CNY 0.1 to 2.0 /m^2.

5.2.1.6 Application Examples

The waterfront buffer zone has been applied in the comprehensive environmental improvement project of Funiuxi in Dadukou District, Chongqing. This project mainly focuses on local species and constructs a new multi-level "tree shrub herb" three belt buffer system, retention system, and gabion system for the shoreline system. The buffer zone system based on trees, shrubs, and herbs has biodiversity, good landscape effects, and the ability to reduce and purify pollutants; This buffer zone system stabilizes the riverbank through gabion belts, adopts different plant combinations for different levels of riverbank belts, intercepts pollutants layer by layer in retention channels, reduces flow velocity, reduces particulate matter, and utilizes different types of vegetation to form a multi-level filtration and absorption system, while also having good landscape and biodiversity functions.

After the project was completed and put into operation in December 2014, it significantly reduced the input of pollution load, improved the water quality of the river, and alleviated the environmental pollution of secondary rivers in the main city of Chongqing. The main indicators met the requirements for Class V water bodies, and the compliance rates for the permanganate index, NH3-N, dissolved oxygen, and transparency of the assessment sections were 100%, 91.7%, 91.7%, and 83.3%, respectively. The implementation of the project has significantly improved the living environment of residents in the watershed and promoted the development of surrounding real estate.

5.2.2 Artificial Wetland Technology

5.2.2.1 Principles of Artificial Wetland Technology

Artificial wetland is a wetland system designed by humans to fully utilize plants, substrates, and their microorganisms to remove pollutants from runoff or overflow, achieving the goal of controlling rainwater and purifying the quality of rainfall runoff. Rainwater wetlands can generally be divided into two types based on their apparent appearance: surface flow constructed wetlands and subsurface flow constructed wetlands.

5.2.2.2 Structural Characteristics and Design Parameters

i. Technical structure

The composition of artificial wetlands, including substrate, plants, microorganisms (bacteria, fungi, etc.), and animals, plays a positive synergistic role in purifying runoff pollution.

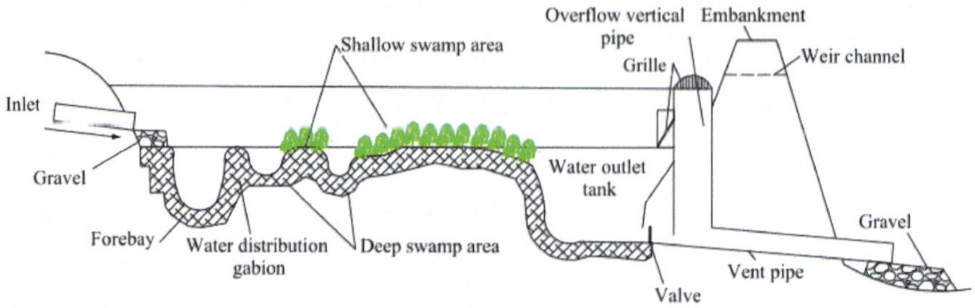

Fig. 5.10 Surface flow artificial wetland [16]

Surface flow constructed wetlands are similar in appearance and function to natural wetlands, distributing water to the surface of the wetland. Rainwater runoff flows on the surface of the constructed wetland, with a shallow water level of generally 10–30 cm and a hydraulic load of 200 $m^3/(104\ m^2 \cdot d)$. The water flow advances in a push flow pattern, forming a layer of surface water flow on the entire wetland surface, which flows out to the terminal and completes the entire purification process. The surface flow artificial wetland is shown in Fig. 5.10.

The subsurface flow constructed wetland system generally consists of four parts: the wetland main body, anti-seepage layer, water distribution device, and water collection device. Among them, the main body of the wetland is composed of a gravel bed planted with plants, with a longitudinal slope set at the bottom of the bed; The setting of impermeable layers can effectively prevent pollutants in wetlands from entering groundwater; The inlet end of the water distribution device constructs a distribution ditch along the bed width, with gravel embedded inside. Rainwater runoff is discharged into the bed from the distribution ditch, and flows horizontally through the lower part of the medium through infiltration, flowing out from the outlet ditch at the other end; The water collection device is equipped with a porous collection pipe at the bottom of the gravel layer at the outlet end, which can be connected to the outlet pipe that can adjust the water level in the bed to control and regulate the water level in the bed. Rainwater runoff pollutants are purified through physical, chemical, and biological processes during the flow of gravel beds. The schematic diagrams of horizontal and vertical subsurface flow constructed wetlands are shown in Figs. 5.11 and 5.12, respectively.

In a horizontal subsurface flow constructed wetland, runoff or overflow flows horizontally from the inlet end of the pool to the outlet end below the surface of the substrate layer. Horizontal subsurface flow constructed wetlands can consist of one or more packed beds. The bed is filled with a matrix, and the bottom of the bed is equipped with an anti-seepage layer to prevent contamination of groundwater. The hydraulic and pollution loads of horizontal subsurface flow constructed wetlands are relatively high, and there is little odor or breeding of mosquitoes and flies. The disadvantage is that the control is relatively complex, and the effect of nitrogen and phosphorus removal is weaker than that of vertical subsurface flow constructed wetlands.

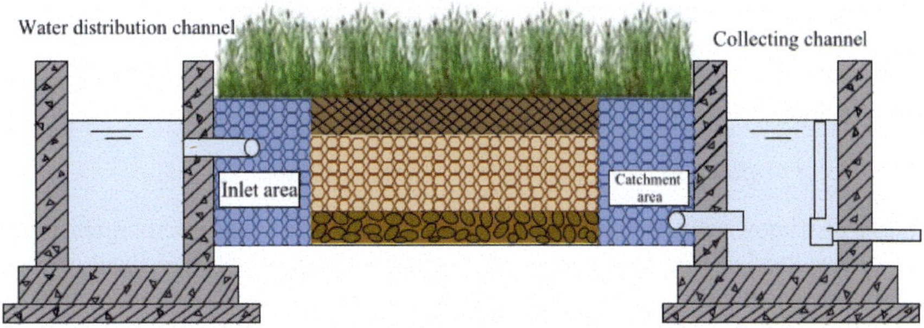

Fig. 5.11 Schematic diagram of horizontal flow subsurface flow constructed wetland [17]

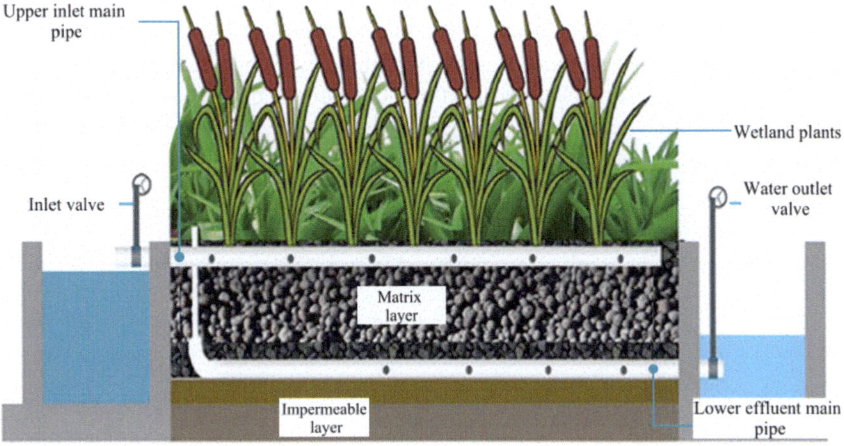

Fig. 5.12 Schematic diagram of vertical flow subsurface flow constructed wetland [18]

In a vertical subsurface flow constructed wetland, runoff or overflow flows vertically through the substrate layer of the pool. The nitrification capacity of vertical subsurface flow constructed wetlands is significantly better than that of horizontal subsurface flow wetlands, and can be used to treat runoff or overflow with high ammonia nitrogen content. However, vertical subsurface flow constructed wetlands have a longer time of drying/flooding, and there is a phenomenon of breeding mosquitoes and flies in summer. Their ability to remove organic matter is weaker than that of horizontal subsurface flow constructed wetlands.

ii. Design requirements

The determination of the designed water volume for artificial wetlands should comply with the relevant provisions of the Outdoor Drainage Design Standard (GB50014). When the project mixes and accepts urban domestic sewage, the design water quality can refer to the relevant provisions in the Outdoor Drainage Design Standards; When the quality of runoff rainwater is different from that of domestic sewage, its design water quality can be determined through investigation.

The inlet water quality of the artificial wetland system should meet the requirements in Table 5.4.

i). *Types of artificial wetland substrates and their selection criteria*

Matrix, also known as filler, is a carrier for the growth of plants and microorganisms in artificial wetlands. It mainly removes pollutants from wastewater through adsorption sedimentation, biofilm on the surface of the matrix, and the combined action of microorganisms. According to the needs, materials with different particle sizes are laid into artificial wetland beds according to a certain thickness. The traditional artificial wetland substrate includes soil, sand, and gravel; In addition, materials with excellent properties such as zeolite, limestone, shale, alum, plastics, and ceramics have been widely used in recent years. The types and physical properties of common substrates in artificial wetlands are shown in Table 5.5. For different characteristic pollutants and removal mechanisms, different substrates can be selected according to Table 5.6.

ii). *Hydraulic Design*

The hydraulic design of subsurface flow artificial wetlands follows Darcy's law in terms of subsurface hydraulics, as expressed in Eqs. (5.4).

$$Q = K_s A S \tag{5.4}$$

where, Q—runoff flow rate, m^3/d;

K_s—Hydraulic conductivity, $m^3/(m^2 \cdot d)$;

A—Total cross-sectional area perpendicular to the water flow, m^2;

Table. 5.4 Requirements for influent water quality of artificial wetland system (unit: mg/L)

Types of artificial wetlands	COD	SS	NH_3-N	TP
Surface flow artificial wetland	≤12	≤100	≤1	≤3
Horizontal subsurface flow artificial wetland	≤20	≤6	≤2	≤5
Vertical subsurface flow artificial wetland	≤20	≤8	≤2	≤5

Table 5.5 Classification and physical properties of media

Type	Effective particle size D10 (mm)	Porosity (%)	Conductivity [$m^3/(m^2{\cdot}d)$]
Coarse sand	2	32	1000
Gravel	8	35	5000
Fine gravel	16	38	7500
Middle gravel	32	40	10,000
Gravel	128	45	100,000

Table. 5.6 Main substrates for removing different types of pollutants and their water purification mechanisms

Pollutant types	Main matrix	Removal mechanism
Organic matter	Coal ash residue, anthracite, bioceramics, gravel, soil, etc	The adsorption and sedimentation of the substrate; The adsorption and absorption of biofilms on the substrate surface and the absorption and degradation of organisms
Nitrogen (inorganic nitrogen: NH3-N, NO2-, NO3-, etc., and organic nitrogen: amino acids, urea, amines, etc.)	Zeolite, crushed stone, fly ash, fine brick chips, and fly ash combination, etc	Matrix adsorption; Nitrification and denitrification of substrate surface organisms
Phosphorus (soluble phosphate, solid mineral phosphate, solid organic phosphate)	Blast furnace slag, steel slag (containing more calcium, iron, and aluminum elements) and shale combination, grass ash, crushed stone, and soil combination, etc	Matrix adsorption; Metabolism of microorganisms on the substrate surface

S—Hydraulic gradient, m/m.

Due to the physical limitations of actual wetland systems, Darcy's law is not entirely applicable to subsurface flow constructed wetlands. Darcy's law assumes laminar flow conditions, but the hydraulic conditions of subsurface constructed wetlands may change when large or rough gravel is used as the substrate. When the hydraulic design gradient is high, turbulence may form in the coarse gravel. Darcy's law also assumes that the inflow and outflow rates are consistent, but in reality, the inflow and outflow rates of subsurface wetlands will change due to factors such as precipitation, evaporation, and leakage. In addition, short flows are formed due to differences in porosity or construction quality. These factors limit the application of Darcy's law in subsurface flow constructed wetlands, but it still serves as an acceptable model for the design of subsurface flow constructed wetland systems. If the substrate in the subsurface flow constructed wetland is small to medium gravel (<4 cm), and the construction quality is improved to control the short flow to the maximum extent, the system design relies on the minimum hydraulic gradient, and the design flow rate adopts $(Q_{in} + Q_{out})/2$, Darcy's law can still provide a close approximation simulation for the hydraulic conditions of the subsurface flow wetland.

ii. Aspect ratio

The aspect ratio of subsurface flow wetlands should be controlled below 3:1, and the length of regular subsurface flow wetlands should be 20–50 m. For irregular subsurface flow wetlands, the issue of uniform water distribution and collection should be considered.

iii). *Slope*

The hydraulic slope of the slope subsurface flow wetland should be between 0.5 and 1%.

iv). *Water inlet structure*

The inlet device of the subsurface flow wetland with inlet structure often uses perforated pipes, which are installed in two main forms: surface and underground. Among them, the water inlet perforated pipes installed on the surface are all equipped with adjustable outlets. Adjust the position based on the settlement of the inlet perforated pipe to ensure even distribution of rainwater runoff, and also provide the maximum possibility for subsequent position adjustment and maintenance; Underground installation utilizes 2–3 valve outlets to avoid the formation of algal mucosa on surface rocks, which can lead to system blockage.

Usually, crushed stones with a particle size of 8–15 cm are used in the inlet area to maintain sufficient hydraulic gradient in the bed to ensure rapid filtration, prevent blockage and algae growth. In warm and sunny weather conditions, plants or facilities should be used for shading in the water outlet area; In cold weather, insulation measures should be taken to prevent frost cracking when distributing water on the ground.

v). *Effluent Structure*

The outlet structure of subsurface flow constructed wetlands mainly includes underground outlet forms, dam box or dam gate structures. Among them, the underground perforated pipe outlet form is one of the most commonly used forms. Place the underground perforated pipe in a shallow trench, below the bed bottom, to complete the discharge of runoff in the bed and ensure sufficient hydraulic gradient; Underground perforated pipes can also be placed above the bed bottom or at the top of the bed, but it is prone to surface overflow. In most cases, the outlet is directly connected to the final discharge pipe or to a concrete channel. It is recommended to use an adjustable outlet to ensure sufficient hydraulic gradient; Connect perforated pipes to adjustable outlets to provide reliability and flexibility for subsurface wetlands.

5.2.2.3 Scope of Application and Advantages and Disadvantages

i. Applicable scope

Artificial wetlands are suitable for construction in areas with low population density, fully utilizing the surrounding terrain characteristics and setting them according to local conditions; It can also be built on vacant land next to residential buildings; Or use ponds and landscape pools in parks for renovation.

ii. Advantages and disadvantages

Advantages: Artificial wetlands can fully utilize wetland plants, soil, and their microorganisms to remove pollutants from runoff rainwater, maintaining a high hydraulic load; The removal effect of pollutants such as BOD_5, SS, and coliform is significant, and it also has strong phosphorus removal ability. The effluent effect is stable, effective, and reliable; Energy saving, low operating costs, simple operation, and no need for complex self-control systems for control; There are fewer mechanical, electrical, and automatic control equipment, and the workload of equipment management is also relatively low; It can purify rainwater runoff pollutants and beautify the environmental landscape.

Disadvantages: Artificial wetlands require a large land area, have weak resistance to harsh weather conditions, and their purification ability is greatly affected by the maturity of crops, requiring control of mosquito and fly breeding.

Subsurface wetlands have good insulation effects, and the treatment effect is less affected by climate and season. Proper management during operation can effectively prevent the breeding of mosquitoes and flies and the generation of odors, but the investment is much larger than surface flow wetlands.

5.2.2.4 Operation and Maintenance Requirements

i. Timely adjustment of water level

According to various extreme conditions such as rainstorm, flood, drought and icing period, the water level can be adjusted without backwater at the inlet end and flooding at the outlet end; When short flow occurs, water level adjustment is also necessary.

ii. Plant management and maintenance

After planting plants in artificial wetlands, it is necessary to fill them with water. In order to promote the development of plant roots, water level regulation should be carried out in the initial stage; After the establishment of the plant system, it should ensure continuous provision of sewage to ensure the density and healthy growth of aquatic plants; According to the growth status of plants, management should be carried out for missing seedlings, weed removal, timely harvesting, and pest control. It is not advisable to use herbicides, insecticides, etc.; For large-scale artificial wetland wastewater treatment projects, consideration should be given to the installation of devices that utilize plant bioenergy.

iii. Low temperature environment operation measures

Take insulation measures for artificial wetlands to ensure that the water temperature is not lower than 4 °C; Regularly conduct permafrost depth tests on artificial wetlands to understand the operational status of the wetland system; Strengthen pre-treatment to reduce the pollution load of artificial wetland systems.

iv. Anti blocking measures

Control the concentration of suspended solids in rainwater runoff entering artificial wetlands; Regularly initiate dredging; Appropriate use of intermittent operation mode; Partially replace the substrate of the artificial wetland system.

5.2.2.5 Technical Processing Effect

When the influent concentration is high, the constructed wetland system has a good ability to withstand impact loads on COD. Under different influent load conditions, the constructed wetland has a certain removal effect on COD; When the influent COD concentration is low, hydraulic load has a certain impact on the purification effect of wetlands. When the hydraulic load is 1.543 $m^3/(m^2{\cdot}d)$, the effluent COD concentration is higher than when the hydraulic load is 0.285 $m^3/(m^2{\cdot}d)$.

The artificial wetland system can achieve a removal rate of over 75% for ammonia nitrogen under different hydraulic impact loads, but its ability to withstand impact loads is poor. The variation of hydraulic load can significantly affect the micro ecological environment inside the artificial wetland system, affecting the growth of nitrifying and denitrifying bacteria. In addition, the scouring effect of the impact load on the substrate may also be one of the reasons for the unsatisfactory removal effect of ammonia nitrogen.

The phosphorus removal rate under different impact loads can reach over 80%. The purification principle of phosphorus in subsurface wetlands is mainly adsorption and precipitation. Higher hydraulic loads will have a scouring effect on the surface of the filler, shortening the settling time of particles. Therefore, the removal rate of phosphorus decreases significantly at high loads. The treatment rate at a hydraulic load of 1.543 $m^3/(m^2{\cdot}d)$ is 9.6% lower than that at a hydraulic load of 0.285 $m^3/(m^2{\cdot}d)$.

5.2.2.6 Application Examples

The application of artificial wetland runoff pollution control technology in the runoff pollution control project of Chongqing Yuanbo Garden's lake tributary explored the effects of different types of combinations of artificial wetlands and biological filters on the upstream overflow sewage treatment of Yuanbo Garden's lake tributary. An ecological restoration technology for the overflow sewage treatment of Yuanbo Garden's lake tributary, mainly based on artificial wetland technology, was proposed (see Fig. 5.13). The process flow of the project is shown in Fig. 5.14. The project covers an area of 1552.18 square meters, with a sewage treatment capacity of 800 m^3/day. The influent water quality has a COD of 46.42 mg/L, BOD_5 of 20 mg/L, TN of 4.15 mg/L, TP of 0.17 mg/L, and a retention time of 19 h. The water quality after treatment with artificial wetlands is better than the Class IV standard in the Environmental Quality Standards for Surface Water (GB3838).

Fig. 5.13 Location of demonstration zone

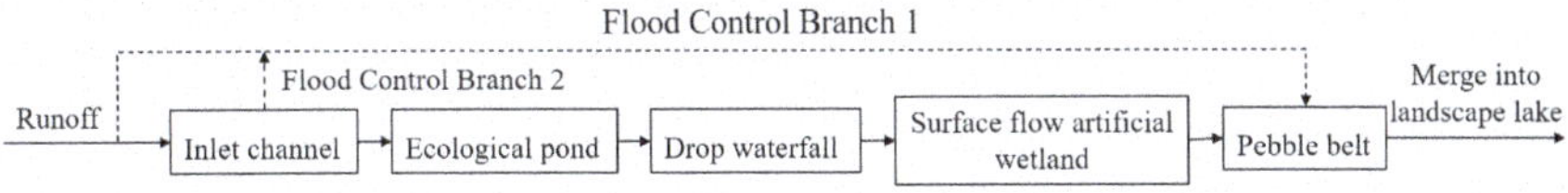

Fig. 5.14 Demonstration zone process flow

5.2.3 Pond Wetland Purification Combination Technology

5.2.3.1 Principle of Pond Wetland Purification Combination Technology

Pond wetland technology is a water pollution ecological purification technology designed and constructed based on the material cycle and purification principles of natural ecosystems, fully utilizing the triple synergistic effects of physics, chemistry, and biology in the ecosystem. Among them, the purification principles of pond technology mainly include physical sedimentation, storage, interception, and biological purification; The mechanisms of interception and removal of pollutants in wetland technology mainly include sedimentation, plant absorption, decomposition of dead branches and leaves, as well as adsorption, interception, filtration, ion exchange, complexation reaction, and microbial action of soil matrix.

5.2.3.2 Structural Characteristics and Design Parameters

i. Technology Combination Mode

The combination mode of pond wetland purification technology can be divided into three types: series, parallel, and hybrid.

i) *Series Combination Mode*

In the control of runoff pollution, pond wetland technology usually improves the removal rate and anti-interference ability of the system in a series combination mode, as shown in Fig. 5.15.

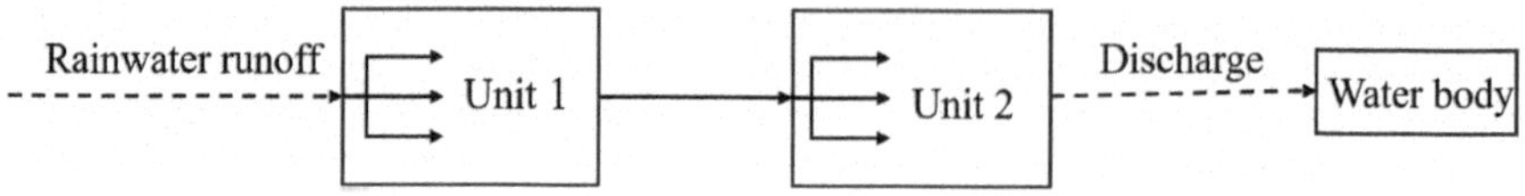

Fig. 5.15 Schematic diagram of series combination mode. (Unit 1 and Unit 2 represent ponds or wetlands in the system)

After the rainwater runoff is purified step by step through a series system, the pollutants in the runoff are gradually reduced, and the purified runoff is directly discharged into the surrounding water bodies. This series combination mode is specifically manifested as multi pond series system, multi-level wetland series system, and pond wetland series system, which can fully leverage the advantages of each technology, improve the treatment efficiency of the system, and have high removal efficiency for various types of pollutants.

ii) *Parallel Combination Mode*

The parallel combination mode is mainly manifested in the combination of different types of technologies and the parallel application of the same technology combination, as shown in Fig. 5.16. This combination method can fully leverage the advantages of various types of wetlands and compare the purification effects and characteristics of different types of wetlands. Runoff pollution passes through a unified water distribution system and flows parallel through various system units to obtain purified runoff, which is then collected by a collection system and discharged into surrounding water bodies. This parallel combination method usually adopts a unified water distribution, and the hydraulic load is basically consistent, which is conducive to comparing the purification characteristics of different types of wetlands.

iii) *Hybrid combination mode*

The hybrid combination mode usually refers to the control system adopting a series combination mode as a whole and a parallel combination mode locally. These two combination modes coexist and have various forms, as shown in Fig. 5.17.

Rainfall runoff is first treated by a series system, then purified by a parallel system, and finally discharged into surrounding water bodies. This parallel combination system is also part of a series system as a whole.

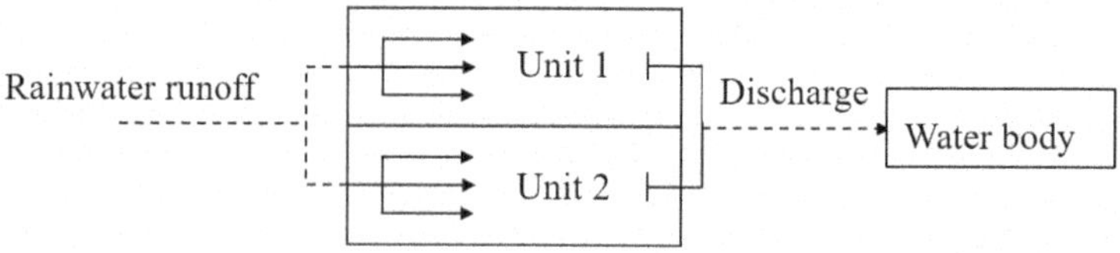

Fig. 5.16 Schematic diagram of parallel combination mode

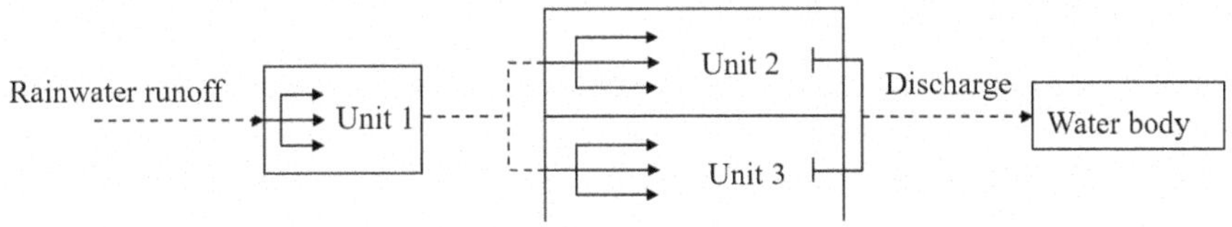

Fig. 5.17 Schematic diagram of hybrid combination mode

ii. Design requirements

In the actual engineering application process of pond wetland purification combination technology, the following four aspects need to be paid attention to.

i) *Selection of technology types*

The selection of technology types should be based on the climate, terrain, topography, water pollution characteristics, and land use conditions of the watershed to choose appropriate pond and wetland technologies. The interception efficiency of pond technology is high, the storage capacity is large, and the land occupation is relatively small, which has certain landscape value. It is generally used in the front end of the runoff end purification system; Wetland technology has a variety of types, high pollutant removal efficiency, and good landscape effects, making it the most widely used ecological engineering technology, usually used in the backend.

ii) *Selection of technology combination mode*

The selection of technology combination mode is based on the pollution load of the initial rainwater or overflow sewage to be treated, selecting a single pond and wetland technology or a multi-level technology combination mode. The anti-interference ability of a single pond and wetland technology is poor, and it is greatly affected by seasonal and other factors. The effluent quality of pollutants is also unstable, making it difficult to achieve the goal of controlling pollutants in the watershed. The combination of multi-stage pond and wetland technology combines the advantages of these two technologies, and the removal efficiency of pollutants is relatively stable, making it the most common application mode.

iii) *Selection of substrate materials*

The selection of substrate materials is one of the key issues to consider in the design and construction process of ponds and wetlands. Matrix materials have strong adsorption capacity for pollutants in runoff and are one of the main ways for ponds and wetlands to remove pollutants. When selecting matrix materials, it is advisable to choose materials from local or nearby areas with strong absorption capacity and relatively abundant resources.

iv) *Plant species selection*

Plants are an important component of ponds and wetlands, providing landscape value and biodiversity for the watershed. There are differences in the absorption capacity of pollutants among different plant species. Pond and wetland technologies should try to select local dominant species with strong ability to remove characteristic pollutants and high biomass, while taking into account species diversity, landscape, and seasonal compatibility issues.

5.2.3.3 Scope of Application and Advantages and Disadvantages

i. Applicable scope

The application area of the pond wetland purification combination technology is very wide. It can not only be applied to the control of runoff pollution in rural areas, small towns, urban tourist areas, and areas with relatively abundant land resources such as urban parks, squares, and green spaces, but also to industrial areas, commercial areas, and residential areas with relatively scarce land resources in cities through technological improvements.

ii. Advantages and disadvantages

Advantages: The pond wetland purification combination technology can efficiently remove rainwater runoff pollution, improve and restore ecosystems, increase biodiversity, provide biological habitats, and enhance the landscape and ecological value of the watershed; It has the advantages of simple construction, low maintenance and operation costs, low energy consumption, high pollutant removal rate, good effluent quality, and simple operation.

Disadvantages: The combination technology of pond wetland purification usually requires a significant amount of land resources, resulting in high construction costs and susceptibility to degradation and blockage issues.

5.2.3.4 Operation and Maintenance Requirements

The degradation and blockage of pond and wetland systems have always been a limiting factor for the promotion of these two technologies, with particulate matter being the main cause of their blockage. In the process of controlling rainwater runoff pollution, the concentration of particulate pollutants is high, and front-end pretreatment processes need to be equipped. Common pre-treatment processes include source reduction technologies such as permeable pavement, soil improvement, vegetation greening, and vegetation filter belts; Process and end control technologies such as grid, sedimentation tank, retention pond and rainstorm tank.

5.2.3.5 Technical Processing Effect

The pond wetland combination technology has a high retention rate for runoff pollutants under different rainfall conditions. The smaller the rainfall, the higher the retention rate. The highest retention rates for TSS, COD, TN, and TP can reach 92.9%, 96.0%, 85.7%, and 80.9%.

5.2.3.6 Application Examples

The combination technology of pond wetland purification has been applied in the comprehensive environmental improvement project of Funiuxi in Dadukou District, Chongqing. The project is located downstream of the Funiuxi Railway Bridge in Jiansheng Town. This project integrates multi-level runoff pollution control technologies such as retention channels, low-level ponds, multi ponds, and artificial wetlands. The project was completed in January 2015, with a scale of 1000 m^3/d and an area of 1450 m^2. The treatment object is the initial rainwater runoff. During the operation of the project, the surface runoff is reduced by about 50%, and the removal rates of SS, COD, NH_3-N, TN, and TP in the initial surface runoff are above 75%, 45%, 50%, 30%, and 45%, respectively. The treated effluent water quality meets the Class V standard in the Environmental Quality Standards for Surface Water (GB3838) and is used for ecological replenishment of the Funiuxi River; The engineering investment is 260 yuan/m^2, and the operating cost is about 0.02–0.03 yuan/m^3; The investment in multi-level runoff pollution control technologies such as retention channels, low-level ponds, multi ponds, and artificial wetlands constructed in this project is only 50% of that of similar technologies in China; It has the function of improving regional biodiversity and beautifying the coastal landscape.

References

1. Technical guidelines for urban drainage systems and runoff pollution control. (draft proposal, in Chinese).
2. Zhang, W., & Jiang, J. (2020). Application of Mashai rivulet storage reservoir in initial rainwater runoff pollution control in Fuzhou. *China Municipal Engineering, 06*, 52–54. (in Chinese).
3. Wang, J., Peng, L., & Li, H., et al. (2021). A rainwater runoff control device based on pipe network regulation and storage. Beijing: CN112647582A. (in Chinese).
4. Wang, W., & Wang, J. (2021). Pipe network runoff purification device based on the principle of swirling sand settling. Beijing: CN112717493A. (in Chinese).
5. National Major Scientific and Technological Project for Water Pollution Control and Treatment: Research and engineering demonstration of key technologies for construction, renovation, and operation control of urban rainwater and wastewater pipeline networks with high interception rate of mixed flow system (2008ZX07317–001–01, in Chinese).
6. Zhang, Y. (2018). *Study on optimization method of runoff pollution control measures based on urban ecological stormwater management*. Tianjin University. (in Chinese).

7. National Major Scientific and Technological Project for Water Pollution Control and Treatment: Research and demonstration of water environment pollution control and quality improvement technology in old urban areas (2008ZX07313–001, in Chinese).
8. Xu, Z., Wang, W., Li, H., et al. (2010). Treatment technique of combined sewer overflows. *Environmental Engineering, 28*(S1), 153–156. (in Chinese).
9. Li, Y., Zhang, L., Geng, X., et al. (2020). Research on the application of sponge city concept on highways. *Soil and Water Conservation Application Technology, 03*, 27–29.
10. Guo, E., Chen, L., Sun, R., et al. (2015). Effects of riparian vegetation patterns on the distribution and potential loss of soil nutrients: A case study of the Wenyu River in Beijing. *Frontiers of Environmental Science & Engineering, 9*(2), 279–287.
11. Li, P., Cui, B., Fu, W., et al. (2013). Effects of riparian buffer vegetation types and widths on pollutant removal. *Journal of Nanjing Forestry University (Natural Science Edition), 37*(06), 47–52. (in Chinese).
12. Haycock, N. E., & Muscutt, A. D. (1995). Landscape management strategy for the control of diffuse pollution. *Landscape and Urban Planning, 31*(1–3), 313–321.
13. Osborne, L. L., & Kovacic, D. A. (1993). Riparian vegetated buffer strips in water-quality restoration and stream management. *Freshwater Biology, 29*(2), 243–258.
14. Lee, K. H., Isenhart, T. M., & Schultz, R. C. (2003). Sediment and nutrient removal in an established multi-species riparian buffer. *Journal of Soil and Water Conservation, 58*(1), 1–8.
15. Hefting, M. M., Clement, J. C., Bienkowski, P., et al. (2005). The role of vegetation and litter in the nitrogen dynamics of riparian buffer zones in Europe. *Ecological Engineering, 24*(5), 465–482.
16. Jilin People's Publishing House. (2016). *Jilin province construction standardization management office technical guidelines for Sponge City construction in Jilin Province.* (in Chinese).
17. http://www.zjsce.com/210228-18068/246068.html.
18. http://www.jnzezhong.com/yyal/qlrgsd.html.

Chapter 6
Rainfall Runoff Pollution Control Technology Process Package, Application Models and Technology Package

After years of accumulation, especially in the past 15 years of research, China has made great progress in the single technology of urban rainfall runoff pollution control, but a single technology often cannot resolve the pollution control problems caused by rainfall runoff in a region (or confluence area), and a number of technologies are required to be reasonably combined to form a complete set of technologies to effectively solve the problem of runoff pollution in a specific area. On the basis of the introduction of individual technologies in Chaps. 3 –5 of this book, this chapter summarizes the research progress of the Major National Water Pollution Control and Treatment Science and Technology "Water Special" on urban rainfall runoff pollution control, and establishes technical process packages, application modes and complete sets of technologies according to the needs of urban rainfall runoff pollution control problems, combined with the individual technical characteristics and the demonstration project or engineering demonstration of "Water Special", to provide technical support for urban runoff pollution control.

6.1 Build Ideas

In the past 15 years, many technical achievements of the "Water Special" in rainfall runoff pollution control technology were summarized, and after evaluation of their technical readiness, technological innovation and comprehensive technical performance, 72 technical achievements with high technical performance and engineering application were screened. On this basis, the 72 technical achievements were summarized into 4 technical links in the whole chain treatment technology system of urban rainfall runoff pollution control (diagnosis and evaluation technology, source reduction technology and facilities, process control technology and facilities, and back-end treatment technology and facilities). The technical link is further divided into supporting technology, supporting technology point and single technology, and the

D. Sun and F. Qi, *Urban Rainfall Runoff Pollution Control Technology and Application*,
https://doi.org/10.1007/978-981-96-4174-1_6

technical topology diagram is shown in Fig. 1.7. Among them, the diagnostic evaluation technology contains 18 individual technologies, the source reduction technology and facility contains 27 individual technologies, the process control technology and facility contains 17 individual technologies, and back-end governance technologies and facility contains 10 individual technologies.

According to the technical topology structure, four process packages were formed, including diagnosis and evaluation technology, source reduction technology, process control technology and back-end governance technology. On this basis, combined with the type of urban underlying surface, drainage system, runoff and pollutant migration path, as well as the specific runoff pollution control needs and goals, the source reduction technology, process control technology and back-end treatment technology were selected from the process package to form a technical application mode with certain applicable scenarios and conditions. Finally, according to the current situation, planning and development goals of urban infrastructure construction in China and the runoff pollution control objectives, a complete set of rainfall runoff pollution control technologies suitable for different drainage systems (combined flow system, diversion system and mixed flow system) in China's cities and towns is proposed. The process package, technology application mode and the construction process of the complete set of technology are shown in Fig. 6.1.

6.2 Urban Rainfall Runoff Pollution Control Technology Package

As mentioned above, urban rainfall runoff pollution control consists of the following four technical process packages.

6.2.1 Diagnostic Evaluation Technology Process Package

This process package is required to quantitatively express the urban rainfall runoff process and the pollutant load and migration therein, simulate and predict the runoff and pollution load, analyze the pollution load and source of urban runoff, serve the monitoring of runoff pollutants and the efficiency of treatment facilities, effectively guide the planning, design, construction, management and maintenance of urban runoff pollution control facilities, and realize a sustainable water cycle in the development area.

There are three sub-process packages in the diagnostic evaluation technology package, namely the pollution analysis technology sub-process package, the system scheme design technology sub-process package and the monitoring and evaluation technology sub-process package.

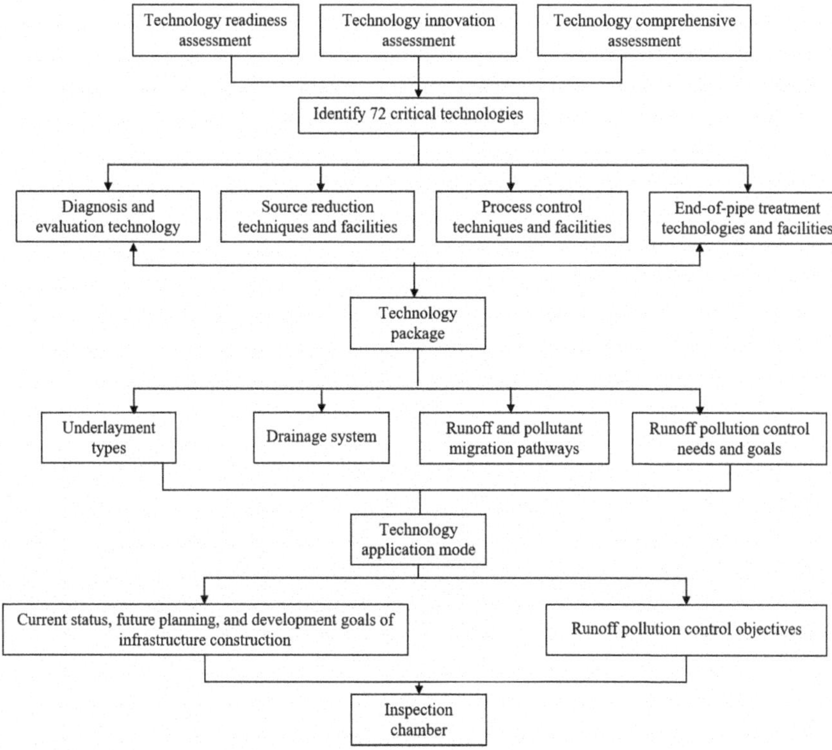

Fig. 6.1 Construction flow chart of technical process package, application mode and complete set of technology

Among them, the sub-process package of pollution analysis technology includes six technologies, including the estimation methods of rainfall runoff pollution to the receiving water body, the urban non-point source pollution load in typial mountain cities, the prediction model of stormwater production-confluence-flow of low-impact development in green buildings, water quality and yield process expression of urban surface runoff based on dynamic runoff coefficient and pollutant reduction coefficient, the simulation and prediction of runoff infiltration and pollution reduction of grass planting ditch, and simulation of rainwater runoff reduction by infiltration roads based on regional soil geological characteristics.

The system scheme design technology sub-process package includes the best comprehensive operation of initial rainwater pollution control based on "concentration control" rainwater regulation and storage, the applicability of urban rainwater retention and utilization, the applicability of overflow pollution control of urban drainage system, the optimization of retention factor based on improved SWMM and unit investment environmental benefits, the synchronous process monitoring of water quantity and quality and model parameter determination method of urban non-point source pollution, the overflow capacity test and optimization setting of urban road rainwater outlets, and the design of drainage system with integrated interception-storage-treatment.

The monitoring and evaluation technology sub-process package includes four technologies: rain garden filler permeability test equipment and key technologies, efficiency evaluation of regulation and storage system based on the combination of on-site monitoring and system simulation, analysis and evaluation model of LID on water quality, and the evaluation method of LID on rainy overflow pollution reduction. The specific technology name, source of achievement, use and scope of application are shown in Table 6.1.

6.2.2 Source Reduction Technology Process Package

The process package is required to realize rainwater infiltration, retention, storage, purification and resource utilization through green ecological facilities, and achieve the purpose of rainwater soil infiltration, groundwater recharge, rainwater runoff peak cutting, rainwater purification and comprehensive utilization.

The source reduction technology process package includes three sub-process packages, namely the penetration technology sub-process package, the retention technology sub-process package and the utilization technology sub-process package.

Among them, the infiltration technology sub-process package includes the the construction of infiltration pavement to strengthen rainwater infiltration and purification, enhanced infiltration of surface runoff by the pavement, the purification of surface runoff formed from impermeable underlying surface, new OGFC pavement structure and optimization, enhanced infiltration of surface runoff by permeable road surface, green infiltration by cellular concrete with plant, percolation purification by tree pool with combined multi-media, planting-maintenance-rapid restoration of green land plants based on rainwater shock and water level change, porous substrate grass planting ditch, control of non-point source pollution in migration section in mountain cities, source area biological seepage reduction in mountain urban surface runoff, the construction of shallow trenches achieving strengthen rainwater infiltration and purification, soil infiltration and emission reduction.

The sub-process package of detention storage technology includes green roof construction, roof runoff diversion purification, applicability green roof source pollution control and interception, green building community rainwater wetland runoff control, green building community stepped green space interception runoff, urban green space multi-functional storage—retention emission reduction—water quality assurance, automatic abandonment of snow melting agent biological retention zone, biological retention system suitable for northern urban roads, new biological retention pool (two-phase biological retention system with synchronous nitrogen and phosphorus removal function), garden-style rainwater collection and infiltration, new rain garden to eliminate the utilization of stormwater runoff from urban stormwater, and a rapid treatment technology and application of diatomaceous earth in the initial rainwater ecological retention.

Table 6.1 Technical information sheet for pollution control of urban rainfall runoff

Technical link	Supporting technology	Individual technologies	The name of the project	Application	Scope of application
Diagnostic evaluation techniques	Pollution analysis technology	Estimation Techniques for Pollution Load from Rainfall Runoff into Receiving Water Bodies	Development and Engineering Demonstration of Functional Restoration and Water Quality Improvement Technologies for Landscape Water Bodies in Central Districts of Tianjin	Monitoring and simulating runoff samples to identify the types of pollutants in different rainfall events, calculating the concentration and total amount of pollutants in the receiving water body	Pollution of rainfall runoff
		Techniques for Nonpoint Source Pollution Load in Mountainous Cities	Research and demonstration on urban water pollution control and comprehensive improvement of water environment technology	Establishing a runoff pollution load model for the rainfall process in mountainous urban watersheds, calculating the concentration of SS during the rainfall runoff process, and assessing the generation of runoff pollution loads in each sub-watershed of the basin	Design of rainwater systems for mountainous Cities
		Green Building and Community Low Impact Development Stormwater System Pollution Generation and Prediction Technology	Research and demonstration of stormwater system for green building and low-impact development of community	Dynamic calculation of the entire process of rainfall runoff, extraction of GIS geographical data, spatial computation and spatial analysis, and connection with China's design standard system and specifications	Rainwater -building and community rainwater quality and quantity simulation and sponge city facility performance assessment

(continued)

Table 6.1 (continued)

Technical link	Supporting technology	Individual technologies	The name of the project	Application	Scope of application
		Techniques for Expressing Urban Surface Runoff Volume and Water Quality Processes Based on Dynamic Runoff Coefficients and Pollutant Reduction Coefficients	Research on urban surface runoff reduction and non-point source pollution reduction technologies	Process of urban surface runoff water quantity and quality	Urban surface runoff water quality process
		Simulation and Prediction Technology for Runoff Infiltration and Pollution Reduction Based on Grassed Swales	Technology integration and demonstration of landscape water body function restoration and water quality improvement in central urban area of Tianjin	Simulation and prediction of pollution reduction of rainfall runoff by grassed swales	Residential areas, commercial areas, parks, and roadsides
		Simulation techniques for reducing rainwater runoff through permeable pavements suitable for soil and geological characteristics	Technology integration and demonstration of landscape water body function restoration and water quality improvement in central urban area of Tianjin	Permeable pavement runoff reduction mathematical model can be used to calculate the runoff reduction rate of permeable bricks and permeable concrete pavements	Northern water-scarce cities, impervious surfaces

(continued)

Table 6.1 (continued)

Technical link	Supporting technology	Individual technologies	The name of the project	Application	Scope of application
	System scheme design technology	Optimal Integrated Operation Technology for Initial Rainwater Pollution Control Based on "Concentration Control" Rainwater Retention	Development and Engineering Demonstration of Techniques for Functional Restoration and Water Quality Improvement of Landscape Water Bodies in the Central Area of Tianjin	The initial rainwater with peak pollutant concentrations at the diversion outlet can be directed into storage tanks and sewage treatment plants, significantly reducing the pollution level of the initial rainwater	Initial rainwater pollution control
		Applicability of Urban Rainwater Retention and Utilization Technologies	Urban Areas of the Chaohu Lake Basin Technologies and Demonstration for Rainwater and Wastewater Collection, Treatment, and Runoff Pollution Control	Improving the original soil, enhancing the facility's ability to treat rainwater runoff, install perforated drainage pipes at the bottom of the facility to collect filtered rainwater, connect them to a rainwater inlet shaft, and discharge into the municipal stormwater system, thereby achieving the goal of reducing rainwater runoff	Rainwater runoff pollution control

(continued)

Table 6.1 (continued)

Technical link	Supporting technology	Individual technologies	The name of the project	Application	Scope of application
		Applicability of Urban Drainage System Overflow Pollution Control Technologies	Urban Areas of the Chaohu Lake Basin Rainwater and Wastewater Collection, Treatment and Runoff Pollution Control Technologies and Demonstration	For rainwater runoff pollution control	Rainwater runoff pollution control
		Optimization Technology for Interception Multiples Based on Improved SWMM and Unit Investment Environmental Benefits	Chao Lake Basin Urban Area Rainwater and sewage collection, treatment and runoff pollution control technology and demonstration	Used to simulate water quality and quantity during urban heavy rain periods	Simulating water quality and quantity during heavy rain in a city
		Synchronous monitoring of water quantity and quality in urban non-point source pollution and methods and techniques for determining model parameters	Research on Urban Surface Runoff Control and Nonpoint Source Pollution Reduction Technologies	For calculating the scale of urban surface runoff reduction and control related projects	Calculate the scale of engineering related to urban surface runoff reduction and pollution control
		Testing and Optimization Techniques for the Overflow Capacity of Urban Road Storm Drains	Research on Urban Drainage System Overflow Pollution Reduction and Runoff Control Technology	For the design of rainwater outlets in drainage systems	Design of Drainage System Rainwater Inlets

(continued)

Table 6.1 (continued)

Technical link	Supporting technology	Individual technologies	The name of the project	Application	Scope of application
		Key technologies for the design of integrated drainage systems with interception, storage, and treatment capabilities	Research on Urban Drainage System Overflow Pollution Reduction and Runoff Control Technology	Designing and application of overflow pollution control and rainwater storage tanks for drainage systems	Control of Overflow Pollution in Drainage Systems and Design and Application of Rainwater Retention Basins
	Monitoring technology	Rain garden fill material permeability coefficient testing equipment and key technologies	Development and Engineering Demonstration of Functional Restoration and Water Quality Improvement Technologies for Landscape Water Bodies in Tianjin Central Urban Area	To test the infiltration coefficient of rain garden fillers	Urban impervious surfaces
		Assessment Technique for the Efficiency of Retention Systems Combining Field Monitoring and System Simulation	Comprehensive Pollutant Reduction and Water Quality Assurance Key Technologies in the Main Urban Area of Kunming	Utilizing a combination of online monitoring and computer modeling methods, a comprehensive assessment and consideration of the individual and integrated efficiency of detention ponds are conducted at the system level	All regulation and storage systems, supporting the planning, design, operation, and performance evaluation of regulation and storage systems

(continued)

Table 6.1 (continued)

Technical link	Supporting technology	Individual technologies	The name of the project	Application	Scope of application
		Analysis and evaluation model and technology of LID's impact on water quality	Research on Urban Surface Runoff Control and Nonpoint Source Pollution Reduction Technologies	The model is not only capable of simulating the current situation but also provides users with a complete set of solutions	LID model construction for site
		LID Assessment Technology for Reducing Pollution from Stormwater Overflow	Research on Pollution Reduction and Runoff Control Technologies for Urban Drainage System Overflow	Applied to the assessment of stormwater overflow pollution reduction in Tinlin drainage systems	Patch-based LID planning, optimized design, and source control of stormwater overflow pollution in high-density urban areas with combined sewer overflow systems
Source reduction technology	Infiltration technology	Enhanced Rainwater Infiltration and Purification with Permeable Pavement Construction Technology	Tianjin Central Urban Area Landscape Water Body Function Restoration and Water Quality Improvement Technology Development and Engineering Demonstration	Using permeable pavements and pervious asphalt to enhance rainwater infiltration and purification, thereby achieving the goal of reducing stormwater runoff	Northern Water-scarce Cities Non-permeable Surfaces

(continued)

Table 6.1 (continued)

Technical link	Supporting technology	Individual technologies	The name of the project	Application	Scope of application
		Surface Runoff on Roads Infiltration Promotion Technology	Chongqing Main City Drainage System Safety and Urban Non-point Source Pollution Control Technology Research and Integrated Demonstration	Various adsorbent materials are buried in the shallow soil to enhance the infiltration of surface runoff on roads	Enhancing surface runoff infiltration in areas not suitable for deep excavation
		Impervious Surface Runoff Treatment Technology	Green Building and Community LID Rainwater System Research and Demonstration	Achieving staggered discharge of rainwater through the flow paths and collection times of rooftops, roads, and squares, reducing the operational load on subsequent treatment structures, and enhancing the effectiveness of pollution control and flow interception	Peak flow reduction, pollutant reduction, and flow delay
		New OGFC Pavement Structure and Optimization Technology	Urban Road and Open Space LID Rainwater System Research and Demonstration	Using recycled OGFC (Open-Graded Friction Course) mixture to design high porosity, well-draining permeable pavements that reduce stormwater runoff from urban road motor lanes, non-motorized lanes, and sidewalks	Urban road motor lanes, non-motorized lanes, and sidewalks

(continued)

Table 6.1 (continued)

Technical link	Supporting technology	Individual technologies	The name of the project	Application	Scope of application
		Pervious Pavement Infiltration Promotion Technology	Hefei Binhu New District LID and Water Environment Remediation Technology Research and Engineering Demonstration	Using natural minerals (quartz sand, clay) as materials for pervious paving to increase the infiltration of rainwater on the road surface	Runoff volume control and pollution reduction in newly developed urban areas
		Plant-Grown Permeable Concrete Green Infiltration Technology	Urban Road and Open Space LID Rainwater System Research and Demonstration	Permeable paving made from plant-grown permeable concrete as a substrate, used to reduce runoff from multifunctional stormwater detention facilities' banks and vegetation belts	Banks and vegetation belts of multifunctional stormwater detention facilities
		Composite Multi-medium Infiltration and Filtration Purification Tree Pit Technology	Urban Road and Open Space LID Rainwater System Research and Demonstration	Using high adsorption capacity materials such as slag as artificial media in tree pits to purify urban road runoff	Sponge city transformation of urban roads in old city areas with insufficient green space
		Green Space Planting, Maintenance, and Rapid Repair Technology Based on Rainwater Impact and Fluctuating Water Levels	Urban Road and Open Space LID Rainwater System Research and Demonstration	Designed for low impact development rainwater facilities in cities with similar climate and soil conditions like Beijing and Shenzhen	Urban LID Rainwater Facility Design

(continued)

Table 6.1 (continued)

Technical link	Supporting technology	Individual technologies	The name of the project	Application	Scope of application
		Porous Matrix Grass Swale Technology	Hefei Binhu New District LID and Water Environment Remediation Technology Research and Engineering Demonstration	Using natural minerals (quartz sand, clay, gravel, pumice) as the substrate for grass swales to control runoff flow	Runoff volume control and pollution reduction in newly developed urban areas
		Integrated Technology for Controlling the Migration of Urban Non-point Source Pollution in Mountainous Cities	Chongqing Liangjiang New District Urban Water System Construction Technology Research and Demonstration	Maintaining the balance of urban water cycles, protecting receiving water bodies from secondary pollution, and partially replenishing groundwater	Pollution control of runoff from roads with slopes of 5–10% in urban areas
		Bio-infiltration and Reduced Runoff Technology for Source Areas of Urban Surface Runoff in Mountainous Cities	Chongqing Liangjiang New District Urban Water System Construction Technology Research and Demonstration	The surrounding landscape green spaces retain and promote infiltration of runoff from impervious surfaces, reducing urban surface runoff while controlling the initial runoff pollution from impervious surfaces	Runoff from impervious surfaces such as roads and parking lots in mountainous urban areas

(continued)

Table 6.1 (continued)

Technical link	Supporting technology	Individual technologies	The name of the project	Application	Scope of application
		Infiltration Trench Construction Technology for Enhanced Rainwater Infiltration and Purification	Tianjin Central Urban Area Landscape Water Body Function Restoration and Water Quality Improvement Technology Development and Engineering Demonstration	Used for the infiltration and purification of rainwater on the ground	Northern water-scarce cities Non-permeable surfaces
		Soil Augmentation and Runoff Reduction Technology	Key Technology Research and Engineering Demonstration for Rainwater Runoff Pollution Control and Ecological Utilization in River Network Cities	Using a combination of different media to enhance soil infiltration and collect rainwater for in-situ infiltration and purification	Control and utilization of rainwater runoff in areas with poor soil infiltration performance such as residential communities, urban roads, and landscape green spaces
	Stagnation technology	Green Roof Construction Technology	Chongqing Main City Drainage System Safety and Urban Non-point Source Pollution Control Technology Research and Integrated Demonstration	Used for transforming urban rooftops into green roofs	Urban Roof Transformation

(continued)

Table 6.1 (continued)

Technical link	Supporting technology	Individual technologies	The name of the project	Application	Scope of application
		Roof Runoff Diversion and Purification Technology	Chongqing Main City Drainage System Safety and Urban Non-point Source Pollution Control Technology Research and Integrated Demonstration	Effectively purifies roof runoff by separating it into different qualities, reducing roof runoff	Urban Roof Runoff Control
		Applicable Green Roof Source Pollution Control and Intercept Flow Technology	Green Building and Community LID Rainwater System Research and Demonstration	Used to reduce the peak flow of roof runoff, pollutants, and delay the time for runoff to reach its peak	Peak Runoff Reduction, Pollutant Reduction, and Runoff Delay
		Green Building Community Rainwater Wetland Runoff Control Technology	Research and Demonstration of LID Rainwater Systems in Green Buildings and Communities	Relying on rainwater gravel wetland devices to control the volume of rainwater runoff, including the reduction and delay of rainwater runoff peaks, as well as the absorption of the total runoff volume	Rainwater - Runoff Delay, Total Runoff Control, Runoff Pollution Reduction
		Stepped Green Space Runoff Moderation Technology in Green Building Communities	Research and Demonstration of LID Rainwater Systems in Green Buildings and Communities	Used for rainwater runoff delay, total runoff volume control, and runoff pollution reduction	Rainwater - Runoff Delay, Total Runoff Volume Control, Runoff Pollution Reduction

(continued)

Table 6.1 (continued)

Technical link	Supporting technology	Individual technologies	The name of the project	Application	Scope of application
		Urban Green Space Multifunctional Storage and Detention - Retention Reduction - Water Quality Assurance Technology	Research and Demonstration of LID Rainwater Systems in Urban Roads and Open Spaces	Applied in the banks and vegetation belts of multifunctional storage facilities to regulate and purify runoff rainwater	Banks and vegetation belts of multifunctional storage facilities
		Automated De-icing Agent Bypass Function of Bioretention Swale Technology	Research and Demonstration of LID Rainwater Systems in Urban Roads and Open Spaces	Utilizing elevation control and filtration structures to achieve automatic infiltration and bypass of de-icing agent-containing meltwater, with filtering soil microorganisms capable of biodegrading the de-icing requirements	Control and utilization of urban road runoff in cold northern regions, achieving automatic infiltration and bypass of de-icing agent-containing meltwater
		Bioretention System Technology Suitable for Northern Urban Roads	LID Rainwater System Comprehensive Demonstration and Assessment	Can be used to protect the integrity of ecosystems, collect rainwater for reuse or to replenish groundwater, and remove pollutants through various means, reducing the pollutant load of stormwater runoff	Northern Road Bioretention Strips

(continued)

Table 6.1 (continued)

Technical link	Supporting technology	Individual technologies	The name of the project	Application	Scope of application
		New Type of Bioretention Pond: A Two-Stage Bioretention System with Simultaneous Nitrification and Denitrification (SND) Capabilities	LID Rainwater System Comprehensive Demonstration and Evaluation	The new type of bioretention pond with simultaneous nitrogen and phosphorus removal capabilities can protect the integrity of ecosystems, collect rainwater for reuse or to replenish groundwater, and remove pollutants through various means, reducing the pollutant load in stormwater runoff	Removing nitrogen and phosphorus from stormwater runoff
		Garden-Style Rainwater Collection and Infiltration Technology	Hefei Binhu New District LID and Water Environment Remediation Technology Research and Engineering Demonstration	By constructing grassed swales, it achieves the purification of rainwater quality, control of runoff volume, and reduction of pollution	Runoff volume control and pollution reduction in newly developed urban areas
		Innovative Rain Garden Technology for Urban Storm Runoff and Flood Utilization	Key Technology Research, Reclamation Treatment of Rain and Sewage in Water-Scarce Cities, and Different Water Use Pathways Engineering Demonstration	Used for the design of rain gardens and to provide a theoretical basis for their future design and operation through in-depth monitoring and research on the performance of rain gardens, reducing stormwater runoff, removing pollutants, and considering ecological landscape effects	Road Stormwater Runoff Pollution Control

(continued)

Table 6.1 (continued)

Technical link	Supporting technology	Individual technologies	The name of the project	Application	Scope of application
		Initial Rainwater Ecological Retention and Diatomite Rapid Treatment Technology and Application	Hefei Binhu New District LID and Water Environment Remediation Technology Research and Engineering Demonstration	Using composite diatomite to treat initial rainwater	Initial Rainwater - Rapid Purification
Process control technology	Utilization Technology	Road Surface Rainwater Collection, Storage, Purification, and Utilization System and Technology	Key Technology Research and Engineering Demonstration for Regeneration Treatment of Rain and Sewage in Water-Scarce Cities and Different Water Use Pathways	Capable of collecting road surface rainwater for purification and reuse, thereby reducing road stormwater runoff pollution	Road Stormwater Runoff Pollution Control
		Parking Space Rainwater In-Situ Purification and Water Recycling Technology	Research and Demonstration of Regional Optimized Operation of Sewage Treatment Systems and Urban Runoff Pollution Reduction Technologies	Capable of recycling car wash wastewater or rainwater, and the purified water can be used for car washing or irrigating green spaces. If car washing is done directly in the parking space, the wastewater can be in-situ infiltrated and purified, which will not generate runoff and can also allow the wastewater to be purified and reused	Removal of Pollutants from Rainwater and Sewage

(continued)

Table 6.1 (continued)

Technical link	Supporting technology	Individual technologies	The name of the project	Application	Scope of application
	Shunt technology	Spatiotemporal Quality-Differentiated Collection and Treatment Technology for Stormwater Runoff	Research and Demonstration of Comprehensive Protection Technologies for Water Environment in Rapidly Urbanizing New Districts	Improves regional flood discharge issues while effectively reducing the non-point source pollution from stormwater runoff in rapidly developing urban areas	Removal of pollutants from stormwater runoff
		Wireless Broadcast-Based Initial Rainwater Diversion Technology	Research and Demonstration of LID Rainwater Systems in Urban Roads and Open Spaces	Indirectly judges the quality of the collected water based on the amount of rainfall and issues 'collect' or 'divert' commands wirelessly to control the opening and closing of electric valves for rainwater collection or control of runoff pollution	Initial Rainwater Diversion
		Combined Sewer Overflow Rainwater Interception and Diversion Control Device and Key Technology	Chongqing Main City Heavy Pollution River Water Pollution Control and Water Quality Improvement Technology Research and Demonstration	Can intercept rainwater from combined sewer overflows and separate initial rainwater from domestic sewage	Control of initial rainwater runoff pollution at combined flow discharge outlets in old urban areas of mountainous cities

(continued)

Table 6.1 (continued)

Technical link	Supporting technology	Individual technologies	The name of the project	Application	Scope of application
		Combined Sewer Overflow (CSO) End-of-Pipe Integrated Treatment Technology	Key Technology Research and Engineering Demonstration for Rainwater Runoff Pollution Control and Ecological Utilization in River Network Cities	Aiming for water environment capacity to meet standards, optimizing storage facilities within the drainage system and end-of-pipe ecological purification facilities to achieve efficient purification of overflow rain and sewage	End-of-pipe treatment for CSO pollution in plain river network areas, downstream drainage管网, or where sewage treatment plants have insufficient treatment capacity
	Pollution interception technology	High-Efficiency Grit Chamber and Key Technology for Stormwater Inlets	Research and Demonstration of LID Rainwater Systems in Urban Roads and Open Spaces	Installing pollution interception devices at stormwater inlets to retain pollutants in stormwater runoff	Efficient pollution interception at stormwater inlets on urban roads
		Stormwater Inlet Decontamination Device and Application Technology	LID Rainwater System Comprehensive Demonstration and Evaluatio	Installing decontamination devices at stormwater inlets to effectively remove some pollutants from stormwater runoff	LID Facilities
		High-Efficiency Adsorptive Purification Strip Technology with Grit Chamber for Road Stormwater	Research and Demonstration of Regional Optimized Operation of Sewage Treatment Systems and Urban Non-point Source Reduction Technologies	Used for intercepting pollutants in road stormwater	Removal of Pollutants from Stormwater Runoff

(continued)

Table 6.1 (continued)

Technical link	Supporting technology	Individual technologies	The name of the project	Application	Scope of application
		Automatic Purification Stormwater Manhole and Pollution Interception Technology	Research and Demonstration of LID Rainwater Systems in Urban Roads and Open Spaces	Filters and settles larger particulate matter in rainwater, reducing the amount of sediment in subsequent rainwate	Stormwater Suspended Solids (SS) Filtration
		Intelligent Stormwater Check Well Pollution Interception Device and Technology	Research and Demonstration of LID Rainwater Systems in Urban Roads and Open Spaces	Enables stormwater inlets to have multiple functions such as retaining pollutants, reducing runoff, and replenishing groundwater, effectively intercepting debris entering stormwater inlets and pipes, ensuring the maximum drainage capacity of stormwater pipes	Stormwater Inlet
		Vortex Sand Sedimentation Technology for Stormwater Pipes	Research and Demonstration of LID Rainwater Systems in Urban Roads and Open Spaces	Suitable for removing heavy particulate matter from rainwater in stormwater pipes	Removal of Heavy Particulate Matter from Rainwater
		Initial Rainwater Dedicated Pipe Storage and Detention Technology	Integrated Demonstration of Urban Water Pollution Process Control and Comprehensive Water Environment Improvement Technologies	Separates and treats initial rainwater to effectively eliminate runoff pollutants and alleviate the operational pressure on sewage treatment plants	Intercepts initial rainwater of about 5 mm from roads

(continued)

Table 6.1 (continued)

Technical link	Supporting technology	Individual technologies	The name of the project	Application	Scope of application
		Separate System Rainwater Drainage System End Floating Media Filtration Pollution Control Technology	Key Technology Research and Engineering Demonstration for Rainwater Runoff Pollution Control and Ecological Utilization in River Network Cities	Under the guidance of the overflow weir, the flow passes through the floating filter layer before exiting the outfall pipe. Due to the interception by the filter layer surface and the adhesion to the surface of the filter media particles, suspended solids (SS) are retained, and the rainwater is thereby purified	Pretreatment for separate system rainwater discharge outlets up to DN1000 and end-of-pipe ecological purification facilities
		Urban Overflow Pollution Reduction and Sewer Sediment Control Technology	Chaohu City Urban Water Environment Quality Improvement Research and Comprehensive Demonstration	Used for reducing urban overflow pollution and controlling sediment in drainage pipes	Control of overflow pollution at mixed stormwater inlets or combined sewer overflow outlets
	Regulating storage technology	Urban Combined Sewer System Overflow Reduction Technology	Key Technology Research and Engineering Demonstration for Water Quality Improvement and Protection of the Nanfei River in Hefei City	Used for reducing overflows in urban combined sewer systems	Control of combined sewer overflow pollution in urban areas

(continued)

Table 6.1 (continued)

Technical link	Supporting technology	Individual technologies	The name of the project	Application	Scope of application
		Overflow Pollution Control Detention Pond Design Technology Based on Kunming Rainfall and Drainage Characteristics	Key Technologies for Comprehensive Pollution Reduction and Water Quality Assurance in the Main Urban Area of Kunming	Integrated design technology for combined sewer detention ponds, based on the analysis of rainfall and drainage characteristics in the main urban area of Kunming, including scale calculation, inflow, emptying, and sediment removal mode selection	Design of combined sewer stormwater and wastewater detention ponds for runoff pollution control
		Hydrological Model-Based Design Method and Technology for Initial Rainwater Detention Ponds	Key Technologies for Comprehensive Pollution Reduction and Water Quality Assurance in the Main Urban Area of Kunming	Using hydraulic models to design initial rainwater detention ponds	Construction of initial rainwater detention ponds
		Initial Rainwater Detention Pond Design Technology Based on Kunming's Rainfall and Water Resource Utilization Characteristics	Key Technologies for Comprehensive Pollution Reduction and Water Quality Assurance in the Main Urban Area of Kunming	Determining the capacity algorithm for separated stormwater initial rainwater detention projects in the Kunming area, proposing a technical framework system primarily focused on rainwater detention, supplemented by other utilization measures	Design, construction, management, and maintenance of separated stormwater initial rainwater detention and utilization projects

(continued)

Table 6.1 (continued)

Technical link	Supporting technology	Individual technologies	The name of the project	Application	Scope of application
Back-end governance techniques	Physicochemical treatment technology	Pump Station Rainwater Enhanced Coagulation, Sedimentation, and Filtration Purification Technology	Technological Development and Engineering Demonstration for the Restoration of Functionality and Water Quality Improvement of Landscape Water Bodies in the Central Urban Area of Tianjin	Setting up coagulation facilities at pump stations to effectively remove suspended solids and colloids from runoff rainwater through coagulation	Purification treatment and reuse of rainwater runoff pollutants
		On-Site Treatment Technology and Equipment for Initial Rainwater Swirl Separation and High-Density Clarification Treatment	Urban Rain and Sewage Collection, Treatment, and Runoff Pollution Control Technology and Demonstration in the Chaohu Basin	Utilizing the centrifugal force created by high-speed tangential flow to effectively separate particulate pollutants	Pollutant Removal
		Initial Rainwater Non-point Source Pollution Control Hydraulic Swirl - Rapid Filtration Technology	Research and Demonstration of Water Pollution Control and Quality Improvement Technologies in Old Urban Areas	Using the adhesion between suspended particles and filter media particles, as well as physical screening, to retain suspended solids in rainwater and purify it	Interception, storage, treatment, and discharge of initial rainwater from roads

(continued)

Table 6.1 (continued)

Technical link	Supporting technology	Individual technologies	The name of the project	Application	Scope of application
	Ecological treatment technology	Composite Flow Constructed Wetland Treatment System and Technology	Key Technology Research and Engineering Demonstration for Regeneration Treatment of Rain and Sewage in Water-Scarce Cities and Different Water Use Pathways	Utilizing wetland plants, soil, and microorganisms to remove pollutants from rain and sewage, achieving flood control and purification of rainwater runoff	Purification and utilization of urban sewage and rainwater runoff from different functional areas
		Steep Slope Terraced Wetland Purification Technology in Mountainous Areas	Research and Demonstration of Water Pollution Control and Water Quality Improvement Technologies for Heavily Polluted Rivers in the Main City of Chongqing	Utilizing wetland plants, soil, and microorganisms to remove pollutants from rain and sewage water, achieving flood control and purification of rainwater runoff	Steep slope areas in mountainous regions

(continued)

Table 6.1 (continued)

Technical link	Supporting technology	Individual technologies	The name of the project	Application	Scope of application
		Separate System Rainwater Drainage System End Seepage-Storage Combined Ecological Synergistic Pollution Control Technology	Key Technology Research and Engineering Demonstration for Rainwater Runoff Pollution Control and Ecological Utilization in River Network Cities	Used for the purification of rainwater at the end of the separate system rainwater drainage system discharge outlets	End of separate system rainwater drainage system discharge outlets up to DN1000
		Three-Zone (Herbaceous-Shrubs-Trees) System Ecological Buffer Strip Technology	Research and Demonstration of Water Pollution Control and Water Quality Improvement Technologies for Heavily Polluted Rivers in the Main City of Chongqing	Used for steep mountain river banks to reduce pollutants in initial rainwater runoff	Reduction of pollutants in initial rainwater runoff from steep mountain river banks
		Multi-Pond System Ecological Buffer Strip Technology	Research and Demonstration of Water Pollution Control and Water Quality Improvement Technologies for Heavily Polluted Rivers in the Main City of Chongqing	Used for relatively gentle riverbank areas in mountainous cities to remove pollutants from initial rainwater runoff	Purification of initial rainwater in mountainous cities

(continued)

Table 6.1 (continued)

Technical link	Supporting technology	Individual technologies	The name of the project	Application	Scope of application
		Rainwater Recharge Landscape Water Body Water Quality Assurance and Rainwater Detention Technology	Research and Demonstration of LID Rainwater Systems in Green Buildings and Communities	Integrating artificial wetlands and fountain oxygenation facilities to treat landscape water bodies primarily replenished by rainwater, the water body is regulated by different water level detention volumes to reduce peak flow and delay peak timing of rainwater runoff in green buildings and communities	Rainwater - Pollutant Reduction and Water Volume Regulation
		Urban Non-point Source Pollution Control Water Body Purification and Ecological Coupling Restoration Technology	Research and Comprehensive Demonstration for Improvement of Urban Water Environment Quality in Chaohu City	Used for oxygenation, degradation of organic matter, and efficient nitrification/denitrification processes, while effectively enhancing the water body's self-purification ability, and improving urban aquatic ecology and water landscape	Quality enhancement and ecological restoration of impaired urban landscape water bodies

The utilization technology sub-process package includes two technologies: pavement rainwater harvesting, storage and purification utilization system and parking space rainwater *in-situ* purification, water storage and reuse. The specific technology name, source of achievement, use and scope of application are shown in Table 6.1.

6.2.3 Process Control Technology Process Package

The process package is required to realize the diversion, interception, storage and purification of rainwater runoff from the surface to the drainage pipeline stage, effectively guide the transformation of rainwater outlets, initial rainwater separation and purification, pipe network interception, rainwater storage and on-site treatment of overflow pollution, and realize the reduction of rainwater runoff and the purification of pollutants in the rainwater collection link.

The process control technology process package includes three sub-process packages, namely the sub-process package of diversion technology, the sub-process package of pollution interception technology and the sub-process package of transfer and storage technology.

Among them, the sub-process package of diversion technology includes four technologies, such as spatiotemporal collection and treatment of rainwater runoff, initial rainwater abandonment based on wireless broadcast, overflow rainwater interception and diversion control device and key technology, and CSO terminal comprehensive treatment of combined pipe network.

The sub-process package of sewage interception technology includes 9 technologies, such as high-efficiency sewage interception device and key technology at rainwater outlet, decontamination device and application of rainwater outlet, high-efficiency adsorption and purification belt with high-efficiency sewage interception type rainwater grate, automatic purification rainwater inspection well and sewage interception, intelligent sewage interception device of rainwater inspection well, swirl grit for rainwater pipeline, initial rainwater special pipe storage and storage, floating medium filtration pollution control at the end of the diversion rainwater drainage system, and urban overflow pollution reduction and sediment reduction control of drainage pipelines.

The sub-process package of the regulation and storage technology includes four technologies, such as the overflow reduction of urban combined flow system, overflow pollution control reservoir design based on Kunming rainfall drainage characteristics, initial rainwater reservoir design method based on hydraulic model, and initial rainwater reservoir design based on Kunming's rainfall and water resource utilization characteristics. The specific technology name, source of achievement, use and scope of application are shown in Table 6.1.

6.2.4 Back-End Treatment Technology Process Package

This process package is required to realize the efficient purification of pollutants from the overflow of combined pipe networks and the discharge of rainwater pipes by diversion.

The back-end treatment technology process package have two sub-process packages, namely physical and chemical treatment technology sub-process package and ecological treatment technologysub-process package.

Among them, the physical and chemical treatment technology sub-process package include three technologies, such as coagulation and precipitation, swirl separation, and filtration. The ecological treatment technology sub-process package includes 7 technologies, including composite flow constructed wetland treatment, cascade wetland purification in mountainous steep bank slope belt, ecological collaborative pollution control combining end infiltration and storage of diversion rainwater drainage system, ecological buffer zone of three-belt (herb-shrub-arbor) system, ecological buffer zone of multi-pond system, water quality assurance and rainwater regulation and storage of rainwater recharge landscape water body, and urban non-point source polluted water purification and ecological coupling restoration. The specific technology name, source of achievement, use and scope of application are shown in Table 6.1.

6.3 Application Mode of Urban Rainfall Runoff Pollution Control Technology

In view of the serious problem of urban runoff pollution in China, there are the following five technical needs in the current urban rainfall runoff pollution control technology.

(1) Green LID engineering technology. This can effectively reduce the pollution load of runoff at the source. However, at present, there are many types of underlying surfaces in cities and towns in China, and there is a lack of LID engineering technology application mode for different types of underlying surfaces.
(2) The new type of sewage interception rainwater gutter inlet. This is an important facility for collecting rainwater and synchronously realizing runoff pollution purification. Traditional rainwater gutter inlet have problems such as easy blockage and difficult dredging, resulting in low water collection efficiency and poor purification capacity, and it is necessary to transform and refine the management of rainwater outlets to achieve gray-green combination.
(3) Overflow pollution control technology. China's combined pipe network covers a wide area and the rainy combined drainage flow is large, exceeding the operating load of sewage treatment plants. Therefore, there is a risk of overflow pollution and appropriate overflow pollution control technology and joint scheduling

strategies between rainwater collection systems and urban sewage treatment plants are urgently needed.

(4) Initial rainwater control technology. The rapid development of urbanization in China, leading to the proportion of underlay hardening increasing year by year. The diversion rainwater pipe network showed problems such as misconnection, damage and sediment of pollutant pipes, resulting in high erosion effect in the early rainy days, high initial rainwater pollution load. Moreover, many cities have the problem of direct discharge of rainwater into the river, and it is urgent to effectively solve the problem of abandonment, purification, storage, terminal treatment and combined application technology that can effectively solve the problem of initial rainwater pollution.

(5) Engineering technology application model and complete set of technology. The uneven speed of urban development in China has brought about great differences in urban infrastructure conditions. It is difficult to realize the transformation of rainwater sewage diversion in the combined drainage system of the old city. In many urban areas, there are mixed areas of combined and diversion systems. Therefore, it is necessary to give engineering technology application models and complete sets of technologies suitable for different urban areas and drainage systems in China.

In view of the above problems, this section summarizes the research results of "water special" and forms the LID technology model, the technology model of the rainwater gutter inlet transformation and management, the technology model of the combined drainage system overflow regulation and storage, and the technology model of the initial rainwater purification of the diversion drainage system.

6.3.1 Technical Model of LID Runoff Pollution Control

With the rapid development of urbanization in China, the demand for land for urban expansion has changed the type of urban undersurface, the ground is gradually covered by hard materials such as cement and asphalt, the proportion of impervious surface is becoming higher and higher, and the water stagnancy and permeability of the underlying surface are becoming worse and worse. The city's original natural water cycle process has been disrupted, resulting in increased probability of flooding disasters and serious pollution of stormwater runoff. LID technology can solve these problems at the source to a certain extent. On the one hand, LID technology can reduce part of the runoff and pollution load from the source, replenish groundwater, and improve the utilization rate of rainwater; on the other hand, it can reduce the pressure on stormwater network transportation and sewage treatment plant operation. The LID technical models provided in this chapter include: building community runoff pollution control technology model, urban road path flow pollution control technology model, ecological parking runoff pollution control technology model, urban green space runoff pollution control technology model and rainwater comprehensive utilization technology model.

6.3.1.1 Technical Model of Runoff Pollution Control in Building Community

i. Objective

In response to the problem of runoff pollution in building areas, this technology model uses LID facilities to reduce rainfall runoff from roofs and roads in urban building districts and public building areas.

ii. Technical Composition and Technical Process

This technology mode consists of eight technologies, including rain garden, sunken green space, grass planting ditch, ecological tree pool, permeable pavement, green roof, rainwater outlet technology and rainwater break interface, and the technical process is shown in Fig. 6.2.

Roof rainwater is purified through the retention and infiltration of green roof technology, and excess rainfall runoff is further purified by rainwater disconnection technology from the roof to the ground in pebble ditches/rain gardens/grass ditches/sunken green spaces. Part of the rainwater on the pavement of the community seeps into the ground through the permeable pavement, part flows into the rainwater inlet, and the other part is further purified by the LID facility that adds functional decontamination media.

iii. Application Efficiency Analysis

This technology model has been applied in the first phase project of demolition and resettlement in Wanzhangpo Area, Guangming New District, Shenzhen. The project adopts an integrated application of green building and community as runoff pollution control technologies, including green roof, rain garden, rainwater tree pond, grass planting ditch, sunken green space, permeable paving and rainwater disconnection.

The project is located at the south of Huayu Road, at the east of Niushan Road, and at the west of Guangming Avenue, including 1–5 residential buildings, covering an area of 2.75 hm^2 and a construction area of 64,700 m^2. The engineering process technical process is shown in Fig. 6.3, and the engineering images are shown in

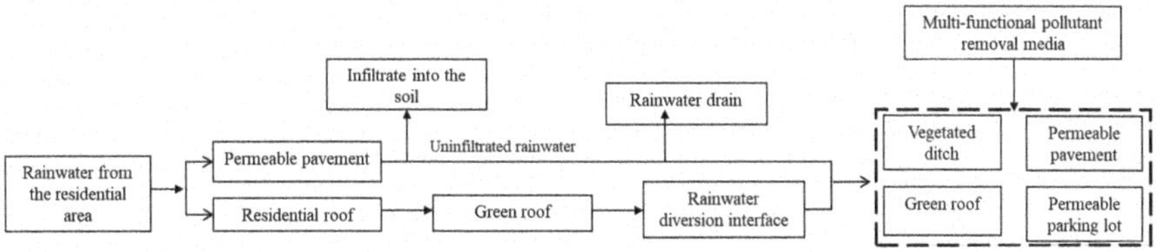

Fig. 6.2 Technical mode process of building community

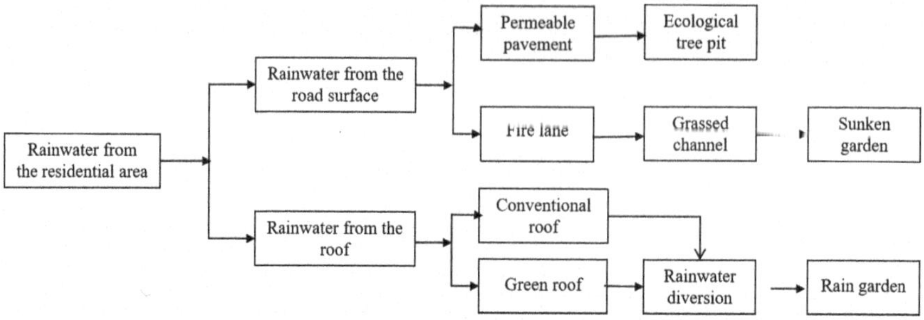

Fig. 6.3 Process technical process of the first phase project of runoff pollution control project for demolition, relocation and resettlement in Wanzhangpo area

Fig. 6.4. The runoff generated by permeable paving can enter the ecological tree pond for retention and purification, and the runoff generated by the fire lane can enter the concave green space through the grass planting ditch for retention and purification. Runoff from green and ordinary roofs can be disconnected through rainwater risers to rain gardens for purification. Applying this set of technical models, the impervious area of the community has been reduced by 20%, the runoff coefficient has been reduced to 0.27, the average annual rainfall peak has been reduced by more than 40%, and the pollutant load into river has been reduced by more than 70%.

6.3.1.2 Technical Model of Pollution Control of Urban Road Surface Runoff

i. Objective

This technology model significantly reduces runoff by reducing rainfall from urban pedestrian roads and motor vehicle roads by using LID facilities to reduce rainfall runoff from urban pedestrian roads and motor vehicle roads.

ii. Technical Composition and Technical Process

The technical model consists of seven technologies, including rain garden, sunken green space, grass planting ditch, ecological tree pool, permeable paving, permeable asphalt and rainwater overflow.

The technical process of urban road technology mode is shown in Fig. 6.5. Part of the road rainfall runoff is infiltrated into the ground through permeable paving technology or permeable asphalt technology, and when the infiltration is saturated or the rainfall runoff volume is too large, the excessive rainfall runoff will flow into the pebble ditch/stepped green space/sunken green space/grass planting ditch/ecological tree pond and other LID facilities on both sides of the road where functional decontamination media is added, and when the rainwater exceeding the treatment capacity of the facility it will overflow into the drainage pipe network through the overflow. When parking spaces are built on both sides of the road, rainfall runoff

a) Permeable paving

b) Eco tree pool

c) Grass ditch

d) Rainwater breaks

Fig. 6.4 Demolition and relocation of Wanzhangpo area in the first phase project of runoff pollution control project

at this time can also be collected into the parking space rainwater purification and storage facilities.

iii. Application Efficiency Analysis

This technology mode has been applied in the park road of Shenzhen Guangming New District. Guangming New District Park Road is located on the south and west sides of Xincheng Park, the north of Huaxia Road, the east of Guangqiao Road, arc-shaped around Xincheng Park, with a total project scale of 5.1 ha. Park Road is a collector streets as a municipal level II road, the width of the red line of the road is 40 m, the road motor vehicle lane is two-way four lanes, the central green width is 9 m, the roadside green width is 3 m, the bicycle lane is 1.5 m wide, and the sidewalk width is 3.5 m. Motorways and bicycle lanes are asphalt, and sidewalks are paved with permeable bricks. The total length of the road is 2.2 km, and the road is divided into two parts, of which the green belt (1.4 km) in the north bid section is transformed

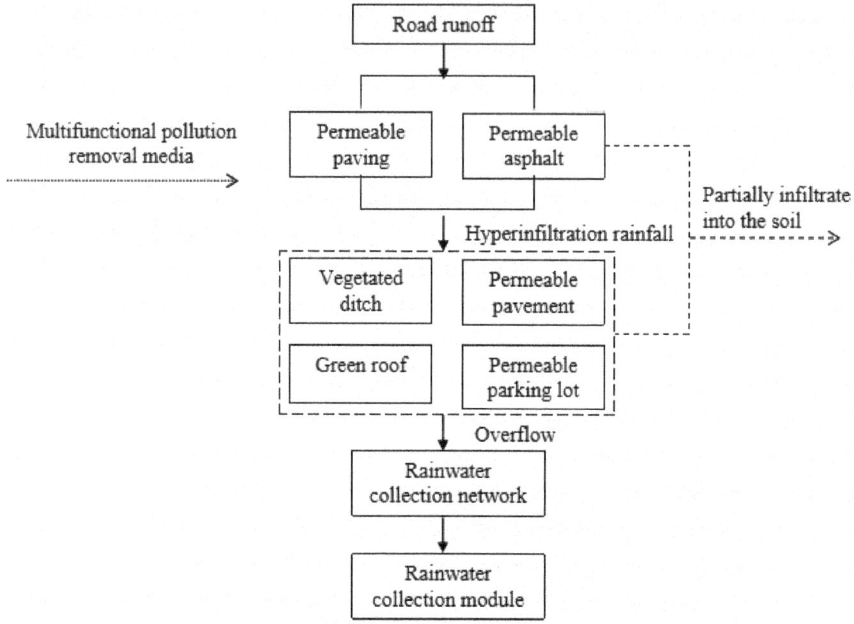

Fig. 6.5 Technical process of urban road path flow pollution control technology

into a biological detention belt, and the south bid section (the 0.8 km) is a normal design road, and the two lot roads have the same road cross-section except for the detention belt.

The technical process of this projects is shown in Fig. 6.6. Part of the runoff rainwater on sidewalks, motorways and bike lanes infiltrates into the ground through permeable brick pavements and permeable asphalt pavements. When the infiltration is saturated or the rainfall runoff volume is too large, the excess rainfall runoff flows into the bioretention zone through the road curb opening. Stormwater overflows are provided in the bioretention zone as drainage channels for runoff stormwater that exceeds the treatment capacity of the biological retention zone. Through the rainfall runoff retention, emission reduction, pollution control and seepage promotion of this process, the total annual runoff control rate of the park road area has reached more than 60%, and the annual runoff pollution load reduction has reached 40%, which has improved the ecological landscape effect of the park road and the comprehensive drainage capacity of the road, and the actual image of the project is shown in Fig. 6.7.

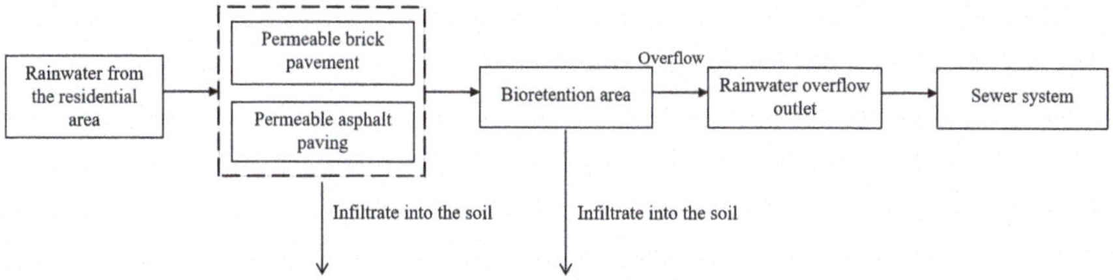

Fig. 6.6 Technical process of surface runoff control engineering of park in Guangming New District, Shenzhen

Fig. 6.7 Actual view of engineering biological retention zone and rainwater overflow

6.3.1.3 Ecological Parking Runoff Pollution Control Technology Model

i. Objective

In view of the problems of traditional parking that are easy to accumulate water and breed mosquitoes and flies in rainy days, ecological parking technology can absorb the runoff rainwater accumulated in the parking space and eliminate the pollutants in the rainy runoff, greatly improving the parking itself and the surrounding environment.

ii. Technical Composition and Technical Process

This technology model consists of three technologies, including permeable brick, packing filter layer and underground reservoir, and the treatment technical process is shown in Fig. 6.8. After the rainwater falls in the parking, it first infiltrates through the permeable bricks with good permeability on the surface of the parking, and then it was purified to a certain extent through the packing filter layer, and enters the underground reservoir with the rainwater collection pipe at the bottom of the parking and the pipeline under the green underground. The purified rainwater in the reservoir

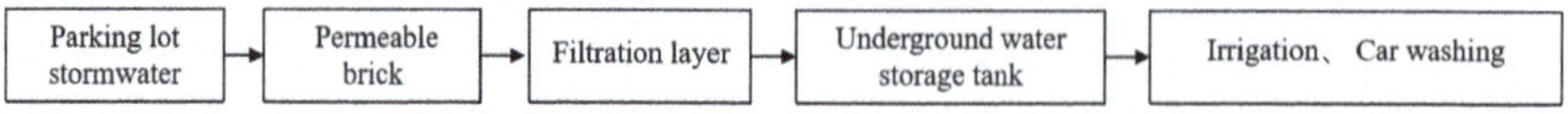

Fig. 6.8 Ecological parking runoff pollution control technology model and its technical process

Fig. 6.9 Application of ecological parking runoff pollution control technology model

can be used for landscaping, landscape water replenishment and reuse water such as car washes.

iii. Application Efficiency Analysis

This technical model is applied to the parking on the south side of Shangxian River Sub-district of Wuxi City, with a construction area of 40 m^2, a water storage volume of 4 m^3, an effective volume of 3.8 m^3, and the length and width of each parking space are 6 m and 3 m respectively. Purified rainwater from the bottom of the ecological parking is channeled through a permeable channel between the two parking spaces to a water storage tank buried in the green ground on the side. The water storage tank is a combined water tank made of plastic module material, and the module volume can be flexibly selected according to the water storage needs. The average removal efficiency of SS, COD, NH_4-N, TN and TP in runoff were 89.3%, 66.4%, 86.8%, 43.6% and 83.3%, respectively, which significantly improved the environment quality of the parking itself and its surroundings, reduced runoff pollution, and realized the comprehensive utilization of rainwater, and the actual project scene is shown in Fig. 6.9.

6.3.1.4 Urban Green Space Runoff Pollution Control Technology Model

i. Objective

Aiming at the problems of poor soil permeability and high shallow groundwater table, this technology model mainly increases the soil infiltration performance, improves

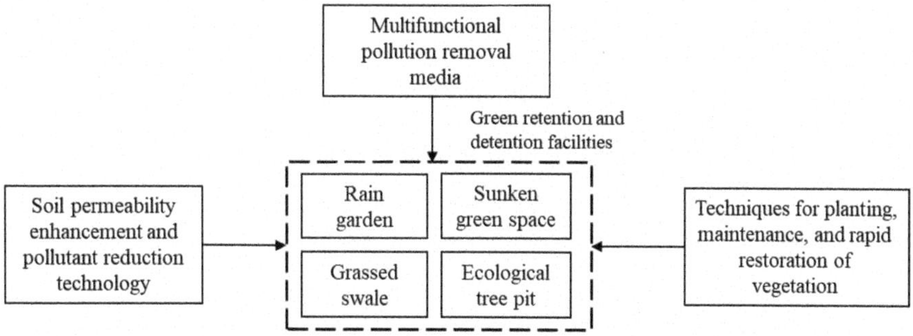

Fig. 6.10 Technological process of urban green space runoff pollution control technology

the rainwater infiltration efficiency and infiltration volume, and reduce rainfall runoff, through soil infiltration and emission reduction technology. Secondly, after using soil infiltration and emission reduction technology, plant planting and maintenance technology and a series of LID technologies, it was achieved that the improvement of landscape effect and the effective reduction of rainfall runoff and runoff pollutants.

ii. Technical Composition and Technical Process

This technical mode including 6 key technologies, including rain garden, sunken green space, grass planting ditch, ecological tree pond, soil infiltration and emission reduction, plant planting maintenance and rapid restoration, and the technical process is shown in Fig. 6.10. Urban rainfall runoff first increases rainwater infiltration through soil infiltration and emission reduction technologies. After the infiltration, the runoff and runoff pollution are further reduced by adding green detention facilities with multi-functional decontamination media. Using plant planting, maintenance and rapid restoration technology, maintains green stagnant storage.

iii. Application Efficiency Analysis

This technology mode has been applied in the key technical engineering of rainwater control and utilization of the ecological greenway network of Huancheng East Road in Jiaxing City, and the technological process is shown in Fig. 6.11. This technology mode reduces the pollution of stormwater runoff of the greenway itself and the surrounding at the source, and further effectively intercepts the pollutants in the rainwater before discharging into the receiving water body to protect the surface water environment. The project eliminated rainfall volumne greater than 20 mm, the total amount of external stormwater runoff reduction efficiency is more than 30%, and the pollution load reduction efficiency of SS and COD is greater than 30% (Fig. 6.12).

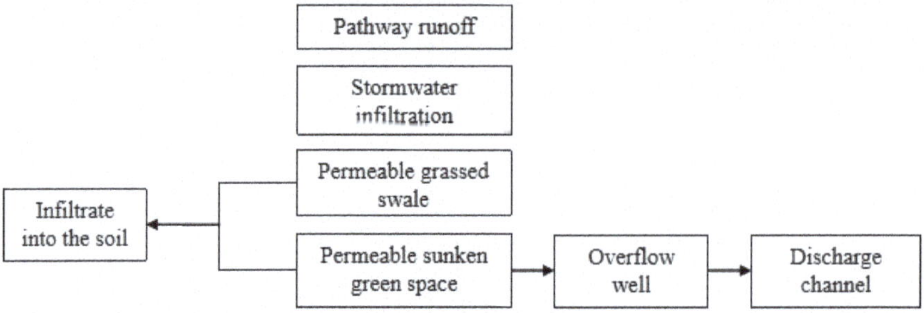

Fig. 6.11 Rainwater control and utilization engineering treatment process of ecological greenway network of Huancheng East Road, Jiaxing City

Fig. 6.12 Jiaxing City Huancheng East Road ecological greenway network rainwater control and utilization project real scene

6.3.1.5 Technical Model of Comprehensive Utilization of Rainwater

i. Objective

In view of the low comprehensive utilization rate of rainwater in the traditional fast rainwater drainage method, this technology mode collects rainfall runoff and purify it centrally by a physical and chemical method, and finally reuses it for community greening and watering, road flushing and toilet flushing.

ii. Technical Composition and Technical Process

The comprehensive utilization treatment technology mode of rainwater consists of eight technologies, including green roof, ecological parking, permeable pavement, concave green space, grass planting ditch, rainwater overflow, swirl grit and filtration and disinfection, and the technical process is shown in Fig. 6.13. After rainfall falls on LID facilities such as sunken green space, grass planting ditch, green roof, permeable parking, permeable pavement and permeable plaza with multi-functional decontamination medium, it is purified and stored through concave green space and grass planting ditch, and the runoff exceeding its capacity is transported to the rainwater collection module through the rainwater collection pipe network, and finally the runoff is physicochemical reused by swirling grit and filtration and disinfection.

iii. Application Efficiency Analysis

This technology model has been applied in the general public sports center of Guangming New District, Shenzhen, covering an area of 61,885 m^2 and a total construction area of 20,221 m^2. Guangming New District General Public Sports Center is located at the intersection of Guangqiao Road and Huaxia Road, north of Guangming New

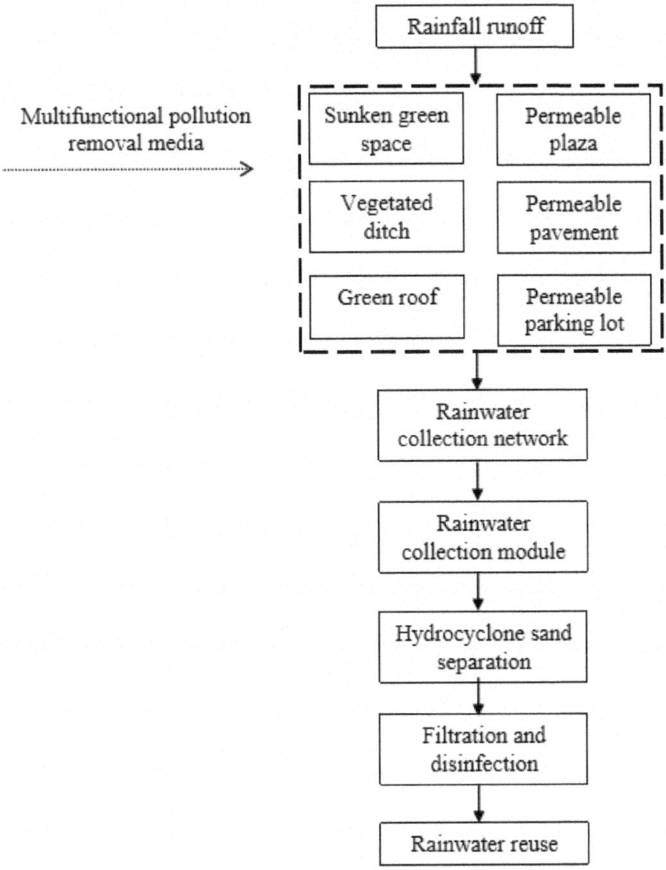

Fig. 6.13 Rainwater comprehensive utilization treatment technology model and its techical process

Town Park. The project adopts a number of rainwater runoff emission reduction facilities to form a comprehensive rainwater utilization system including green roof, sunken green space, grass planting ditch, permeable parking lot and rainwater collection module and other technical facilities, the technical processis shown in Fig. 6.14, and the actual project scene is shown in Fig. 6.15. Among them, the source emission reduction technology includes low-impact development facilities such as green roofs, permeable plazas, permeable parking lots, and sunken green spaces. Process emission reduction technology includes grass planting ditches and rainwater harvesting networks for rainwater diversion and control. End treatment technology includes rainwater harvesting modules and treatment equipment collecting and purifying rainwater and re-overflowing rainwater beyond the design standards into the municipal stormwater network. The project achieves a comprehensive control rate of 75% of annual runoff, a 50% reduction in annual runoff pollutants, a reduction of 37–47% in runoff peaks, and a rainwater reuse of about 6,000 m^3/a, which is used for greening irrigation and road flushing, and achieves annual water saving of about 20,000 CNY.

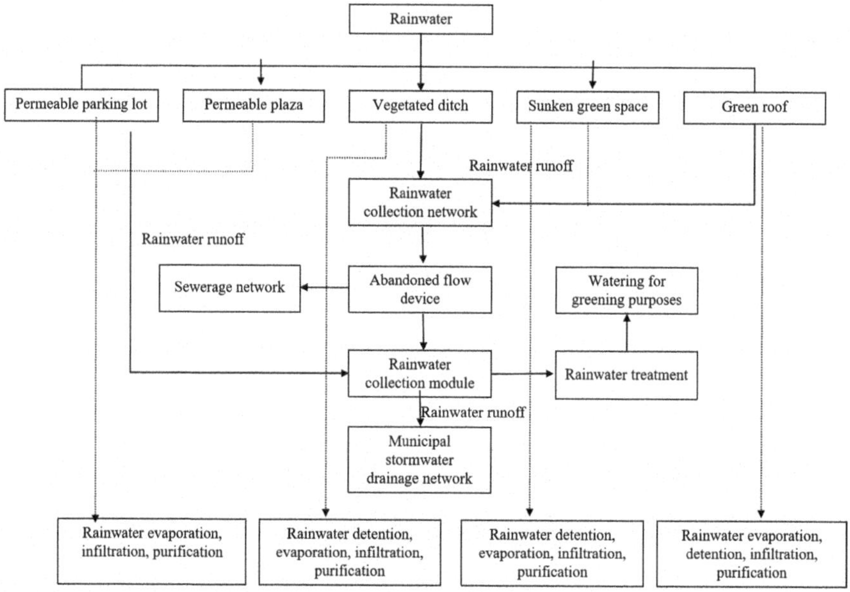

Fig. 6.14 Technical model and technical processof rainwater comprehensive utilization of general public sports center in Guangming New District, Shenzhen

a) Grass ditch

b) Green roof

c) Permeable Plaza

d) Rainwater purification device

Fig. 6.15 Rainwater comprehensive utilization project of general public sports center in Guangming New District, Shenzhen

6.3.2 Technical Model of Stormwater Gutter Inlet Renovation and Management

6.3.2.1 Objective

In view of the problems of easy blockage and difficult dredging of traditional rainwater gutter inlet, the technical mode of rainwater gutter inlet transformation and management effectively reduces the deposition of initial rainwater pollutants in rainwater wells, and greatly reduces the management difficulty of rainwater wells.

6.3.2.2 Technical Composition and Technical Process

This technology mode consists of four technologies, including sewage interception technology, stream abandonment, collection and purification technology and maintenance technology, and the treatment process is shown in Fig. 6.16. Rainfall runoff enters the dumping device after the interception through the rainwater gutter inlet and the sewage hanging basket, and infiltrates the groundwater through the seepage

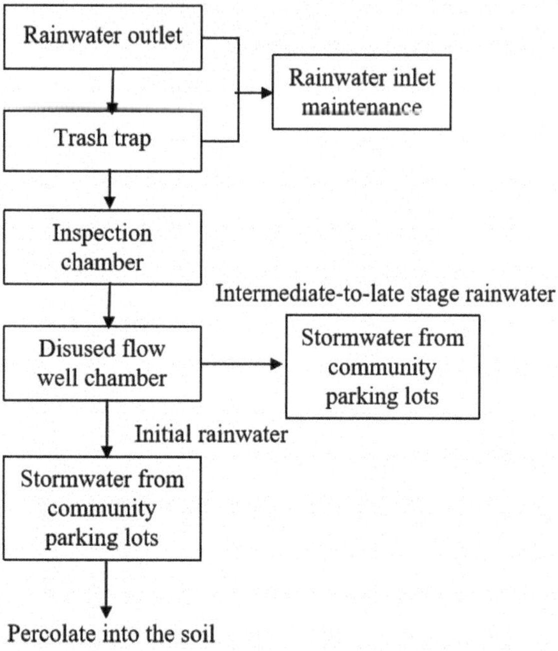

Fig. 6.16 Technical process of stormwater gutter inlet renovation and management model

pipe to replenish the groundwater, and the rainwater is transported to the rainwater collection module in the middle and late stages.

6.3.2.3 Application Efficiency Analysis

This technology model has been applied in the auxiliary road of Beijing Airport Expressway and the residential area under the jurisdiction of Zuojiazhuang Street in Beijing, among which the rainwater outlet sewage interception device is shown in Fig. 6.17. Runoff rainwater first passes through the sewage filtration device to intercept the pollutants of larger particles in the rainwater, and then enters into the rainwater well waste chamber for storage, and the initial runoff rainwater enters into the quartz sand layer, gravel layer and surrounding soil through the seepage pipe connected at the bottom of the waste chamber. When the middle and late runoff rainwater enters into the water depth in the rainwater well waste chamber reaches the height of the outlet pipe, the middle and late runoff rainwater accumulated in the rainwater well waste chamber flows out through the outlet pipe and enters the rainwater pipe network.

This technology model successfully intercepts at least 90% of the garbage entering stormwater gutter inlets and stormwater pipes, saving more than 90% of pipeline maintenance costs. Under the conditions of small rainfall recurrence period and

Fig. 6.17 Rainwater gutter inlet sewage interceptor

large rainfall recovery period, the removal efficiency of SS was in the range of 50%–80% and 40%–60%, and the runoff reduction rate was 25%–42% and 16%–25%, respectively.

6.3.3 Technical Model of Overflow Regulation and Storage of Combined Drainage System

6.3.3.1 Reservoir Technology Mode

i. Objective

In view of the problems of large drainage flow in rainy days, water pollution caused by direct discharge into the river beyond the operating load of the sewage treatment plant, this technology mode greatly alleviates the pollution loads of the overflow to the river by transporting the overflow sewage to the reservoir for treatment by physical and ecological facilities before discharging it into the river.

ii. Technical Composition and Technical Process

The technology mode of storage regulation is composed of three technologies, such as sewage interception well, storage reservoir and *on-site* treatment, and the treatment process is shown in Fig. 6.18. The combined drainage system is transported to the reservoir through the sewage interception well, and the combined drainage is discharged into the river after *on-site* treatment on rainy days, and transported to the sewage treatment plant for the purification when there is no rain and then discharged into the river.

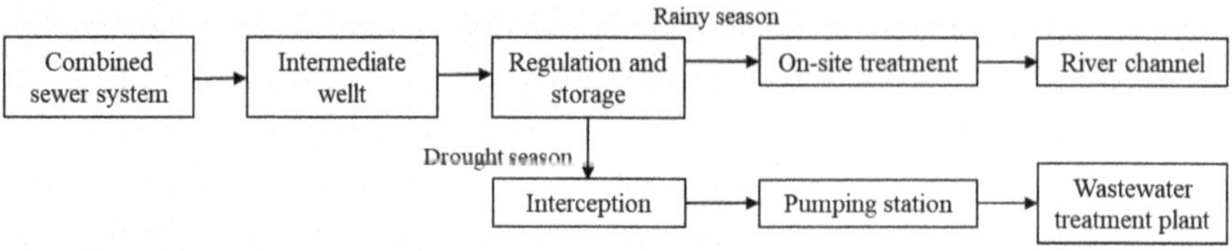

Fig. 6.18 Technical processof the technology mode of the reservoir

iii. Application Efficiency Analysis

This technology mode has been applied in the CSOs storage project of Shangcun Pumping Station in Zhonglou District, Changzhou City. The service area of the pumping station is mainly an unrenovated urban village using inline combined drainage, with a service area of about 55,000 m^2, a population of about 2,750 people, and a sewage capacity of about 400 m^3/d. In the Shangcun area, the excess rain and sewage being combined in the Shangcun area, was stored in the reservoir, and when it was sunny, the sewage was intercepted by the pumping station and entered the sewage treatment plant. If the amount of rainfall is heavy on a rainy day, the part that exceeds the storage capacity will be purified *on-site* and then discharged into the water body. The volume of the early rain storage tank is 45 m^3, the volume of the *on-site* processing fast filter tank is 35 m^3, and the processing capacity is 70 m^3/h.

6.3.3.2 Technical Mode of Joint Operation of Storage Pond-Sewage Treatment Plant

i. Objective

In view of the problem that the operating load of the single sewage treatment plants is vulnerable to influenced when treating combined sewage, this technology model can use the storage pond to purify the rainwater that exceeds the load of the sewage treatment plant in the rainy season, and reduce the running pressure of the sewage treatment plant.

ii. Technical Composition and Technical Process

The technical model includes four important technical links, including rainwater gutter inlets, rainwater wells, storage pond and rainy season purification of urban sewage treatment plants, and the treatment process is shown in Fig. 6.19. In rainy days, drainage enters the combined pipe network through rainwater gutter inlets, rainwater wells and other facilities, and then was transfers into the sewage treatment plant for the purification and was discharged into the river. When the rainwater sewage volume exceeding the treatment load of the sewage treatment plant, it is transported to the storage pond, and then transported to the sewage treatment plant for storage pond when there is no rain, and then is discharged after *on-site* purification in the rainy season.

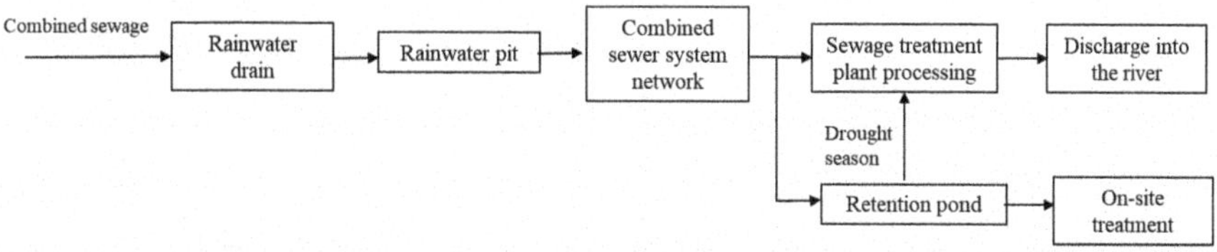

Fig. 6.19 Treatment process of the joint operation technology mode of the reservoir-sewage treatment plant

iii. Application Efficiency Analysis

This technology mode is applied to the joint control and operation project of rainwater sewage storage system and sewage treatment plant in the north area of Kunming City, involving 7 storage ponds and two sewage treatment plants (as the fourth and the fifth sewage treatment plant in Kunming City). The project selects the global control strategy for the optimization of the northern drainage area of Kunming, through numerical simulation method, realizesing that the joint operation technology mode of drainage pipe network, storage control facilities and sewage treatment plant, and the engineering process is shown in Fig. 6.20. In rainy days, the storage pond stores part of the rainwater to relieve the operating load of the sewage treatment plant. If there is still pressure in operation at this time, the first-level intensive treatment can be used to quickly purify and discharge the rainwater, and then effectively reduce the pollution load that discharged into the environmental water body in the urban area on rainy days, and the rainwater stored in the storage pond is transported to the sewage treatment plant for the storage pond when there is no rain.

Based on this project, three rainfall sites were selected for field testing from the 2014 rainy season. The results show that the storage pond designed and operated by applying the optimization strategy can reduce the overflow by 11.4%, SS overflow load by 9.7%, COD overflow load by 15.5%, NH_4-N overflow load by 19.6%, TN overflow load by 17.1% and TP overflow load by 16.1%. The long-term operation monitoring results of Kunming No. 5 Sewage Treatment Plant, show that the average removal efficiency of SS, COD, NH_4-N, TN and TP in combined sewage sewage

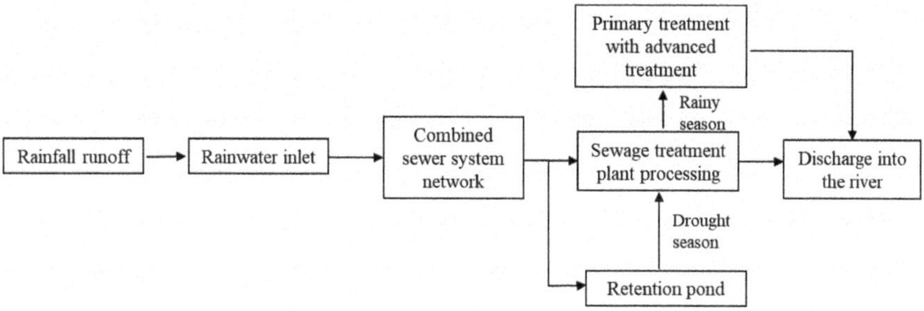

Fig. 6.20 Technical process of joint operation of storage pond-sewage treatment plant in the main urban area of Kunming

a) Stage 1 intensive treatment reaction cell b) First-level intensive treatment effluent

Fig. 6.21 Actual view of first-level enhanced treatment technology of Kunming No. 5 sewage treatment plant in the rainy season

in the rainy season were 87.2%, 84.3%, 42.5%, 28.5% and 67.9%, respectively, by the rainy season response technology developed by the first-level enhanced sewage treatment plant (Fig. 6.21).

6.3.4 Technical Model of Rainwater Purification in the Initial Stage of the Diversion Drainage System

6.3.4.1 Technology Mode Depended on Rainwater Split Based on Water Quality

i. Objective

Aiming on the purification of initial rainwater with high pollution loads before its discharing into the surface water, this technology mode finished the initial rainwater quality diversion and split-flow by rainwater gutter inlet and collection well, achieving that the separation of initial rainwater with high pollution loads and the middle and late stage rainwater with good water quality. Among them, the initial rainwater is purified and discharged into the receiving surface water, or enters into the storage facility for further purification and then enters into the river to achieve pollution purification of the initial rainfall runoff. Rainwater with better water quality is simply purified and then comprehensively reused or directly into the river.

ii. Technical Composition and Technical Process

This technology mode consists of rainwater gutter inlet technology, rainwater inspection well, rainwater flow abandonment technology, rainwater transfer and storage facilities, initial rainwater purification facilities and other technologies, and the technicial process is shown in Fig. 6.22. The rainfall runoff entering into the rainwater

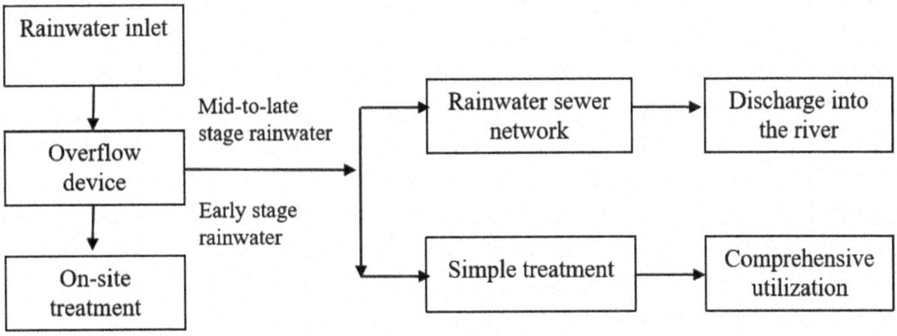

Fig. 6.22 Technology process of mode depended on rainwater split based on water quality

collection port and the rainwater collection well, is further separated from the rainwater quality in the middle and late stages through the rainwater flow abandonment technology. Among them, the initial rainwater is discharged into the river after being transferred and stored, and the rainwater in the middle and late stages is directly discharged into the river through the diversion rainwater pipe network or comprehensively utilized after simple treatment.

iii. Application Efficiency Analysis

This set of technical models has been applied in Xincheng Park, Guangming New District, Shenzhen. The south gate of Xincheng Park is equipped with a broadcast-type initial rainwater drainage device, which is composed of a host with two sub-mechanisms, the distance between the main machine and the sub-machine is about 100 m, and the wireless connection. The sub-machine controls two DN300 electric valves to switch the direction of water flow. When the rainwater from upstream is switched by the discarding device, and the preset discharge standard is not met, the initial rainwater is discharged to the green space for retention and purification; When the rainfall reaches the preset value of abandonment, the rainwater is directly discharged downstream and into the downstream water body, thereby controlling the non-point source pollution caused by rainwater. After one year of operation test, the equipment is in good operation, and this technology has been promoted in batches in the Jinan Second Ring South Road project.

6.3.4.2 Modulation and Purification Technology Mode

i. Objective

The technical model mainly collects the initial rainwater through the reservoir, and uses physical and biological treatment facilities to purify the initial rainwater to solve the problem of high pollution of the initial rainwater.

ii. Technical Composition and Technical Process

The technology mode of storage and purification is composed of rainwater gutter inlet, reservoir, coagulation and sedimentation, filtration, constructed wetland, ecological buffer zone and wetland park, the treatment process is shown in Fig. 6.23, the diversion rainwater enters into the reservoir after passing through the rainwater gutter inlet and the inspection well, and then the rainwater is further purified and discharged into the surface water by biological treatment facilities such as waterfront buffer belt, wetland park and constructed wetland after being treated by biological treatment facilities such as coagulation, sedimentation and filtration in the rainy season. During the dry season, the rainwater stored in the reservoir is transported to the sewage treatment plant for treatment and discharged into the surface water body.

iii. Application Efficiency Analysis

This technical model was applied in the initial rainwater treatment project of Tangxi River in Hefei Binhu New Area, and the technical process is shown in Fig. 6.24. Hefei Binhu New Area targeted 10 rainwater gutter inlets on Luzhou Avenue in Tangxi River basin to intercept the initial rainwater through drainage renovation and setting up interception wells. Laying interception pipes along the river to uniformly collects the initial rainwater intercepted at each gutter inlet. On the north bank of the Tangxi River and the east of Huizhou Avenue, south of Zhongshan Road and west of Guangxi Road, a wetland-type reservoir was built to recharge the collected

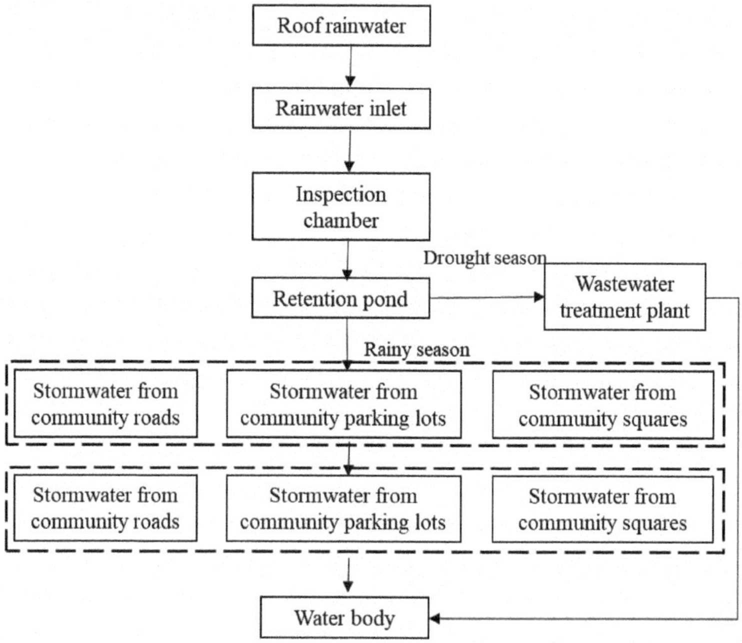

Fig. 6.23 Technical process of storage and purification technology mode

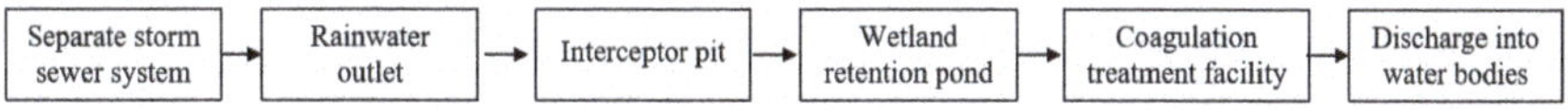

Fig. 6.24 Engineering process of Tangxi River storage and purification technology

a) Wetland storage facilities   b) Coagulation and purification facilities

Fig. 6.25 Actual image of rainwater treatment project in the initial stage of Tangxi River transfer and purification technology

initial rainwater, where the reservoir is equipped with a complete set of diatomite coagulation and purification facilities, which purifies the initial rainwater and then discharges it into the Tangxi River. The initial rainwater treatment project of Tangxi River was operated from 2016, with annual COD and TP reduced by 120–200 t and 1–1.5 t respectively; the construction cost was 0.73 million CNY/t water, and the operating cost was 0.2–0.3 CNY/t water. This effectively alleviated the influence of initial rainwater on the water quality of Tangxi River and achieved obvious economic, social and environmental benefits.

The actual image of the project is shown in Fig. 6.25. At present, the project is running stably, and the key core technologies have been popularized and applied in the lower reaches of the Tangxi River in Hefei City and the initial rainwater treatment project in the Shifenli River Basin of Hefei City.

6.3.4.3 Technical Mode of Regulation and Storage by a Special Pipeline

i. Objective

The technical mode of regulation and storage by a special pipeline realizes the rainwater purified dividually of the initial rainwater and middle and late rainwater to reduce river pollution.

ii. Technical Composition and Technical Process

This technology model consists of three key technologies, including rainwater gutter inlet, rainwater inspection well and early rainwater storage special pipe, and the specific process is shown in Fig. 6.26. Among them, the rainwater runoff from the road entering into the stormwater gutter inlet preferentially, flows into the initial

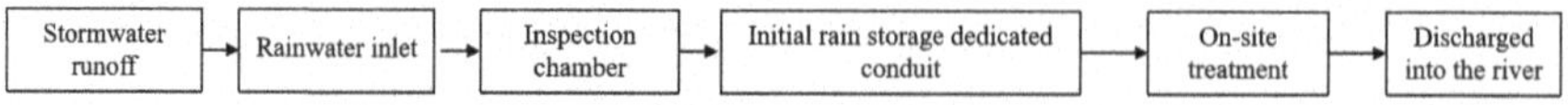

Fig. 6.26 The techical processof the storage adjustment technology mode by the special pipeline

rainwater storage pipe at the bottom of the inspection well, and the initial rainwater is stored after the rain being discharged into the river channel after on-site treatment such as cyclone separators and fast filters.

iii. Application Efficiency Analysis

The storage technology mode achieved by special pipeline is applied to the initial rainwater special storage project of Hengtang River West Road (Zhulin North Road - Beitanghe East Road) in Changzhou City, with a laying length of 1.16 km, the spokes of the engineering service road is 32 m (including motorized roads, non-motorized roads and sidewalks), the service area is about 19,648 m^2, and the engineering process is shown in Fig. 6.27. The initial runoff rainwater on the road preferentially flows into the special pipe installed under the rainwater well, and the initial runoff rainwater in the special pipe after the rain is treated by a cyclone separator and fast filter and discharged into the wetland park.

In addition, the integrated technology that couples the storage technology mode with the *on-site* processing technology has built four promotion projects in Changzhou Hengtang River West Road, Lihua Pumping Station, Zhulin Pumping Station and Jinling Pumping Station. The site before and after the implementation of the project is shown in Fig. 6.28. The promotion project adopts the above technological flow, and the processing capacity is 45, 160, 60 and 70 m^3/h, respectively, with a total of 335 m^3/h. Among them, the removal efficiency of COD, TP and NH_4-N were 56.6%, 55.9% and 64.9%, respectively.

In view of the rainwater retention in the later stage, three treatment projects were built in Hengtang River West Road Pumping Station, Shangcun Pumping Station and Chaoyang Pumping Station, respectively, using sand filtration + special pipe storage + sponge water storage module technology, high-speed fiber filtration technology and rain garden technology, with treatment capacities of 45, 100 and 104 m^3/h, respectively, totaling 249 m^3/h; The service area is 5000, 55,000 and 2441 m^2 respectively, for a total service area of 62,441 m^2.

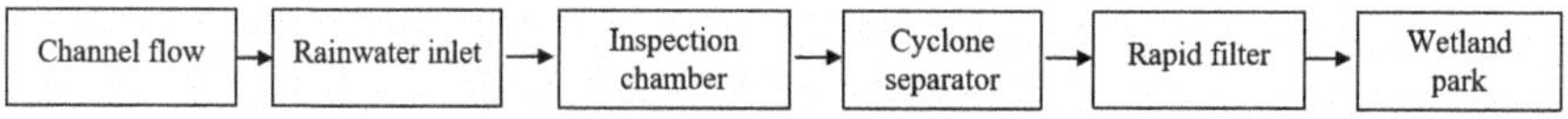

Fig. 6.27 Technical Processof Hengtang Hexi Road special storage project in Changzhou City

Fig. 6.28 Before and after the implementation of the special management and storage project of Hengtang Hexi Road, Changzhou City

6.4 Complete Set of Technologies for Controlling Urban Rainfall Runoff Pollution

6.4.1 Complete Set of Runoff Pollution Control Technology of Combined Drainage System

6.4.1.1 Objective

There are two problems in the combined drainage system: first, the combined pipe network cannot separate rainwater from municipal sewage, and the heavily polluted rainfall runoff enters into the sewage treatment plant with domestic sewage, resulting in high pressure on the sewage treatment plant; Second, when the rainfall is huge, it is easy to form rainwater overflow. In view of the above problems, this complete set of technology effectively controls the rainwater of the combined drainage system through LID sewage interception, rainwater well interception, physical and chemical sewage interception and ecological sewage interception, and removes solid pollutants and dissolved pollutants in it.

6.4.1.2 Technical Composition and Technical Process

The technical mode consists of rain garden, sunken green space, grass planting ditch, ecological tree pool, pebble ditch, permeable paving, rainwater outlet, reservoir, coagulation and sedimentation, filtration, constructed wetland, ecological buffer zone and wetland park, etc., and the technical process is shown in Fig. 6.29.

In areas with LID facilities, rainwater from different underlying surfaces enters into the LID facility, part of which is naturally infiltrated to replenish groundwater, and the other part enters into the sewage interception type, which intercepts the sewage through the rainwater gutter inlet and enters into the combined pipe network,

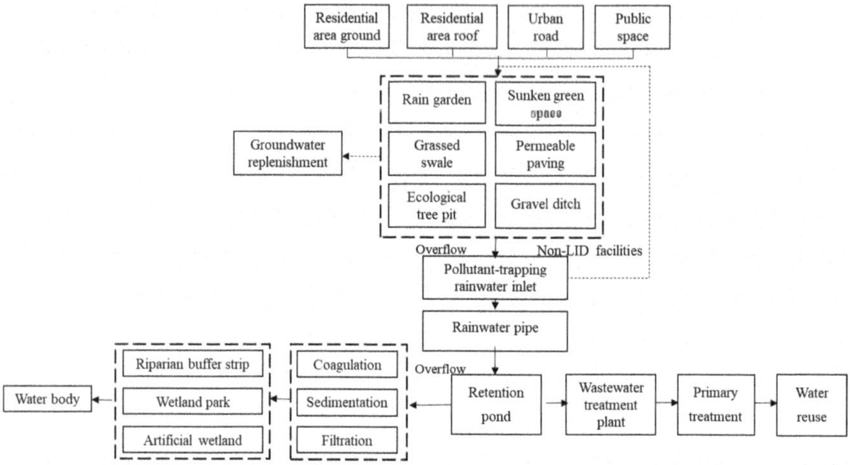

Fig. 6.29 Technical model and processof urban runoff pollution control technology of combined drainage system

and then transported to the sewage treatment plant through the sewage interception main pipe for further purification. The part exceeding the treatment loads of the sewage treatment plant, is transported to the reservoir for physicochemical and biological treatment and then discharged into the surface water.

In areas where there are no LID facilities, rainwater formed from different underlying surfaces directly enters the sewage interception type stormwater gutter inlet for sewage interception, and the intercepted rainwater enters into the combined pipe network and is transported to the sewage treatment plant through the interception main pipe for the further purification. Rainwater that exceeds the treatment loads of the sewage treatment plant is transported to the reservoir, where it is discharged into the receiving surface water after physicochemical and biological treatment.

6.4.1.3 Application Efficiency Analysis

The source reduction technology in this technology model has been applied in the runoff pollution control project of the third phase of the community in Nanhuayuan, Jiaxing Province, and the engineering process is shown in Fig. 6.30. In order to effectively reduce the runoff pollution of the third phase of Funan Garden and improve the overall water environmental quality, the LID concept was adopted. The rainwater pipe disconnection technology is adopted to introduce the rainwater of the building roof into the sunken green space and rainwater garden set up in the surrounding area for infiltration and purification. Transform roads, parking plants and squares in the community into infiltration paving to increase rainwater infiltration, achieving excess rainwater into sunken green spaces and rain gardens for infiltration and purification. Rainwater overflows that exceed the storage capacity of sunken green spaces and rain

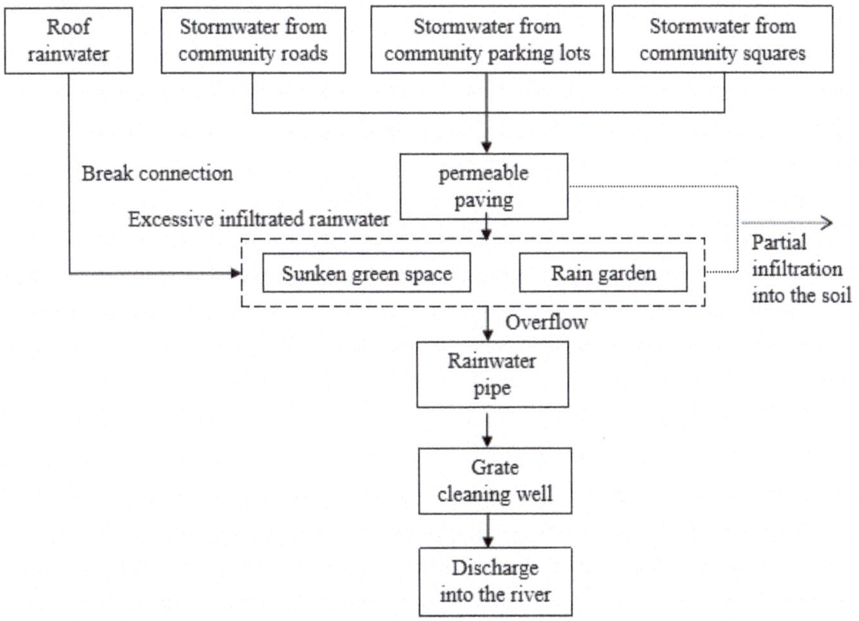

Fig. 6.30 Runoff pollution control process of Funan Garden Phase III

gardens, entering into stormwater pipes, are filtered through grid decontamination wells and discharged into rivers. After the implementation of the project, the total annual runoff control rate of the third phase of Funan Garden increased from 36% before the renovation to 84.6%, reducing 1825 t of rainwater into the river every year, realizing the integration of runoff pollution control and landscape effect in the community. The actual image of runoff pollution control in Funan Garden is shown in (Figs. 6.31 and 6.32).

a) Sunken green space b) Rain garden

Fig. 6.31 Runoff pollution control project of Phase III of Funan Garden

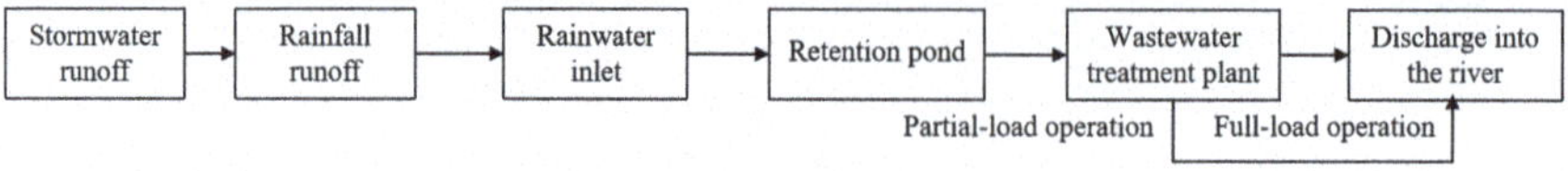

Fig. 6.32 **Technical** process of rainwater sewage storage project

This complete set of technology has also been partially applied in Kunming rainwater sewage storage technology project, and the engineering process is shown in Fig. 6.32. The project has three sub-projects, namely the Haiming River Reservoir, the Wulong River Reservoir and the Baiyun Road Reservoir, with a total scale of 44,100 m^3. Three rainwater reservoirs are connected to the 10th sewage treatment plant, the third sewage treatment plant and the fifth sewage treatment plant. When the rainfall runoff is small and the reservoir is not running at full capacity, all the collected rainfall runoff is transported to the sewage treatment plant for purification. When the reservoir is running at full capacity, excess rainfall runoff is discharged directly into the city's inland river. After online monitoring, the average annual reduction of SS, COD, TN and TP in rainfall runoff was 338.35 t, 421.04 t, 49.57 t and 8.13 t, respectively. The Baiyun Road reservoir is shown in Fig. 6.33.

This complete set of technology has also been partially applied in the CSOs storage and purification project of Changzhou Shangcun Pumping Station. The engineering process is shown in Fig. 6.34. In Shangcun area, the sewage in the sunny day is intercepted by the pumping station into the municipal sewage pipe network, and the excess rain sewage combined with the wastewater enters into the reservoir for storage on rainy days, and is pumped to the sewage treatment plant after the rainy day, greatly reducing the overflow on rainy days. If the rainfall is heavy on a rainy day, the drainage exceeding the storage capacity is purified by the suspension fast filter device and discharged into the surface water, which greatly reduces the loads of CSOs pollutants and protects the receiving water body. The effective volume of the reservoir is nearly 800 m^3, and the storage capacity is 1055.78 m^3/h. The diameter of the floating fast filter is 2.4 m and the scale is 102.7 m^3/h, which can

Fig. 6.33 Baiyun road reservoir images

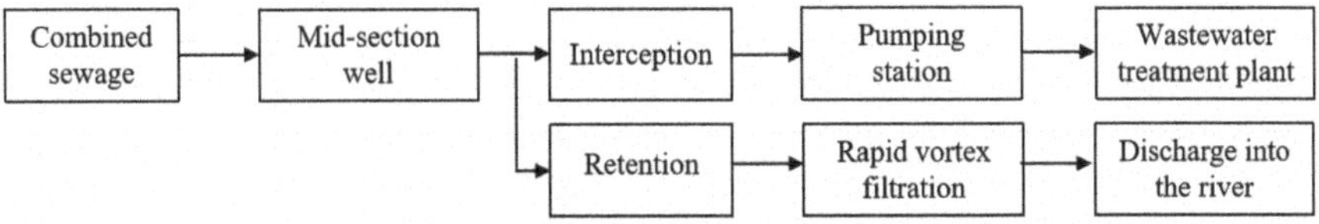

Fig. 6.34 Technical process of CSOs storage and purification engineering at Shangmura pumping station

Fig. 6.35 Suspension fast filter device image

realize automatic backwashing and its device is shown in Fig. 6.35. According to the third-party monitoring and evaluation report of the project, the SS load reduction performance in the six sample collecting reached more than 30%. The entire project can be intelligently operated and remotely monitored for interception, storage and rapid treatment, which is suitable for promotion and use in areas where the combined drainage system is adopted.

6.4.2 Complete Set of Runoff Pollution Control Technology for Diversion Drainage System

6.4.2.1 Objective

The construction of a diversion rainwater drainage system in the new urban area, and the change of the combined flow system to the diversion system in the old city to achieve the diversion of rain sewage, is one of the important means to solve the overflow pollution of the combined system. However, in the early stage of rainfall, the amount of rainfall water is large and the flow rate is fast, which has a strong erosion effect on the underlying surface and rainwater drainage pipes, carrying a large number of pollutants and pipe sludge, and directly entering into the river. This will bring a high pollution load, causing the water quality decrease of the receiving water, and even black and odorous. In order to effectively solve the problem of runoff pollution in the diversion drainage system, it is necessary to integrate the source, process and terminal technologies to control the urban rainfall runoff of the diversion drainage system, through methods such as surface LID interception, rainwater well interception, storage and storage, enhancing operation of sewage treatment plants, physical and chemical interception and ecological interception, and remove the pollutants carried by rainfall runoff.

6.4.2.2 Technical Composition and Technical Process

The complete set of runoff pollution control technology of diversion drainage system consists of rain garden, sunken green space, grass planting ditch, ecological tree pond, pebble ditch, permeable paving, rainwater outlet, reservoir, coagulation and sedimentation, filtration, constructed wetland, ecological buffer zone, wetland park and rainy season response of sewage treatment plant, etc., the process is shown in Fig. 6.36.

Part of the rainwater entering into the LID facility, is naturally infiltrated to replenish groundwater or stored in the LID facility for physical and chemical treatment to achieve comprehensive utilization; the other part enters into the sewage interception type rainwater gutter inlet, and enters into the rainwater pipeline after the interception through the rainwater gutter inlet; and part of the overflowing rainwater is transported to the reservoir for physical and biological treatment and then discharged into the water body, and the other part is transported to the sewage treatment plant for first-level intensive treatment and recycling in the rainy season.

Rainwater that does not enter into the LID facility directly, enters into the sewage interception type rainwater gutter inlet for sewage interception, and after interception,

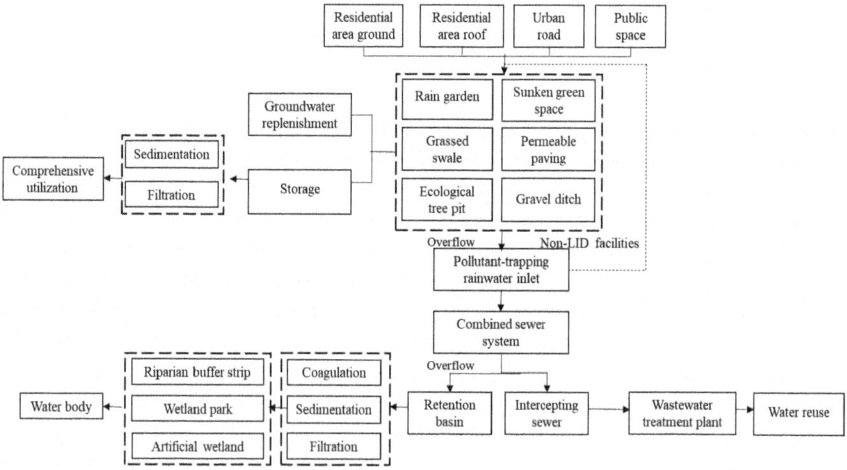

Fig. 6.36 Technical model and process of urban runoff pollution control in diversion system drainage system

it enters inmto the rainwater pipeline, and part of it is transported to the reservoir for physical and biological treatment before being discharged into the water body, and the other part is transported to the sewage treatment plant for intensive treatment and recycling in the rainy season.

6.4.2.3 Application Efficiency Analysis

In the first phase of the demolition and resettlement project of Wanzhangpo Area of Guangming New District, Shenzhen, the urban runoff pollution control project of the diversion system drainage system was built, mainly focusing on the source reduction project. The project covers an area of 2.75 ha, with a construction area of 64,700 m^2, involving 5 residential buildings (1# ~ 5#). The project is bordered by Huayu Road to the south, Niushan Road to the east, and Guangming Avenue to the west.

The project realizes the organic combination of roof rainfall runoff pollution control, road square rainfall runoff control and public green space runoff control, and the process is shown in Fig. 6.37. Among them, the runoff generated by the fire lane on the north side of the residential building 2# and 3#, can enter into the sunken green space through two grass planting ditches, and be discharged into the rainwater pipe after the detention and purification. Construction of permeable paving between residential buildings 3# and 4# achieves the runoff infiltration and interception, and the remaining runoff enters into the ecological tree pond, and is discharged into the rainwater pipeline after further retention and purification. The roof of the main building of 1# ~ 5# residential building is an ordinary roof, and the runoff generated can be disconnected to the rainwater garden through the rainwater riser, and be discharged into the rainwater pipeline after the detention and purification. The podium roof of the 1# ~ 5# residential building is built with a green roof, which intercepts the rainwater, and the runoff generated can be disconnected to the rainwater garden through the rainwater riser, and be discharged into the rainwater pipe after the detention and purification. The performance of the first phase of the demolition and resettlement project in Wanzhangpo area and the actual image after the completion are shown in Fig. 6.38, and the project has reduced SS by more than 60%.

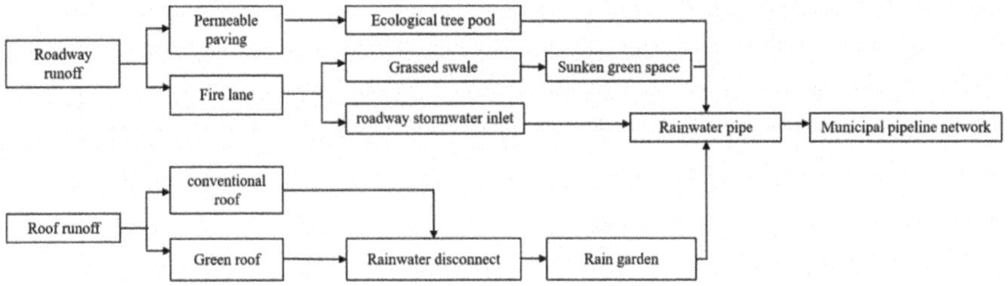

Fig. 6.37 Runoff pollution control process of the first phase of demolition, relocation and resettlement in Wanzhangpo area

a) Eco tree pool b) Grass ditch

c) Rain garden d) Rainwater breaks

Fig. 6.38 Contribution effect and actual image of runoff pollution control of the first phase of demolition, relocation and resettlement in Wanzhangpo area

This technology model has also been applied in the five diversion rainwater reservoir projects in the west area of the main urban area of Kunming and the enhanced primary project of the fifth sewage treatment plant in the rainy season, mainly reflected in the runoff pollution process storage treatment and the back-end sewage treatment plant treatment. The engineering process is shown in Fig. 6.39, the rainwater runoff is collected into the rainwater reservoir through the rainwater pipe network, and then entered into the sewage treatment plant through the rainwater pipeline for enhanced primary treatment, and the treated rainwater is discharged into the Kunming urban inland river, and finally into Dianchi Lake as recharge water.

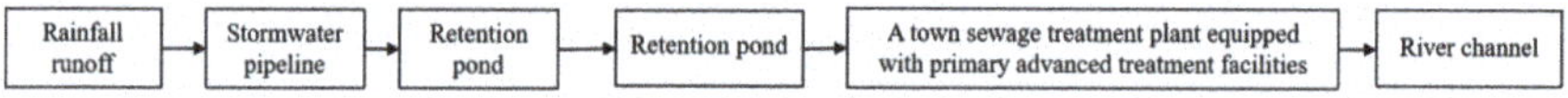

Fig. 6.39 Technical process of rainwater transfer and storage and enhanced primary treatment in the main urban area of Kunming City

The five diversion rainwater reservoirs in the western area are Kunyi Middle School, Qimugou, Laoyunlianghe, Wulong River and Xiaolugou Reservoir, with a total storage scale of 56,000 m^3. According to the typical annual characteristic rainfall of 837.76 mm in Kunming, the total annual characteristic rainfall of surface runoff rainfall (continuous rainfall greater than 3 mm) is 837.76 mm, and the total COD load of the western area can be reduced by 671.9 t COD/a, and the operation effect is good.

In order to cope with the rainy season shock load, No. 5 Sewage Treatment Plant in Kunming, built a rainy season treatment structure based on enhanced primary treatment of sewage treatment plant, with a scale of 100,000 m^3/d. When the liquid level in front of the influent lifting pump of the sewage treatment plant exceeds 5.6 m, the rainy season coping mode will be activated. The influent water exceeding 1.3 times the design flow rate of the biochemical section, will enter into the rainy season to cope with the treatment structure. When the structure was running, the PAC dosage was 75 mg/L, the PAM dosage was 75 μg/L, and the hydraulic residence time was 1 h. The average removal rates of SS, COD, NH_4-N, TN and TP in rainy season diverted rainwater reached 87.2%, 84.3%, 28.5%, 42.5% and 67.9%, respectively, and the operating cost was about 0.6 yuan/m^3 considering construction, equipment, chemicals, electricity consumption, labor and other factors.

References

1. Liu, L., Huang, O., Guo, L., et al. (2016). Research on hydraulic of urban road gully. *Water Supply and Drainage, 52*(S1), 12–16. (in Chinese).
2. Duan, B., Zhao, J., Gao, Y., et al. (2013). Control effect of green roof on roof runoff pollutions. *Environmental Science and Technology, 36*(9), 57–59.
3. Yi, S., Wang, X., Kang, W., et al. (2017). Storage and purification efficiency of improved depressed green spaces on residential stormwater runoff. *China Water & Wastewater, 33*(5), 134–138. (in Chinese).
4. Kong, L., Kang, W., Tan, S., et al. (2016). Design and application of step green land in green buildings community. *China Water & Wastewater, 32*(1), 90–93. (in Chinese).
5. Xu, P., Si, S., Zhang, J., et al. (2015). Study on the control effects of permeable asphalt road and road retention on water quality and quantity of runoff, Shenzhen. *Water Supply and Drainage, 51*(11), 64–69. (in Chinese).
6. Wang, X. (2015). *Study on characteristics of runoff pollution and integrated control technologies in rapid urbanization region*. Tsinghua University. (in Chinese).
7. Wang, X. (2016). Practice and exploration for sponge city construction in Jiaxing City. *China Water & Wastewater, 32*(14), 33–35.
8. Wang, X., Wang, Y., Ren, X., et al. (2016). Application and simulation-based evaluation of a low impact development system in a sports center in Shenzhen. *Water Supply and Drainage, 52*(5), 91–96. (in Chinese).
9. Zhu, H. (2016). *Research of intercepting and treatment technologies for first flush rainwater from urban roads*. Tsinghua University. (in Chinese).
10. Guo, F., Li, Z., Wu, Y., et al. (2015). Effectiveness analysis of enhanced primary treatment in WWTP during the rainy season in Kunming. *China Water & Wastewater., 31*(09), 11–14. (in Chinese).

Chapter 7
Typical Cases of Urban Rainfall Runoff Pollution Control Engineering

By organically combining source reduction technologies and facilities, process control technologies and facilities, as well as backend treatment technologies and facilities, a total of four technological processes, four application modes, and three complete sets of technologies have been integrated. Building upon this foundation, this chapter presents seven typical case studies of urban rainfall runoff pollution control, highlighting the utilized technological processes, application modes, and the application status of complete sets of technologies.

7.1 Based on Source Analysis, the Rainwater and Sewage Storage and Primary Enhanced Treatment Project

7.1.1 Case Background

The Dianchi Lake Basin, located in Kunming City, is located at the watershed of the Yangtze River, Pearl River, and Red River, and is classified as a severely water-scarce area. According to the distribution characteristics of water systems in Kunming City, urban rivers ultimately flow into the Caohai and Waihai of Dianchi Lake. In recent years, Kunming City has intensified its efforts to comprehensively rectify the inflow rivers into Dianchi Lake. The pollutant concentration in these rivers has shown a decreasing trend year by year, but still does not meet Class V surface water standards. The pollution load discharged into Dianchi Lake from urban rivers is the main external source of pollution, accounting for over 70% of the total pollutant load in the Dianchi Lake Basin in terms of COD, TN, and TP transported to the lake annually. Among them, the pollution load from rainfall runoff into the rivers is relatively high. Therefore, addressing the pollution from rainfall runoff in the main

D. Sun and F. Qi, *Urban Rainfall Runoff Pollution Control Technology and Application*,
https://doi.org/10.1007/978-981-96-4174-1_7

urban area of Kunming is crucial for improving the water environment quality of Dianchi Lake.

The "Research and Demonstration of Key Technologies for Comprehensive Pollution Reduction and Water Quality Assurance in the Main Urban Area of Kunming (2011ZX07302-001)" project, which was launched under the National Twelfth Five-Year Plan for Water in 2011, focuses on overall pollutant reduction in the main urban area of Kunming. The research is focused on the design and optimization of rainwater and sewage storage systems within the drainage networks, exploring measures for urban sewage treatment plants equipped with diversion-based drainage systems to cope with the challenges posed by the rainy season. The project involves the construction of storage tanks prior to the overflow channels of the main drainage routes within the second ring road, as well as the establishment of rain season treatment structures in sewage treatment plants. These measures significantly reduce the frequency of overflow discharge from combined sewer networks into rivers and the total pollutant load, thereby reducing the overall pollution load discharged into the rivers. This project provides key technological support and feasible strategies for addressing rainfall runoff pollution in the main urban area of Kunming, supporting the comprehensive pollutant reduction and water quality assurance technology system and serving as a demonstration and reference for water environment management in other lakefront cities in China.

7.1.2 Process Route and Key Technologies

7.1.2.1 Process Route

Firstly, carry out analysis of urban runoff pollution sources in Kunming; Based on the analysis results, optimize the design and construction of rainwater and sewage storage tanks, construct a first level strengthening project for urban sewage treatment plants to cope with rainy seasons, and achieve plant network linkage. The process route is shown in Fig. 7.1. Among them, the overflow rainwater and sewage from the combined pipe network enter the sewage treatment plant for routine treatment through the storage tank; The initial rainwater collected by the diversion system pipe network enters the sewage treatment plant for primary enhanced treatment through the rainwater storage tank. After treatment, the two types of sewage are discharged into the inland rivers of Kunming City and eventually merged into Dianchi Lake as make-up water.

According to this process route, 17 regulating and storage tank projects have been constructed in the main urban area of Kunming City, as well as the first level enhanced treatment project of Kunming Fifth Sewage Treatment Plant during the rainy season.

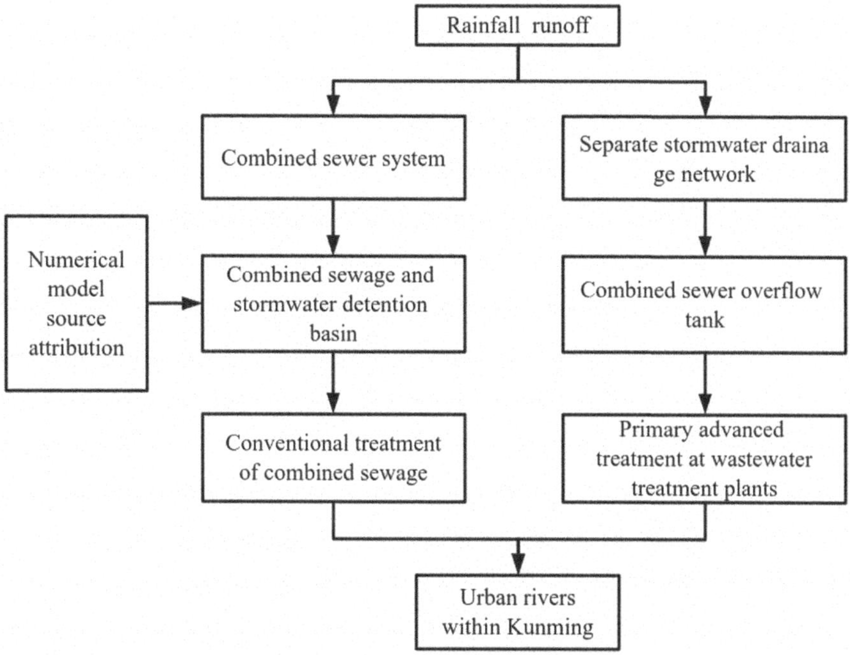

Fig. 7.1 Process route for urban runoff pollution control project in Kunming's main urban area

7.1.2.2 Key Technologies and Applications

Based on the above-mentioned projects, four key technologies have been developed and verified. These include the design technology for overflow pollution control and storage tanks based on rainfall drainage characteristics, the design technology for initial rainwater storage tanks based on rainfall and water resource utilization characteristics, the efficiency evaluation technology for storage systems combining on-site monitoring and system simulation, and the technology for enhanced primary treatment of urban sewage treatment plants during the rainy season.

i. Design Technology of Overflow Pollution Control and Storage Tank Based on Rainfall Drainage Characteristics

(i) *Related Design Parameters*

This technology combines the scale calculation methods of rainwater storage tanks used in the United States and Japan for overflow pollution control, proposing to convert the dry season sewage volume into equivalent rainfall intensity; Based on rainfall statistical data, calculate the interception rate, which is simple and easy to master.

Rainfall design: The rainfall characteristics parameters required for the design of combined sewer overflow storage facilities are mainly rainfall intensity and cumulative rainfall, which are determined based on local rainfall data for the past decade or more. By using rainfall intensity as the x-axis and cumulative rainfall as the y-axis, a rainfall duration curve is obtained, and the corresponding rainfall amount for a given rainfall intensity is calculated through numerical simulation.

Sewage volume design: The sewage volume within the service scope of combined sewer overflow storage facilities should be determined according to the local sewage special planning. When conditions permit, adjustments can be made based on measured data. The dry season sewage volume is converted into equivalent rainfall intensity and calculated according to Eq. (7.1).

$$i_{HW} = Q_{HW}/(10\psi \cdot A \cdot 24) \tag{7.1}$$

where, i_{HW}—rainfall intensity equivalent to the sewage volume QHW, mm/h;

Q_{HW}—average dry flow sewage volume on clear days, m^3/d;

ψ—runoff coefficient;

A—confluence area, hm^2.

Interception design: The interception factor and rainfall intensity in the combined sewer drainage system are interrelated. The interception volume for different interception factors corresponds to the rainfall amount for their respective rainfall intensities, while the overflow volume represents the rainfall amount exceeding that particular rainfall intensity. The details are shown in Table 7.1.

Design of regulation and storage capacity: The effective capacity of the combined overflow regulation and storage facility should be comprehensively considered based on regulation and storage objectives, dry flow sewage volume, pipeline layout, downstream sewage system surplus, and surrounding environment, can be determined

Table 7.1 Rainfall capture and overflow volume in Kunming's main urban area combined sewer system at various diversion ratios

Diversion ratio	Corresponding rainfall intensity (mm/h)	Annual average rainfall (mm/year)	Retention volume (mm/year)	Overflow volume (mm/year)	Retention volume to rainfall ratio (%)
0	0	828.6	0	828.6	0
1	0.52	828.6	239	589.6	28.84
1.5	0.78	828.6	307	521.6	37.05
2	1.04	828.6	355	473.6	42.84
3	1.58	828.6	425	403.6	51.29
4	2.08	828.6	471	357.6	56.84
5	2.6	828.6	508	320.6	61.31
8	4.16	828.6	587	241.6	70.84
10	5.2	828.6	624	204.6	75.31
15	7.8	828.6	692	136.6	83.51

using mathematical models. If a mathematical model is not used, the effective volume of the combined overflow storage facility should be determined based on local rainfall characteristics, environmental capacity of the receiving water body, interception multiple of the drainage system, dry flow sewage volume of the system, service area of the drainage system, and surplus of the downstream sewage system, and can be calculated according to Eq. (7.2).

$$V = 3600 \cdot t_i(n - n_0) \cdot Q_{dr} \cdot \beta \tag{7.2}$$

where, *V*—effective volume of storage facilities, m^3;

Ti—The inflow time (h) of the regulation and storage facilities should be 0.5 to 2 h. When the overflow sewage quality of the combined drainage system on rainy days has no obvious initial effect in a single rainfall event, the upper limit should be taken; On the contrary, the lower limit can be taken;

N—the interception multiple during the operation of the regulation and storage facilities;

N_0—original interception multiple of the drainage system;

Q_{dr}—Dry flow sewage volume before the closure well, m^3/s;

B—The safety factor can be taken as 1.1–1.5.

(ii) *Location and Scale of Technology Application*

The "Overflow Pollution Control and Storage Tank Design Technology Based on Rainfall Drainage Characteristics" has been widely applied in the Hai Ming River storage tank, Wu Long River storage tank, Bai Yun Road storage tank, and Shanghai Xin Wanping Pumping Station storage tank. The storage volumes of these four tanks are 28,000 m^3, 11,000 m^3, 9,100 m^3, and 9,000 m^3 respectively.

i. Design Technology for Initial Rainwater Storage Tanks Based on Rainfall and Water Resource Utilization Characteristics

(i) *Related Design Parameters*

This technology proposes an algorithm for designing the capacity of rainwater storage facilities in the early stages of diversion systems. For areas with conditions, the effective volume of storage facilities should be determined using a mathematical model after comprehensive consideration of storage objectives, pipeline network layout, rainwater utilization, and site environmental factors; For areas that do not meet the conditions of the mathematical model at the moment, the calculation method with rainfall thickness as the main control index should be used for the regulation and storage capacity. Based on the rainfall pattern, runoff characteristics, and water resource utilization characteristics of Kunming area, reference values for controlling rainfall thickness for different regulation and storage scales are provided.

Rainfall design: The rainfall characteristic parameters required for the design of the rainwater storage tank in the early stage of the diversion system are mainly rainfall intensity and accumulated rainfall. They should be determined based on local rainfall data over the past 10 years, with rainfall intensity as the horizontal axis and accumulated rainfall as the vertical axis, to obtain the rainfall curve. Through numerical simulation, the given rainfall intensity can be obtained.

Runoff: calculated according to Eq. (7.3).

$$Q = \psi \cdot i \cdot A \tag{7.3}$$

where, Q—Design flow rate, L/s;

I—Design rainstorm intensity, L/(s · hm^2).

The runoff calculated according to (7.3) is the basic data for designing rainwater drainage facilities, collecting and discharging pipelines (channels).

Design of regulation and storage capacity: In areas with conditions, the effective capacity of diversion system regulation and storage facilities should be determined using a mathematical model after comprehensive consideration of regulation and storage objectives, pipeline layout, rainwater utilization, and surrounding environment.

If a mathematical model is not used, the effective volume of the diversion system storage tank should be calculated using different formulas based on local rainfall characteristics, environmental capacity of the receiving water body, service area of the rainwater system, classification of storage facilities (peak reduction, pollution control, etc.), and degree of utilization. When used for pollution control, the initial rainwater storage capacity should be calculated according to Eqs. (7.4).

$$Q_y = 10 \times H_y \times \psi \times A \tag{7.4}$$

where, Q_y—Control amount of runoff pollution, m^3;

H_y—Design controlled rainfall thickness, mm; Can take 4–8; In order to better control urban runoff pollution, it is recommended to take a high or low value when conditions permit;

The rainfall runoff coefficient can be calculated according to Eq. (7.5):

$$\psi = 0.05 + 0.009I \tag{7.5}$$

where, I-proportion of impermeable area within the catchment area,%;If the proportion of impermeable area is 80%, then $I = 80$.

Controlling rainfall thickness: The most important parameter in initial rainfall regulation and storage facilities is rainfall thickness, which directly affects the effectiveness of pollution control. The main factors considered include: pollution status of the underlying surface, catchment time, water quality control requirements, etc. If conditions permit, it is recommended to first calculate and analyze based on the measured results; In the absence of measured results, considering the importance of Dianchi Lake protection and regional characteristics, it is recommended to take the following upper limit values as much as possible.

① Single building roof: The convergence time is very small, and the water quality is generally good, so it is recommended to use 1–3 mm;
② Small watershed area: The catchment time is very small, and it is recommended to use 3–5 mm, but for areas with poor water quality, it needs to be appropriately increased;
③ Middle watershed area: If the convergence time is less than 25 min, it is recommended to use 5–15 mm; If the convergence time is greater than 25 min, it should be determined after actual measurement;
④ Comprehensive area: determined after actual measurement.

(ii) *Location and Scale of Technology Application*

"Design technology of initial rainwater storage tank based on rainfall and water resources utilization characteristics" has been applied in the storage tank of Kunming No.1 Middle School, Qimu Ditch, Laoyunliang River, Wulong River and Xiaolugou in the western area of Kunming, with a total storage capacity of 56,000 m^3.

iii. A Technology For Evaluating The Efficiency of Regulating And Storage Systems by Combining Field Monitoring And System Simulation

(i) *Related Design Parameters*

This technology constructs a complete set of efficiency evaluation index system for regulating and storage tanks from three aspects: interception capacity, drainage capacity, and utilization capacity, as shown in Fig. 7.2. Among them, the interception capacity mainly evaluates the storage/interception capacity of the regulation and storage facilities for the runoff generated by rainfall, as well as the ability to reduce pollution loads; The drainage capacity mainly assesses whether the runoff peak caused by rainstorm after the addition of regulation and storage facilities will cause system overflow and the load capacity of the pipe network system; The utilization capacity of the regulation and storage pool mainly evaluates the inflow and storage capacity of the regulation and storage facilities caused by the runoff generated by rainfall after the addition of regulation and storage facilities, and quantitatively evaluates the utilization efficiency of the regulation and storage facilities.

The performance evaluation index system for rainwater and sewage storage facilities in this key technology includes 20 indicators, and the meanings of each indicator are shown in Table 7.2.

After determining the evaluation indicators, calculate the operational efficiency evaluation index of the regulation and storage facilities, so that the calculation value is limited to 0–1. Based on the classification of evaluation indicators for the operational efficiency of storage facilities, determine the operational efficiency level of storage facilities. The highest level of efficiency is level five, with a first level efficiency of 0–0.2, a second level efficiency of 0.2–0.4, a third level efficiency of 0.4–0.6, a fourth level efficiency of 0.6–0.8, and a fifth level efficiency of 0.8–1. The specific evaluation effectiveness levels are shown in Table 7.3.

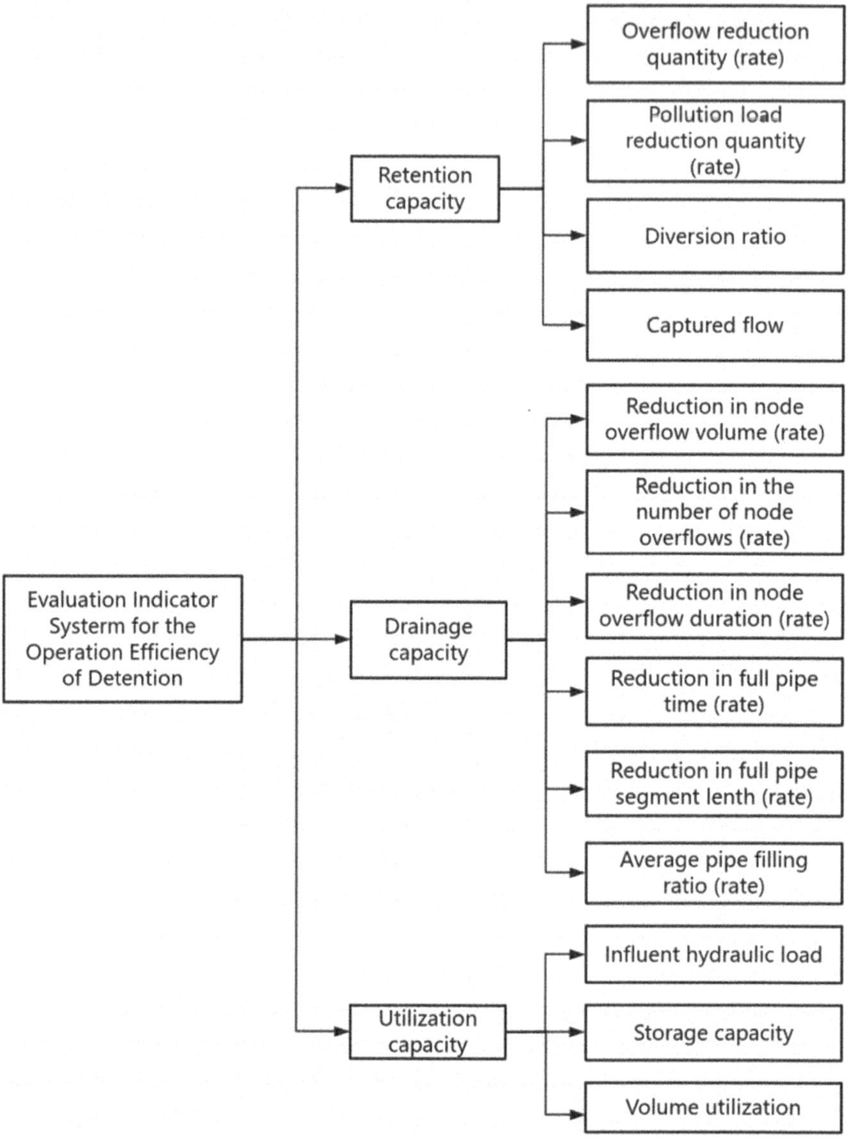

Fig. 7.2 Evaluation indicator system for detention facility efficiency

(ii) *Location and Scale of Technology Application*

The technology of combining on-site monitoring and system simulation to evaluate the effectiveness of the regulation and storage system has been applied in 17 regulation and storage tanks in the main urban area of Kunming, with a total regulation and storage capacity of nearly 220,000 m^3. Set up rainfall scenarios with a return period

Table 7.2 Meaning of detention facility efficiency evaluation indicators

Number	Assessment indicators	Meaning of indicators
1	Overflow reduction volume, m^3	The difference between overflow volume at the combined sewer system outfall before and after the operation of the rainwater and wastewater detention facility
2	Overflow reduction rate,%	The ratio of overflow reduction volume to the total overflow volume at the combined sewer system outfall before the operation of the rainwater and wastewater detention facility
3	Overflow load reduction, kg	The difference between the total pollution load at the combined sewer system outfall before the operation of the rainwater and wastewater detention facility and the total pollution load at the outfall after the operation of the facility
4	Overflow load reduction rate,%	The ratio of overflow load reduction to the total pollution load at the outfall before the operation of the detention facility
5	Eaptured flow volume, m^3	The maximum rainfall volume that the drainage system within the service area of the rainwater and wastewater detention facility can handle without overflowing, with the condition that the drainage system does not overflow as the boundary
6	Diversion ratio	The ratio of the combined sewer flow intercepted before overflow at the outfall to the dry season sewage flow before overflow
7	Reduction in node overflow volume, m^3	The difference between the sum of overflows at individual nodes before the operation of the rainwater and wastewater detention facility and the sum of overflows at individual nodes after the operation of the facility
8	Node overflow reduction rate,%	The ratio of the reduction in node overflow volume after the operation of the detention facility to the sum of node overflow volumes before the operation of the rainwater and wastewater detention facility
9	Reduction in node overflow volume, in units	The difference between the total number of node overflows before the operation of the rainwater and wastewater detention facility and the total number of node overflows after the operation of the facility
10	Reduction rate in the number of node overflows,%	The ratio of the reduction in node overflow volume to the number of node overflows before the operation of the rainwater and wastewater detention facility
11	Reduction in node overflow duration, h	The difference between the sum of overflow durations at individual nodes before the operation of the rainwater and wastewater detention facility and the sum of overflow durations at individual nodes after the operation of the facility

(continued)

Table 7.2 (continued)

Number	Assessment indicators	Meaning of indicators
12	Reduction rate in node overflow duration,%	The ratio of the reduction in node overflow duration to the sum of overflow durations at individual nodes before the operation of the rainwater and wastewater detention facility
13	Reduction in full pipe time (h)	The difference between the sum of full pipe operation time for individual pipe segments before the operation of the rainwater and wastewater detention facility and the sum of full pipe operation time for individual pipe segments after the operation of the facility
14	Reduction rate in full pipe time,%	The ratio of the reduction in full pipe time to the sum of full pipe operation time for individual pipe segments before the operation of the rainwater and wastewater detention facility
15	Reduction in full pipe segment length, km	The difference between the sum of lengths of full pipe operation for pipe segments before the operation of the rainwater and wastewater detention facility and the sum of lengths of full pipe operation for pipe segments after the operation of the facility
16	Reduction rate in full pipe segment length,%	The ratio of the reduction in length of full pipe operation for pipe segments to the sum of lengths of full pipe operation for pipe segments before the operation of the rainwater and wastewater detention facility
17	Average pipe filling Level	The average filling level of all pipe segments within the service area of the rainwater and wastewater detention facility
18	Influent hydraulic load, $m^3/(h \cdot m^2)$	The ratio of captured flow volume to the inflow time and the bottom area of the rainwater and wastewater detention facility
19	Storage capacity, m^3	The product of the storage depth of the rainwater and wastewater detention facility and the corresponding cross-sectional area of the pool at a given time
20	Volume utilization efficiency,%	The ratio of the storage volume to the maximum capacity of the rainwater and wastewater detention facility

Table 7.3 Detention efficiency classification

Detention efficiency index range	Efficiency level	Physical significance of efficiency levels
$0<\eta \leq 0.2$	First level	The ratio of the values of the detention basin efficiency assessment indicator system after the construction of the detention basin to the corresponding baseline before the construction of the detention basin
$0.2<\eta \leq 0.4$	Second level	
$0.4<\eta \leq 0.6$	Third level	
$0.6<\eta \leq 0.8$	Forth level	
$0.8<\eta \leq 1$	Fifth level	

of 0.25 years, a return period of 0.75 years, a return period of 1 year, a return period of 2 years, a return period of 3 years, and a return period of 5 years, respectively, to track and monitor the operational efficiency of 17 individual storage tanks in the main urban area of Kunming, and simulate and evaluate the overall operational efficiency of each tank. The evaluation scenario conditions and results are shown in Tables 7.4 and 7.5, respectively.

Based on the performance evaluation of individual storage tanks, the main urban area of Kunming is divided into three drainage areas: the north area, the west area, and the south area, for the comprehensive performance evaluation of rainfall storage tanks. Among them, the northern area includes 7 storage ponds, including Baiyun Road, Golden Avenue, Hetaoqinggou, Xuechang North Gully, Xuefu Road, Ma Xian Gully, and Yuantong Gully; The western area includes 5 storage ponds, including Laoyunliang River, Xiaolugou, Qimugou, Kunyi Zhonghe and Wulong River; The southern area includes five storage ponds, namely Cailian River, Lanhuagou, Daguan River, Haiming River, and Mingtong River. The comprehensive operational efficiency evaluation results of the three drainage areas under different rainfall scenarios are shown in Tables 7.6, 7.7 and 7.8.

From Table 7.6, it can be seen that under the 0.25 year rainfall scenario, the comprehensive operational efficiency evaluation index of the storage tank in the northern area is 0.85, reaching level five efficiency; The comprehensive energy efficiency under the 0.75 year and 1 year rainfall scenarios is level four, with energy efficiency indices of 0.75 and 0.73, respectively; 2. Under the 5-year rainfall scenario, the energy efficiency level is Level III.

From Tables 7.7, it can be seen that under the 0.25 year rainfall scenario, the comprehensive operational efficiency evaluation index of the western area's storage and regulation pond is 0.16, which is a first level efficiency; The comprehensive energy efficiency under the 0.75 year rainfall scenario is Level 3, with an energy efficiency index of 0.41; 1. Under the scenario of rainfall with a return period of 2, 3, and 5 years, the efficiency level is Level 2.

Table 7.4 Event-based rainfall scenario information

Design standards	Total rainfall, mm	Average rainfall intensity, mm/h	Peak rainfall intensity, mm/h
Rains once every three months	9.01	3	22.73
Rains once every nine months	26.3	8.77	66.31
Rains once every year	30.82	10.27	77.73
Rains once every two years	41.73	13.91	105.22
Rains once every three years	48.11	16.04	121.31
Rains once every five years	56.14	18.71	141.57

Table 7.5 Individual detention basin efficiency assessment

Detention basin	Assessment criteria	Rains once every three months	Rains once every nine months	Rains once every year	Rains once every two years	Rains once every three years	Rains once every five years
Baiyun Road	Overflow reduction rate,%	–	100.00	95.70	44.82	19.48	26.19
	Pollution load reduction rate,%	–	100.00	95.70	44.82	19.48	26.19
	Volume utilization rate,%	–	100.00	100.00	100.00	100.00	100.00
	Composite index	0.000	1.000	0.968	0.586	0.396	0.446
	Efficiency level	–	Level 5	Level 5	Level 3	Level 2	Level 3
Golden Tower	Overflow reduction rate,%	100.00%	79.82	67.16	51.15	33.00	22.02
	Pollution load reduction rate,%	100.00%	79.82	67.16	51.15	33.00	22.02
	Volume utilization rate,%	90.38%	95.63	95.63	95.63	100.00	100.00
	Composite index	0.976	0.838	0.743	0.623	0.498	0.415
	Efficiency level	Level 5	Level 5	Level 4	Level 4	Level 3	Level 3
Hetaojinggou	Overflow reduction rate,%	–	100.00	95.70	83.20	58.48	39.36
	Pollution load reduction rate,%	–	100.00	95.70	83.20	58.4%	39.36
	Volume utilization rate,%	–	100.00	100.00	100.00	100.00	100.00
	Composite index	–	1	0.968	0.874	0.689	0.545

(continued)

Table 7.5 (continued)

Detention basin	Assessment criteria	Rains once every three months	Rains once every nine months	Rains once every year	Rains once every two years	Rains once every three years	Rains once every five years
	Efficiency level	–	Level 5	Level 5	Level 5	Level 4	Level 3
Xiaochang North Ditch	Overflow reduction rate,%	–	82.93	67.14	39.95	38.97	25.56
	Pollution load reduction rate,%	–	82.93	67.14	39.95	38.97	25.56
	Volume utilization rate,%	–	100.00	100.00	100.00	100.00	100.00
	Vomposite index	0.000	0.872	0.754	0.55	0.542	0.442
	Efficiency level	–	Level 5	Level 4	Level 3	Level 3	Level 3
Xuefu Road	Overflow reduction rate,%	–	99.53	85.86	28.04	25.25	11.22
	Pollution load Reduction rate,%	–	99.53	85.86	28.04	25.25	11.22
	Volume utilization rate,%	–	100	100	100	100	100
	Composite index	0.000	0.996	0.894	0.46	0.439	0.334
	Efficiency level	–	Level 5	Level 5	Level 3	Level 3	Level 2
Maxian Gully	Overflow reduction rate,%	–	80.70	68.65	65.70	55.72	49.93
	Pollution load reduction rate,%	–	80.70	68.65	65.70	55.72	49.93
	Volume utilization rate,%	–	100	100	100	100	100

(continued)

Table 7.5 (continued)

Detention basin	Assessment criteria	Rains once every three months	Rains once every nine months	Rains once every year	Rains once every two years	Rains once every three years	Rains once every five years
	Composite index	0.000	0.855	0.765	0.743	0.668	0.624
	Efficiency level	–	Level 5	Level 4	Level 4	Level 4	Level 4
Yuantong Gully	Overflow reduction rate,%	–	100.00	88.33	59.28	26.44	4.28
	Pollution load reduction rate,%	–	100.00	88.33	59.28	26.44	4.28
	Volume utilization rate,%	–	100.00	100.00	100.00	100.00	100.00
	Composite index	0.000	1	0.912	0.695	0.448	0.282
	Efficiency level	–	Level 5	Level 5	Level 4	Level 3	Level 2
Laoyunliang River	Overflow Reduction rate,%	100.00	100.00	100.00	35.95	10.82	8.55
	Pollution load reduction rate,%	100.00	100.00	100.00	35.95	10.82	8.55
	Volume utilization rate,%	9.07	12.15	15.28	100.00	100.00	100.00
	Composite index	0.773	0.78	0.788	0.52	0.331	0.314
	Efficiency level	Level 4	Level 4	Level 4	Level 3	Level 2	Level 2
Xiaota Gully	Overflow reduction rate,%	100	100	100	32.31	33.63	14.13
	Pollution load reduction rate,%	100	100	100	32.31	33.63	14.13

(continued)

Table 7.5 (continued)

Detention basin	Assessment criteria	Rains once every three months	Rains once every nine months	Rains once every year	Rains once every two years	Rains once every three years	Rains once every five years
	Volume utilization rate,%	0	4.28	9.93	86.64	100	100
	Composite index	0.75	0.761	0.775	0.459	0.502	0.356
	Efficiency level	Level 4	Level 4	Level 4	Level 3	Level 3	Level 2
Qimu Gully	Overflow reduction rate,%	100.00	100.00	100.00	34.85	18.34	3.71
	Pollution load reduction rate,%	100.00	100.00	100.00	34.85	18.34	3.71
	Volume utilization rate,%	0.00	0.00	14.44	100.00	100.00	100.00
	Composite index	0.75	0.75	0.786	0.511	0.388	0.278
	Efficiency level	Level 4	Level 4	Level 4	Level 3	Level 2	Level 2
Kunming No. 1 High School	Overflow reduction rate,%	83.30	73.17	81.57	27.65	12.35	8.17
	Pollution load reduction rate,%	83.30	73.17	81.57	27.65	12.35	8.17
	Volume utilization rate,%	100.00	100.00	100.00	100.00	100.00	100.00%
	Composite index	0.875	0.799	0.862	0.457	0.343	0.311
	Efficiency level	Level 5	Level 4	Level 5	Level 3	Level 3	Level 3
Wulong River	Overflow reduction rate,%	42.44	32.17	25.88	23.89	18.15	16.16

(continued)

Table 7.5 (continued)

Detention basin	Assessment criteria	Rains once every three months	Rains once every nine months	Rains once every year	Rains once every two years	Rains once every three years	Rains once every five years
	Pollution load reduction rate,%	42.44	32.17	25.88	23.89	18.15	16.16
	Volume utilization rate,%	100.00	100.00	100.00	100.00	100.00	100.00
	Composite index	0.568	0.491	0.444	0.429	0.386	0.371
	Efficiency level	Level 3	Level 3	Level 3	Level 3	Level 2	Level 2
Cailian River	Overflow reduction rate,%	100.00	64.16	53.82	30.78	14.52	11.88
	Pollution load reduction rate,%	100.00	64.16	53.82	30.78	14.52	11.88
	Volume utilization rate,%	14.23	100.00	100.00	100.00	100.00	100.00
	Composite index	0.786	0.731	0.654	0.481	0.359	0.339
	Efficiency level	Level 4	Level 4	Level 4	Level 3	Level 2	Level 2
Lanhua River	Overflow Reduction rate,%	–	52.48	47.10	42.93	26.17	42.73
	Pollution load reduction rate,%	–	52.48	47.10	42.93	26.17	42.73
	Volume utilization rate,%	–	42.11	100.00	100.00	100.00	100.00
	Composite index	–	0.499	0.603	0.572	0.446	0.57
	Efficiency level	–	Level 3	Level 4	Level 3	Level 3	Level 3

(continued)

Table 7.5 (continued)

Detention basin	Assessment criteria	Rains once every three months	Rains once every nine months	Rains once every year	Rains once every two years	Rains once every three years	Rains once every five years
Daguan River	Overflow reduction rate,%	100.00	100.00	99.15	82.93	62.88	56.39
	Pollution load reduction rate,%	100.00	100.00	99.15	82.93	62.88	56.39
	Volume utilization rate,%	14.38	100.00	100.00	100.00	100.00	100.00
	Composite index	0.786	1	0.994	0.872	0.722	0.673
	Efficiency level	Level 4	Level 5	Level 5	Level 5	Level 4	Level 4
Haiming River	Overflow reduction rate,%	–	100.00	68.89	43.48	35.47	25.64
	Pollution load reduction rate,%	–	100.00	68.89	43.48	35.47	25.64
	Volume utilization rate,%	–	21.64	39.29	100.00	100.00	100.00
	Composite index	–	0.804	0.615	0.576	0.516	0.442
	Efficiency level	–	Level 5	Level 4	Level 3	Level 3	Level 3
Mingtong River	Overflow reduction rate,%	100.00	100.00	92.14	50.36	35.33	35.06
	Pollution load reduction rate,%	100.00	100.00	92.14	50.36	35.33	35.06
	Volume utilization rate,%	3.55	20.94	35.45	100.00	100.00	100.00
	Composite index	0.759	0.802	0.78	0.628	0.515	0.513
	Efficiency level	Level 4	Level 5	Level 4	Level 4	Level 3	Level 3

Table 7.6 Efficiency level results of north zone detention basins

Assessment criteria	Rains once every three months	Rains once every nine months	Rains once every one year	Rains once every two years	Rins once every three years	Rains once every five years
Overflow reduction rate,%	100.00	73.80	67.86	37.24	29.94	20.52
Pollution load reduction rate,%	100.00	73.80	67.86	37.24	29.94	20.52
Full-pipe segment length reduction rate,%	0.58	1.21	0.70	0.42	0.22	0.39
Full-pipe time reduction rate%	0.58	1.21	0.70	0.42	0.22	0.39
Reduction rate of overflow events at nodes%	100.00	55.56	38.89	22.70	16.67	3.40
Reduction rate of overflow time at nodes%	43.30	46.00	43.30	36.36	4.00	1.80
Reduction rate of overflow volume at nodes%	100.00	87.70	71.75	3.45	2.71	1.31
Volume utilization rate,%	61.70	91.50	100	100	100	100
Composite index	0.85	0.75	0.73	0.52	0.47	0.41
Efficiency level	Level 5	Level 4	Level 4	Level 3	Level 3	Level 3

Table 7.7 The effectiveness level results of the Western District detention basin

Assessment criteria	It rains once every three months	It rains once every nine months	It rains once every one year	It rains once every two years	It rains once every three years	It rains once every five years
Overflow reduction rate,%	–	35.13	30.98	16.23	6.46	5.71
Pollution load reduction rate,%	–	35.13	30.98	16.23	6.46	5.71
Full-pipe segment length reduction rate,%	17.79	9.88	4.50	1.72	0.73	1.08
Full-pipe time reduction rate%	27.86	25.20	25.35	11.32	7.65	4.80
Reduction rate of overflow events at nodes%	66.67	42.86	29.03	13.48	12.38	7.63
Reduction rate of overflow time at nodes%	60.00	33.00	12.50	3.70	2.00	1.33
Reduction rate of overflow volume at nodes%	56.25	41.24	24.33	4.29	3.26	3.24
Volume utilization rate,%	40.68	57.63	58.00	100	100	100
Composite index	0.16	0.41	0.37	0.39	0.33	0.32
Efficiency level	Level 1	Level 3	Level 2	Level 2	Level 2	Level 2

Table 7.8 The effectiveness level results of the Southern District detention basin

Assessment criteria	It rains once every three months	It rains once every nine months	It rains once every one year	It rains once every two years	It rains once every three years	It rains once every five years
Overflow reduction rate,%	100	71.00	63.00	45.00	30.00	32.00
Pollution load reduction rate,%	100	71.00	63.00	45.00	30.00	32.00
Full-pipe segment length reduction rate,%	13.67	5.38	9.28	6.63	5.69	6.03
Full-pipe time reduction rate%	27.86	27.54	23.34	8.04	4.80	3.59
Reduction rate of overflow events at nodes%	60.00	45.83	29.63	13.92	14.58	9.26
Reduction rate of overflow time at nodes%	78.57	32	11.02	4.00	3.00	1.33
Reduction rate of overflow volume at nodes%	56.25	41.24	24.33	4.29	3.26	3.24
Volume utilization rate,%	3.36	38.65	61	100	100	100
Composite index	0.68	0.58	0.58	0.57	0.47	0.48
Efficiency level	Level 4	Level 3	Level 3	Level 3	Level 3	Level 3

From Table 7.8, it can be seen that the comprehensive efficiency index of the storage pond in the southern region is the highest under the 0.25 year rainfall scenario, with a value of 0.68, which is a fourth level efficiency, while it is a third level efficiency under other rainfall scenarios.

On the basis of conducting comprehensive performance evaluations of individual and individual rainfall storage tanks, further conduct comprehensive performance evaluations of annual rainfall storage tanks. Using rainfall data from August 7, 2013 to August 8, 2014, the total rainfall was 1103.8 mm. The rainfall was concentrated from June to October, with a cumulative rainfall of 913 mm, accounting for 82.7% of the total annual rainfall. Based on this data simulation, the operational efficiency of the regulation and storage facilities on a year-round scale is evaluated. The comprehensive operational efficiency evaluation results of the regulation and storage tanks in three drainage areas under the annual rainfall scenario are shown in Table 7.9.

From Table 7.9, it can be seen that the comprehensive efficiency indices of the storage facilities in the northern, western, and southern regions within the annual rainfall boundary conditions are 0.253, 0.21, and 0.197, respectively, with efficiency levels of Level 2, Level 2, and Level 1.

iv Primary enhanced treatment technology for urban sewage treatment plants during rainy season.

(i) *Related Design Parameters*

The fundamental principle of this technology is to treat as much rainwater and runoff collected during the rainy season as possible in the sewage treatment plant. A new rainy season response treatment structure is constructed in the sewage treatment plant, and chemical coagulation strengthening technology is used to treat the collected runoff and rainwater. The added chemicals are poly aluminum chloride (PAC) and polyacrylamide (PAM).

When the liquid level in front of the inlet lift pump of the sewage treatment plant exceeds 5.6 m, the rainy season response mode will be activated. Inflow exceeding 1.3 times the design flow rate of the biochemical section will enter the rainy season response treatment structure. The dosage of PAC is 75 mg/L, the dosage of PAM is 75 μ g/L, and the hydraulic retention time is 1 h. The three-year operation data shows that the average removal rates of SS, COD, BOD_5, NH_3-N, TN, and TP in the combined wastewater during the rainy season can reach 87.2%, 84.3%, 90%, 28.5%, 42.5%, and 67.9%, respectively. Taking into account factors such as construction,

Table 7.9 Annual Rainfall Detention Basin Comprehensive Efficiency Assessment

Drainage zone	Overflow reduction rate,%	Overflow frequency reduction rate,%	Efficiency index	Efficiency level
North Zone	22.68	27.96	0.253	Level 2
West Zone	13.40	28.30	0.21	Level 2
South Zone	18.10	21.24	0.197	Level 1

equipment, chemicals, electricity consumption, and labor, the operating cost is 0.6 yuan/m^3.

(ii) *Location and Scale of Technology Application*

The first level enhanced treatment technology for urban sewage treatment plants during the rainy season has been applied in the fifth sewage treatment plant in Kunming, with a design scale of 185,000 m^3/d. The main biochemical process adopts A^2O, and the effluent quality meets the first level A standard of the "Pollutant Discharge Standard for Urban Sewage Treatment Plants". In order to cope with the impact load during the rainy season, the sewage treatment plant has constructed a treatment structure for the rainy season, with a treatment capacity of 100,000 m^3/d.

7.1.3 Application Effectiveness

7.1.3.1 Engineering Effectiveness

Kunming City has applied the above technology to construct three regulating and storage tanks in the second ring area, which has a good effect on reducing the initial rainwater and overflow pollution load under light rain scenarios. According to statistics, from 2014 to April 2018, the cumulative reduction of combined rainwater wastewater from the Haiming River, Wulong River, and Baiyun Road storage tanks was 1.8018, 69.89, and 6.7933 million cubic meters, respectively. The total annual reduction of COD, SS, TN, and TP in the three storage tanks reached 505.88 tons, 2821.32 tons, 140.86 tons, and 19.61 tons, respectively.

This technology has been popularized and applied in the regulation and storage tank of Kunming No.1 Middle School, the regulation and storage tank of Qimu Ditch, the regulation and storage tank of the old Yunliang River and the regulation and storage tank of Xiaolugou in the western area of Kunming City. The operation effect is good, and the total pollutant load can be reduced by 671.9 t COD/year; Seven regulation and storage ponds will be constructed in the Panlong River Basin. Under the once in a year rainfall scenario, the annual reduction of COD, TN, and TP will reach 91.2 tons, 4.77 tons, and 0.53 tons, respectively.

7.1.3.2 Economic Performance

Based on this technology, a total of 17 rainwater storage tanks have been built in the main urban area of Kunming, with a total storage capacity of 212,400 m3, a service area of 54.03 km^2, and an investment of 1.819 billion yuan. The construction cost of a unit volume rainwater storage tank and the construction cost of a unit service area are 8564 yuan and 33.67 yuan respectively, and the operation and maintenance cost is approximately 1.23 yuan/m^3.

7.1.3.3 Environmental Effects

Kunming rainwater storage tanks all adopt underground storage construction technology, and through the combination of storage tanks and water purification plants, the collection of rainfall runoff and water quality purification in different areas are achieved. The upper part of the storage pool is usually a green space or park, with fitness and leisure facilities built to coordinate with the surrounding environment, improving the environmental friendliness of the gray infrastructure and having high environmental benefits.

7.1.4 Technical Highlights and Increments

a. The technical highlights of this case are mainly reflected in three aspects: At the beginning of the design of the rainwater storage tank in this case, advanced model simulation technology was used to provide more scientific guidance on the volume design, construction location, and construction quantity of the storage tank.
b. The rainwater storage tank in this case achieved linkage between multiple rainwater storage tanks and between rainwater storage tanks and water purification plants through the application of online monitoring systems in actual operation, effectively reducing the phenomenon of full load operation of rainwater storage tanks and water purification plants during the rainy season.
c. This case utilizes primary enhanced treatment technology to purify the rainwater collected in the rainwater storage tank, effectively reducing the runoff pollution load.

The technological increment of this case is reflected in the proposed new method for designing rainwater and sewage storage tanks, the development of enhanced treatment technology for sewage treatment plants during the rainy season, and the construction of a joint operation mode between rainwater and sewage storage and sewage treatment plants. Compared to traditional rainwater and sewage storage tank construction and operation technologies, the overall efficiency is improved by more than 20%.

7.2 Rainfall Runoff Pollution Control Project in Steep Slope Area of Mountain City

7.2.1 Case Background

Chongqing is a typical mountainous city with undulating terrain and large slope. The pollution load of rainfall runoff is large and the runoff erosion effect is strong. With the rapid development of economy and society, the urbanization construction of Chongqing has developed rapidly. Some natural ecosystems such as natural green space and forest land have been replaced by hardened pavement or roof, and the proportion of impermeable ground is increasing. The problem of runoff pollution is increasing day by day, the peak of urban rainstorm is obviously increased, the problem of rainstorm disaster is prominent, and the capacity of existing urban drainage system is obviously insufficient.

On July 17, 2007, Chongqing suffered a heavy rainstorm and flood disaster with a recurrence period of 115 years, and the urban waterlogging was very serious. After the rainstorm, there were 165 severely affected places or regions in the main city of Chongqing. Urban waterlogging is accompanied by rainfall runoff pollution, resulting in increasingly serious water environment problems in Chongqing.

Under the background of rapid urbanization, ensuring the safe operation of urban drainage system, effectively reducing the pollution load of urban rainfall runoff, gradually improving the overall efficiency and safety stability of pollutant reduction of water environment facilities, improving the quality of urban water environment and promoting the health of urban water ecological environment have become the overall objectives and requirements of water quality and pollution control in the Three Gorges Reservoir area.

Under the background of rapid urbanization, ensuring the safe operation of urban drainage systems, effectively reducing urban rainfall runoff pollution loads, gradually improving the overall efficiency and safety stability of water environmental facilities for pollutant reduction, improving urban water environmental quality, and promoting the health of urban water ecosystems have become the overall objectives and requirements for water quality and pollution control in the Three Gorges Reservoir Area.

This case is based on the national water project "Chongqing Liangjiang New Area Urban Water System Construction Technology Research and Demonstration (2012ZX07307-001)".

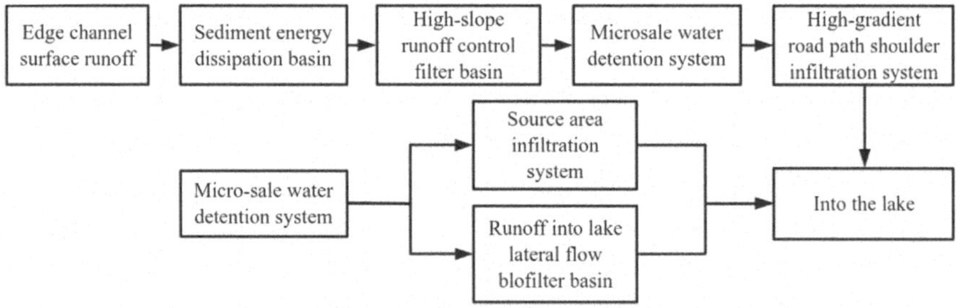

Fig. 7.3 Process flowchart of rainfall runoff pollution control project in the mountainous urban area of Jiangnan Park, Chongqing Garden Expo Park

7.2.2 *Process Route and Key Technology*

7.2.2.1 Process Route

Based on the above topics, the rainfall runoff pollution control project in the large slope area of Jiangnan Garden of Chongqing Garden Expo Park was constructed. The process flow is shown in Fig. 7.3. The surface runoff from the upper reaches of the Jiangnan Garden outside the Garden Expo Garden first enters the precipitation energy dissipation pool, which accepts rainwater runoff inside the Garden Expo Garden through the rainwater side ditch. After the runoff is treated by the precipitation energy dissipation pool, the water flow rate slows down and some sediment is also precipitated. Part of the surface runoff after energy dissipation and precipitation enters the combined modular large slope runoff control filter system, and enters the micro waterscape system for detention after infiltration treatment. Another part of the runoff directly enters the micro-waterscape system for retention; when the amount of rainwater entering the micro-waterscape system exceeds the designed storage capacity of the micro-waterscape, the micro-waterscape system will overflow. The overflow water enters the large slope road runoff shoulder belt infiltration system, the source area infiltration system and the runoff into the lake. The lateral flow biofilter system quickly infiltrates into Longjing Lake through the underground infiltration drainage channel.

7.2.2.2 Key Technologies and Applications

i. Related Design Parameters

(i) *Combined modular large slope runoff control filter technology*

This technology consists of stagnant water layer, planting matrix layer, stagnant layer/water storage layer and surface plants. The whole system profile and real scene are shown in Figs. 7.4 and 7.5.

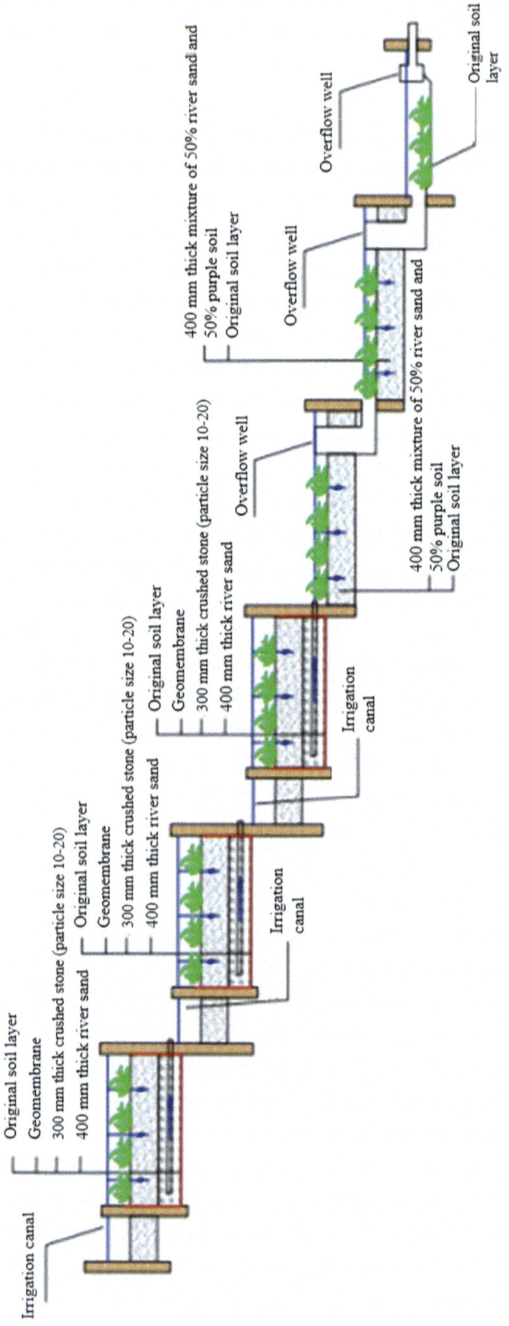

Fig. 7.4 Cross-sectional diagram of combined modular high-gradient runoff control filter bed system

Fig. 7.5 Realistic view of combined modular high-gradient runoff control filter bed system

The stagnant water layer: the stagnant water layer provides a temporary storage and retention space for surface runoff to realize the settlement of particulate pollutants. Considering the factors such as landscape, function and safety, the height of the stagnant water space is set to 500 mm.

The planting matrix layer: the planting matrix layer is mainly river sand with different proportions of purple soil, the permeability coefficient changes step by step, the maximum is 42 cm/h, and the depth is set to 400 mm.

Retention layer/water storage layer: the retention layer prolongs the residence time of urban surface runoff, plays a role in temporary storage and retention of heavy pollution runoff in the initial stage of urban surface, and undertakes the main nitrogen removal function. The retention layer is composed of gravel with a particle size of 20–40 mm and a design depth of 300 mm. Geotextiles are laid on the surface of the retention layer to prevent the downward migration of soil particles in the upper planting matrix layer and the occurrence of clogging. In order to improve the phosphorus removal capacity, slag was added to the retention layer of the last filter chamber as a filler to adsorb soluble phosphorus in surface runoff.

Surface plants: plant configuration selection of tall grass herbaceous plants Miscanthus, Yudaicao and Pennisetum.

(ii) *Micro-Waterscape Stagnation Technology*

The technical profile and real scene are shown in Figs. 7.6 and 7.7. The average water depth of the miniature waterscape is 120 cm, and the hydraulic retention time is 2.5 h. Pebbles with a particle size of 50–200 mm were laid at the bottom of the pond, and the thickness of the pebble layer was 10–30 cm. Potted water lily, acorus tatarinowii and water bamboo taro are the main plants, and canna, arundo donax and brown bamboo are arranged around the pool. The plant spacing of canna is 0.4–0.6 m.

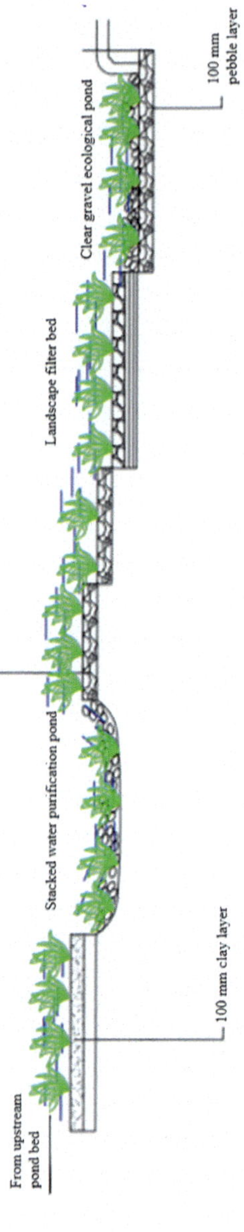

Fig. 7.6 Cross-sectional diagram of micro water feature retention technique

Fig. 7.7 Realistic view of micro water feature retention technique

The slope between the pool and the surrounding road is a vegetation filter belt. The vegetation belt is a cluster of Arundo donax and Brown bamboo. It is dispersed and mixed at a density of 6–8 plants/m^2, and the lawn grass is densely planted at the gap. The planting specifications are generally 12–18 g/m^2.

(3) Large slope road runoff shoulder infiltration technology

According to the distribution pattern of urban roads and green belts, the 'shoulder infiltration technology of large slope road runoff road shoulder' is arranged along the shoulder side of the road, and the technical profile and real scene are shown in Figs. 7.8 and 7.9. The service area of the cascade infiltration system is controlled at 5%–10% of the surface area of the service road. The width of the vegetation filtration area should be controlled at 60–120 cm, the slope should be controlled within 3: 1, and the bottom of the retention infiltration area should be smooth, with an average width of 300 cm. The height of aquifer is 500 mm; the planting matrix layer was 80% river sand composite 20% purple soil with a thickness of 300 mm, and the permeability coefficient was 40 cm/h. The middle layer adopts coarse river sand, and the depth is designed to be 100 mm, so that the surface runoff continues to penetrate and prevents the top planting soil layer from blocking the bottom gravel layer. The reservoir and drainage layer in the infiltration area is composed of gravel (Φ30–40 mm), the depth is designed to be 200 mm, and the D110 perforated pipe is laid in the last pond. Plants choose iris, canna and other hygrophytes.

(iv) *Penetration Promotion Technology in Source Area Under Complex Geological Conditions*

As shown in the technical cross-section (Fig. 7.10) and the actual scene image (Fig. 7.11), the bottom width of the source area infiltration promotion technology is

Fig. 7.8 Cross-sectional diagram of high-gradient road runoff route shoulder infiltration technique

Fig. 7.9 Realistic view of high-gradient road runoff route shoulder infiltration technique

designed to be 3000 mm, with a length-to-width ratio of 2:1 and a longitudinal slope of 1.4%. The height of the stagnant water space is set at 150 mm. The planting substrate layer is 500 mm thick, composed of 80% river sand and 20% purple soil, with a permeability coefficient of 40 cm/h. The storage and drainage layer is composed of gravel (Φ30–40 mm), with a depth of 300 mm. The storage and drainage layer provides temporary storage for infiltrated surface runoff and extends the infiltration time of the surface runoff into the deep soil. Perforated drainage pipes (D110) are installed in the gravel layer, with a slope of 3‰. When selecting vegetation, considerations include native species, well-developed root system, tolerance to flooding and drought, as well as landscape aesthetics, with the choice of zebra grass.

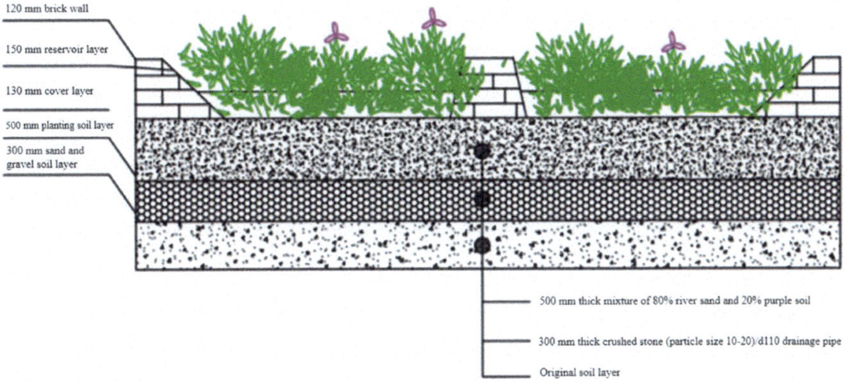

Fig. 7.10 Cross-sectional diagram of source area infiltration promotion technique under complex geological conditions

Fig. 7.11 Realistic view of source area infiltration promotion technique under complex geological conditions

(v) *Purification Technology of Lateral Flow Biofilter Enhanced by Runoff into the Lake*

As shown in the technical cross-section (Fig. 7.12) and the actual scene image (Fig. 7.13), the hydraulic load design for the lateral flow constructed wetland for intensified runoff treatment in the lake is 0.8 $m^3/(m^2 \cdot d)$. The planting soil layer consists of a combination of river sand and purple soil. According to the treatment process, the proportion of river sand gradually decreases, resulting in a gradual decrease in permeability rate. The maximum permeability coefficient is 51 cm/h, while the minimum is 4.0 cm/h. The filling material mainly consists of gravel and coarse sand. The thickness of the filling material is 300 mm at its thickest, and the thickness of the filling layer gradually decreases along the direction of the water flow. The surface vegetation consists of moneywort and iris, with a planting density of 20 plants/m^2.

(vi) *Technology Application Location and Scale*

The "source area infiltration promotion technology" under complex geological conditions, the "shoulder infiltration technology for steep slope pathway flow", the "combined modular steep slope runoff control filter module technology", the "micro water feature retention technology", and the "lateral flow constructed wetland for intensified runoff treatment in the lake" have been applied in the Chongqing Garden Expo Park Jiangnan Garden mountainous urban area rainfall runoff pollution control project. The respective scales are 630 m^2, 410 m^2, 875 m^2, 875 m^2, and 238 m^2. At the same time, these technologies have been widely promoted and applied in the Yuelai New City water environment restoration project.

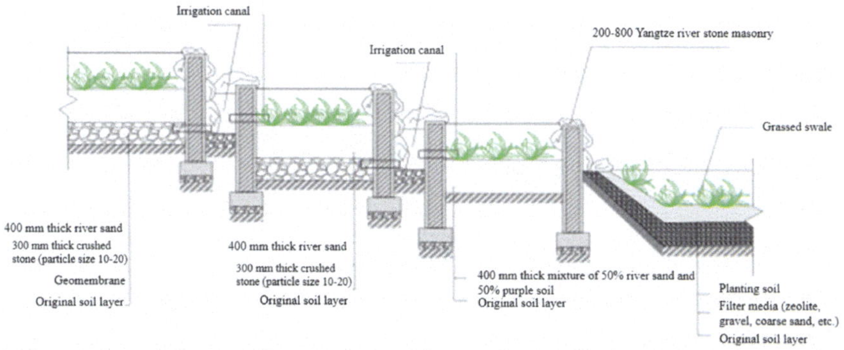

Fig. 7.12 Cross-sectional diagram of enhanced lateral flow biofiltration technology for runoff into lake purification

Fig. 7.13 Realistic view of enhanced lateral flow biofiltration technology for runoff into lake purification

7.2.3 Application Effectiveness

7.2.3.1 Technical Effectiveness

Based on the monitoring of the Chongqing Garden Expo Park Jiangnan Garden mountainous urban area rainfall runoff pollution control project in 2015, the total annual rainfall was 1270.1 mm. Rainfall events with less than 30 mm did not produce any outflow, and all surface runoff from upstream areas was intercepted. There were 66 days with rainfall greater than 5 mm, and 9 days with rainfall exceeding 30 mm. The project reduced the direct inflow into Longjing Lake by 83.7% of the total runoff volume. The project achieved average removal rates of 44.3% for SS, 44.7%

for COD, 74.3% for TN, and 41.8% for TP in the runoff. This translates to an annual reduction of pollution load by 6.03 t for SS, 6.93 t for COD, 1.43 t for TN, and 0.04 t for TP.

7.2.3.2 Economic Performance

Based on the engineering construction, two types of products have been developed: the initial runoff enhancement treatment and later runoff collection purification integrated equipment, and new high-performance runoff infiltration promotion materials. These products are suitable for controlling surface source pollution in mountainous cities, promoting water quality improvement and functional enhancement in mountainous urban water environments, and have broad application prospects.

7.2.3.3 Environmental Effectiveness

The implementation of the rainfall runoff pollution control project in the mountainous urban area has effectively reduced the surface source pollution load entering Longjing Lake within the Garden Expo Park. It has enhanced the flow of water in the bay area, strengthened the self-purification capacity of the water, and promoted the improvement of water quality in the lake. The water quality of the lake can reach Class IV water quality standards. Currently, the Garden Expo Park has become an important choice for leisure activities for Chongqing residents, as well as a destination for tourists from other places and team building activities.

7.2.4 Technical Highlights and Technical Increments

In response to the challenges of complex terrain, steep slopes, intense rainfall runoff, and initial prominence of surface source pollution in typical mountainous watersheds, a "source segment-migration segment" control technology system for surface source pollution in mountainous urban areas has been developed. This system effectively slows down the flow velocity in areas with steep slopes, increases hydraulic retention time, and removes pollutants from the runoff effectively. It has been demonstrated in the engineering project of Longjing Lake within the Chongqing Garden Expo Park and also applied in the water environment restoration project in Yuelai New City.

7.3 Demonstration Project of Initial Rainwater Purification Gray-Green Multi-Stage Regulation and Storage in Binhu New Town

7.3.1 Case Background

The Lakeside New District is located on the shore of Chaohu Lake, the fifth largest freshwater lake in China. It serves as the financial, commercial, administrative, cultural, tourism, life service, and research and development center of Hefei City. It is also the water gateway for Hefei City to connect with the Yangtze River and integrate into the Yangtze River Delta. The main water systems in the Lakeside New District include Nanfei River, Shiwuli River, and Tangxi River. Among them, Nanfei River and Shiwuli River are more seriously polluted, with water quality ranging from Class V to inferior Class V, far from meeting the water quality standards required by the "Functional Zoning of Water Environment in Hefei City" and still far from achieving the goals of comprehensive pollution control in Chaohu Lake.

Due to the lack of clean water supply, poor hydraulic conditions, limited water environmental capacity, and weak self-purification ability in most of the water channels in the Lakeside New District, pollutants entering the rivers are prone to accumulate and amplify their pollution effects in the water body. The incomplete rain and wastewater drainage system results in the direct discharge of urban sewage and a large amount of mixed rain and wastewater into urban rivers such as Tangxi River and Nanfei River, which eventually flows into Chaohu Lake. This is one of the main reasons why the pollution in urban rivers and Chaohu Lake has not fundamentally improved.

With the significant construction efforts in recent years, the Lakeside New District has established a comprehensive drainage system covering the urban areas, including both separate and combined sewer systems. With the goal of "complete interception and full treatment," the Lakeside New District has gradually constructed a large number of pollution interception pipelines and pumping stations along the rivers. Through measures such as pollution interception and control, the control rate of point source pollution in the central urban area has greatly improved, leading to some improvement in the overall water environment of the city. However, effective control measures for urban rainfall runoff pollution are still lacking, which has exacerbated the contradiction between runoff pollution and river water environment pollution, posing a threat to the achievements of water environment management.

The annual rainfall in the Lakeside New District is close to 1000 mm, mainly concentrated during the flood season. In the early stage of development, the Lakeside New District faced severe rainwater and overflow pollution issues, which can be primarily summarized as follows: (1) The current combined sewer system still faces challenges such as low interception ratios and high overflow frequency. The phenomenon of mixed rain and wastewater in the separate sewer system is widespread, with a considerable amount of sewage being discharged into the rivers

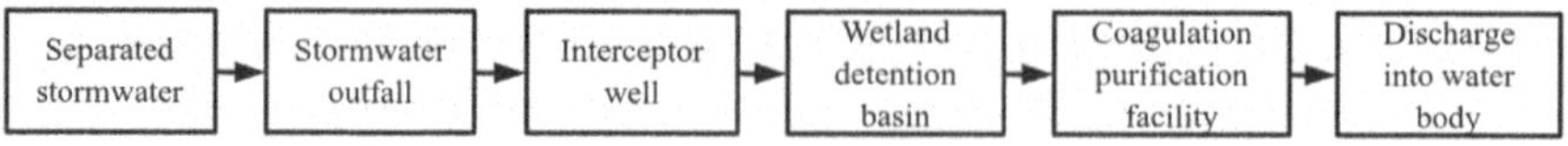

Fig. 7.14 Process route of rainwater purification engineering in Hefei Binhu new area (Initial Stage)

through stormwater outlets, intensifying the level of water pollution. (3) The existing drainage system has difficulties in effectively controlling surface runoff pollution, and its operational management and monitoring methods are lagging behind. Consequently, it is evident that in the context of gradual improvement in point source pollution control, runoff pollution will become one of the key factors limiting the water environmental quality in the Lakeside New District.

This case is based on the project "Research on Low Impact Development and Water Environment Improvement Techniques and Engineering Demonstration in the Lakeside New District of Hefei City (2011ZX07303-001)," which was initiated in 2011 as part of the National Water Special Program.

7.3.2 Process Route and Key Technology

7.3.2.1 Process Route

Based on the above-mentioned project, the initial rainwater purification project was established in the Lakeside New District of Hefei. The process route is shown in Fig. 7.14. In this project, Hefei City focused on the 10 stormwater outlets along Luzhou Avenue in the Tangxi River, and implemented interception measures by reconstructing the outlets and installing interception wells to intercept the initial rainwater. A system of interception pipes was laid along the river to collect the intercepted initial rainwater from each outlet. On the northern bank of the Tangxi River, east of Huizhou Avenue, south of Zhongshan Road, and west of Guangxi Road, wetland-type detention ponds were constructed to store the collected initial rainwater. These detention ponds were equipped with complete sets of diatomaceous earth coagulation and purification facilities to treat and purify the initial rainwater before it is discharged into the Tangxi River.

7.3.2.2 Key Technologies and Applications

Based on the above project, two key technologies were developed and applied: the "wet pond detention technology" and the "rapid treatment technology for initial rainwater ecological retention using diatomaceous earth".

i. Wet Pond Storage Technology

(i) *Related Design Parameters*

The wet pond detention technology mainly relies on the wetland-type detention pond to achieve the purpose of runoff detention. The vertical structure and the actual scene of the wet pond are shown in Figs. 7.15 and 7.16. In this case, a green space excavation wetland-type detention pond was constructed on the northern bank of the Tangxi River. The pond has a top dimension of 70 m × 70 m, a maximum depth of 7 m, and a slope ratio of 1:2. Its effective storage capacity is 20,000 m^3. Due to the high groundwater level along the Chaohu Lake, anti-seepage measures were taken in the wet pond. After completing the excavation of the wet pond, the soil was compacted and a geotechnical membrane was installed to prevent seepage. A 50–100 cm compacted clay protective layer was applied on top of the membrane. Considering the landscape effect, the water level in the wet pond is set at 0.2–0.4 m below the ground elevation. The buried depth of the interception pipe before entering the detention pond is approximately 6.0–8.0 m, requiring the installation of a lift pump room. The maximum flow rate of the interception main pipe is 2.53 m^3/s. The pump room is equipped with three sets of 84.4 L/s submersible axial flow pumps and two sets of 417 m^3/h submersible sewage pumps. In order to prevent the impurities from clogging the pumps during the initial rainfall, three sets of 900 L/s pulverizing type bar screens were installed in the front section of the pump room for debris removal.

(ii) *Technology application location and scale*

The technology has been applied in the initial rainwater purification project in Hefei Binhu New Area. The project covers an area of about 700 m^2, the design scale is 20000 m^3, and the effective water depth is 1.5 m.

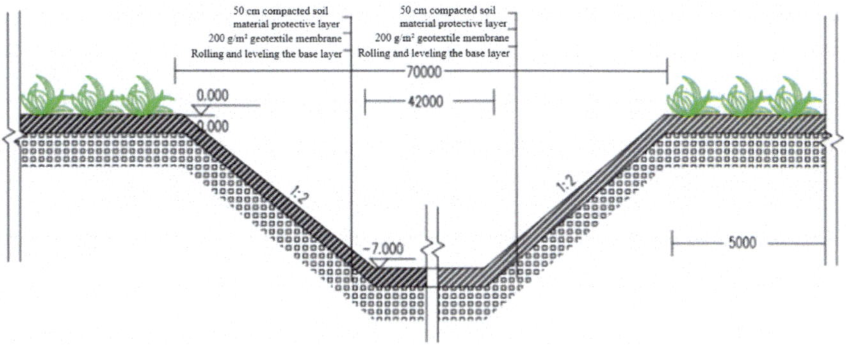

Fig. 7.15 Schematic diagram of vertical structure of wet pond

Fig. 7.16 Realistic view of wet pond

i. Rapid Treatment Technology of Initial Rainwater Ecological Retention Diatomite

(i) *Related Design Parameters*

The design capacity of the diatomite rapid treatment facility for ecological retention of initial rainwater is 5000 t/d. The complete set of fully automatic operation equipment is adopted to realize the unattended automatic control of dosing, sludge discharge, start/stop and so on.

The underground early-stage rainwater ecological retention diatomaceous earth rapid treatment facility pit is built together with the lift pump room, and it is constructed in a buried form with a glass top cover for easy visitation. The specific treatment process is as follows: the initial rainwater is pumped from the detention pond by a water intake pump and enters the mixing tank (as seen in the actual scene in Fig. 7.17). Modified diatomaceous earth is added to the mixing tank to achieve rapid mixing with the initial rainwater, and the mixture flows into the grid flocculation tank. In the flocculation tank, hydraulic stirring is employed to promote particle collisions and agglomeration, forming larger flocculated particles. The mixture then evenly enters the upward-flow sloping tube settling tank through the distribution zone. The sloping tube settling tank is equipped with hexagonal honeycomb tubes, which facilitate the separation of sediment and water. The clarified water flows out into the Tangxi River.

Fig. 7.17 Diatomaceous earth coagulant mixing tank

The purification treatment of initial rainwater using this technology involves a dosage of 300–500 kg/day of chemicals and a hydraulic retention time of approximately 50 min. The sludge from the grid flocculation tank and sloping tube settling tank is discharged into the sludge collection trough and periodically released into the municipal drainage network.

(ii) *Technology Application Location and Scale*

This technology has been applied in the initial rainwater purification project in Binhu New District of Hefei. The project covers an area of about 50 m^2 and the design scale is 5000 m^3/d.

7.3.3 Effectiveness of Application

7.3.3.1 Technical Effectiveness

The COD removal rate of pollutants in this project is 60%–90%, the total phosphorus removal rate is 70%–90%, and the annual pollution load into the river is reduced by more than 30%.

7.3.3.2 Economic Performance

The construction cost of storage facilities for the initial rainwater treatment project of Tangxi River rainwater discharge outlet in Hefei City is 120 yuan/m3; compared with the traditional storage tank, the construction cost is saved by more than 80%. In addition, the operation cost of diatomite flocculation phosphorus removal technology is 0.2 yuan/m^3.

7.3.3.3 Environmental Effectiveness

The project plant area is coordinated with the surrounding landscape. In order to increase the landscape effect, a series of low impact development technologies and wet pond technologies are also applied in the project area, which realizes the perfect combination of landscape and engineering, effectively reduces the impact of initial rainwater on the water quality of Tangxi River, and achieves obvious economic, social and environmental benefits.

7.3.4 Technology Highlights and Technology Increments

The technological highlights of this typical case are mainly reflected in two aspects: (1) By utilizing diatomaceous earth coagulation to remove phosphorus, efficient removal of phosphorus from rainfall runoff is achieved. (2) After diatomaceous earth phosphorus removal technology, efficient removal of SS and settled phosphorus from rainfall runoff is achieved through the utilization of sloping tube settling technology.

This technology integrates rainwater detention ponds, diatomaceous earth coagulation for phosphorus removal, sloping tube settling technology, rainwater pumping stations, and ecological pond technology, achieving the organic unity of the entire system and ensuring its efficient operation. The removal rate of TP from collected rainfall runoff and Tangxi River water remains stable at over 90%, playing a crucial role in controlling pollution in Tangxi River and Chaohu Lake.

7.4 Overflow Pollution Control Project in the Old Urban Area of Taihu Lake Basin

7.4.1 Case Background

Changzhou City is located in the Yangtze River Delta plain, with a dense population, a developed economy, a high urbanization rate, and an increasingly prominent contradiction between economic development and water environment protection. With the rapid socio-economic development in Changzhou City, the emission rate of pollutants far exceeds the rate of control, resulting in the discharge of pollutants greatly exceeding the water's pollution-carrying capacity. A large amount of pollutants enter water bodies through various channels, exacerbating the pollution load.

Non-point source pollution, due to its dispersibility, suddenness, and high impact, has gradually become a focal point in water pollution control efforts.

CSOs, as a typical non-point source pollution, has become one of the important factors restricting water pollution in old urban areas. CSOs pollutants come from various sources, including domestic wastewater, stormwater runoff, and pipe sediments, making their pollution complex due to the diversity of origins. Among them, the inflow of stormwater runoff is the fundamental cause of CSOs pollution in combined sewer networks. Therefore, the development of simple and effective urban stormwater runoff pollution control technologies is of great significance in reducing the pollution load of stormwater runoff and further utilizing rainwater resources efficiently.

This case study is based on the research and demonstration projects initiated under the National Water Special Program in 2008, namely "Research and Demonstration on Pollution Control and Quality Improvement Techniques in Old Urban Areas" and the project launched in 2012, "Integrated Demonstration of Process Control and Comprehensive Improvement Techniques for Urban Water Pollution."

7.4.2 Process Route and Key Technology

7.4.2.1 Process Route

Based on the aforementioned projects, the CSOs pollution interception rapid filtration purification project and the stormwater non-point source pollution collection and treatment project have been constructed separately.

i. Combined Overflow Pollution Interception Rapid Filtration Project

The process route of the project is shown in Fig. 7.18. In this route, sewage in the Shangcun area is intercepted by a pumping station during sunny days. After being treated at the sewage treatment plant, it enters the municipal sewage pipe network. Excess stormwater and sewage during rainy days are stored in a detention basin to reduce stormwater overflow. The portion of stormwater and sewage that exceeds the storage capacity is treated by a suspended quick filtration device before being discharged into water bodies, thus reducing pollutant concentrations in combined sewer overflows (CSOs), improving river water quality, and preserving the water environment quality of the receiving water body.

ii. Separate Collection and Treatment Project of Rainwater Surface Source Pollution

The engineering process route is shown in Fig. 7.19. Initially, road runoff during rainfall is directed to a dedicated pipe located beneath the stormwater well. When the pipe is completely filled, the mid-to-late stage road runoff then enters the diversion pipe, achieving the separation of different quality levels of road stormwater runoff. The initial stage runoff in the dedicated pipe is treated by a swirl separator and rapid filtration pool before being discharged into the wetland park. The mid-to-late stage

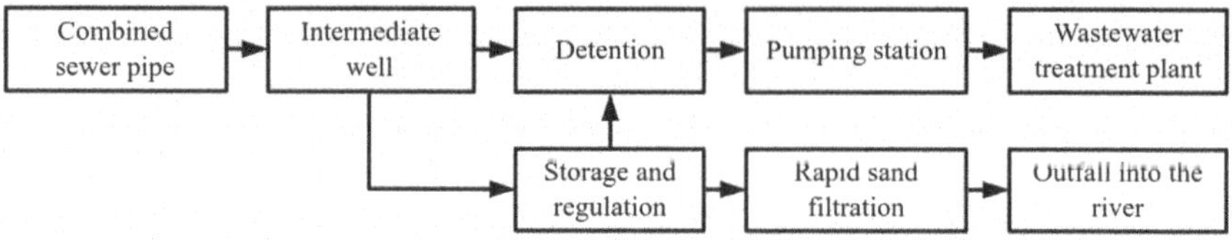

Fig. 7.18 Technical route of combined sewer overflow pollution interception and rapid filtration purification engineering

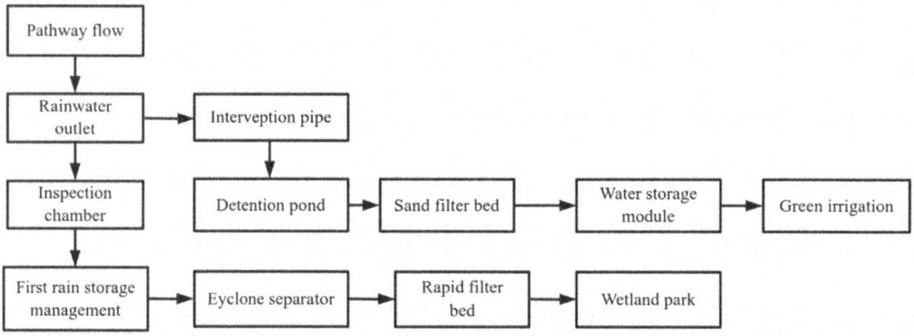

Fig. 7.19 Technical route for collection and treatment of Stormwater nonpoint source pollution

stormwater in the diversion pipe is treated by a sand filtration pool before entering the storage module for irrigation of green spaces within the pumping station. Any stormwater that exceeds the storage capacity is discharged into the wetland park.

7.4.2.2 Key Technologies and Applications

i. Combined Overflow Pollution Hydrocyclone-Rapid Suspension Filtration Technology

(i) *Related Design Parameters*

The suspended rapid filtration pool consists of three filter chambers, labeled A, B, and C, each with a height of 1.0 m. The types and thicknesses of filter materials in each chamber can be adjusted. On sunny days, the sewage in the detention tank is transported by diversion pumps 1 and 2 to the Dahongqi Road pumping station and then to the sewage treatment plant for processing. During rainy days, as upstream stormwater and sewage continuously flow in, when the liquid level in the detention tank rises to the set height of 1, the inlet pump of the suspended rapid filtration pool automatically starts, and the device begins to treat combined sewer overflows (CSOs), discharging the filtered effluent directly into the Tongji River. When the liquid level in the detention tank continues to rise to the set level 2 (higher than the set level 1), the inlet gate of the pumping station closes, and the emergency gate opens. Rainwater and sewage no longer enter the detention tank, and the suspended rapid filtration pool ceases operations. However, the diversion pumps and stormwater and sewage lifting pumps continue to work, continuously transporting the stored CSOs in the detention tank to the Dahongqi Road pumping station, and then to the sewage treatment plant for processing.

At the bottom of the detention tank, a backwashing system has been installed, consisting of 10 perforated pipes with a diameter of 15 mm and a spacing of 1.0 m. These pipes are connected to a backwashing pump. When sediment dredging is required, the backwashing pump can be activated to use the backwashing pipes to agitate the accumulated pollutants and stagnant water at the bottom of the detention tank. With the assistance of diversion pumps 1 and 2, the stirred mixture is then transported to the Dahongqi Road pumping station and subsequently sent to the sewage treatment plant for processing.

(ii) *Technology application location and scale*

The "Combined Sewer Overflow Polluted Hydraulic Vortex-Rapid Suspended Filtration Technology" has been applied at the Shangcun Pumping Station in the Zhonglou District of Changzhou City. The service area of this pumping station mainly consists of unreconstructed urban villages that adopt a direct discharge combined sewer system. The service area covers an approximate area of 55,000 m^2 with a population of around 2,750 people. The sewage volume is approximately 400 m^3/d. The effective volume of the detention tank is close to 800 m^3, and its detention capacity reaches 1055.78 m^3/h. The suspended rapid filtration pool has a diameter of 2.4 m and a rapid processing capacity of 102.7 m^3/h, with automatic backwashing capability (Fig. 7.20).

i. Initial Rainwater Management Storage Technology

(i) *Related Design Parameters*

In the demonstration project, the connecting pipes for rainwater inlets are made of high-density polyethylene (HDPE) double-walled corrugated pipes, while the initial

Fig. 7.20 Engineering site of combined overflow pollution hydraulic vortex-rapid suspended filtration technology

rainwater collection pipes use polyethylene (PE) pipes. According to the characteristics of the implemented roads, specific pipe diameters of 280, 315, and 450 mm have been set, with corresponding pipe lengths of 457, 547, and 156 m. The total length of the specialized pipes laid is 1.16 km, serving an area of 19,648 m^2. The project allows for the diversion and quality control of rainwater on Xitang River West Road, with an initial rainwater volume of 5 mm. The specialized pipes utilize hydraulic control for diversion (Fig. 7.21).

(ii) *Technology Application Location and Scale*

The "Initial Rainwater Specialized Pipe Storage and Retention Technology" has been applied at the Hengtang River West Road Pumping Station in Changzhou City. The specialized pipes have been laid over a length of 1.16 km. The demonstration project serves a road network with a total width of 32 m, including motor vehicle lanes, non-motorized vehicle lanes, and sidewalks, covering an area of approximately 19,648 m^2. Additionally, this technology has been promoted and applied at the Lihua Pumping Station, Zhulin Pumping Station, and Jinling Pumping Station in Changzhou City. The processing capacities of these three promotion projects are 160, 60, and 70 m3/h, respectively (Fig. 7.22).

7.4.3 Application Effectiveness

7.4.3.1 Technical Effectiveness

The "CSOs Polluted Water Hydraulic Cyclone-Rapid Suspended Filtration Technology" achieves removal rates of 95%, 50%, 80%, 30%, and 70% for SS, COD, NH_3-N, TN, and TP, respectively. The "Initial Rainwater Specialized Pipe Storage and Retention Technology" achieves removal rates of 56.6%, 55.9%, and 64.9% for COD, TP, and NH_3-N, respectively.

7.4.3.2 Economic Performance

The "CSOsPolluted Water Hydraulic Cyclone-Rapid Suspended Filtration Technology" has an investment construction cost of 100 yuan/m2 and an operational maintenance cost of 0.05 yuan/m3. On the other hand, the "Initial Rainwater Specialized Pipe Storage and Retention Technology" has an investment construction cost of 5000 yuan/m^2 and an operational maintenance cost of 0.5 yuan/m^3.

7.4.3.3 Environmental Effectiveness

The "Combined Overflow Polluted Water Hydraulic Cyclone-Rapid Suspended Filtration Technology" has been applied at the Shangcun Pumping Station in the

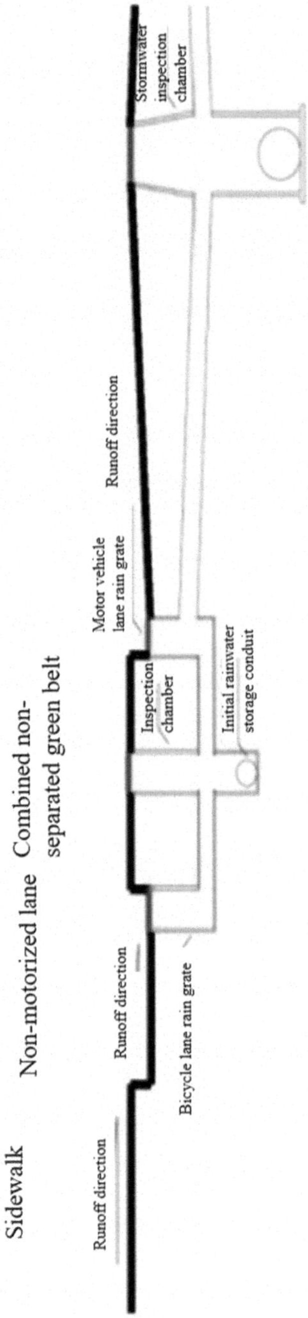

Fig. 7.21 Schematic diagram of initial rainwater storage conduit storage technology

Fig. 7.22 Initial rainwater conduit

Zhonglou District of Changzhou City. It relies on the storage and retention of the detention basin and on-site treatment to address overflow pollution that exceeds the capacity of the wastewater treatment plant. This technology effectively controls the discharge volume of combined overflow during rainfall, significantly reducing the pollution caused by overflow wastewater entering the receiving water body.

The "Initial Rainwater Specialized Pipe Storage and Retention Technology" effectively addresses the issue of initial runoff rainwater pollution within and around the West Hutang River Road Pumping Station. On one hand, it demonstrates good pollutant removal effectiveness for pollutants in rainfall runoff. On the other hand, the treated rainwater can be utilized for park greening irrigation and supplementing wetland water, significantly enhancing the utilization rate of rainwater.

7.4.4 Technology Highlights and Technology Increments

To address the issue of combined overflow pollution, the "Combined Overflow Polluted Water Hydraulic Cyclone-Rapid Filtration Technology" has been developed. This technology significantly improves the diversion ratio of the pipeline network and reduces the impact of water flow on the network by incorporating detention basins without requiring extensive modifications to the existing pipeline network. The key breakthrough lies in the rapid filtration technology for suspended particles, which features a higher filtration rate, eliminates the need for a separate backwashing system, occupies less space, and requires simple maintenance. The filtration rate of the rapid filtration technology can reach 20 m/h, providing an average removal rate of over 45% for SS at high filtration speeds. When the particle size exceeds 150 μm, the removal effect is essentially 100%.

To address the issue of initial road rainwater, the "Initial Rainwater Specialized Pipe Storage and Retention Technology" has been developed. This technology enables the complete separation of initial rainfall, providing different design parameters for various road conditions. During usage, the specialized pipe for the "upstream" does not affect the "downstream", and it is characterized by simple maintenance, low cost, and thorough separation of rainwater. Additionally, the design parameters of this technology can be modified according to specific circumstances.

7.5 Comprehensive Demonstration Project of Multi-Element Low-Impact Development and Utilization in New Urban Areas

7.5.1 Case Background

With the development of the economy, urbanization in Wuxi City has accelerated. Taking the example of the emerging urban area, Taihu New City, in Wuxi, the rate of ground hardening has significantly increased after the construction of the new city. Improper disposal of garbage, oil leakage from motor vehicles, tire wear, exhaust emissions, air deposition, the loss of fertilizers in green areas, and varying degrees of leakage from production areas in certain enterprises have led to the accumulation of various pollutants on different urban impervious surfaces. During rainfall, the initial runoff creates a scouring effect, carrying a large amount of accumulated pollutants from impervious surfaces, which are directly discharged through the stormwater pipe network, severely affecting the water quality of urban water bodies, particularly slow-flowing water bodies such as lakes and reservoirs. Urban stormwater runoff pollution has become one of the significant pollution sources in the water environment of Taihu New City in Wuxi. Therefore, to improve the ecological environment in Wuxi and protect and restore the aquatic ecosystem, this case relies on low-impact development and green infrastructure to effectively reduce the pollution load of stormwater runoff in Taihu New City.

This case is based on the project "Research and Demonstration of Regional Optimization of Wastewater Treatment System Operation and Urban Non-point Source Reduction Techniques" initiated in 2011 as part of the National Water Special Project.

7.5.2 Process Route and Key Technology

7.5.2.1 Process Route

Based on the above project, the Shangxian River wetland landscape greening and non-point source pollution control project were constructed, and the process flow is shown in Fig. 7.23. Rainwater carrying surface pollutants enters the high-efficiency interception-type rainwater screen after which it passes through the rainwater collection pipe and enters the surface of the high-efficiency adsorption purification zone. It then infiltrates through the filler layer composed of zeolite and activated carbon, reaching the bottom collection system composed of cobblestones. In the collection system, the treated runoff is discharged and discharged into the lake nearby. The rainwater on the green space flows and infiltrates on the lakeside slope, undergoing underground flow filtration and reaction in a discrete high-efficiency 1adsorption barrier column. During this process, the purified water is discharged into the water body. After rainfall on the parking lot, it first infiltrates through permeable bricks with good permeability and then collects in an underground reservoir after undergoing infiltration and purification through the filler layer. This water can be used for car washing and plant irrigation.

Based on the above project, three key technologies were developed, namely, "Road Rainwater High-Efficiency Adsorption Purification Zone Technology with High-Efficiency Interception-Type Rainwater Screen," "Underground Flow Filtration Reaction Barrier Technology," and "On-Site Purification and Water Storage Reuse Technology for Parking Lots."

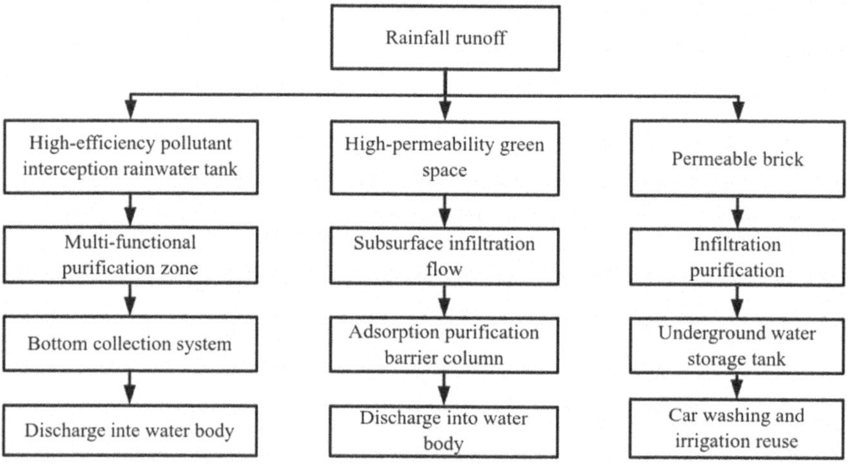

Fig. 7.23 Process flow of Shangxian River Wetland Landscape Greening and Nonpoint Source Pollution Control Project2. Key technologies and applications

i. Related Design Parameters

(i) *The Technology of Road Rainwater high-efficiency Adsorption Purification Zone with high-efficiency interception-type Rainwater Screen.*

The technology primarily consists of a high-efficiency interception-type rainwater screen and an adsorption zone, as depicted in Fig. 7.24 for the structural representation. The rainwater screen is made of stainless steel and can be directly installed into existing cement rainwater screens along the roadside. The upper layer serves as the retention zone, while the lower layer functions as the sedimentation zone. The main body of the high-efficiency adsorption zone utilizes the existing sunken grassed swales on both sides of the road, replacing the original soil in the swales with zeolite and activated carbon that exhibit higher pollutant removal performance. This aims to remove pollutants from the road runoff. The adsorption zone comprises layers of existing permeable rectangular bricks at the top, followed by sand, a mixture of activated carbon and zeolite, and is supported by cobblestones. It effectively purifies the road runoff.

(ii) *the Technology of Underground Flow Filtration Reaction Barrier*

The structural representation of the technology can be seen in Fig. 7.25. In this design, each barrier pillar is 0.6 m long with a diameter of 100 mm. It is surrounded by a mesh structure made of fine-pored iron wire mesh, while the interior is filled with a mixture of activated carbon, zeolite, and other efficient adsorption carriers. The overall barrier adopts a discrete pillar design, with the barrier pillars arranged in two rows near the water's edge. The spacing between pillars from left to right is 1 m, and the spacing from front to back is 0.5 m. This technology utilizes physical adsorption, chemical precipitation, and biological reactions to treat the infiltrated flow from green spaces, preventing direct pollution of the water body by the runoff from the green areas.

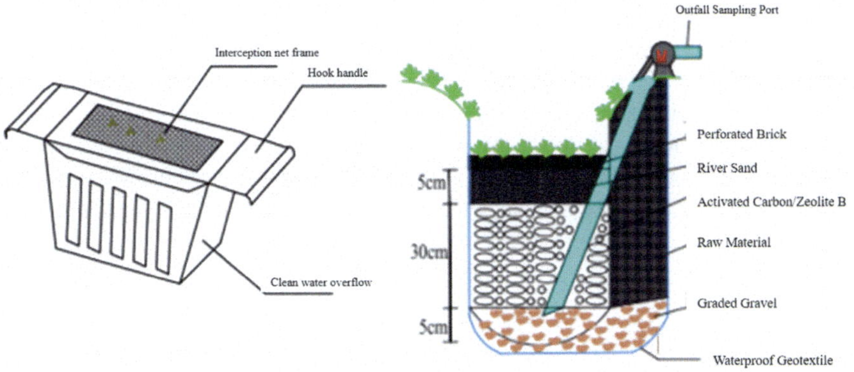

Fig. 7.24 Schematic diagram of road Stormwater high-efficiency adsorption and purification belt technology structure with high-efficiency pollution interception type rainwater grate

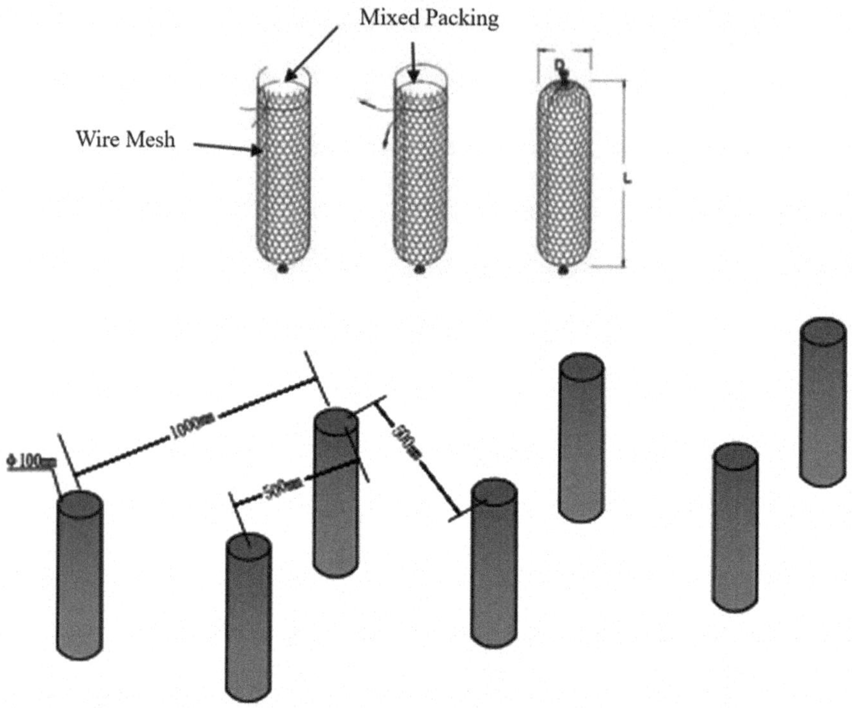

Fig. 7.25 Schematic diagram of subsurface flow reactive barrier technology structure

(iii) *In-Situ Purification and Storage Technology for Parking Lot Stormwater Reuse*

The plan view and sectional view of this technology can be seen in Figs. 7.26 and 7.27, respectively. Each parking space in the parking lot has a length and width of 6.0 m and 3.0 m, respectively. Purified rainwater from the ecological parking lot is directed through permeable pipes located between two parking spaces and is then transported into a combination of underground plastic storage tanks buried on the side of the green area. The modular volume of the storage tanks can be flexibly selected based on the water storage needs. These storage tanks are easy to install, have a high load-bearing capacity, and do not support mosquito breeding or the growth of algae.

The top surface of the parking lot is made of permeable sand bricks, while the lower layers consist of fill filter layer, support layer, and drainage layer. During the construction of the parking spaces, the total excavation depth is 0.6 m.

i. Technology Application Location and Scale

The "Road Efficient Adsorptive Purification Zone with High-Efficiency Pollutant Interception Rainwater Grate" technology has been applied in the northern part of Phase I of Shangxian River, with a total length of 150 m. The roads in the service area are equipped with high-efficiency pollutant interception rainwater grates to control the entry of pollutants.

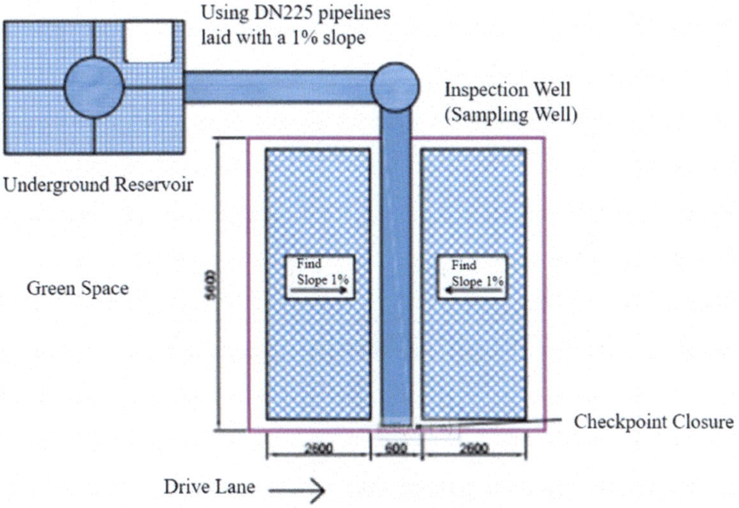

Fig. 7.26 Plan view of parking lot in-situ purification, storage, and reuse technology for parking spaces

The "Underground Infiltration Filtration Reaction Barrier Technology" has been applied in Phase II of Shangxian River, Wuxi Taihu New Town, with a total length of 200 m.

The "In-Situ Purification and Water Recycling Technology for Parking Lots" has been applied in the southern part of Phase I of Shangxian River, with a construction area of 40 m^2 and a storage module volume of 4.0 m^3 (Fig. 7.28).

7.5.3 Application Effectiveness

7.5.3.1 Technical Effectiveness

Based on the wetland landscape greening and non-point source pollution control project in Shangxian River, the central water system of Wuxi Taihu New Town, with a service catchment area of 1.0 km2, the reduction of SS and TP in the demonstration area exceeds 70%, while the reduction of COD exceeds 50%.

7.5.3.2 Economic Performance

The construction investment cost of this project is 1,958 yuan/m2, with an operational and maintenance cost of 2.2 yuan/d.

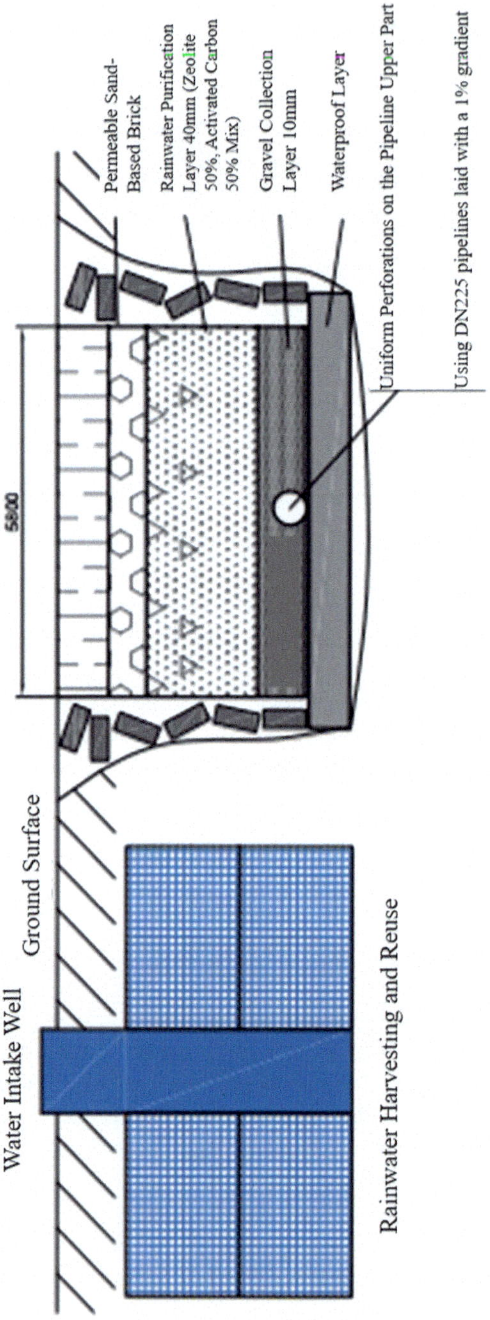

Fig. 7.27 Cross-sectional view of parking lot in-situ purification, storage, and reuse technology for parking spaces

Fig. 7.28 Construction drawing of Shangxian River Wetland landscape greening and non-point source pollution control project

7.5.3.3 Environmental Effectiveness

After the implementation of the project, the original stagnant and malodorous roadside ditches have been transformed into highly permeable absorption zones. They effectively control runoff pollution from the road, while also beautifying the environment and preventing mosquito breeding.

7.5.4 Technology Highlights and Technology Increments

The road efficient adsorption purification belt technology with efficient sewage interception rainwater grate 'adopted in Wuxi City has realized the flow reduction and water quality purification of road runoff through process blocking. After the first interception of rainwater by the rainwater grate, the runoff rainwater is intercepted twice by the purification belt. Compared with the traditional rainwater grate and purification belt, the removal effect of pollutants in runoff rainwater is better.

The underground subsurface flow infiltration reaction barrier technology 'treats the green space seepage through physical adsorption, chemical precipitation and biological reaction, and blocks the green space runoff directly into the water body. This technology not only considers the removal of surface runoff pollutants, but also has a good removal of runoff pollutants that penetrate below the surface.

The in-situ purification and reuse technology of water storage in parking lot has realized the in-situ treatment and reuse of runoff. The runoff rainwater is purified by the permeability of the permeable brick through the purification layer, and the purified rainwater is accumulated in the water storage tank. The rainwater in the water storage tank can be used to wash the car or irrigate the green space. Compared with the traditional technology, it can absorb the runoff rainwater accumulated on the parking space and eliminate the pollutants in the rainwater runoff, greatly improve the parking space itself and the surrounding environment, realize the comprehensive utilization of rainwater resources and save water.

The engineering application of the above key technologies has solved the problem of runoff rainwater pollution in Wuxi, and greatly reduced the pollution degree of Shangxian River in Wuxi Taihu New Town.

7.6 Demonstration of Runoff Pollution Control in Low Impact Development of River Network Cities

7.6.1 Case Background

Located in the northeast of Zhejiang Province and the hinterland of the Hangjiahu Plain in the Yangtze River Delta, Jiaxing City is one of the important cities in the Yangtze River Delta. It is located at the intersection of rivers, seas, lakes and rivers, and strangles the throat of the Taihu Lake Corridor. At the same time, as a typical Jiangnan water town, the terrain is flat and the water network is dense. It is a typical plain river network area with high groundwater level, poor soil permeability, large and uneven rainfall.

In the process of urbanization, a large number of urban areas have been replaced by hardened surfaces, and the population has continued to gather in urban areas. The accumulation of population and the continuous hardening of the underlying surface have led to the production of a large number of rainwater runoff pollutants, which has become an important source of urban water environmental pollutants. In addition, the poor water quality of the upper reaches of the urban inland river and the small flow velocity in the plain river network area lead to the poor self-purification ability of the urban inland river, the continuous improvement of water quality, and the urgent improvement of water environmental pollution.

This case is based on the project of 'Key Technology Research and Engineering Demonstration of Rainwater Runoff Pollution Control and Ecological Utilization in River Network Cities' launched by the National Water Special Project in 2011.

7.6.2 Process Route and Key Technology

7.6.2.1 Process Route

Based on the above topics, four projects were constructed, including urban roads, residential quarters, combined overflow storage and greenway rainwater runoff pollution control and ecological utilization.

i. Urban Road Rainwater Runoff Pollution Control and Ecological Utilization Project

The process route of urban road rainwater runoff pollution control and ecological utilization project is shown in Fig. 7.29. The runoff rainwater on the road is collected by the rainwater inlet and transported by the rainwater pipeline to the floating medium filtration device installed at the end of the rainwater pipeline. After the filtration and precipitation of the device, it is discharged into the river channel.

ii. Rainwater Runoff Pollution Control And Ecological Utilization Project in Resdential District

The process route of rainwater runoff pollution control and ecological utilization project in residential area is shown in Fig. 7.30. Runoff generated by green roofs and ordinary roofs can be disconnected through rainwater risers to the soil-improved rainwater garden for infiltration and purification. Runoff generated by permeable paving squares and roads flows into the grassed swale nearby, and then passes through the grassed swale to transport it to the soil-improved sunken green space for infiltration and purification. After purification, rainwater is directly discharged into the river channel through open channels or underground channels.

iii. Ecological Purification Project of Combined Overflow Regulation And Storage

The process route of the combined overflow storage ecological purification project is shown in Fig. 7.31. During the dry season or light rain, the sewage is directly intercepted by the pumping station to the Chengdong reclaimed water plant. In the rainy season, the storage tank undertakes the sewage beyond the treatment capacity of the sewage plant. After the pretreatment of the terminal floating matter filtration device, it is treated in situ by the ecological purification facility of the subsurface flow constructed wetland, so as to realize the in-situ reduction and purification of the combined overflow pollution and improve the water quality of the receiving water body.

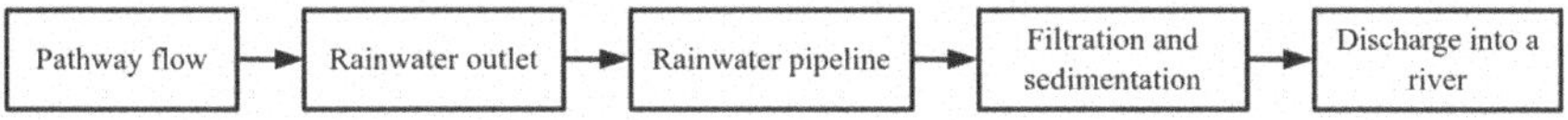

Fig. 7.29 Process route of urban road Stormwater runoff pollution control and ecological utilization project

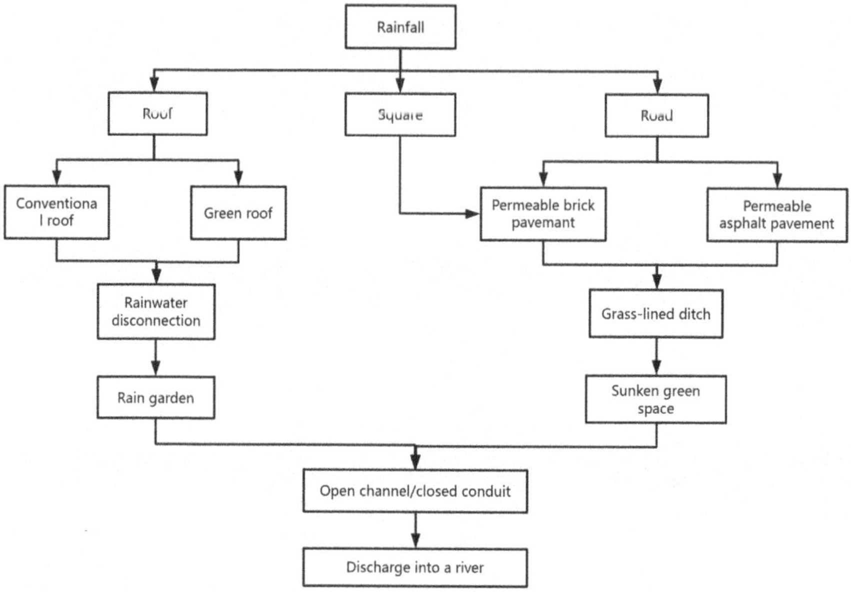

Fig. 7.30 Process route of residential area Stormwater runoff pollution control and ecological utilization project

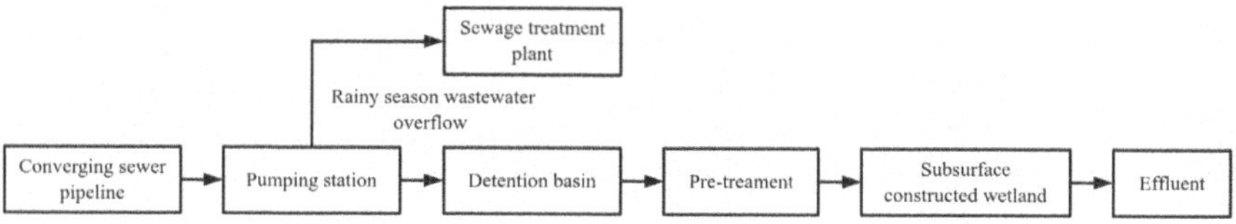

Fig. 7.31 Combined sewer overflow (CSO) storage and ecological purification engineering process route

iv. Urban Greenway Rainwater Runoff Pollution Control And Ecological Utilization Project

The process flow of urban greenway rainwater runoff pollution control and ecological utilization project is shown in Fig. 7.32. The rainwater runoff is discharged into the rainwater pond or rainwater wetland through the discharge port, and the suspended matter in the rainwater runoff is removed by precipitation. Under the action of water level difference, rainwater runoff gradually infiltrates into the river channel through infiltration weir or infiltration dam, and soluble pollutants can be intercepted or filtered during infiltration.

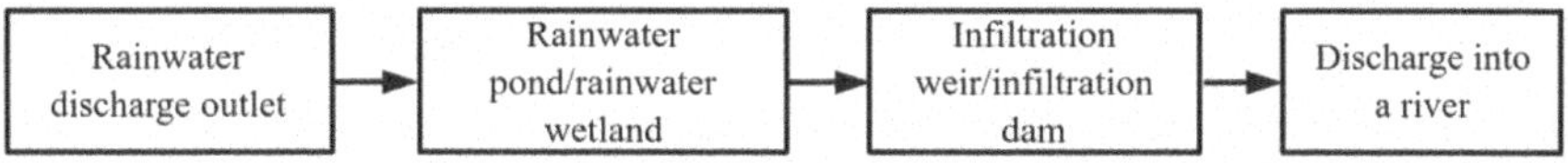

Fig. 7.32 Urban Greenway Stormwater runoff pollution control and ecological utilization engineering process route

7.6.2.2 Key Technologies and Applications

Based on the above demonstration project, four key technologies were developed, including 'floating medium filtration technology at the end of split rainwater system', 'soil improvement and infiltration reduction technology', 'integrated treatment technology of combined overflow sewage' and 'ecological synergistic pollution control technology combined with infiltration and storage at the end of split rainwater drainage system'.

i. Filtration Technology of Floating Medium at The End of Split Rainwater System

(i) *Related Design Parameters*

The filter device formed by this technology is designed as a weir type, tilted and placed, with an inclination angle of about 60° to the horizontal direction. The device is shown in Fig. 7.33. The inclined setting, on the one hand, increases the filtration area, and secondly facilitates the precipitation of the filtered SS. After the water flow passes through the weir filter, SS is intercepted, and the purified water is directly discharged from the outlet. When the SS is saturated, the loosening rod is relaxed, the floating particles expand, and the SS enters the sludge area, thereby achieving filtration and purification. The thickness of the filter medium is 20–30 cm. The head loss is 18–20 cm at the design flow rate. The particle size of lightweight filter material is 5–10 mm. The removal rate of SS was 60%–80%.

(ii) *Technology Application Location*

This technology has been applied in Wenchang Road, Jiaxing City. The rainwater system of Wenchang Road directly discharges rainwater runoff pollution and a small amount of mixed sewage, which poses a great threat to the pollution of the surrounding receiving water body. By using this technology, SS and other pollutants are intercepted and filtered by floating particles, and then the floating particles expand under the action of buoyancy, and the SS precipitation intercepted and filtered between particles is removed to realize non-dynamic backwashing (Fig. 7.34).

ii. Soil Improvement, Infiltration and Emission Reduction Technology

(i) *Related Design Parameters*

Aiming at the characteristics of poor permeability and high shallow groundwater level of Jiaxing clayey soil, the key technologies with soil improvement and water quality control as the main objectives are proposed. Plant straw, sand, slag, zeolite

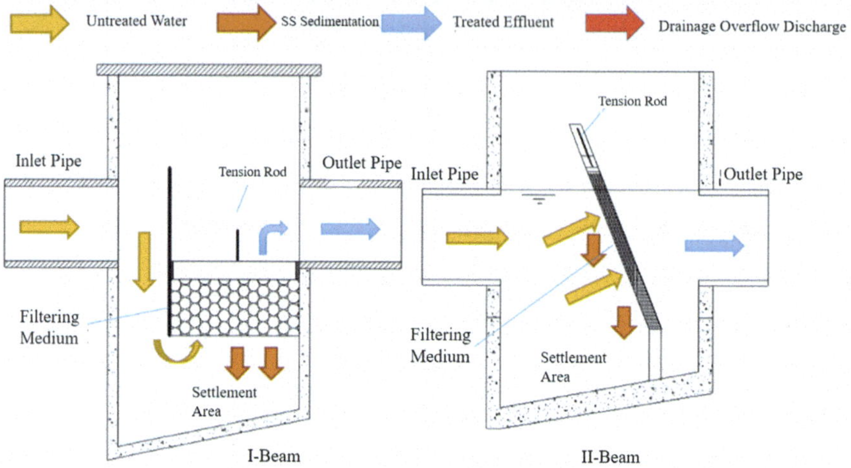

Fig. 7.33 Schematic diagram of diversion rainwater system end floating medium filtration technology device

and ceramsite were selected as soil improvement media to study the improvement of soil permeability and water purification effect of the above soil improvement media on clayey soil, as shown in Table 7.10. The technology is applied to Orchard Road, Shihe Town, Jiaxing, using clay + medium sand + bark, and clay + medium sand combination, as shown in Fig. 7.35.

(ii) *Technology Application Location and Scale*

This technology has been applied in Jiaxing Jiangshuigang Ecological Greenway and Shihe Town. Among them, the scale of Jiangshuigang Greenway project is about 0.5 ha, and the scale of Jiaxing Shihe Town project is 90642 m^2 (Fig. 7.36 and 7.37).

i. Comprehensive Treatment Technology of Combined Overflow Sewage Terminal

(i) *Related Design Parameters*

The total effective storage volume of the reservoir is 20000 m^3, and the inlet end of the reservoir is regulated by an electric gate. There are 6 corridors in the storage tank, and the storage tank is washed by vacuum washing equipment. The constructed wetland ecological purification facilities are built on the upper part of the storage tank, with an area of about 2500 m^2. The hydraulic retention time is 2 days, the hydraulic load is 0.125 m^3/(m^2·d), and the design influent flow is 300 m^3/d. The engineering profile is shown in Fig. 7.38.

(ii) *Technology Application Location and Scale*

The technology has been demonstrated on the north side of the plant boundary of Chengdong Reclaimed Water Plant in Jiaxing City. The service area is 2.4 km^2,

Fig. 7.34 Real scene of diversion rainwater system end floating medium filtration technology device

Table 7.10 Removal efficiency of different composition combinations of pollutants and permeability coefficient

Cla + Expanded clay + Zeolite		Project	COD	TN	TP	SS	NH_3-N
Soil thickness: 30 cm	2:1:1	Average removal rate,%	91	65	83	98	53
		Permeability coefficient	2.58 m/d				
Clay + Zeolite + Medium sand		Project	COD	TN	TP	SS	NH3-N
Soil thickness: 40 cm	2:1:1	Average removal rate,%	96	73	82	99	60
		Permeability coefficient	2.07 m/d				

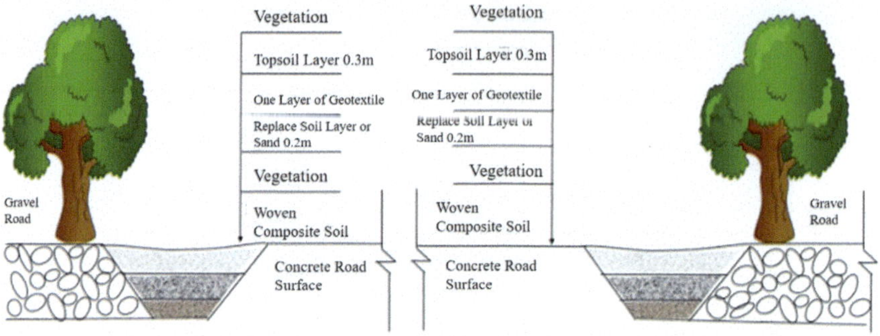

Fig. 7.35 Application of soil improvement and permeability enhancement and pollution reduction technology on Guoyuan Road in Shihe Town

Fig. 7.36 Engineering application in Shihe Town

and the total effective regulation volume is 20000 m^3. The technology has been popularized and applied in Chengnan Park and Wenhua Park, including 7000 m^3 storage tank in Chengnan Park and 9000 m^3 storage tank in Wenhua Park (Fig. 7.39).

iv. Ecological Synergistic Pollution Control Technology Combined With Infiltration And Storage at The End of Split Rainwater System

(i) *Related Design Parameters*

The ecological synergistic pollution control technology combined with infiltration and storage at the end of the split rainwater drainage system couples the detention and sedimentation functions of ecological facilities such as rainwater ponds and

Fig. 7.37 Jiangshuigang road engineering application

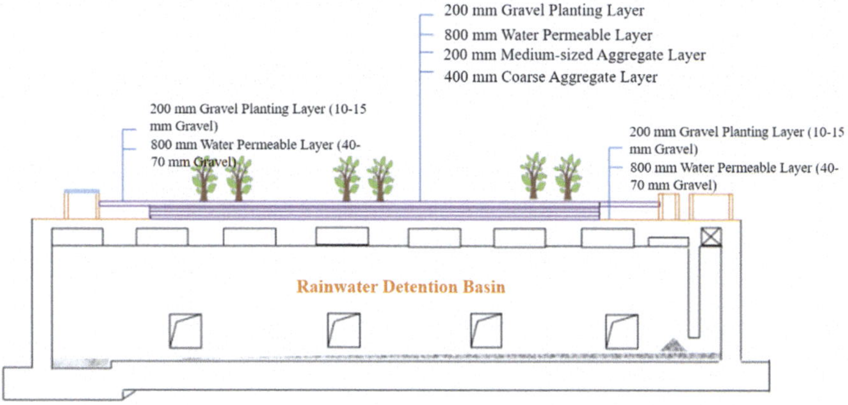

Fig. 7.38 Cross-sectional diagram of comprehensive treatment project at the end of confluent overflow sewage

rainwater wetlands with the interception and purification functions of infiltration weirs, dams and banks to achieve synergistic purification, stabilize the purification of soluble and insoluble pollutants, and avoid the blockage of ecological infiltration facilities and the instability of the effluent quality of rainwater ponds. During the design return period, the sedimentation surface load of the detention tank is more than 15 $m^3/(m^2 \cdot d)$. The width of filler is 0.7–1.0 m. The particle size of the filler is 8–15 mm; the hydraulic retention time should be 1–3 d (Fig. 7.40).

Fig. 7.39 Real scene of comprehensive treatment project at the end of confluent overflow sewage

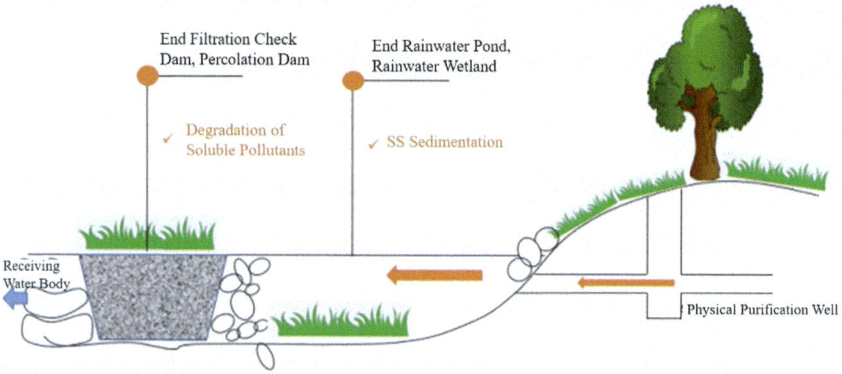

Fig. 7.40 Schematic diagram of ecological synergistic pollution control technology with combined infiltration and storage at the end of diversion rainwater system

(ii) *Technology Application Location and Scale*

This technology has been applied in the ecological greenway of Jiaxing Botanical Garden, and the area of ecological measures at the end of the construction is 4400 m^2 (Fig. 7.41).

7.6.3 Application Effectiveness

7.6.3.1 Technical Effectiveness

The average removal rates of SS and COD in Wenchang Road runoff pollution control project was 77.0% and 70.5%, respectively. The average reduction rate of the total amount of rainwater runoff discharged from the runoff pollution control in Shihe

Fig. 7.41 Real scene of botanical garden ecological greenway engineering

Town was 32.0%, the control rainfall was 28.6 mm, and the average reduction rate of SS was 89.6%. The reduction rates of SS, COD, NH_3-N, TN and TP were 33%–94%, 34%–87%, 70%–95%, 64%–87% and 20%–70%, respectively. The removal rates of SS, COD, TN and TP were 35.0%–98.2%, 68.5%–98.4%, 35.2%–92.6% and 60.

7.6.3.2 Economic Performance

The investment and construction cost of "pollution control technology of floating medium filtration at the end of split rainwater system" is 3000 yuan/seat (civil construction without wells). The investment and construction cost of "soil improvement, infiltration and emission reduction technology" is 300 yuan/m^2, and the annual operation and maintenance cost is 8.0 yuan/m^2. The investment and construction cost of "combined overflow sewage treatment technology" is 800 yuan/m^2 for ecological wetland, 2500 yuan/m^3 for storage tank, and 50 yuan/d for operation and maintenance. The investment and construction cost of the 'ecological synergistic pollution control technology combined with infiltration and storage at the end of the split rainwater drainage system' is 300 yuan/m^2, and the operation and maintenance cost is 8 yuan/(m^2·a).

7.6.3.3 Environmental Effectiveness

The developed technology provides strong support for the pilot construction of sponge city in Jiaxing. In the pilot area of 18.44 km^2, the COD load of rainwater runoff can be reduced by 1621.5 t/a, and the COD of combined sewage overflow can be reduced by 118.45 t/a, which achieves the pollutant reduction index of regional water environment quality improvement.

7.6.4 Technology Highlights and Technology Increment

Aiming at the problems of large rainfall, poor soil permeability and insufficient self-purification capacity of water bodies in plain river network cities, Jiaxing City has constructed a complete set of technologies for rainwater runoff pollution control in river network cities with 'source reduction, end interception and control, and river network self-purification'. From 2015 to 2018, this technology was adopted in the sponge pilot area of 18.44 km^2 in Jiaxing, and four types of demonstration projects were built, including urban roads, residential areas, ecological purification of combined regulation and storage in the middle of the city, and urban greenways. The goal of "light rain without water accumulation, heavy rain without waterlogging, and water body without black and odor" was achieved.

7.7 Comprehensive Demonstration Project of Runoff Pollution Control in Low Impact Development Area

7.7.1 Case Background

As the source of rainwater runoff, urban roads and open space, green buildings and communities are low-impact development and construction. It is of great value to realize the in-situ infiltration, stagnation, storage, purification, utilization and drainage of rainwater resources, realize the sustainable circulation of urban rainwater, accelerate the transformation of urban rainwater drainage methods, effectively control flood disasters and runoff pollution, alleviate the shortage of urban water resources in China, and improve the ecological carrying capacity of urban water. Urban roads and open space land generally account for 20–30% of the total urban construction land, of which urban road land generally accounts for 15–20%, but the pollution load generated by urban road rainwater runoff accounts for 40–60% of the total urban rainwater runoff pollution load. In addition, open spaces such as urban park green space are the main carriers of urban ecological functions. In addition to providing rest and entertainment functions, they are also the main places for animal and plant habitats and urban rainwater runoff storage and consumption. Therefore, it is of great significance to reduce the total loss of rainwater runoff in urban roads and open spaces, green buildings and residential areas and its pollution to the water environment.

Shenzhen Guangming New District is an important economic development zone in Shenzhen. In January 2010, the Shenzhen Municipal Government and the Ministry of Housing and Urban–Rural Development signed the "Cooperation Framework Agreement on Building a National Low-carbon Ecological Demonstration City" and established the urban development strategy of creating a low-carbon ecological demonstration city. Combined with Shenzhen 's natural conditions and urban development characteristics, Shenzhen will be built into a social and economic prosperity and vitality, comfortable and pleasant living and production environment, resource and energy efficiency significantly improved, carbon dioxide emissions remain low, low-impact development civilization concept deeply rooted in the hearts of the people, urban complex ecological system healthy and harmonious, has an important demonstration role in the country and has an advanced level of low-impact development cities in the world. Based on this background, the low impact development demonstration project of Guangming New District is established.

This case is based on the national water project launched in 2011 'low impact development of rainwater system comprehensive demonstration and evaluation (2010ZX07320-003)', 'urban road and open space low impact development of rainwater system research and demonstration (2010ZX07320-002)' and 'green building and community low impact development of rainwater system research and demonstration (2010ZX07320-001)' three topics.

7.7.2 Process Route and Key Technology

7.7.2.1 Process Route

Based on the above topics, four types of demonstration projects are constructed, which are residential areas, public buildings, park green space and road rainfall runoff pollution control demonstration projects.

i. Wanzhangpo Area Resettlement Phase I Runoff Pollution Control Project

The process route of the first-stage runoff pollution control project in Wanzhangpo District is shown in Fig. 7.42. The runoff generated by the permeable pavement can enter the ecological tree pool and be discharged into the rainwater pipeline after further retention and purification. The runoff generated by the fire lane can enter the sunken green space through the grass planting ditch, and is discharged into the rainwater pipeline after retention and purification. The runoff generated by the green roof and the ordinary roof can be disconnected to the rainwater garden through the rainwater standpipe, and discharged into the rainwater pipe after retention and purification.

ii. Runoff Pollution Control Project of Mass Sports And Cultural Center

The process route of the runoff pollution control project of the mass sports and cultural center is shown in Fig. 7.43. Rainwater from parking lots, squares and roofs is collected into grassed swales and sunken green spaces for retention and purification, and excessive runoff rainwater is collected by rainwater pipe networks. The initial rainwater with more serious pollution is abandoned into the sewage pipe network, and the middle and late runoff rainwater enters the rainwater collection module for collection and storage. After treatment, it is used for greening and irrigation. The middle and late runoff rainwater overflow exceeds the storage capacity of the

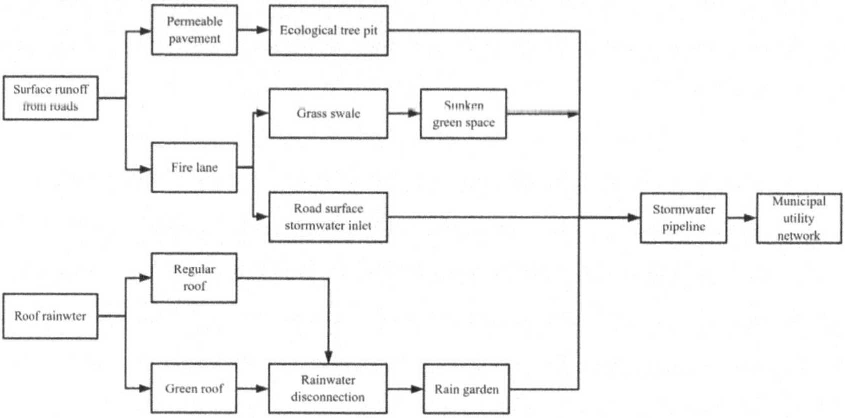

Fig. 7.42 Process route for Phase I of runoff pollution control project in the Wanzhangpo area for demolition and resettlement

rainwater collection module and is discharged into the municipal rainwater pipe network.

iii. Runoff Pollution Control Project of Low Impact Development in Xincheng Park

The process route of runoff pollution control project for low impact development of Xincheng Park is shown in Fig. 7.44. After the collection of green rainwater and road rainwater, some of them can enter the dry stream through the grass planting ditch for detention and purification, and the runoff rainwater exceeding the treatment capacity of the dry stream can be discharged into the municipal rainwater pipe network through the rainwater overflow port set in the dry stream. When there is too much rainwater entering the grassed swale, the runoff rainwater can be stored in the modular reservoir through the sewage interception rainwater inlet in the grassed swale, and used for greening and irrigation after the rainy season.

iv Runoff Pollution Control Project of Low Impact Development of Park Road

The process route of runoff pollution control project in low impact development of park road is shown in Fig. 7.45. A part of the runoff rainwater on sidewalks, motorways and bicycle lanes infiltrates into the ground through permeable bricks or permeable asphalt pavements. When the infiltration capacity is saturated or the rainfall runoff is too large, excessive rainfall runoff flows through the road edge stone openings into the bioretention zone or sunken green space for detention and purification. Runoff rainwater exceeding the treatment capacity of the bioretention zone can be discharged into the municipal rainwater pipe network through the rainwater overflow outlet set up in the bioretention zone.

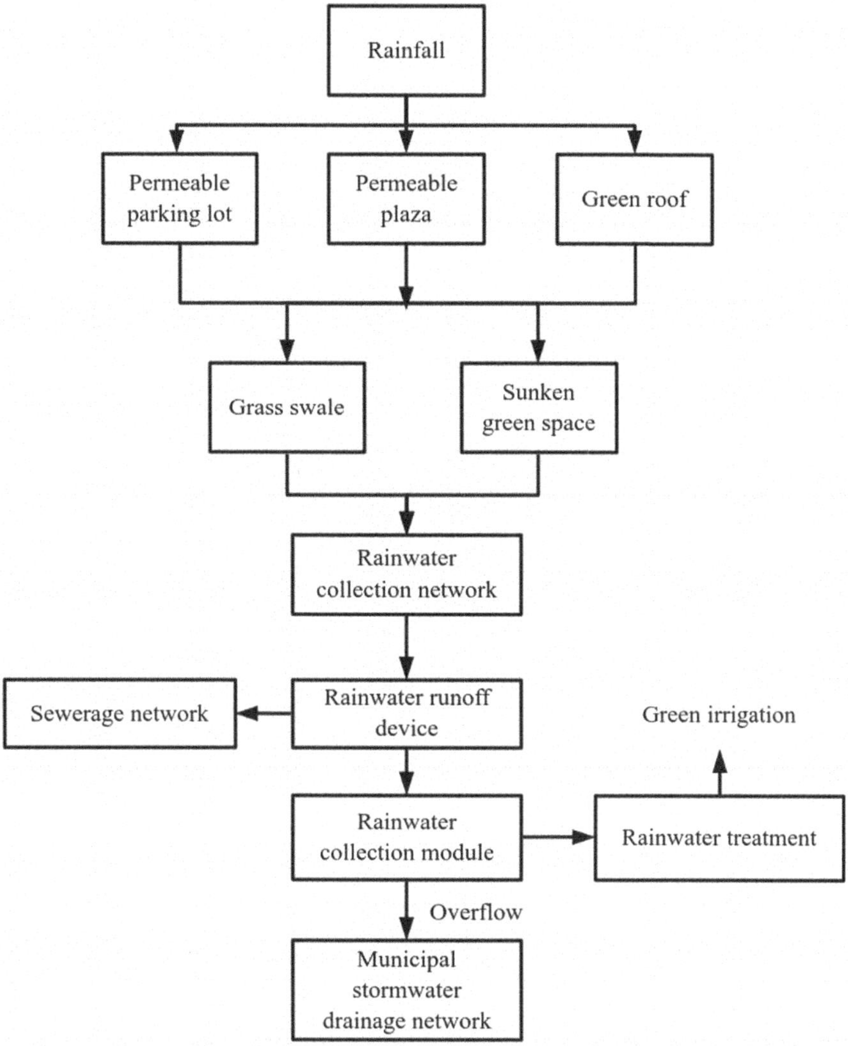

Fig. 7.43 Stormwater pollution control engineering process route for the mass sports and culture center

7.7.2.2 Key Technologies and Applications

Based on the above projects, a total of five technologies were developed, namely 'sunken green space technology', 'rain garden technology', 'green roof technology', 'grassed swale technology' and 'permeable pavement technology'.

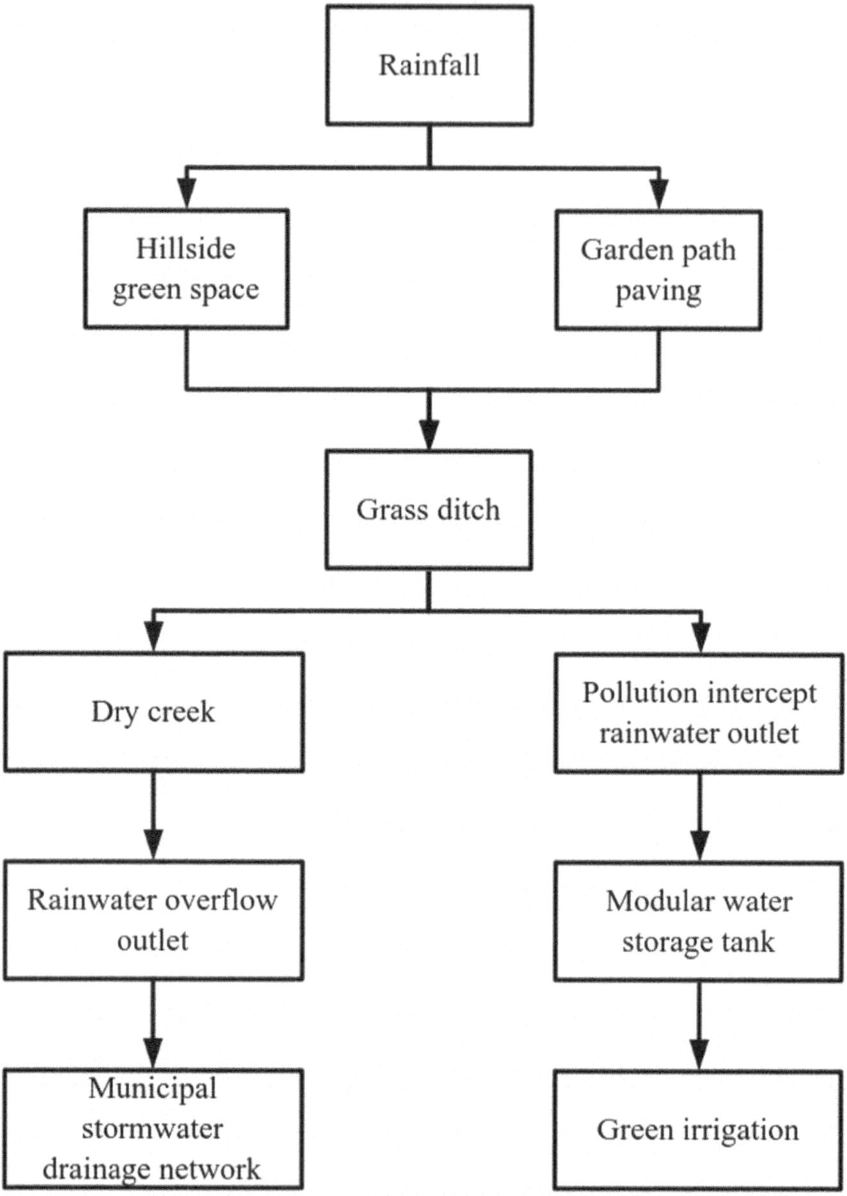

Fig. 7.44 Low-impact development Stormwater pollution control engineering process route for Xincheng Park

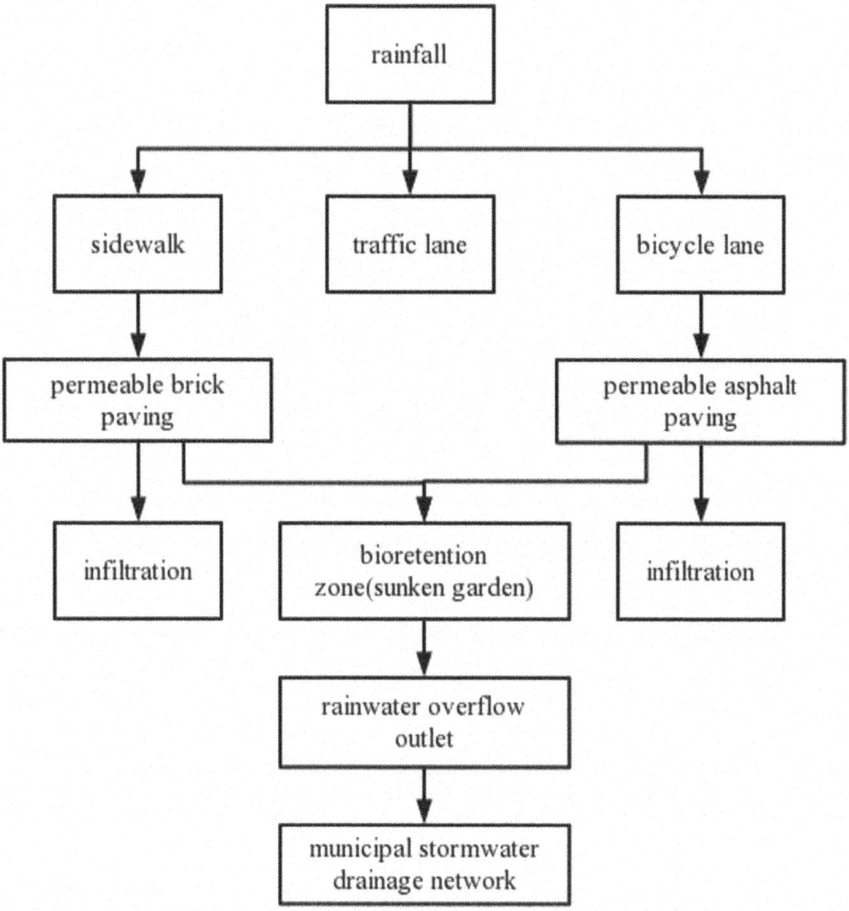

Fig. 7.45 Low-impact development Stormwater pollution control engineering process route for park road

i. Concav Green Space Technology

(i) *Related Design Parameters*

In the construction of green space, the technology requires the elevation of green space to be lower than the surrounding ground, so as to facilitate the inflow of surrounding rainwater runoff. The technology fully accumulates rainwater in the concave space in the application area, which significantly increases the infiltration time of rainwater, and can purify the accumulated rainwater through the green space, which has a great effect on the reduction of pollutants. In addition, the seepage water can also play a role in supplementing groundwater. The specific parameters are as follows: Taiwan grass is selected as the vegetation layer; the sand ratio of the

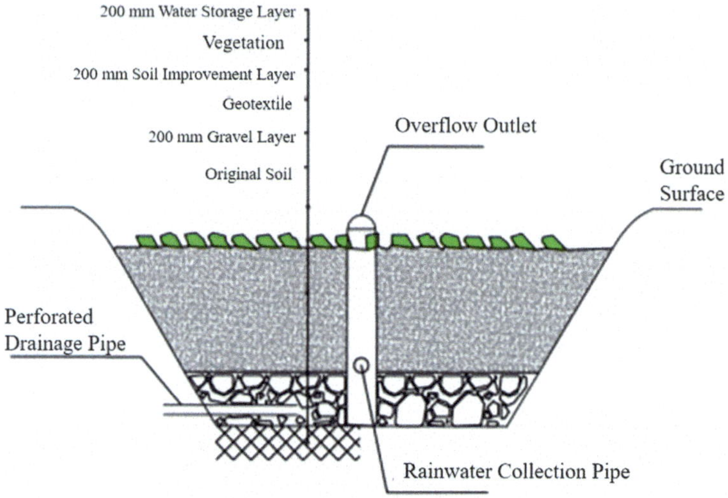

Fig. 7.46 Sectional view of concave green space

improved soil layer is 1/2; the pebble layer is selected with Φ200–400 mm pebble (Fig. 7.46).

(ii) *Technology Application Location and Scale*

This technology has been applied in the concave green space on the west side of Wanzhangpo Children's Paradise, with a scale of about 786 m^2 (Fig. 7.47).

i. Rain Garden Technology

(i) *Related Design Parameters*

This technology is mainly used in low-lying areas. It can reduce the peak flow of surface runoff by storing and purifying runoff rainwater through plants, soil and microbial systems. It can also reduce runoff pollution through adsorption, degradation, ion exchange and volatilization. The specific parameters are as follows: Autumn maple is selected as the flower bed vegetation layer, and Taiwan grass is selected as the shallow ditch; the sand ratio of the improved soil layer is 1/2, and the lower part is laid with geotextile. The pebble layer selects Φ40-60 mm pebbles, which can not only dissipate energy, but also facilitate the uniform inflow of rainwater into the green space. The PVC pipe with a diameter of 20 mm was selected for the perforated drainage pipe. The surface layer of the flower bed is 3% slope, and the inverted parabolic section is taken from the ephemeral ditch. The width is about 400 mm, the depth is about 50 mm, and the longitudinal slope is about 3 ‰ (Figs. 7.48 and 7.49).

(ii) *Technology Application Location and Scale*

This technology has been applied in Block B of Wanzhangpo Resettlement Housing Project, with a length of about 48 m and a width of about 8.2 m (Fig. 7.50).

Fig. 7.47 Real scene of concave green space

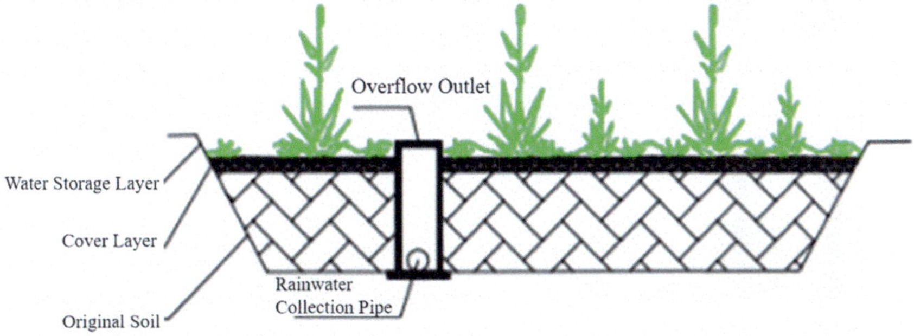

Fig. 7.48 Rain garden device structure diagram

i. Green Roof Technology

(i) *Related Design Parameters*

Vegetation selection of native plants in Shenzhen Taiwan grass, strong growth, cold and drought tolerance; hDPE concave and convex drainage plate is selected for the drainage layer; the protective layer adopts modified asphalt root-resistant puncture waterproofing membrane, which has both root-blocking and waterproofing functions. The Φ30–40 mm pebbles are selected in the drainage ditch (Fig. 7.51).

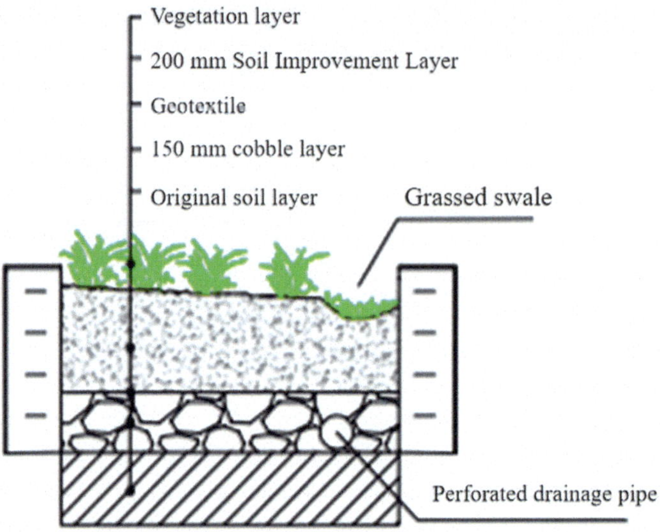

Fig. 7.49 Rain garden cross-section diagram

Fig. 7.50 Rain garden photo

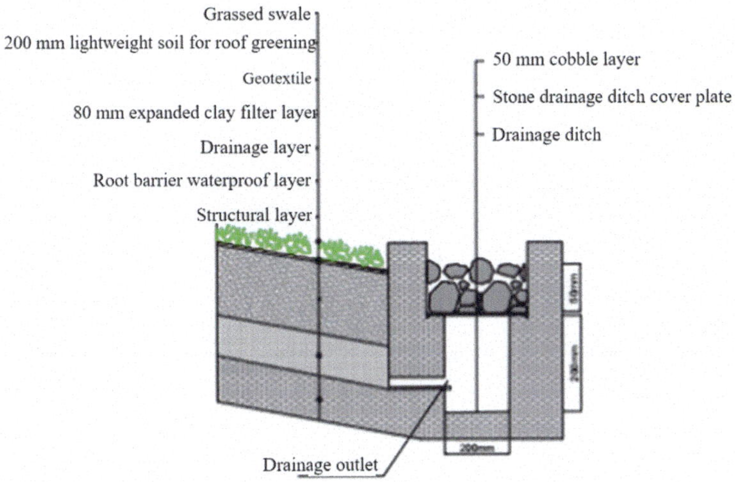

Fig. 7.51 Green roof cross-section diagram

(ii) *Technology Application Location and Scale*

This technology has been applied to the roof of the mass sports center, with a total area of about 12,500 m^2 (Fig. 7.52).

Fig. 7.52 Green roof photo

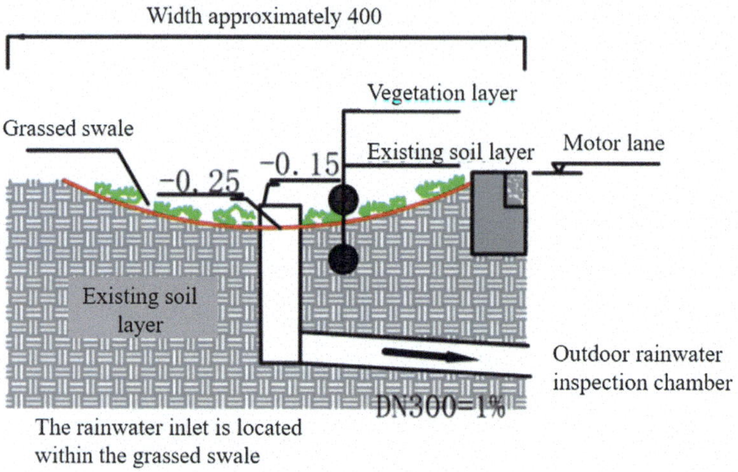

Fig. 7.53 Grassed swale cross-section diagram

iv. Grassed Ditch Technology

(i) *Related Design Parameters*

According to the actual terrain slope, the technology is mainly applied to the side of the green space, collecting part of the surface runoff into the transfer road and the runoff of the higher green space. The structure from top to bottom is water storage layer, vegetation layer, soil layer and gravel layer. The relevant parameters are as follows: the vegetation selects the local plant Cyperus alternifolius in Shenzhen, which is conducive to growth; the inverted parabolic section is selected, and the width is about 400 mm. Considering the flat terrain of the demonstration area, the longitudinal slope of the grassed swales is 3 ‰ (Fig. 7.53).

(ii) *Technology Application Location and Scale.*

This technology has been applied in Xincheng Park, with a total area of about 2263 m^2 (Fig. 7.54).

v. Permeable Pavement Technology

(i) *Related Design Parameters*

This technology is mainly used in paving small pavement bricks on street pavements, reserving gaps between bricks, so that rainwater can penetrate into the ground from the gaps, which can delay runoff and reduce water volume. The effect is remarkable. The structure of permeable brick pavement is shown in Fig. 7.55, and the structural layer is composed of permeable brick surface layer, leveling layer, base layer and soil base layer. The relevant parameters are as follows: the width of the indirect seam of the brick seam of the permeable brick surface layer is 3 mm, the brick

Fig. 7.54 Grassed swale photo

seam is filled with sand and rolled; the surface layer of grass planting brick contains 40 mm planting soil, and Taiwan grass is selected as the plant. The medium sand is used in the leveling layer, and the medium sand requires Φ0.3–5 mm graded sand; the permeable brick pavement base adopts continuous graded gravel, and the grass-planting brick pavement adopts Φ15–25 mm graded gravel; permeable brick pavement has no bottom base, grass planting brick bottom base adopts lime-fly ash gravel; the soil foundation is compacted, and the compactness is greater than or equal to 93%.

(ii) *Technology Application Location and Scale*

This technology has been applied in Guangming New District Park Road, with a total area of about 2500 m^2 (Fig. 7.56).

7.7.3 Application Effectiveness

7.7.3.1 Technical Effectiveness

The first-stage runoff pollution control project of demolition and resettlement in Wanzhangpo District has reached more than 70% of the total annual runoff control

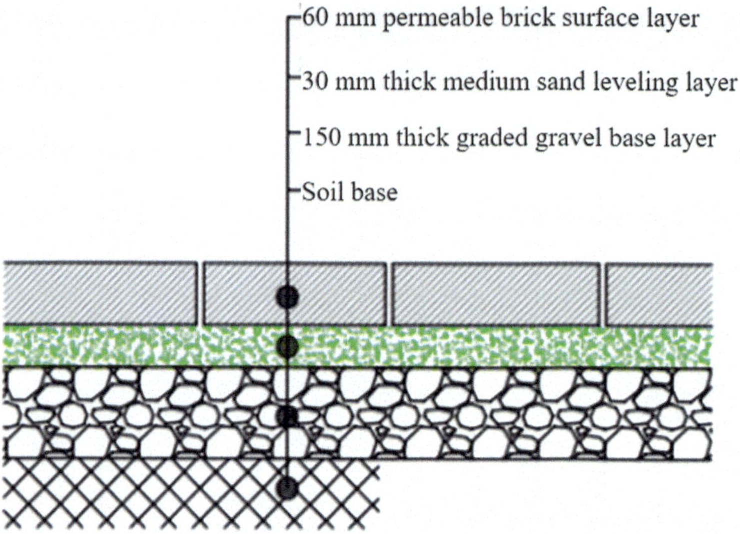

Fig. 7.55 Permeable pavement structure diagram

rate, and the runoff SS reduction rate is higher than 49%. The runoff pollution control project of the mass sports and cultural center has achieved an annual runoff comprehensive control rate of more than 70%, SS reduction of more than 50%, and a comprehensive effect of runoff peak reduction of 37% to 47%. The annual total runoff control rate of the low impact development runoff pollution control project in Xincheng Park can reach more than 85%. The total annual runoff control rate of the low impact development runoff pollution control project of the park road is 66.3%–69.2%, and the SS reduction is 68.0%–90.0%.

7.7.3.2 Economic Performance

Relying on the construction of urban rainfall runoff pollution control project, three products of new plastic module combined pool, seepage drainage plate and initial rainwater abandonment device have been developed and applied in the market. Among them, the new plastic module combination pool 3 years to achieve sales of 61,300 m^2, total sales of 1.84 million yuan; the seepage drainage board has been applied in more than 65,000 m^3 in the actual project, with a total sales of 97.5 million yuan and an average annual collection and utilization of rainwater of more than 1.4 million m^3. The initial rainwater discarding device has achieved sales of 150 sets, with a total sales value of 5.25 million yuan.

Fig. 7.56 Permeable pavement sidewalk photo

7.7.3.3 Environmental Effectiveness

In the construction of urban rainfall runoff pollution control project in Guangming New District of Shenzhen, a variety of low impact development technologies including green roof, permeable pavement, grass planting ditch, sunken green space and rainwater garden are adopted. The application of these technologies has effectively reduced the rainfall runoff pollution in Shenzhen, improved the utilization rate of rainwater, improved the urban environment and greatly reduced the risk of urban waterlogging. During the rainfall period after typhoon 'Miaobai' landed, there was no obvious water accumulation in the demonstration project.

7.7.4 Technology Highlights and Technology Increment

In the construction of urban rainfall runoff pollution control project in Guangming New District of Shenzhen, single technologies such as green roof, permeable pavement, grass planting ditch, concave green space and rainwater garden were adopted, forming a relatively concentrated and contiguous low-impact development comprehensive demonstration area, covering typical construction projects such as residential areas, public buildings, park green space and roads, realizing the large-scale and integrated application of regional low-impact development.

Chapter 8
Performance Evaluation of Key Technology Application for Urban Rainfall Runoff Pollution Control

The application of urban rainfall runoff pollution control technology in China is still at an early stage, and there is a lack of informed decision-making in selecting key technologies. To enhance the integration of urban rainfall runoff pollution control in infrastructure planning, design, construction, and implementation, a comprehensive assessment of current key technologies is urgently required. This assessment will analyze the challenges associated with these technologies and provide a scientific foundation for their improvement, optimization, and effective utilization. This chapter presents an assessment framework for key technologies in urban rainfall runoff pollution control and their performance in application. Its objective is to evaluate the challenges and identify directions for improving these technologies in practical implementation.

8.1 Evaluation Indicator System

When assessing a specific class of objects, it is essential to clarify the aspects that will be the focus of the assessment. This involves developing assessment indicators based on the characteristics of the objects being evaluated. Additionally, the significance of each indicator's influence on the objects should be analyzed to determine their respective weights. Finally, an appropriate assessment method is selected to conduct the evaluation [1]. This section presents an evaluation system for key technologies in urban rainfall runoff pollution control.

D. Sun and F. Qi, *Urban Rainfall Runoff Pollution Control Technology and Application*,
https://doi.org/10.1007/978-981-96-4174-1_8

8.1.1 Evaluation Index

A. Principles for Selecting Evaluation Index

The selection of assessment indicators is the first step in evaluating the performance of technology applications. The rationality of these indicators significantly impacts the accuracy and reliability of the outcomes. To comprehensively, objectively, scientifically, systematically, and accurately assess the performance of rainfall runoff pollution control technology, it is essential to establish appropriate assessment indicators based on specific principles, as outlined below.

i. Scientific

The accuracy and reliability of assessment results are enhanced by the scientific and rational selection of indicators. Therefore, the chosen assessment indicators should have clear concepts and connotations, a solid theoretical foundation, and the ability to reflect the essential characteristics of the subject being evaluated.

ii. Systematic

Assessment indicators should ensure reasonability, coordination, and unity, forming a cohesive structure that systematically and comprehensively reflects the overall situation of the assessed object. They must avoid causal or overlapping relationships, with well-defined boundaries and distinct differentiation at the same level.

iii. Comprehensive

The assessment indicators should provide a comprehensive and holistic reflection of the overall state of the technology being evaluated. Depending on the subject, these indicators should encompass various aspects of urban rainfall runoff pollution control technologies and projects, including technical, economic, environmental, social, and management dimensions. They should capture the technical performance, management mechanisms, as well as the ecological, environmental, economic, and social benefits derived from their application. While having a greater number of indicators can be beneficial, an excessive quantity may result in an overwhelming workload. Therefore, the indicators should strike a balance being both comprehensive and representative, to minimize the assessment workload and reduce errors.

iv. Objective

The indicators within the assessment system have clear objectives and are aligned with specific goals. The selection of these indicators directs the assessment target to actively progress in accordance with the objectives of the chosen criteria. This approach provides guidance for future development, capitalizes on strengths, and addresses areas of improvement.

v. Operational Plausibility

The content of the indicators should be concise and intuitive, typically aligned with the existing national statistical indicator system. They should correspond to specific objectives, with data that is easily collectible, observable, and measurable. This ensures the authenticity, continuity, feasibility, and operability of the data and statistics.

vi. Combination of Qualitative and Quantitative Analysis

To achieve a comprehensive and integrated assessment of the evaluation object, the selection of assessment indicators should follow the principle of combining qualitative and quantitative aspects. Qualitative indicators should be appropriately quantified and standardized to enable consistent measurement and analysis.

These construction principles apply not only to the application performance assessment indicators of key technologies discussed in this chapter, but also to the engineering performance assessment indicators covered in Chaps. 9 and 10.

B. Construction of Evaluation Index

Technology assessment indicators are essential tools for evaluating technological solutions. When evaluating key urban rainfall runoff pollution control technologies, hese indicators can be complex and difficult to quantify. To ensure scientific rigor, comprehensiveness, and systematization, the Analytical Hierarchy Process (AHP) is employed to develop performance assessment indicators for these critical technologies.

In accordance with the aforementioned principles, the selection of indicators for assessing urban rainfall runoff pollution control technologies is based on a thorough review of the literature and real-world engineering applications. These indicators cover multiple aspects, including their meaning, classification, function, benefits, and subsequent maintenance and management. To establish a comprehensive assessment framework, indicators are selected across four key dimensions: technical performance, economic performance, management performance, and benefit performance. The resulting structure of assessment indicators is presented in Fig. 8.1.

The hierarchical structure of the performance assessment framework consists of three levels. The first level (A) includes four sub-systems: technical performance, economic performance, management performance, and benefit performance. The second level (B) forms the criterion layer, where each sub-system is defined based on its distinct characteristics, resulting in a total of nine assessment subjects. The third level (C) consists of specific indicators that directly reflect the overall performance of each technology under assessment [3].

In this Chapter, we have selected four first-level indexes,—technical performance, economic performance, management performance, and benefit performance—as the key assessmentcriteria for technologies used to control urban rainfall runoff pollution. These four indicators correspond to the four subsystems described earlier. Technical performance reflects the appropriateness of technology selection, while economic performance assesses the cost-effectiveness of the technology. Management performance evaluates the level of management required for technology's implementation, and benefit performance measures the environmental, social, and economic benefits derived from its application.

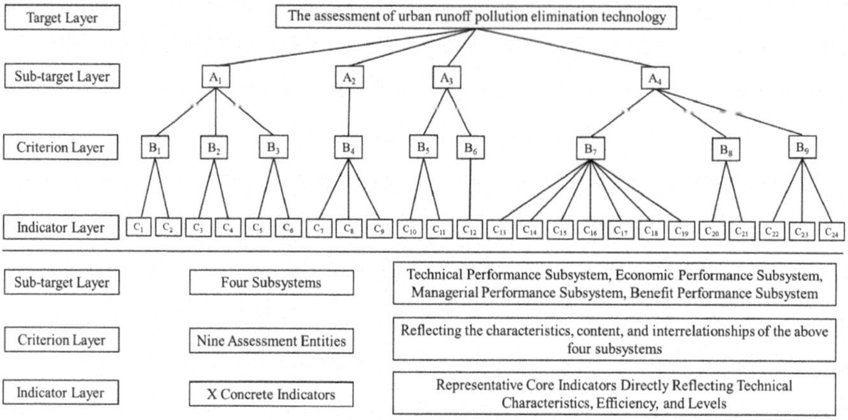

Fig. 8.1 Structure of the performance assessment framework for key technologies in urban rainfall runoff pollution control

In the technical performance subsystem, the assessment indicators primarily focus on the reliability, operability, and timeliness of the technology. These indicators should accurately reflect the performance of the key technology in controlling rainfall runoff pollution. In the economic performance subsystem, the cost of the technology is a key indicator that reflects its economic viability. The management performance subsystem includes indicators related to personnel and technology management, which assess the level of management required for effective technology implementation. Finally, in the benefit performance subsystem, three indicators are considered: environmental, economic, and social benefits, which together provide a comprehensive assessment of the technology's overall impact. The performance evaluation index system for the application of key technologies in controlling urban rainfall runoff pollution is presented in Table 8.1. While some of the indexes in this system can be quantitatively expressed, others can only be qualitatively described.

i. Meaning of Quantitative Indicators

(i) Technical Maturity

This refers to the current level of general usability achieved by the key technology, reflecting its maturity.

(ii) Useful Life

It refers to the operating or service life of the technology or equipment under stable conditions. The longer the operating time or service life, the more economically viable the technology or equipment becomes. Therefore, technologies with longer lifespans should be prioritized.

Table 8.1 Performance assessment indicator system for key technologies in urban rainfall runoff pollution control

Sub-target level	Standardized level	Index layer
Technical performance A_1	Reliability B_1	Technology maturity C_1
		Operational stability C_2
	Operability B_2	Site area(m^2)C_3
		Difficulty of construction C_4
	Timeliness B_3	Useful life (years) C_5
		Frequency of maintenance (times/year) C_6
Economic performance A_2	Technical cost B_4	Unit construction cost (rmb/m^2) C_7
		Operation and maintenance costs (%) C_8
		Personnel staff cost (rmb) C_9
Managing Performance A_3	People management B_5	Number of personnel (persons) C_{10}
		Professional Requirement C_{11}
	Technology management B_6	Difficulty of operation and management C_{12}
Comprehensive benefits A_4	Environmental benefit B_7	TN reduction rate(%)C_{13}
		TP reduction rate(%)C_{14}
		COD reduction rate(%)C_{15}
		SS reduction rate(%)C_{16}
		Reduction rate of total runoff(%)C_{17}
		Peak runoff reduction rates(%)C_{18}
		Utilization of rainwater resources C_{19}
	Economic benefit B_8	Cost of water savings from reuse of effluent(rmb)C_{20}
		Save money on municipal wastewater treatment(rmb)C_{21}
	Social benefit B_9	Reduce pressure on municipal wastewater treatment C_{22}
		Mitigating flooding in towns C_{23}
		Recharge groundwater C_{24}

(iii) Maintenance Frequency

This refers to the number of times per year that technology or equipment requires maintenance or inspection to ensure proper operation and minimize the risk of failure. A lower maintenance frequency indicates reduced effort and cost for ongoing maintenance and management.

(iv) Unit Construction Cost

This refers to the cost per square meter or cubic meter, consumed by the technology before it is officially put into use. It includes design, civil construction, equipment, and other related expenses. A lower unit construction cost implies a lower upfront investment, providing a financial advantage when compared to technical with similar technical performance.

(v) Operation and Maintenance Costs

These costs encompass the expenses associated with operating and maintaining the technology or equipment after its initial use. They include raw material costs, power consumption, and other relevant expenses, typically expressed as a percentage of the construction cost. Lower operation and maintenance costs indicate.

$$\text{Operation and Maintenance Costs} = \frac{\text{Total operating costs}}{\text{Total construction cost}} \times 100\% \tag{8.1}$$

(vi) Number of Personnel

This refers to the number of operators required to manage and operate the technology or equipment, including technical management and maintenance personnel. Under the same technical performance conditions, a lower number of required operators suggests simpler management and maintenance, making the technology a more favorable option.

(vii) Reduction Rates of SS, COD, TN, TP, Total Runoff, and Peak Runoff

The control of pollutants is a crucial indicator when evaluating key technologies for managing urban rainfall runoff pollution. Currently, the primary focus of these technologies is on effectively reducing the total runoff load, achieving some degree of runoff peak shaving, and decreasing the emissions of TSS, COD, TN, and TP emissions. Hence, this chapter selects these indicators as comprehensive benefit metrics for assessing the effectiveness of key technologies in controlling urban rainfall runoff pollution.

ii. Implications of Qualitative Indicators

(i) Operational Stability

Operational stability refers to the technology's capacity to withstand external conditions. It indicates how well the technology maintains consistent performance and functionality over time, even under challenging or changing conditions.

(ii) Site Area

Site area refers to the physical space the technology occupies. Technologies with a large footprint may face spatial constraints during installation or practical application.

(iii) Difficulty of Construction

Difficulty of construction refers to the complexity involved in the construction, installation, and commissioning of the technology or equipment.

(iv) Personnel Costs

Personnel costs refer to the salary expenses of technical managers and staff involved in the construction, operation, and maintenance of the equipment.

(v) Professional Requirements

Professional requirements refer to the expertise needed by technical managers and maintenance personnel for the operation and maintenance management of the equipment, depending on the complexity of daily operations.

(vi) Difficulty of Operation and Management

Difficulty of operation and management reflects the workload for technical managers and staff in daily management operations, as well as the complexity involved in operating and managing the equipment.

(vii) Rainwater Resource Utilization

Rainwater resource utilization measures how effectively technical equipment can promote the reuse of rainwater.

(viii) Water Saving Cost of Water Reuse

Water saving cost of water reuse represents the cost savings achieved by reusing rainwater (e.g., irrigation, car washing, etc.) after treatment with technical equipment.

(ix) Saving Municipal Wastewater Treatment Costs

Saving municipal wastewater treatment costs refers to the reduction in the cost of treating rainwater by diverting it from the municipal pipeline network to the sewage treatment plant.

(x) Reduce the Pressure of Municipal Wastewater Treatment

Reducing the pressure of municipal wastewater treatment involves decreasing the volume of rainwater runoff entering the sewage treatment plant through the use of rainfall runoff treatment technology, thus alleviating the treatment burden on the plant.

(xi) Improvement of the Quality of the Water Environment

Improving the quality of the water environment refers to the technology's ability to reduce the discharge of rainfall runoff and pollutants into receiving water bodies through processes such as retention, filtration, infiltration, and purification processes, thereby enhancing the water quality in urban areas.

(xii) Alleviation of Urban Flooding

Alleviating urban flooding involves reducing surface runoff through measures such as rainfall retention, evaporation, and infiltration during the operation of rainfall runoff pollution control technology.

(xiii) Groundwater Recharge

Groundwater recharge refers to the capability of runoff pollution control technology to facilitate the infiltration of surface rainwater runoff into the ground, thereby replenishing groundwater resources.

8.1.2 Determination of Weights for Evaluation Index

Different levels of indicators within the assessment system have varying degrees of influence on the overall assessment results. To emphasize the importance of key technology indicators in rainfall runoff pollution control, it is essential to assign weights to each assessment indicator. This section outlines the methodology and process for calculating these weights.

A. Analytic Hierarchy Process

One method for determining the weights is hierarchical analysis, which combines both quantitative and qualitative analysis in multi-program or multi-objective decision-making processes. This method involves breaking down complex problems into objective, guideline, and program levels, and comparing the degree of contribution of the indicators at each level. By constructing judgment matrices and performing the necessary calculations, the relative importance of each programs' weights can be derived, providing a basis for decision-making [4]. Hierarchical analysis modeling comprises four steps [5].

i. Establishment of a Hierarchical Structural Model

The decision-making problem is first decomposed into multiple levels and elements, such as goals, guidelines, and programs. A hierarchical structural model is then constructed based on the interrelationships and dependencies between these factors.

ii. Construction of Judgment Matrices

The relationships between factors are analyzed, and two-by-two comparative judgments are made on the contribution of each element in a specific level to the relevant element in the previous level. This process involves constructing two-by-two comparative judgment matrices.

iii. Hierarchical Single Sorting and Consistency Test

The judgment matrix obtained in the previous step is used to calculate the maximum eigenvalue and the corresponding eigenvector W. This eigenvector represents the relative weight of each element in a certain level, normalized against the relevant elements in the previous level. A consistency test is then performed.

iv. Hierarchical Total Ordering and Consistency Test

Using the relative weights obtained in the previous step, the relevant elements at the previous level are weighted. This calculation is performed from the bottom up to determine the total ordering weights of each level, especially for the lowest element at the target level (top) level. A consistency test is also carried out.

The hierarchical analysis method, based on the principles outlined above, has been widely applied in various fields of ecological and environment pollution management. This includes air pollution control, water environment safety assessment, water pollution source evaluation, research on water quality indicators and protection measures, as well as the development of ecological and environment quality assessment index systems [4].

B. Cluster Analysis

Hierarchical Clustering Methods are a statistical analysis method used for classification based on multiple indicators. The fundamental concept is as follows: initially, each research object is considered its own class. Then, the distance between each class is calculated, and the two closest classes are merged into a new class. This merging process is repeated until all objects are combined into one class, with each iteration reducing the number of classes by one. The final classification can be determined based on specific requirements or a set distance threshold. The systematic clustering method typically comprises six steps, outlined as follows.

i. Data Standardization

When sample data vary widely in magnitude or units, standardization is necessary.

(i) Initialization

Each sample is considered as an individual class, resulting in the creation of n classes. The distance between any two samples among these n classes is calculated.

(ii) Merge Closest Classes

The two classes with the closest distance are merged into a new class.

(iii) Iterative Merging

The distance between the new class and the remaining classes is calculated, and the two classes with the closest distance are subsequently merged into another new class.

(iv) Repeated Merging

Repeat the fourth step until only one class remains.

(v) Clustering Diagram and Results

Generate a clustering diagram to determine the final classification results. The systematic clustering method is widely used in the field of ecological and environmental governance. It is primarily employed for comprehensive assessments of atmospheric quality, groundwater quality, engineering geological environment quality, water quality analysis in environmental monitoring, and in the study of pollution zoning and causal factors in towns and regions, among other technical fields.

C. Calculation of Index Weight

The systematic clustering method typically consists of six steps. In recent years, the hierarchical analysis method has gained widespread usage for calculating weights of technology assessment indicators. This method involves inviting relevant experts to score and evaluate each indicator based on their perceived influence on the technology. However, it is important to note that for qualitative indicators, the experts' scoring plays a decisive role, leading to strong expert subjectivity. Due to variations in experts' knowledge, professional backgrounds, focus areas, and personal preferences, inconsistent assessment results and even significant discrepancies may arise.

To effectively capture each expert's scoring and obtain more reasonable, accurate, and reliable assessment results, this section combines the hierarchical analysis method with the systematic clustering method. This integration aims to reduce the subjectivity in expert scoring, enhance consensus among experts, and better reflect the characteristics of group decision-making. The specific steps are as follows: first, design an expert consultation form according to the hierarchical analysis method requirementsl next invite experts to evaluate and score the indicators; then, process the expert data to derive individual evaluation results; after that, applying the systematic clustering method to analyze the experts' evaluation results and determine weight coefficients for each expertl and finally, perform a weighted summation based on the evaluation results and weight coefficients to optimize the weights of the evaluation indexes, integrating the opinions of all experts. The main calculation process is outlined below.

i. Development of the Expert Scoring Table

The expert scoring table is designed to evaluate the indicator system in alignment with the multi-criteria group decision-making model. Experts are invited to make two-by-two comparative judgments on the importance level of the assessment indicators.

Tables 8.2, 8.3, 8.4, 8.5, 8.6, 8.7, 8.8, 8.9 present the expert scoring patterns for the pairwise comparison judgments of performance indicators related to key technology application in urban rainfall runoff pollution control. In these tables, experts assign two-by-two comparative judgment scores to evaluate the contribution level of each indicator relativeto the previous level. Table 8.2 illustrates the expert rating scale for comparative relationships among Level 1 indicators.

Table 8.2 The expert rating scale is used to assess the comparative relationships among level 1 indicators

Evaluation index	Technical performance	Economic performance	Managing performance	Benefit performance
Technical performance				
Economic performance				
Managing performance				
Benefit performance				

Table 8.3 The expert rating scale is utilized to assess the comparative relationship between secondary indicators of technical performance

Evaluation index	Technical reliability	Technical operability	Technical timeliness
Technical reliability			
Technical operability			
Technical timeliness			

Table 8.4 The expert rating scale is utilized to assess the comparative relationships between secondary indicators of management performance

Evaluation index	People management	Technical management
People management		
Technical management		

Table 8.5 The expert rating scale is utilized to assess the comparative relationships between secondary indicators of benefit performance

Evaluation index	Environmental benefit	Economic benefit	Social benefit
Environmental benefit			
Economic benefit			
Social benefit			

When completing the rating scale, experts must follow specific principles for assigning points. Based on the psychological research, which suggests that "the limit of human ability to differentiate information levels is 7 ± 2", the multi-criteria group decision-making model adopts a nine-point scale for evaluating the relative importance of indicators. The meaning of each value on the scale is presented in Table 8.10. Experts are required to complete the scoring form according to the guidelines shown

Table 8.6 The expert rating scale is utilized to assess the comparative relationships among the three levels of indicators within the domain of technical performance

Evaluation index	Tertiary index	
Technical reliability	Technology maturity	operational stability
Technology maturity		
Operational stability		
Technical operability	Floor space (m^2)	Difficulty of construction
Floor space (m^2)		
Difficulty of construction		
Technical timeliness	Useful life (years)	Frequency of maintenance (times/year)
Useful life (years)		
Frequency of maintenance (times/year)		

Table 8.7 The expert rating scale is utilized to assess the comparative relationships among the three levels of indicators within the domain of economic performance

Evaluation index	Tertiary index		
Technology cost	Unit construction cost (rmb/m^2)	Operation and maintenance cost (%)	Personnel staff (rmb)
Unit construction cost (rmb/m^2)			
Operation and maintenance cost (%)			
Personnel staff (rmb)			

Table 8.8 The expert rating scale is utilized to assess the comparative relationships among the three levels of indicators within the domain of management performance

Evaluation index	Tertiary index	
People management	Number of personnel (persons)	Professional requirement
Number of personnel (persons)		
Professional requirement		

in Table 8.2. They assess each indicator listed in the left vertical column of Table 8.9 by comparison it with the indicators listed in the top horizontal row.

ii. Develop a comparative judgment matrix

The expert scoring method was used to construct a comparative judgment matrix of assessment indicators. A total of n valid expert scoring data were obtained by distributing and recycling the scoring forms. Of these, 20% came from engineering managers, 40% from technical researchers, and 40% from engineering designers.

Table 8.9 The expert rating scale is utilized to assess the comparative relationships among the three levels of indicators within the domain of benefit performance

Evaluation Index	Tertiary Index						
Environmental benefit	TN reduction rate (%)	TP reduction rate (%)	COD reduction rate (%)	SS reduction rate (%)	Total runoff reduction rate (%)	Peak runoff reduction (%)	Utilization of rainwater resource
TN reduction rate (%)							
TP reduction rate (%)							
COD reduction rate (%)							
SS reduction rate (%)							
Total runoff reduction rate (%)							
Peak runoff reduction (%)							
Utilization of rainwater resource							
Economic benefit	Cost of water savings from effluent reuse (rmb)				Savings in wastewater treatment costs (rmb)		
Cost of water savings from effluent reuse (rmb)							
Savings in wastewater treatment costs (rmb)							
Social benefit	Reduce pressure on municipal wastewater treatment es			Mitigating flooding in towns and citi		Groundwater recharge	
Reduce pressure on municipal wastewater treatment							

(continued)

Table 8.9 (continued)

Evaluation Index	Tertiary Index						
Environmental benefit	TN reduction rate (%)	TP reduction rate (%)	COD reduction rate (%)	SS reduction rate (%)	Total runoff reduction rate (%)	Peak runoff reduction (%)	Utilization of rainwater resource
Mitigating flooding in towns and cities							
Groundwater recharge							

Table 8.10 Meaning of scale value

Relative importance	Connotation
1	Indicates that two factors are of equal importance compared to each other
3	Indicates that one factor is slightly more important than the other when compared to two factors
5	Indicates that one factor is significantly more important than the other when comparing two factors
7	Indicates that one factor is more strongly important than the other when comparing two factors
9	Indicates the extreme importance of one factor over the other when comparing two factors
2, 4, 6, 8	The middle value of the above two neighboring judgments
$\varphi_{ba} = 1/\varphi_{ab}$	If factor a is compared with factor b to obtain a judgment value of φ_ab, then factor b is compared with factor a to obtain a judgment value of $\varphi_{ba} = 1/\varphi_{ab}$

The judgment matrix was built based on the experts' scoring data, with the matrix elements representing the pairwise comparison relationships between the indicators. For example, for the key technology Level 2 indicators, the experts constructed a judgment matrix according to the scoring in Table 8.2. The scoring data matrix of 10 representative experts is presented in matrix $E_1, E_2, E_3, \cdots, E_{10}$.

$$E_1 = \begin{bmatrix} 1 & 3 & 5 & 1/3 \\ 1/3 & 1 & 3 & 1/5 \\ 1/5 & 1/3 & 1 & 1/7 \\ 3 & 5 & 7 & 1 \end{bmatrix}, E_2 = \begin{bmatrix} 1 & 3 & 5 & 1/5 \\ 1/3 & 1 & 3 & 1/5 \\ 1/5 & 1/3 & 1 & 1/7 \\ 5 & 5 & 7 & 1 \end{bmatrix}$$

,

$$E_3 = \begin{bmatrix} 1 & 3 & 3 & 1/3 \\ 1/3 & 1 & 3 & 1/5 \\ 1/3 & 1/3 & 1 & 1/5 \\ 3 & 5 & 5 & 1 \end{bmatrix}, E_4 = \begin{bmatrix} 1 & 3 & 4 & 1/3 \\ 1/3 & 1 & 2 & 1/6 \\ 1/4 & 1/2 & 1 & 1/7 \\ 3 & 6 & 7 & 1 \end{bmatrix}$$

,

$$E_5 = \begin{bmatrix} 1 & 5 & 3 & 1/5 \\ 1/5 & 1 & 1/2 & 1/7 \\ 1/3 & 2 & 1 & 1/5 \\ 5 & 7 & 5 & 1 \end{bmatrix}$$

,

$$E_6 = \begin{bmatrix} 1 & 3 & 5 & 1/3 \\ 1/3 & 1 & 3 & 1/5 \\ 1/5 & 1/3 & 1 & 1/5 \\ 3 & 5 & 5 & 1 \end{bmatrix}$$

,

$$E_7 = \begin{bmatrix} 1 & 3 & 3 & 5 \\ 1/3 & 1 & 1 & 3 \\ 1/3 & 1 & 1 & 3 \\ 1/5 & 1/3 & 1/3 & 1 \end{bmatrix}, E_8 = \begin{bmatrix} 1 & 2 & 3 & 7 \\ 1/2 & 1 & 3 & 5 \\ 1/3 & 1/3 & 1 & 3 \\ 1/7 & 1/5 & 1/3 & 1 \end{bmatrix},$$

$$E_9 = \begin{bmatrix} 1 & 1/2 & 1/3 & 1/3 \\ 2 & 1 & 1/2 & 1/2 \\ 3 & 2 & 1 & 1 \\ 3 & 2 & 1 & 1 \end{bmatrix}, E_{10} = \begin{bmatrix} 1 & 5 & 3 & 1 \\ 1/5 & 1 & 1/3 & 1/5 \\ 1/3 & 3 & 1 & 1/3 \\ 1 & 5 & 3 & 1 \end{bmatrix}.$$

iii. Calculation of Weight Vector

The comparative judgment matrix obtained from each expert's scoring is used to determine the weight vector of the first-level indicators for the consistency test. To illustrate the calculation process, we will use E_1 as an example.

The weight vector for the first-level indicators obtained from Expert 1 is shown in Table 8.11.

$$W = (0.263, 0.118, 0.055, 0.564)^T$$

iv. Consistency Test

Table 8.11 The procedure for calculating the vector of weights for Level 1 indicators

Judgment matrix					In-line rides	Consecutive product to the nth power	Normalization
E_1	A1	A2	A3	A4	$M_i = \prod_{i=1}^{n} a_{ij}$	$\overline{W_i} = \sqrt[n]{M_i}$	$W_i = \overline{W_i} / \sum_{i=1}^{n} \overline{W_i}$
A1	1	3	5	1/3	5	1.495	0.263
A2	1/3	1	3	1/5	0.2	0.669	0.118
A3	1/5	1/3	1	1/7	0.01	0.312	0.055
A_4	3	5	7	1	105	3.201	0.564

Note: n-number of orders of the judgment matrix; a_{ij}-individual elements of the judgment matrix; W_i-the ith element of the weight vector

From the previous step, we obtained the weight vector for the first-level indices as $W = (0.263, 0.118, 0.055, 0.564)^T$. Subsequently, the consistency test for the judgment matrix is conducted.

The consistency test of the judgment matrix involves calculating the Consistency Index *CI* value and the Consistency Ratio *CR* value using Eqs. (8.2) and (8.3), respectively.

$$CI = \frac{\lambda_{max} - n}{n - 1} \tag{8.2}$$

$$CR = \frac{CI}{RI} \tag{8.3}$$

If the Consistency Ratio (CR) is less than 0.10, the judgment matrix is considered to have satisfactory consistency. If the CR is greater than or equal to 0.10, adjustments are required to achieve satisfactory consistency.

(i) Calculation of the Maximum Characteristic Root λ_{max}

Let λ_{max} be the maximum characteristic root of the judgment matrix A, and let W be the corresponding eigenvector, Thus, we have the equation AW $= \lambda_{max}$W. Therefore, λ max can be obtained using Eq. (8.4).

$$\lambda_{max} = \frac{1}{n} \sum_{i=1}^{n} \frac{(AW)_i}{w_i} \tag{8.4}$$

By substituting the judgment matrix E and the weight vector W into Eq. (8.4), we obtain .$\lambda_{max} = 4.117$

(ii) Calculation of the Judgment Matrix Consistency Index (CI)

Using Eq. (8.2), we calculate CI as $CI = \frac{\lambda_{max} - n}{n-1} = \frac{4.117-4}{4-1} = 0.039$

(iii) Calculation of Stochastic Consistency Ratio (CR)

At this stage, it is necessary to introduce the average stochastic consistency index, *RI* for judgment matrices. The values of *RI* for judgment matrices of order 1 to 15 are shown in Table 8.12.

Since n = 4, looking up Table 8.12 shows that $RI = 0.89$. Calculated from Eq. (8.3):

$$CR = \frac{CI}{RI} = \frac{0.039}{0.89} = 0.044 < 0.1$$

The judgment matrix passes the consistency test as its consistency ratio is less than 0.10, indicating satisfactory consistency. Therefore, the vector of key technology level 1 index weights for expert 1 is $W = (0.263, 0.118, 0.055, 0.564)^T$. This implies that the weights of level 1 indexes for technical performance, economic performance, managerial performance, and benefit effectiveness are $w_1 = 0.263$, $w_2 = 0.118$, $w_3 = 0.055$ and w_4=0.564 respectively, for the application performance of key rainfall runoff pollution control technology for urban low-impact development.

Using the same procedure, scoring data from other relevant experts are processed to obtain their respective indicator weights and consistency test results. The calculation results for 10 representative experts are presented in Table 8.13.

v. Optimization of indicator weights

The hierarchical analysis method relies on expert scoring to determine indicator weights. However, for qualitative indicators, the subjective factors of experts can lead to significant deviations in results, as shown in Table 8.13. To address this issue, the systematic clustering method is introduced to analyze the scoring data of each expert by clustering. This approach enables obtaining more reasonable, accurate, and reliable technical assessment results, while optimizing the weights of the assessment indicators. In this study, the assessment result of each expert are treated as vectors. The cosine of the angle between these vectors is used to define the degree of consistency between the assessment results of two experts. This criterion is employed for clustering analysis [6]. Assuming that the feature vectors, i.e., assessment results of two experts at the same level, are $X = (\times 1, \times 2, \ldots xn)$ and $Y = (y1, y2, \ldots yn)$, the degree of consistency of these two experts' assessment results, d_{xy}, can be expressed using Eq. (8.5).

$$d_{xy} = \cos\theta_{xy} = \frac{(X, Y)}{|X| \cdot |Y|} = \frac{\sum_{i=1}^{n} x_i y_i}{\sqrt{\sum_{i=1}^{n} x_i{}^2 \sum_{i=1}^{n} y_i{}^2}} \tag{8.5}$$

$$0 < \cos\theta_{xy} < 1$$

Table 8.12 Average consistency indicator RI value

n	1	2	3	4	5	6	7	8	9	10	11	12	13	14	15
RI	0	0	0.52	0.89	1.12	1.26	1.36	1.41	1.46	1.49	1.52	1.54	1.56	1.58	1.59

Table 8.13 Summary of the weighted values of the indicators identified by the 10 experts

Indicators		1	2	3	4	5	6	7	8	9	10
Level 1A	Technical performance A1	0.263	0.222	0.248	0.248	0.223	0.275	0.520	0.483	0.109	0.391
	Economic performance A2	0.118	0.113	0.126	0.101	0.059	0.123	0.201	0.314	0.189	0.067
	Managing Performance A3	0.055	0.053	0.073	0.064	0.102	0.062	0.201	0.144	0.351	0.151
	Benefit Performance A4	0.564	0.613	0.554	0.587	0.616	0.540	0.078	0.059	0.351	0.391
Technical performance A1	Technical Reliability B1	0.429	0.54	0.387	0.249	0.435	0.740	0.333	0.714	0.25	0.637
	Technical operability B2	0.428	0.297	0.443	0.594	0.487	0.094	0.333	0.143	0.25	0.105
	Technical timeliness B3	0.143	0.163	0.169	0.157	0.078	0.167	0.333	0.143	0.5	0.258
Economic performance A2	Technology costs B4	1	1	1	1	1	1	1	1	1	1
Management Performance A3	People management B_5	0.75	0.25	0.25	0.25	0.5	0.833	0.25	0.333	0.5	0.833
	technical management B_6	0.25	0.75	0.75	0.75	0.5	0.167	0.75	0.667	0.5	0.167
Benefit Performance A4	Environmental benefit B_7	0.258	0.211	0.2	0.23	0.279	0.279	0.637	0.279	0.25	0.714
	Economic benefit B_8	0.105	0.084	0.2	0.648	0.072	0.072	0.258	0.649	0.25	0.143
	social benefit B_9	0.637	0.705	0.6	0.122	0.649	0.649	0.105	0.072	0.5	0.143
Technical Reliability B1	Technology maturity C_1	0.5	0.67	0.5	0.5	0.5	0.75	0.5	0.5	0.5	0.75
	Operational stability C_2	0.5	0.33	0.5	0.5	0.5	0.25	0.5	0.5	0.5	0.25
technical operability B2	Floor space C_3	0.25	0.25	0.25	0.33	0.167	0.875	0.75	0.833	0.667	0.75
	Difficulty of construction C_4	0.75	0.75	0.75	0.67	0.833	0.125	0.25	0.167	0.333	0.25
Technical timeliness B3	Useful life C_5	0.5	0.75	0.75	0.67	0.25	0.125	0.75	0.833	0.5	0.75
	Frequency of maintenance C_6	0.5	0.25	0.25	0.33	0.75	0.875	0.25	0.167	0.5	0.25
technology cost B4	Unit construction cost C_7	0.429	0.54	0.429	0.429	0.481	0.515	0.2	0.6	0.25	0.429
	Operation and maintenance cost C_8	0.428	0.163	0.429	0.143	0.405	0.097	0.6	0.2	0.25	0.429

(continued)

Table 8.13 (continued)

Indicators		1	2	3	4	5	6	7	8	9	10
	Personnel staff cost C_9	0.143	0.297	0.143	0.429	0.114	0.388	0.2	0.2	0.5	0.143
People management B5	Number of personnel C_{10}	0.75	0.25	0.25	0.25	0.25	0.125	0.75	0.667	0.75	0.25
	Professional Requirement C_{11}	0.25	0.75	0.75	0.75	0.75	0.875	0.25	0.333	0.25	0.75
Technical management B6	Difficulty of operation and management C_{12}	1	1	1	1		1	1	1	1	1
Environmental benefitB7	TN reduction rate (%)	0.057	0.063	0.059	0.066	0.041	0.036	0.176	0.086	0.176	0.286
	TP reduction rate (%)	0.057	0.063	0.059	0.066	0.041	0.143	0.176	0.124	0.176	0.286
	COD reduction rate (%)	0.057	0.089	0.059	0.066	0.041	0.202	0.176	0.045	0.176	0.113
	SS reduction rate (%)	0.057	0.17	0.059	0.068	0.041	0.103	0.176	0.119	0.176	0.113
	Total runoff reduction rate (%)	0.312	0.37	0.24	0.283	0.405	0.298	0.059	0.447	0.059	0.044
	Peak runoff reduction (%)	0.146	0.207	0.201	0.156	0.154	0.202	0.176	0.148	0.059	0.113
	Utilization of rainwater resources	0.312	0.039	0.322	0.295	0.275	0.017	0.059	0.030	0.176	0.044
Economic benefit B8	Cost of water savings from effluent reuse (rmb)	0.75	0.25	0.75	0.75	0.5	0.25	0.5	0.667	0.5	0.75
	Savings in wastewater treatment costs (rmb)	0.25	0.75	0.25	0.25	0.5	0.75	0.5	0.333	0.5	0.25
Social benefit B9	Reduce pressure on municipal wastewater treatment	0.105	0.105	0.2	0.122	0.188	0.405	0.258	0.105	0.4	0.6
	Mitigating flooding in towns and cities	0.258	0.637	0.6	0.32	0.731	0.114	0.637	0.637	0.4	0.2
	Groundwater recharge	0.637	0.258	0.2	0.558	0.081	0.481	0.105	0.258	0.2	0.2

where (X, Y) $= |X| \cdot |Y| \cos\theta_{xy}$ denotes the inner product of two vectors, $d_{xy} = d_{yx}$.

The above formula shows that if $dxy = cos\theta xy = 1$, the assessment results of expert X and expert Y at this level are identical. The degree of consistency between X and Y's eigenvectors increases as dxy approaches 1, indicating greater similarity between the experts' assessment results. When dxy reaches a certain level, these two experts can be classified. Similarly, when $dxy = cos\theta xy = 0$, the assessment results of expert X and expert Y at this level are completely different. As dxy approaches 0, the difference between the two experts' assessment results becomes more significant. Using dxy as a criterion, the expert group's scoring results are systematically clustered and analyzed. The clustering process is not repeated here.

The assessment results issued by categories with a larger number of experts represent the majority opinion and should be given a higher weighting factor, whereas those issued by categories with few experts represent only a minority and should be given a lower weighting factor [7].

Suppose m experts' assessment results have been clustered and analyzed, and Φ_k experts belong to the kth category. The weight λ_k is proportional to the number of experts in the corresponding category, i.e., $\lambda_1 : \lambda_2 : \cdots : \lambda_m = \Phi_1 : \Phi_2 : \cdots : \Phi_m$. Also, $\sum_{k=1}^{m} \lambda_k = 1$. The weight coefficients for the kth expert are shown in Eq. (8.6).

$$\lambda_k = \frac{\Phi_k}{\sum_{i=1}^{m} \Phi_i} \tag{8.6}$$

The weight coefficients of each expert can be calculated sequentially using Eqs. (8.9). Suppose the feature vector of the kth expert's level is $W^{(k)} = ({w_1}^k, {w_2}^k, \cdots, {w_n}^k)^T$. Based on the hierarchical analysis of group decision-making, the feature vector $W = (w_1, w_2, \cdots, w_n)^T$ at this level can be obtained using the same hierarchical matrix of experts. Each factor is then weighted by λ_k, and the weighted average of each factor weight by λ_k gives the eigenvector, as shown in Eq. (8.7).

$$w_i = \sum_{k=1}^{m} \left(\lambda_k \cdot {w_i}^{(k)}\right) \tag{8.7}$$

By applying the cluster analysis method mentioned above, the comprehensive weight vector $W = (w_1, w_2, \cdots, w_n)^T$ can be derived from each expert's opinion on the assessment indicators. At this stage, we obtain the comprehensive weight of each assessment indicator relative to the previous level, known as hierarchical single sort.

For instance, taking the first-level indicators as an example, Table 8.13 illustrates the characteristic vector $W^{(k)}$ derived by the 10 representative experts for these indicators.

$$W^{(1)} = (0.263,\ 0.118,\ 0.055,\ 0.564)^T$$

$$W^{(2)} = (0.222,\ 0.113,\ 0.053,\ 0.613)^T$$

$$W^{(3)} = (0.248,\ 0.126,\ 0.073,\ 0.554)^T$$

$$W^{(4)} = (0.248,\ 0.101,\ 0.064,\ 0.587)^T$$

$$W^{(5)} = (0.223,\ 0.059,\ 0.102,\ 0.616)^T$$

$$W^{(6)} = (0.275,\ 0.123,\ 0.062,\ 0.540)^T$$

$$W^{(7)} = (0.520,\ 0.201,\ 0.201,\ 0.078)^T$$

$$W^{(8)} = (0.483,\ 0.314,\ 0.144,\ 0.059)^T$$

$$W^{(9)} = (0.109,\ 0.189,\ 0.351,\ 0.351)^T$$

$$W^{(10)} = (0.391,\ 0.067,\ 0.151,\ 0.391)^T$$

Equation (8.5) is applied to the aforementioned 10 assessment results in order to calculate the degree of consistency between the pairwise expert assessments. The results are presented in Table 8.14.

Clustering the experts based on the values of the degree of consistency between the two-by-two experts in Table 8.14 results in the 10 representative expert clusters shown in Fig. 8.2.

Table 8.14 The degree of consistency in the assessment results provided by the 10 experts

d_{xy}	1	2	3	4	5	6	7	8	9	10
1	–	0.9961	0.9993	0.9988	0.9892	0.9993	0.5665	0.5414	0.7781	0.9256
2	–	–	0.9962	0.9984	0.9940	0.9922	0.4949	0.4712	0.7761	0.8930
3	–	–	–	0.9985	0.9903	0.9987	0.5680	0.5441	0.8005	0.9246
4	–	–	–	–	0.9949	0.9967	0.5358	0.5053	0.7812	0.9165
5	–	–	–	–	–	0.9854	0.4929	0.4467	0.7966	0.9044
6						–	0.5973	0.5720	0.7854	0.9376
7							–	0.9752	0.5933	0.8047
8								–	0.5663	0.7403
9									–	0.7848
10										–

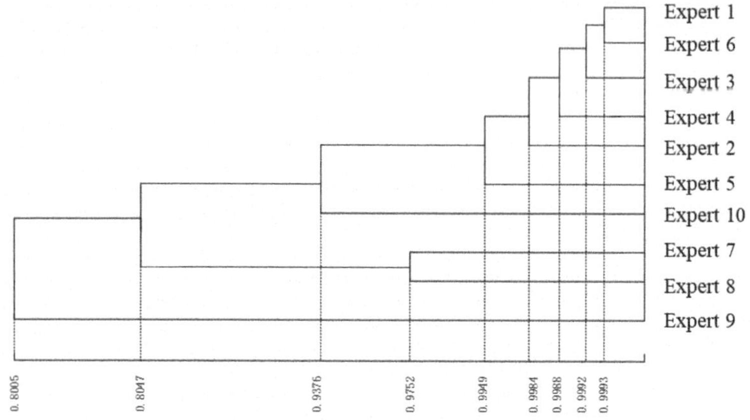

Fig. 8.2 Cluster diagram of expert assessment results

Based on Fig. 8.2, it is evident that categorizing the n experts into three classes is more appropriate: class 1 accounts for 70%, class 2 accounts for 20%, and class 3 accounts for only 10%. The number of experts in each category is as follows: $\Phi_1 = \Phi_2 = \Phi_3 = \Phi_4 = \Phi_5 = \Phi_6 = \Phi_{10} = 7$ By utilizing Eq. (8.6), the weight coefficients for each expert can be determined: $\lambda_1 = \lambda_2 = \lambda_3 = \lambda_4 = \lambda_5 = \lambda_6 = \lambda_{10} = 7/54$.Through these calculations, the feature vector $W^{(k)}$ for each xpert level indicator and their respective expert weight coefficients λ_k can be obtained. Additionally, according to Eq. (8.7), the combined weight vector of the level indicators can be determined as = (0.282, 0.114, 0.092, 0.513).

Similarly, the comprehensive weights of the remaining assessment indicators can be obtained through sequential calculations, where the comprehensive weights represent the intra-group weights of the indicators relative to the previous level. The weights of the indicators at each level are further calculated to obtain the weights of all indicators relative to the target level. This results in the weight vector of each assessment indicator relative to the target level, denoted as $W^* = (w_1{}^*, w_2{}^*, \cdots, w_n{}^*)^T$.

Figure 8.1 illustrates the hierarchical structure of key technology assessment indices for urban rainfall runoff pollution control. It consists of the target layer, sub-target layer, criterion layer, and indicator layer. Assuming the sub-target layer has m layers, the criterion layer has s layers, and the indicator layer has n indicators, the hierarchical ordering of the sub-goal layer to the goal layer is $W'' = (w_1{}'', w_2{}'', \cdots, w_m{}'')^T$. Furthermore, the hierarchical ordering of the criterion layer to the goal layer is $W' = \left(w_{1k}{}', w_{2k}{}', \cdots, w'_{jk}\right)^T, (j = 1, 2, \ldots, s; k = 1, 2, \ldots, m)$.

The hierarchical single ranking of the indicator layer to criterion j of the criterion layer is $W_j = \left(w_{1j}, w_{2j}, \cdots, w_{ij}, \cdots, w_{is}\right)^T$. Thus, the calculation of the hierarchical total ranking matrix of the indicator layer to the target layer, $W^* = (w_1{}^*, w_2{}^*, \cdots, w_n{}^*)^T$, is shown in Eq. (8.8).

Table 8.15 Indicator weights for technical assessment of urban rainfall runoff pollution control

Subgoal level		Standardized level		Indicator layer		
Subgoal Level	Weighting within group w_k''	Criterion	Weighting within group w_{jk}'	Mark	Weighting within group w_{ij}	Weighting relative to the target layer w_i*
Technical performance A_1	0.282	Reliability B_1	0.510	Technology maturity C_1	0.542	0.078
				Operational stability C_2	0.458	0.066
		Operability B_2	0.323	Footprint C_3	0.468	0.043
				Difficulty of construction C_4	0.532	0.048
		Timeliness B_3	0.168	Useful life C_5	0.658	0.031
				Frequency of maintenance C_6	0.342	0.016
Economic performance A_2	0.114	technology cost B_4	1	Unit construction cost C_7	0.466	0.053
				Operation and Maintenance Cost C_8	0.295	0.034
				Staff costs C_9	0.239	0.027
Managing Performance A_3	0.092	People Management B_5	0.489	Number of personnel C_{10}	0.459	0.021
				Professional Requirements C_{11}	0.541	0.024
		Technology Management B_6	0.511	Difficulty of operation and management C_{12}	1	0.047
Benefit Performance A_4	0.513	Environmental Benefit B_7	0.286	TN reduction rate C_{13}	0.075	0.011
				TP reduction rate C_{14}	0.090	0.013
				COD reduction rate C_{15}	0.084	0.012
				SS reduction rate C_{16}	0.091	0.013
				Total runoff reduction rate C_{17}	0.302	0.044

(continued)

Table 8.15 (continued)

Subgoal level		Standardized level		Indicator layer		
Subgoal Level	Weighting within group w_k''	Criterion	Weighting within group w_{jk}'	Mark	Weighting within group w_{ij}	Weighting relative to the target layer w_i*
				Peak runoff reduction rate C_{18}	0.166	0.024
				Rainwater Resource Utilization C_{19}	0.191	0.028
		Economic Benefit B_8	0.184	Cost of water savings from reuse of effluent C_{20}	0.557	0.053
				Saving municipal wastewater treatment costs C_{21}	0.443	0.042
		Social Benefit B_9	0.530	Reduce pressure on municipal wastewater treatment C_{22}	0.201	0.055
				Mitigating flooding in towns and cities C_{23}	0.559	0.152
				Groundwater recharge C_{24}	0.241	0.066

8.1.3 Assessment Criteria

After determining the weights of the key technology assessment indicators for urban rainfall runoff pollution control, it becomes necessary to standardize the individual indicators due to differences in units, orders of magnitude, and other factors. These differences prevent direct calculations or comparisons across indicators. Therefore, the assessment criteria for all indicators must be standardized before analysis and evaluation. Specifically, each of the 24 level-3 indicators in the assessment system is assigned a corresponding scoring standard. The grading and scoring criteria for each indicator are shown in Tables 8.16 and 8.17, respectively.

Table 8.16 Grading and scoring criteria for quantitative indicators

Indicator layer	Indicator grading and scores				
	1	3	5	7	9
Useful life, years	(0,3)	(3,5)	(3,8)	(8,10)	(10,∞)
Maintenance frequency. Times/year	(8,∞)	(6,8)	(4,6)	(2,4)	(0,2)
Unit construction cost. rmb/m^2	(0,600]	(400,600)	(200,400)	[50,200)	(0,50)
O&M costs as a percentage, %	(15,∞)	(10,15)	(5,10)	(2,5)	(0,2)
Number of personnel, persons	5 or above	3 or 4	2	1	0
TN reduction rate,%	[0,10]	(10,30]	(30,50]	(50,70]	(70,100)
TP reduction rate, %	(0,10)	(10,30)	(30,50)	(50,70)	(70,100)
COD reduction rate,%	(0,10)	(10,30)	(30,50)	(50,70)	(70,100)
SS reduction rate,%	(0,10)	(10,30)	(30,50)	(50,70)	(70,100)
Total runoff reduction rate,%	(0,5)	(5,20)	(20,40)	(40,60]	(60,100)
Peak runoff reduction rate,%	(0,5)	(5,20)	(20,40)	(40,60)	(60,100)

8.1.4 Determination of Evaluation Method

This Chapter adopts the comprehensive assessment index method (TAI) as an approach to comprehensively evaluate key technologies for controlling urban rainfall runoff pollution, as depicted in Eq. (8.9).

$$\begin{cases} TAI = \sum_{i=1}^{n} \left(w_i^* \cdot C_i\right) \\ \sum_{i=1}^{n} w_i{}^* = 1 \end{cases} \quad (8.9)$$

where, *TAI*—comprehensive assessment index of control technology;

$w_i{}^*$—Weight of the ith indicator relative to the target layer;

C_i—the scoring value of the ith index.

The comprehensive technology assessment index can be obtained by using Eq. (8.9), after obtaining the weights and scoring values of each assessment indicator. Based on the assessment criteria and value assignment, it can be deduced that$C_i \in [1, 9]$. Furthermore, the sum of weights of each assessment index in Eq. (8.9) is 1, implying that every *TAI* value is within the range of $[1, 9]$. Using these ranges, the technology comprehensive assessment index can be classified into five levels: $[1.0, 2.0]$, $(2.0, 4.0]$, $(4.0, 6.0]$, $(6.0, 8.0]$, $(8.0, 9.0]$, corresponding to very poor, poor, average, good, and very good technologies respectively, as shown in Table 8.18. These levels facilitate the comprehensive assessment of low-impact development technology in the town.

Table 8.17 Grading and scoring criteria for each qualitative indicator

Evaluation of index	Scoring Standard	Point Value
Technology maturity	A. High level of maturity of the technology implemented	(6,9)
	B. Average maturity of technologies implemented	(3,6)
	C. Low maturity of implemented technologies	(1,3)
Operational stability	A. Good operational stability of implemented technologies	(6,9)
	B. General operational stability of implemented technologies	(3,6)
	C. Poor operational stability of implemented technologies	(1,3)
Difficulty of construction	A. Low construction difficulty of the implemented technologies	(6,9)
	B. Medium difficulty of construction of the implemented technologies	(3,6)
	C. Difficulty of construction of the implemented technologies	(1,3)
Footprint	A. Small floor space	(6,9)
	B. Medium footprint	(3,6)
	C. large footprint	(1,3)
Staff costs	A. Lower personnel salary costs	[6,9]
	B. Medium staff salary costs	[3,6)
	C. Higher personnel salary costs	[1,3)
Professional requirements	A. The technologies implemented are largely non-specialized	[6,9]
	B. The technology implemented requires a certain level of expertise	[3,6)
	C. The technology implemented must be professionally staffed	[1,3)
Difficulty of operation and management	A. The technology implemented is less difficult to manage operationally	[6,9]
	B. Difficulty of operational management of implemented technologies is average	[3,6)
	C. Difficulty in the operational management of the technologies implemented	[1,3)
Utilization of rainwater resources	A. High utilization of rainwater resources	[6,9]
	B. Average utilization of rainwater resources	[3,6)
	C. Low utilization of rainwater resources	[1,3)
Cost of water savings from reuse of effluent	A. High effluent reuse rate	[6,9]
	B. Average effluent reuse rate	[3,6)
	C. Low effluent reuse rate	[1,3)

(continued)

Table 8.17 (continued)

Evaluation of index	Scoring Standard	Point Value
Save money on municipal wastewater treatment	A. Higher savings in municipal wastewater treatment costs	[6,9]
	B. Some savings in municipal wastewater treatment costs	[3,6)
	C. Saving a small portion of municipal wastewater treatment costs	[1,3)
Reduce pressure on municipal wastewater treatment	A. Effectively reduce pressure on municipal wastewater treatment	[6,9]
	B. Essentially reduces pressure on municipal wastewater treatment	[3,6)
	C. Almost impossible to reduce pressure on municipal wastewater treatment	[1,3)
Improving the quality of the human environment	A. Effective improvement of the quality of the human environment	[6,9]
	B. Essentially improves the quality of the human environment	[3,6)
	C. Hardly improves the quality of the human environment	[1,3)
Mitigating flooding in towns	A. Effective mitigation of flooding in towns and cities	[6,9]
	B. Basic relief from flooding in the town	[3,6)
	C. Failure to alleviate flooding in the town	[1,3)
Recharge of groundwater	A. Effective groundwater recharge	[6,9]
	B. Some recharge of groundwater	[3,6)
	C. Almost impossible to recharge groundwater	[1,3)

Table 8.18 Grading criteria for the assessment index of urban rainfall runoff pollution control technology

Technology assessment index	(1.0,2.0]	(2.0,4.0]	(4.0,6.0]	(6.0,8.0]	(8.0,9.0]
rank	poorly	mediocre	normal	better	good
Grade Color					

8.2 Analysis and Evaluation of Key Technology Cases

8.2.1 Basic Information of Cases

A. Basic Overview of Beijing Future Science and Technology City

The Beijing Future Science and Technology City is located in the southeastern part of Changping District, Beijing. It is bordered to the east by the Jingcheng Expressway, to the south by the planned Qibei Road, to the west by the boundary of Beiqijia Village, and to the north by the extension of Shunyu Road. The city is approximately 12 km from the Capital Airport and about 15 km from the South Fourth Ring Road. It is also in close proximity to the North Sixth Ring Road in the north, serving as a significant development hub for the expansion of the Beijing Municipal Economy into the Beijing-Tianjin-Hebei region.

The development and construction of Beijing Future Science and Technology City will be executed in two phases. The first phase covers an area of approximately 10 km^2, extending from the western extension of Shunyu Road in the north to the Beijing-Chengdu Expressway and the boundary of Changping District in the east. To the south, it stretches to Qibei South Road, and to the west, it reaches the eastern boundary of the Beiqijia Town Center Group. The land-use planning prioritizes scientific research, education, and green spaces. Out of the total planned land area, 25% (256.1 hm^2) will be allocated for educational, research, and design purposes, 1.0% (10.4 hm^2) for public supporting services, and 5.4% (55.42 hm^2) for residential use, primarily for research and development activities, as well as construction of scientific research and supporting facilities. Protective green space will cover 5.7% (57.89 hm^2), public green space 28.1% (287.4 hm^2), and water areas 8.9% (93 hm^2) of the total planning area.

B. Beijing Future Science City Rainfall Runoff Pollution Control Technology Program

Based on the urban low-impact development classification and zoning method, along with the overall rainwater utilization goals for the region encompassing Beijing Future Science and Technology City, this study decomposes the target rainfall runoff pollution control according to three types of land use: (1) development and construction sites, (2) peripheral urban roads, and (3) public green areas. The parcels are further subdivided by considering the construction and hydrological characteristics of different land types, allowing for the selection and design of appropriate rainwater utilization facilities.

i. Overall Rainfall Runoff Control and Utilization Objectives

After the completion of Beijing Future Science and Technology City, the comprehensive runoff coefficient of rainwater discharged within the planned area of one rainfall in five years will be controlled below 0.43. According to Design Design Code for Rainwater Control and Utilization Engineering (DB11/685), the total annual runoff control rate of the newly developed areas should not be less than 85%, and the corresponding rainfall designed for rainwater utilization is 32.5 mm. Therefore, the overall rainwater control and utilization target for Beijing Future Science and Technology City is as follows: by implementing low-impact development technology, we aim to achieve a multi-year runoff coefficient below 0.43 for a one-in-five-year rainfall event in the developed and constructed land plots. After the development and construction, the average annual discharge runoff should not exceed pre-development levels. Additionally, the average annual runoff coefficient should be ≤ 0.15, and the average annual total runoff control rate (annual rainwater utilization rate) should be ≥ 0.85.

ii. Rainfall Runoff Control and Utilization Target Decomposition

Following the proportions of development and construction sites, public green areas, and urban roads within the development and construction area and adhering to the principles of economic feasibility, safety, reliability, and local conditions, we have determined the rainfall control and utilization targets for development and construction sites, public green areas, and urban roads depicted in Table 8.19.

The general concept of controlling and utilizing rainfall runoff involves the comprehensive implementation of low-impact development principles. Planning and design are tailored to different land areas, including development and construction plots, peripheral urban roads of the plots, and public green areas. Figure 8.4 illustrates this overall concept. Specifically, for development and construction plots, rainwater discharge is strictly limited, with a focus on in-situ consumption, following the principle of "prioritizing infiltration, focusing on stagnant storage and control, and appropriate storage and reuse." This principle guides on-site rainwater management. Regarding the roads surrounding the plots, the emphasis is on permeable pavement control and the proportion of concave green space. Measures such as infiltration, stagnation, seepage storage, self-irrigation, and irrigation are implemented to reduce runoff from urban roads and minimize pollution loads In public green spaces, the main focus is on stagnant water and promoting infiltration tosup

Table 8.19 Breakdown of rainfall runoff control and utilization targets for different site types in Beijing Future Science City

Typology	Annual average Runoff coefficient	Total annual runoff Control rate	5-year rainfall Peak external flow runoff coefficient	Rainwater utilization design rainfall, mm
Development and construction Tectonic plate	≤ 0.13	≥ 0.87	≤ 0.58	rainfall $\leq$ 36 mm, take (time) $\leq$ 1 h
Public center	≤ 0.03	≥ 0.97	≤ 0.1	rainfall $\leq$ 80 mm, take (time) $\leq$ 24 h
City road	≤ 0.4	≥ 0.6	≤ 0.7	rainfall $\leq$ 14 mm, take (time) $\leq$ 9 min
Region-wide	≤ 0.15	≥ 0.85	≤ 0.43	$\leq$ 32.5 h

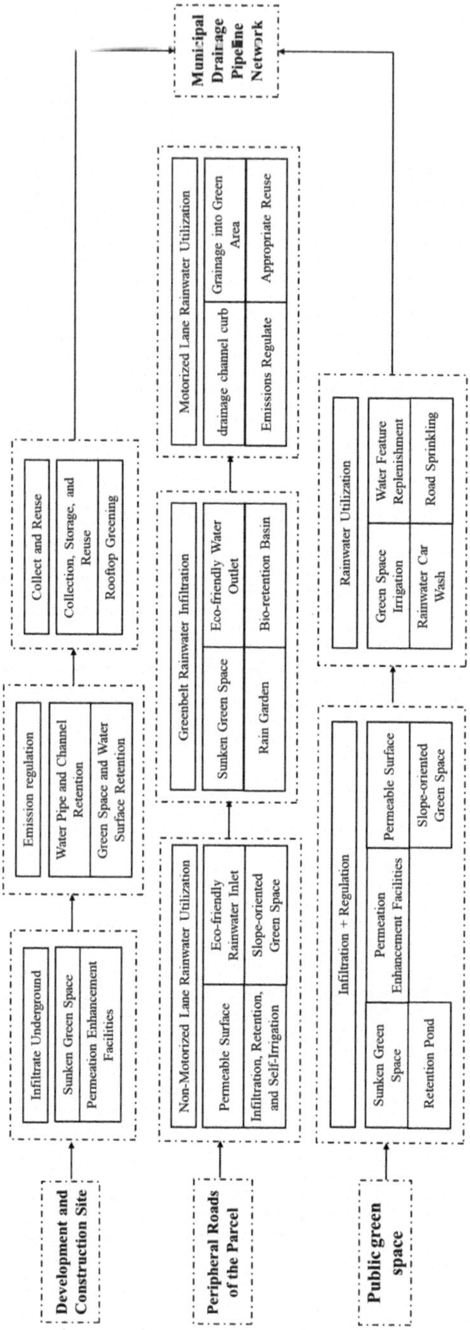

Fig. 8.4 Schematic diagram of the general idea of rainfall runoff control and utilization in Beijing future science and technology city

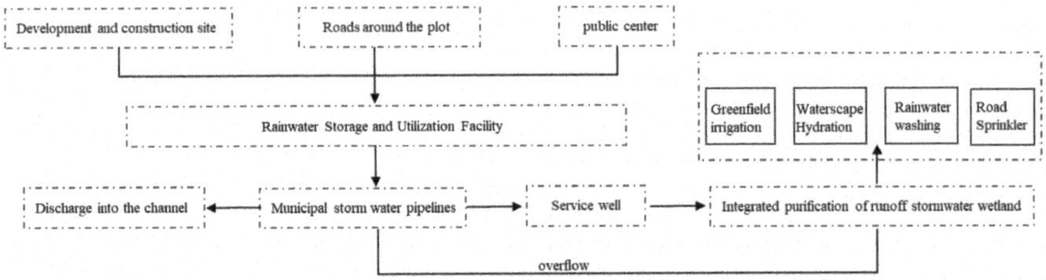

Fig. 8.5 Rainfall runoff control and utilization Zoning in Beijing future science and technology city

port the site's own rainwater consumption. Simultaneously, a portion of the external runoff is stored through infiltration.

To achieve the overall objective of controlling and utilizing rainfall runoff, it is essential to implement a regional program for storage and regulation of such runoff. Given the abundance of planned green areas in the region, these spaces should be considered for their potential role in runoff storage and control. After evaluating several options, the most suitable approach for decentralized storage and control of rainfall runoff across different land parcels is determined. This involves effective collection, storage, and reuse of rainwater, as depicted in Fig. 8.5.

In accordance with local conditions, rainwater storage and utilization facilities are established in each land plot. These facilities are designed to ensure that the discharge of pollutants from rainfall runoff remains within acceptable standards. If rainwater exceeds the set standards, it is directed into the municipal pipeline. Additionally, rainwater tanks are installed in public green spaces, strategically located near the municipal pipeline. These tanks serve to store, purify, and utilizing rainwater for various activities, such as irrigation of green spaces, car washing, and road cleaning.

C. Runoff Pollution Control Demonstration Project for Beijing Future Science City

The Runoff Pollution Control Demonstration Project in Beijing Future Science and Technology City covers both the North Zone (the drainage area into Wenyu River) and the Southeast Zone. These areas serve as the primary sites for the application of urban rainfall runoff pollution control technologies.

i. Construction of rainfall runoff pollution control technology project for towns in the North District.

(i) Building Plot

Rainwater utilization and low-impact development measures have been implemented across six central enterprises: China Guodian, State Grid, Wuhan Iron and Steel Group, Anshan Iron and Steel Group, COMAC, and Shenhua Group. The total area of permeable pavement is 69,990 m^2, the area of recessed green space is 48,772 m^2, and the storage volume of rainwater tank is 2,715 m^3.

(ii) Roads Around the Site

Twelve planned roads are present in the North Zone, with a total length of 13,828 m. The constructed roads extend over an area of 475,923 m^2 and the sidewalks span a total of 104,698 m^2. The sidewalks feature 100% permeable pavement, amounting to the entire 104,698 m^2. The total green area is 137,066 m^2, with an 63,866 m^2area of concave green space. The ecological ditch extends 13,932 m in length, with a net green area volume of 6,386 m^3. The ditch also has a net storage volume of 2,937 m^3, resulting in a unit area net storage volume of approximately 275.1 m^3/hm^2.

(iii) Public Center

A riverfront green space has been created featuring the concave green space area of 88,300 m^2, permeable pavement area of 21,561.4 m^2, and a stagnant storage pond storage volume of 33,097 m^3.

ii. Construction of Urban Rainfall Pollution Control Technology Project in Southeast District.

(i) Building Plot

Four settled central enterprises, including China National Offshore Oil Corporation (CNOOC), China Electronics Corporation (CEC), China Telecom Corporation (CTC), and Weaponry Equipment Corporation (WAC), along with the resettlement residential area, have implemented rainwater utilization and low-impact development facilities. A total construaction area of 50,729 m^2 of permeable pavement, 84,317 m^2 of concave green space, and 2,490 m^3 of rainwater storage volume have been constructed.

(ii) Roads Around the Plot

There are 19 planned roads in the Southeast Zone, with a total length of 18,374 m. The constructed roads cover an area of 783,270 m^2, with sidewalks covering 129,712 m^2. The permeable pavement area also covers 129,712 square meters, achieving a 100% permeable surface rate. The total green area is 151,211 m^2, with the concave green space area as 23,769 m^2. The ecological ditch extends for 8,369 m, with a net storage volume of 2377 m^3 for green area, and the ecological ditch having a net storage volume of 504 m^3. The unit area net storage volume is approximately 45.6 m^3/hm^2.

(iii) Project Overview

Beijing Future Science and Technology City has implemented five urban rainfall runoff pollution control projects and applied seven different technologies for managing rainfall runoff pollution.

(i) Permeable Surface Project with Multiple Types of Technologies

This project is situated in the parking lot on the southern side of the riverfront road section between Lutong West Road and Shenhua Planning Road 4 in the

North District. It aims to implement three types of permeable surface runoff pollution control technologies, namely permeable concrete, permeable bricks, and grass-planting bricks. These technologies are designed to reduce rainfall runoff and control pollution on permeable paved surfaces, as illustrated in Fig. 8.6.

(ii) Undercover Green Space Project

Situated on the northern side of the river center island at the bay of Wenyu River, this project focuses on constructing recessed green spaces, infiltration wells, and other infiltration facilities. These facilities effectively mitigate runoff pollution and control in green areas, as depicted in Fig. 8.7.

(iii) Bioretention Tank Project

This project is situated in the parking lot on the southern side of Riverside Road between Lutong West Road and Shenhua Planning Road 4, in the North District. The project involves constructing bioretention tanks in the green space with relatively low gradients in the eastern section of the parking lot. These tanks aim to reduce runoff and control pollution by facilitating the retention and purification of water, before discharging it into the stormwater pipeline on the southern side, as illustrated in Fig. 8.8.

Fig. 8.6 Different types of permeable surfaces

Fig. 8.7 Schematic of low elevation greenbelt

Fig. 8.8 Schematic of biologic detention pond

(iv) Vegetation Shallow Ditch (Ecological Ditch) Project

The project is situated within the green belt on both sides of the main road in the future science and technology city, in the northern part of the eastern section of the first road, as displayed in Fig. 8.9. The implementation involves constructing concave vegetation shallow ditches, through which road rainwater is retained and purified before being stored in seepage facilities or flowing into the green space. In the event of extreme rainfall, the excess water overflows to the municipal pipeline.

(v) Wetland Test Project for Stormwater Purification

Located north of the Wenyu River, between Lutong West Road Bridge to the east and Shenhua Fourth Road Bridge to the west, this project aims to purify rainwater from select rainwater pipes under the Lutong West Road Bridge. The wetland utilizes the area's concave topography, and diverting rainwater through grass ditches into nearby depressions that are transformed into multi-stage wetlands to purify the water quality. Figure 8.10 depicts the project's layout.

Fig. 8.9 Schematic of a vegetated shallow trench

Fig. 8.10 Schematic of stormwater purification wetland

8.2.2 Assessment Results and Analysis

A. Assessment Results

The field research conducted on the five projects mentioned earlier involved the assessment of seven rainfall runoff pollution control technologies. Using the 24 assessment indicators and established methods outlined in this chapter, comprehensive data and information were collected during the research process. Key technology assessments were performed, and the specific results of the field research are presented in Table 8.20.

The weight of each assessment indicator, as outlined in Table 8.15, along with the survey scores for each technical indicator, presented in Table 8.20, were used to calculate the comprehensive assessment index. This method was employed to carry out the technical assessment, and the results are displayed in Table 21 and Fig. 8.11.

B. Analysis of Indicators

From Table 8.21 and Fig. 8.11, it is evident that the vegetated shallow ditch exhibits the best technical performance, grass brick paving demonstrates the best economic performance, permeable brick paving shows the best management performance, and the bioretention tank achieves the best benefit performance. Equation (8.9) reveals that the superiority of individual technical performance is determined by the weights assigned to the technical indicators and their corresponding scores.

From Tables 8.15 and 8.20, the vegetated swale technique ranks first in terms of maturity and operational stability, second in footprint, ease of construction, and service life, and fourth in maintenance frequency. Although the vegetated shallow ditch has a lower score for maintenance frequency, this indicator carries less weight. However, the other five indicators for this technology have both higher weights and higher scores, making the vegetated shallow ditch the best in terms of technical performance.

Table 8.15 shows that the three-level indicators for economic performance—unit construction cost, operation and maintenance cost, and personnel staffing costs —, have weights of 0.053, 0.034, and 0.027, respectively. Additionally, Table 8.20 indicates that the vegetated brick paving scores 7 in all three categories: unit construction cost, operation and maintenance cost, and personnel staffing costs. Compared to other technologies, the vegetated brick paving ranks second in unit construction cost and

Table 8.20 Summary of case rainfall runoff pollution control technology assessment scores

Assessment indicator/technical title	Pervious concrete paving	Permeable tile paving	Grass Tile Paving	Recessed green space	Bioretention tank	Rainwater wetland	Shallow vegetated gully
Technology maturity	7	7	7	7	6	7	9
operational stability	6	6	6	5	5	7	7
footprint	7	7	7	5	5	4	6
Difficulty of construction	4	3	5	6	5	4	5
useful life	7	9	8	8	7	9	8
Maintenance frequency	6	7	7	3	7	8	4
Unit construction cost	5	5	7	8	5	7	7
Operation and Maintenance Costs	6	7	7	3	7	7	5
Staff costs	5	7	7	4	5	5	6
Number of personnel	7	8	7	7	6	7	6
Professional Requirements	5	7	7	6	4	5	5
Difficulty of operation and management	6	7	5	6	5	6	5
TN removal ratio	5	5	5	3	5	5	5
TP removal ratio	7	7	7	5	5	7	7
COD removal ratio	5	5	5	5	7	9	5
SS removal ratio	9	9	9	9	9	9	9
Total runoff reduction rate	8	8	7	9	5	8	7
Peak runoff reduction rate	5	5	5	9	7	5	3

(continued)

Table 8.20 (continued)

Assessment indicator/ technical title	Pervious concrete paving	Permeable tile paving	Grass Tile Paving	Recessed green space	Bioretention tank	Rainwater wetland	Shallow vegetated gully
Rainwater Resource Utilization	5	5	5	7	6	7	6
Cost of water savings from reuse of effluent	6	6	6	5	7	6	6
Save money on municipal wastewater treatment	6	6	6	5	7	7	5
Reduce pressure on municipal wastewater treatment	6	6	6	5	7	7	5
Mitigating flooding in towns	6	6	6	7	7	7	6
Groundwater recharge	5	5	5	7	9	7	5

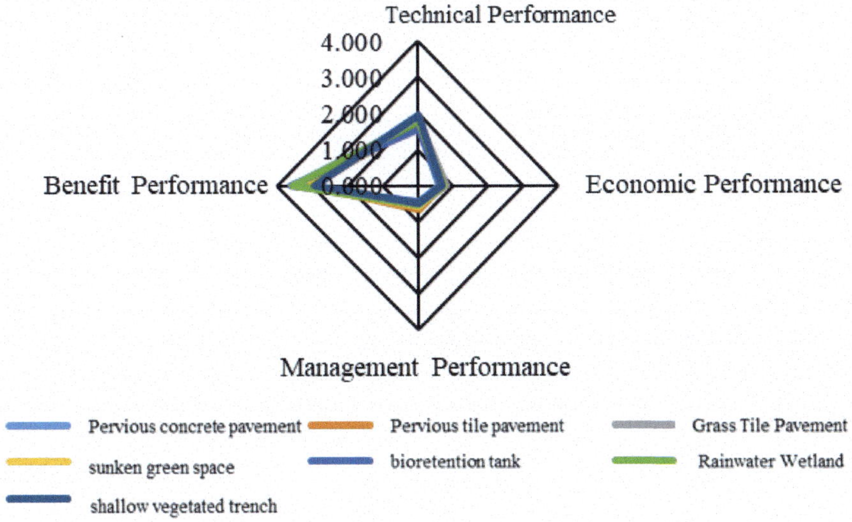

Fig. 8.11 Case rainfall runoff pollution control technology assessment index radar map

Table 8.20 Case rainfall runoff pollution control technology assessment index

	Pervious concrete pavement	Permeable tile pavement	Grass Tile pavement	Sunken green	Bioretention tank	Rainwater wetland	Shallow vegetated gully
technical performance	1.748	1.778	1.843	1.675	1.582	1.779	1.974
economic performance	0.604	0.692	0.798	0.634	0.638	0.744	0.703
Managing Performance	0.549	0.665	0.550	0.573	0.457	0.549	0.481
Benefit Performance	3.077	3.077	3.033	3.359	3.585	3.562	2.916
Total Assessment Index (TAI)	5.978	6.212	6.224	6.241	6.262	6.634	6.074

first in both operation and maintenance cost and personnel staffing costs. Although undercrofts having the same unit construction cost score as grass block paving, they score lower in operation and maintenance cost as well as personnel staffing costs. As a result, grass brick paving exhibit the best economic performance.

According to Table 8.20, bioretention tanks rank first in TN removal rate, water saving cost for effluent reuse, saving municipal wastewater treatment cost, reducing municipal wastewater treatment pressure, alleviating urban flooding, and recharging groundwater. They secure second place in TP removal rate, COD removal rate, peak runoff reduction rate, and stormwater resource utilization. Table 8.15 highlights the greater weight assigned to water saving cost of effluent reuse, saving municipal wastewater treatment cost, reducing municipal wastewater treatment pressure, alleviating urban flooding, and recharging groundwater. Consequently, the bioretention tank exhibits the best benefit performance.

C. Comprehensive Performance Analysis

According to Table 8.21, the prioritized order for the comprehensive assessment of these seven-rainfall runoff pollution control technologies is as follows: rainwater wetland, bioretention tanks, depressed green space, grass tile paving, permeable tile paving, vegetated shallow ditch, permeable concrete paving.

Based on Table 8.18, permeable concrete paving falls into the third category "general," while permeable bricks, grass-block paving, depressed green space, bioretention tanks, rainwater wetland, and vegetated shallow ditch belong to the fourth category "better." These findings align more closely with actual application; although there are slight distinctions among the technologies, with stormwater wetland are the most suitable, while permeable concrete paving is the least suitable. Rainwater wetland tops the rankings primarily due to its superior technical, economic, management, and benefit performance, with scores of 1.779, 0.744, 0.549, and 3.562, respectively, ranking third, second, third, and fourth in each category. In contrast, permeable concrete paving achieves the lowest score due to its inferior technical performance, economic, management, and benefit performance, with scores of 1.748, 0.604, 0.549, and 3.077, placing it fifth, seventh, fourth, and fourth, respectively.

From the above analysis, it is evident that no single technology fully satisfies all requirements in terms of technology, economy, management, and benefit. In engineering applications, a combination of technologies should be selected according to the specific circumstances in order to achieve optimal comprehensive benefits.

References

1. Liang, J. (2010). *Research on Evaluation Methods of Water Pollution Prevention and Control Technology in the Pharmaceutical Industry*. Hebei University of Science and Technology. (in Chinese)
2. Wang, Z. (2011). *Classification Study of Automotive Service Spare Parts Based on AHP and Fuzzy Comprehensive Evaluation Method*. Jilin University. (in Chinese)
3. Qian, C. (2014). *Integration of Common Technologies for Pollution Control in Urban Rivers in Southern China and Evaluation of Engineering Performance*. East China Normal University. (in Chinese)
4. Guo, J., Zhang, Z., & Sun, Q. (2008). Research and application of analytic hierarchy process. *Journal of Safety Science, China, 18*(5), 148–153. (in Chinese).
5. Zhu, J. (2005). *Research and Application of Some Problems in Analytic Hierarchy Process*. Northeastern University. (in Chinese)
6. Wu, Y., Hua, Z., & Zha, Y. (2003). Determination of weights in group decision making and aggregation of judgment matrices in AHP. *Operations Research and Management, 04*, 16–21. (in Chinese).
7. Xu, Z., Guo, H., Yu, Y., et al. (2007). Decision study on construction mode of eco-industrial park based on multi-criteria group decision-making model. *Environmental Science Research, 02*, 123–129. (in Chinese).

Chapter 9
Performance Evaluation of Engineering Application of Integrated Technology for Pollution Control of Rainfall Runoff in Towns and Cities

Chapter 8 provides a comprehensive assessment of key technologies for controlling rainfall runoff pollution in urban areas. In urban rainfall runoff pollution control projects, the effectiveness of individual key technologies is often limited. It usually requires the combination of multiple individual technologies to form an integrated technology for engineering application. During the process of applying integrated technology in engineering projects, it is important to not only focus on technical performance but also consider the economic benefits, ecological and environmental benefits, and long-term maintenance of the integrated technology. In order to effectively evaluate the engineering application effects of integrated technology, this chapter constructs an evaluation method for the performance assessment of integrated technology in controlling urban rainfall runoff pollution. It evaluates the performance of three integrated technology engineering applications: the runoff pollution control project in Wuxi, the overflow pollution control project in Hefei, and the surface runoff pollution purification project in Chongqing. The application effects of integrated technology in controlling runoff pollution are analyzed.

9.1 Performance Evaluation Indicators for Integrated Technology Engineering Applications

9.1.1 Principles for Constructing Evaluation Indicators

The specific principles for constructing the indicator system can be found in Sect. 8.1.1 of Chap. 8. We will not repeat them here.

D. Sun and F. Qi, *Urban Rainfall Runoff Pollution Control Technology and Application*,
https://doi.org/10.1007/978-981-96-4174-1_9

9.1.2 Construction of Evaluation Indicators

The process of constructing evaluation indicators involves the exploration, screening, analysis, and construction of indicators.

Indicator exploration refers to the analysis of evaluation schemes to identify representative measures. This process is often carried out using a comprehensive decomposition method. The overall objective is broken down into relatively specific objectives, establishing a connection with quantitative indicators. These decomposed objectives are typically referred to as criteria or levels. These criteria reflect the structural characteristics of the evaluated object in terms of a particular aspect and the requirements of the overall objective. The criteria are further decomposed one or more times until the resulting indicators can be described in a relatively simple qualitative or quantitative manner. This forms an unfiltered original indicator database.

Indicator screening refers to the rational, effective, and scientific selection of indicators from the original indicator database based on criteria such as comprehensiveness, independence, representativeness, distinctiveness, and operability.

In this study, when selecting performance evaluation indicators for a complete technical engineering application, three aspects of evaluation indicators were formed based on factors such as the purpose, input, output, process, and management of the engineering project. These aspects include economic benefits, ecological environmental benefits, and long-term maintenance of the engineering project. A performance evaluation indicator framework for complete technical engineering was constructed from three levels: the goal level, the criterion level, and the indicator level.

The first level is the target layer, which represents the overall evaluation object of "performance evaluation of integrated technological engineering application" The second level is the standard level, which consists of three subsystems derived from the characteristics of various aspects of complete technical engineering application. These subsystems are the economic benefits subsystem, the ecological environmental benefits subsystem, and the long-term maintenance subsystem of the engineering application. Each subsystem provides a reference direction for the selection of specific indicators in the third level based on its own characteristics. The third level is the specific indicator level, which includes several specific indicators that directly reflect the overall situation of each subsystem. The specific framework structure of the performance evaluation indicator system is shown in Fig. 9.1.

9.1.3 Evaluation Indicators

By analyzing the development process of urban rainfall runoff pollution control technologies in China, as well as the key technologies and their combinations in integrated technical solutions and engineering applications, relevant research and engineering experiences have been summarized. Literature review has been conducted to identify the indicators commonly used in the field of urban rainfall runoff pollution control

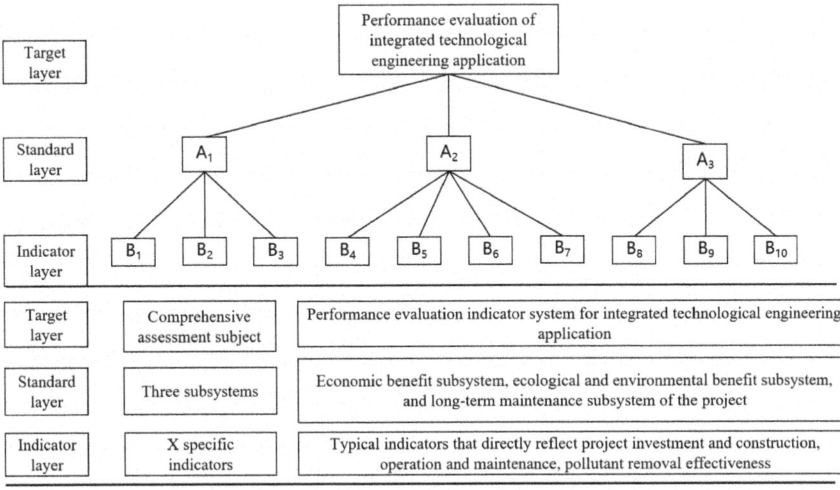

Fig. 9.1 Framework structure of performance evaluation indicators for urban rainfall runoff pollution control integrated technical engineering

technology. Expert consultations in this field have also been carried out. Based on the characteristics of urban rainfall runoff pollution, quantitative, feasible, and applicable performance evaluation indicators have been selected from common environmental assessment indicators. Please refer to Table 9.1 for the specific indicators selected for the performance evaluation indicator system of integrated technical engineering.

In the performance evaluation indicators reflected in Table 9.1, a smaller value is considered more favorable for cost-type indicators, while a larger value is considered more favorable for benefit-type indicators. The specific indicators and their meanings in the standard layer and corresponding indicator layer are as follows.

9.1.3.1 Economic Benefit Index

The economic benefits indicator should reflect the input–output ratio of the integrated technical engineering project and is an important indicator for determining whether the project is suitable for large-scale promotion. As the evaluation objects chosen in this chapter are public welfare engineering projects, universally applicable post-evaluation indicators for engineering economics are adopted to reflect the indicators of engineering economic costs and actual benefits.

The economic benefits of the engineering application in this chapter mainly refer to the "investment construction cost" and "operation and maintenance cost" of the constructed engineering. The indicator of "municipal sewage cost savings" is calculated based on the annual total pollutant control mass of each engineering application, using the conventional cost of pollutant removal in wastewater treatment plants in China [1, 2].

Table 9.1 Evaluation indicators for performance assessment of integrated technical engineering for urban rainfall runoff pollution control

Target layer	Standard layer	Indicator layer	Indicator nature
Complete set of technology engineering application performance	Economic benefit A1	Investment and construction cost B1	Cost type
		Operating and maintenance cost B2	Cost type
		Municipal sewage cost savings B3	Benefit type
	Ecological and environmental benefits A2	The COD purification rate B4	Benefit type
		SS Purification rate B5	Benefit type
		TP Purification rate B6	Benefit type
		NH_4^+-N Purification rate B7	Benefit type
	Engineering long-term maintenance of A3	Water quality stability B8	Cost type
		Technical specification and standard construction B9	Benefit type
		Engineering maintenance and performance appraisal B10	Benefit type

9.1.3.2 Ecological and Environmental Benefits

Urban runoff pollution control primarily aims to reduce the volume of runoff, achieve peak flow control, remove pollutants from runoff, and reduce the pollution load entering receiving water bodies. However, due to significant changes in urban construction and surrounding environments, the ecological environmental goals of urban runoff pollution control have also evolved. Depending on different types of water pollution, regional water environment requirements, and the difficulty of water environment remediation in different areas, the emphasis on ecological environmental benefits varies in different urbanized regions.

Urban rainfall runoff pollution mainly stems from the pollution caused by the surface erosion effects of rainfall. Compared to point source pollution, urban rainfall runoff pollution is characterized by its dispersed sources, multiple origins, and uncertainties. Therefore, the "runoff control rate" is a critical and indispensable indicator.

In addition, according to the "Surface Water Environmental Quality Standards (GB3838)", the purification efficiency of engineering projects for SS, COD, NH_4^+-N, TN, and TP is used as an indicator of ecological environmental benefits.

9.1.3.3 Engineering Long-Term Maintenance Index

Long-term maintenance of engineering projects should reflect aspects related to the sustainable development of comprehensive technical applications. For the comprehensive technical application of urban rainfall runoff pollution control, the indicator of "effluent water quality stability" can directly reflect the stability of the engineering's pollutant reduction efficiency and also reflect the overall reduction effect on pollutant load. It is an important indicator for evaluating the comprehensive technical application of engineering projects. In this chapter, the indicator of "effluent water quality stability" can be represented by the variance value obtained from multiple water quality monitoring during the operation of the project.

From a management perspective, the stable operation of the engineering project also relies on the formulation of standardized technical usage and engineering management regulations. Therefore, "technical specification and standard construction" and "engineering maintenance and performance assessment" are two evaluation indicators. However, these two indicators are difficult to evaluate quantitatively, and detailed information on the specifications and regulations is generally used as internal documents within enterprises and may not be easily accessible. In this chapter, only the submission status of the specifications and regulations at the completion of the project is evaluated as either "yes" or "no" with a score of either "1" or "2". Further quantitative scoring is not performed for detailed content.

9.2 Evaluation Method

Traditional analytic hierarchy process (AHP) and entropy weight method have certain subjectivity in their application process. In this chapter, the TOPSIS (Technique for Order Preference by Similarity to Ideal Solution) method is coupled with the entropy weight method to eliminate the subjectivity of evaluation results and objectively assess the performance of comprehensive technical applications for urban rainfall runoff pollution control.

9.2.1 TOPSIS Method

The TOPSIS method, also known as Technique for Order Preference by Similarity to an Ideal Solution, is an effective multi-objective and multi-criteria decision analysis method. The basic principle of this method is to rank a limited number of evaluation objects based on their distance from the ideal solution (positive ideal solution) and negative ideal solution.

Since there is generally no real meaning for the best or worst solution among a limited number of evaluation objects, the evaluation object that achieves the optimal value for each index in the sample is usually considered as the "best solution", while the object with the worst value for each index is considered as the "worst solution". This approach significantly improves the scientific, accuracy, and operability of performance evaluation and it has been widely applied in performance evaluations across multiple fields.

The modeling steps of the TOPSIS model mainly include the following six steps:

9.2.1.1 Build a Decision Matrix

Suppose there are m evaluation objects, and their set is denoted as D = (D1, D2, ..., Dm). There are n criteria, and their set is denoted as C = (C1, C2, ..., Cn). The target decision matrix is shown in Eq. (9.1).

$$X = (x_{ij})_{m\times n} = x\begin{bmatrix} x_{11} & \cdots & x_{1n} \\ x_{21} & \cdots & x_{2n} \\ \cdots & \cdots & \cdots \\ x_{m1} & x_{m2} & x_{mn} \end{bmatrix} \tag{9.1}$$

9.2.1.2 Dimensionless Standardization

To eliminate the influence of different dimensions in the criteria on the decision-making process, the elements xij are standardized, including both positive and negative criteria. The normalization method is described in Eq. (9.2), and the results are shown in the dimensionless matrix (Eq. 9.3).

$$\begin{cases} Positive criteria: y_{ij} = \dfrac{x_{ij} - \min(x_{ij})}{\max(x_{ij}) - \min(x_{ij})} \\ Negative criteria: y_{ij} = \dfrac{\max(x_{ij}) - x_{ij}}{\max(x_{ij}) - \min(x_{ij})} \end{cases} \tag{9.2}$$

where, x_{ij}—evaluates the value of the object i on the *j*th index;
y_{ij}—evaluates the value of the *j*th index.

$$Y = (y_{ij})_{m\times n} = \begin{bmatrix} y_{11} & \cdots & y_{1n} \\ y_{21} & \cdots & y_{2n} \\ \cdots & \cdots & \cdots \\ y_{m1} & \cdots & y_{mn} \end{bmatrix} \tag{9.3}$$

9.2.1.3 Build a Weighted Decision Matrix

By multiplying the dimensionless matrix Y (Eq. 9.3) with the weights Wj of each criterion, the weighted decision matrix Z = (zij)m × n is obtained, as shown in Eq. (9.4).

$$Z_{ij} = w_j y_{ij} i = 1, 2, \ldots, m; j = 1, 2, \ldots, n \tag{9.4}$$

Where, Z_{ij}—evaluates the value of object i on the j index;

W_j—evaluates the weight vector.

In this chapter, the entropy weight method is used to determine the weights of each criterion. The detailed method for determining the weights is described in Sect. 9.2.2.

9.2.1.4 Determine the Positive Ideal Solution Y^+ Versus the Negative Ideal Solution Y^-

The positive ideal solution Y^+ and the negative ideal solution Y^- for the criteria are determined using Eqs. (9.5) and (9.6) respectively. The positive ideal solution represents the set of optimal values for each criterion, while the negative ideal solution represents the set of worst values for each criterion.

For positive indicators:

$$Y^+ = \max(z_{ij}) \tag{9.5}$$

For reverse indicators:

$$Y^- = \min(z_{ij}) \tag{9.6}$$

$$1 \leq j \leq n; i = 1, 2, \ldots, m$$

9.2.1.5 Computing Distances from Positive and Negative Ideal Solutions D_i^+ and D_i^-

The distance between the *i*th evaluation object and the positive/negative ideal solutions is calculated using Eqs. (9.7) and (9.8) respectively.

$$D_i^+ = \sqrt{\sum\nolimits_{j=1}^{n} (z_j^+ - z_{ij})^2} i = 1, 2, \ldots, m; j = 1, 2, \ldots, n \tag{9.7}$$

$$D_i^- = \sqrt{\sum\nolimits_{j=1}^{n} (z_j^- - z_{ij})^2} i = 1, 2, \ldots, m; j = 1, 2, \ldots, n \tag{9.8}$$

where, z_j^+ and z_j^- refer to the maximum and minimum values in each index, respectively.

9.2.1.6 Calculate the Closeness of the Ideal Solution *C**

The proximity of the *i*th evaluation object to the ideal solution is represented by C* and can be calculated using Eq. (9.9).

$$C^* = \frac{D_i^-}{D_i^- + D_i^+} (i = 1, 2, \ldots, m) \tag{9.9}$$

The evaluation objects are sorted in ascending order based on the value of C*. Since the formula for C* includes the distance between the evaluation object's criterion values and the positive/negative ideal solutions, it represents the proportion of its value to the total distance from the negative ideal solution. Therefore, a higher value of C* indicates a greater distance between the evaluation object and the negative ideal solution, which means it is closer to the positive ideal solution and ranks higher in terms of that criterion [3].

9.2.2 *Optimization of TOPSIS Method with Entropy Weighting*

9.2.2.1 Method of Index Empowerment

The weighting of indicators can reflect the relative importance of evaluation criteria in the entire evaluation model and their level of impact on the evaluation objectives. Therefore, the rationality of indicator weights is closely related to the reliability of the evaluation results. From the established indicator system mentioned above, it can be seen that there are many factors influencing the performance of integrated engineering technologies for urban rainfall-runoff pollution control. It is necessary to use a scientifically reasonable method to determine the weights of each indicator in order to scientifically evaluate the performance of integrated engineering technologies for urban rainfall-runoff pollution control.

There are generally two types of weighting methods for evaluation models: subjective and objective. Common subjective weighting methods include expert scoring and AHP. Subjective weighting methods are based on personal subjective consciousness, that is that the weights of indicators are determined according to the evaluator's personal emphasis on each indicator, mainly relying on the evaluator's personal experience and knowledge. Objective weighting methods measure the weights of various evaluation indicators based on the amount of information that each indicator can reflect or the degree of correlation between indicators. Common objective weighting methods include entropy weighting and data envelopment analysis (DEA).

Due to the limitations of subjective weighting methods and their vulnerability to various subjective factors, this article adopts the entropy weighting method, which is an objective weighting method [4].

9.2.2.2 Entropy Weighting Method

The entropy weighting method is an objective weighting method that determines the weights of indicators based on the amount of information that each indicator can reflect. Entropy originated from physics and was initially used to characterize the disorderliness of thermodynamic systems. Nowadays, the entropy weighting method has been widely applied in various fields such as social economics and engineering technology.

The entropy weighting method primarily assigns weights to indicators based on the principle of entropy. The smaller the entropy value corresponding to an indicator, the greater amount of information it provides. Thus, the indicator is assigned a higher weight, indicating its greater importance in the evaluation. Therefore, the entropy weighting method is suitable for evaluation systems where there are differences in indicator values. In this chapter, the indicators related to the application performance of integrated engineering technologies for urban stormwater runoff pollution control meet this requirement.

Entropy assignment generally includes the following five steps.

i. Build the raw data matrix

Based on Eq. (9.1), construct the original data matrix.

ii. Non-dimensional normalization

Due to the differences in dimension and magnitude of each indicator in the original data matrix x_{ij} for the performance evaluation of integrated technology for urban rainfall-runoff pollution control projects, it is necessary to normalize the original data matrix in order to eliminate the influence of different dimensions and magnitudes on the evaluation results. This is done to construct a standardized decision matrix.

The specific non-dimensional normalization method refers to Eq. (9.2).

iii. Shifted Standardized Data

In order to ensure that the obtained data remains meaningful throughout the calculation process and does not alter the computational results, the non-dimensional normalized data is shifted by one unit, as calculated according to Eq. (9.10).

$$y'_{ij} = y_{ij}+1 \tag{9.10}$$

iv. Calculating Indicator Weights

f_{ij} represents the proportion of the *j*th indicator value of the *i*th object in the total sum of that indicator, calculated according to Eq. (9.11).

$$f_{ij} = \frac{y_{ij}}{\sum_{i=1}^{m} y_{ij}} \tag{9.11}$$

v. Calculating Indicator Entropy

The entropy value of the *j*th indicator is represented by Gj, calculated according to Eq. (9.12).

$$G_j = -\frac{1}{\ln m}\sum_{i=1}^{m} f_{ij}\ln f_{ij} \tag{9.12}$$

vi. Calculating Coefficient of Variation

g_j refers the coefficient of variation for the indicator values Gj. For evaluation indicators, the more distinct the differences in indicator values are, the greater the amount of information they can reflect. Consequently, the weights of the indicators will be higher. This is calculated according to Eqs. (9.9–9.13).

$$g_j = 1 - G_j \tag{9.13}$$

vii. Calculating Indicator Entropy Weight

w_j refers to the entropy weight of the *j*th indicator, which is part of the weight vector mentioned earlier. The entropy weight of an indicator is higher if its corresponding entropy value in the weight vector is smaller. This indicates that the indicator has a greater impact on the overall evaluation result. This calculation is performed according to Eq. (9.14).

$$w_j = \frac{g_j}{\sum_{j=1}^{n} g_j} \tag{9.14}$$

9.2.2.3 TOPSIS Method Improved by Entropy Weight

By combining the TOPSIS method and the entropy weight method, an evaluation model for the application performance of integrated technical engineering in urban rainfall runoff pollution control can be established. The specific evaluation steps are as follows:

(1) Normalizing the original indicator values of the integrated technical engineering application.

(2) Using the entropy weight method to calculate the entropy values of each indicator to determine the weight of each indicator.
(3) Determining the ideal solution and negative ideal solution for each weighted indicator value based on the TOPSIS method.
(4) Calculating the distances between each indicator value and the ideal and negative ideal solutions to obtain the proximity relationship to the negative ideal solution.
(5) Sorting the proximity values from small to large, where a higher proximity value indicates a closer proximity to the ideal solution and a higher comprehensive evaluation score.

9.3 Performance Evaluation of Integrated Technical Engineering Application Case

9.3.1 Basic Information of the Case

9.3.1.1 Wuxi Urban Runoff Pollution Control Project

i. Project Overview

The Wuxi Urban Runoff Pollution Control Project is located in the Taihu New City area of Wuxi City. It utilizes the topographical features and incorporates various control technologies, including efficient intercepting rainwater screens, high-efficiency adsorption purification belts, subsurface flow infiltration reaction barrier walls, and on-site purification and storage for reused parking spaces. These technologies form a complete set of runoff pollution control measures and have been applied in the project to achieve efficient purification of rainwater infiltration processes. For detailed information about the Wuxi Urban Runoff Pollution Control Project and its integrated technologies, please refer to Sect. 7.5.

ii. Operational Effectiveness of the Project

The service area of the runoff pollution control project in Wuxi City is 1160 m^2. The reduction efficiency for SS, COD, TP, and NH_3-N are 90.4, 65.0, 85.0, and 74.0% respectively. The annual reduction loads are 12,567, 2442, 25, and 59 kg, correspondingly.

i) *Application performance of high-efficiency pollutant interception rain screen and high-efficiency adsorption purification strip technology in engineering projects*

During rainfall runoff, road water first enters the rain collection pipe through a fast pollutant interception rain screen. The collected water is then directed to the purification strip, where it undergoes infiltration and purification processes before reaching the bottom layer of the strip. After a brief buffering at the gravel layer, the water is discharged from the facility. The discharged water undergoes separate treatment based on its quality. Water quality that meets the discharge standards is directly released into nearby water bodies, while water with higher levels of contamination needs to be further treated at wastewater treatment plants. The high-efficiency adsorption purification strip achieves average removal efficiency of 89.9% for SS, 58.5% for COD, 80.9% for TP, 54.8% for TN, and 7.6% for NH_4^+-N.

ii) *The application of subsurface flow infiltration and reaction barrier wall technology in the engineering project has been effective*

After rainfall, the runoff flows along the road slope into the surrounding green areas. Due to the existence of infiltration measures, a significant amount of rainwater forms subsurface flow. The subsurface flow carries pollutants such as nitrogen, phosphorus, and trace organic matter into the groundwater. To enter water bodies, these pollutants must first penetrate the barrier wall arranged at the soil–water interface. The facility efficiently adsorbs pollutants carried by the subsurface flow, fixing them within the column. The subsurface flow infiltration and reaction barrier wall achieve average removal efficiency of 92% for SS, 70% for COD, 90.9% for TP, and 57.7% for NH_4^+-N in runoff.

iii) *The application of in-situ purification and storage for water reuse in parking engineering has been effective*

During rainfall in the parking, a portion of the rainwater infiltrates in-situ, passes through filtering media beneath permeable pavement, and enters into the drainage pipe. Another portion generates localized runoff, which enters in-situ the drainage pipe through the filtering effect of permeable pipes between parking spaces. The filtered rainwater is stored in an underground reservoir and can be used for purposes such as irrigation of green areas and car washing. The in-situ purification and storage for water reuse in the parking achieved average removal in-situ of 89.3% for SS, 66.4% for COD, 83.3% for TP, and 86.8% for NH_4^+-N in runoff.

9.3.1.2 The Overflow Pollution Control Project in Hefei City

i. Project Overview

The Hefei City Overflow Pollutant Control Project adopts a combined sewer system overflow control technology, which mainly includes numerical simulation technology for overflow pollution in the combined sewer system, design and operation technology for detention tanks aimed at reducing overflow volume, and optimization operation technology for the combined sewer system in the old city area. The project is located in the old city area of Hefei City, serving an area within the Huancheng Road, with a total area of about 5.2 km^2. The design criteria for detention tanks are based on a rainfall of 8 mm and a detention volume of 6,500 m^3. The detention tank is located in the southeast corner of the Xiaoyaojin Park, with a fully underground structure occupying an area of approximately 1,000 m^2. It consists of a new diversion well, inlet pipe, detention tank, outlet pipe, and discharge pump.

A diversion well is installed at the DN2400 combined sewer main near the Xiaoyaojin existing drainage pump station for the Hefei City Overflow Pollutant Control Project. A bar screen is installed inside the diversion well. Initially, the runoff passes through the bar screen and then enters the detention tank through a new DN600 inlet pipe. A discharge pump is installed inside the detention tank to drain the stored initial rainwater into the sewage main pipe of Huancheng East Road during dry weather, and ultimately discharged after treatment at the Wangxiaoying wastewater treatment plant. The plan dimensions of the detention tank are 37.5 m × 28.8 m, with an effective depth of 7.5 m and a burial depth of approximately 12.05–13.8 m. Gravity inflow and pump pressure boosting are used for discharge. The discharge time of the detention tank is 8–16 h. The detention tank is equipped with 5 sets of gate-flushing devices, 2 discharge pumps, and water metering facilities, as well as supporting electrical, automatic control, deodorization, and ventilation equipment.

ii. Project Application Effect

The working process of the detention tank in the Hefei City Overflow Pollutant Control Project mainly consists of three stages: diversion, storage, and drainage. During the early stages of rainfall when the runoff volume is low, the gates of the detention tank are closed, and the sewage diversion pump intercepts sewage and initial rainwater. When the rainfall intensity exceeds the diversion capacity and the water level inside the pump station approaches the flood safety level, the inlet gate of the detention tank is opened, and the initial rainwater enters the rainwater detention tank. When the detention tank is full, the inlet gate is closed, and self-discharge or forced discharge of the pump station is determined based on the water level of Nanfei River and the water level of the previous tank. After the rainfall ends, when the water level in the front tank of Xiaoyaojin pump station reaches the low water level, the discharge pump is activated to lift and discharge the initial rainwater in the tank to the sewage diversion trunk.

After the completion of the Hefei City Overflow Pollutant Control Project, during the period from June to December 2016 for a total of 214 days (including 83 rainy days), the detention tank operated 28 times, reducing the number of overflow events by 19 times. The reduction rate of overflow volume from the detention tank was 38.3%, and the reduction rate of pollutant load was over 32%. During the operation period, the detention tank reduced the pollutant load of SS, COD, NH_4^+-N, and TN by 34.1, 36.1, 2.0, and 3.0, respectively, corresponding to reduction efficiency of 40, 32, 42, and 39%.

9.3.1.3 Chongqing Surface Runoff Pollution Purification Project

i. Project Overview

The Chongqing Surface Runoff Pollution Purification Project is located in Jiansheng Town, Chongqing. It adopts the technology of initial runoff automatic diversion well, multi-functional pond, and cascade artificial wetland. Jiansheng Town has high building density and a high pollution load from surface runoff, resulting in severe overflow pollution caused by rainfall runoff. The project covers an area of approximately 2500 m^2, with a single treatment capacity of 1000 m^3 for initial stormwater runoff. During the rainy season, the initial surface runoff is automatically diverted into the treatment facilities for storage, treatment, and discharge through the rainwater diversion well.

The process of the Chongqing Surface Runoff Pollution Purification Project is as follows: the initial runoff automatic diversion well separates and stores domestic wastewater, initial runoff, and later-stage runoff, achieving automatic diversion treatment for the three types of sewage. The multi-functional pond achieves both runoff storage and efficient retention and degradation of carbon, nitrogen, and phosphorus pollutants through the aquatic ecosystem inside the pond. The cascade artificial wetland utilizes steep slopes in mountainous river areas, combining sequenced batch wetlands, vertical flow wetlands, and horizontal subsurface flow wetlands to enhance the natural reoxygenation capacity and water regulation capacity of wetlands, improve denitrification efficiency, and form an efficient set of technologies for purifying surface runoff pollution. Among them, the sequenced batch wetland can operate in four different scenarios: "inflow-degradation-outflow-idle", with respective residence times of 0.5, 6, 0.5, and 2 h. Both the second-level vertical subsurface flow wetland and the third-level horizontal subsurface flow wetland in the cascade artificial wetland process operate in an intermittent inflow and outflow manner, corresponding to the operational scenarios of the first-level sequenced batch wetland. During the idle period of the sequenced batch wetland, transpiration is high, which may lead to excessive plant water loss and lodging. Therefore, it is advisable to choose long-growing, aesthetically pleasing, and economically valuable woody plants, such as willows and mangroves (Fig. 9.2).

ii. Project Application Effect

During the rainy season, the combined sewage first enters the initial runoff automatic diversion well, where the initial stormwater runoff is automatically diverted to the multi-functional pond for sediment removal. Then, it goes through the sequenced batch artificial wetland and vertical flow artificial wetland for the removal of organic matter, NH_4^+-N, and TP. Finally, it enters the subsurface flow artificial wetland to enhance the removal of nitrogen and phosphorus nutrients, achieving deep purification of the initial runoff by denitrification and phosphorus removal. The treated water is then safely discharged into the Funiu Creek.

The design of the project ensures that the effluent COD, NH_4^+-N, and TP meet the Class V water quality standards of the "*Surface Water Environmental Quality Standards*" (GB3838). During operation, the annual removal rates of COD, NH_4^+-N, TN, and TP are 82, 94, 65, and 78% respectively. The project has achieved a reduction of 19.4 t/a in COD emissions, 0.9 t/a in NH_4^+-N emissions, 3.28 t/a in TN emissions, and 0.62 t/a in TP emissions.

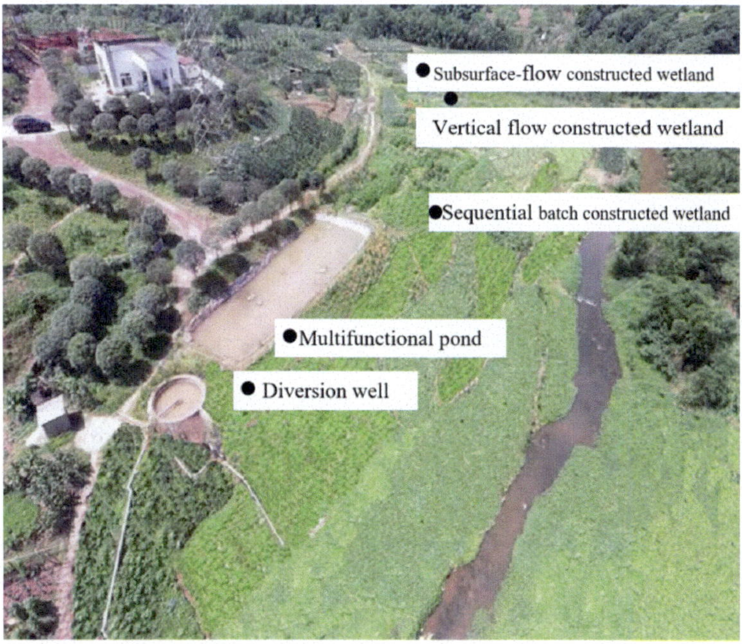

Fig. 9.2 Chongqing surface runoff pollution purification project

9.3.2 Case Evaluation Results and Analysis

Field research and engineering verification were conducted on the three sets of technology projects mentioned above, and their basic information is shown in Table 9.2. In the table, C1 represents the Runoff Pollution Control Project in Wuxi City, C2 represents the Overflow Pollution Control Project in Hefei City, and C3 represents the Chongqing Surface Runoff Pollution Purification Project. Indicators B9 and B10 represent whether there are relevant documents, with "2" indicating the presence of relevant documents and "1" indicating the absence of relevant documents.

9.3.2.1 Determine the Index Weight

Based on the collected basic data of the integrated technology projects, Eqs. (9.1)–(9.3) were applied to standardize the raw data of the indicators in Table 9.2. In order to obtain the weighted values of the indicators in Eq. (9.4), the entropy weighting method Eqs. (9.10)–(9.14) were used to calculate the weight ratios of evaluation indicators B1–B10. Then, the indicator values were multiplied by the weight values to obtain the final weighted values for each indicator. The results obtained from the entropy weighting method and the weighted indicator data are shown in Tables 9.3 and 9.4.

Table 9.2 Original index data of comprehensive evaluation of the engineering application of the integrated technology

Index layer	Wuxi City runoff pollution control project (C1)	Hefei city overflow pollution control project (C2)	Chongqing municipal surface runoff pollution purification project (C3)
Investment and construction cost is B1, ten thousand yuan	88.72	3681.93	196.85
Operation and maintenance cost B2, yuan/m^3	0.089	0.05	0.04
Municipal sewage savings costs B3, yuan/a	1938	45,267	23,553
The COD purification rate, B4,%	65	32	82
SS purifying rate B5,%	90	40	82
TP purifying rate B6,%	85	39	78
NH3-N purifying rate B7,%	74	42	94
Water quality stability B8	0.0083	0.2247	0.0235
Technical specification and standard construction B9	1	2	2
Engineering maintenance and performance appraisal B10	1	1	2

9.3.2.2 Calculating Proximity

Based on the calculated weights for each indicator, according to the evaluation steps of the TOPSIS method improved based on the entropy weighting method, the values for each indicator of the positive ideal solution and the negative ideal solution are determined, as shown in Table 9.5.

Using Eqs. (9.5)–(9.8), the distances between the indicator values of the three engineering applications (C1, C2, and C3) and the positive and negative ideal solutions are calculated respectively. The criterion level distance and total distance from each engineering application case to the positive and negative ideal solutions are shown in Table 9.6.

Based on the distances from the positive and negative ideal solutions in Table 9.6, the indicator proximity/similarity of each engineering application is calculated using Eq. (9.9), resulting in the performance level of each standard layer and overall performance of the engineering applications. Please refer to Table 9.7 for the indicator proximity values (Fig. 9.3).

Table 9.3 Standardized data and weight of comprehensive evaluation index of complete set of technology engineering application

Index layer	Standardized data			Entropy value	Entropy right
	C1	C2	C3		
Investment and construction costs B1	1.0000	0.0000	0.9699	0.9610	0.1023
Operation and maintenance costs B2	0.0000	0.7959	1.0000	0.9644	0.0934
Municipal sewage saves money on costs B3	0.0000	1.0000	0.4989	0.9656	0.0901
COD purifying rate B4	0.6600	0.0000	1.0000	0.9658	0.0897
SSpurifying rate B5	1.0000	0.0000	0.8400	0.9637	0.0952
TPpurifying rate B6	1.0000	0.0000	0.8478	0.9635	0.0956
NH3-Npurifying rate B7	0.6154	0.0000	1.0000	0.9659	0.0893
Stability of the effluent water quality B8	1.0000	0.0000	0.9298	0.9619	0.0998
Technical specification and standard construction B9	0.0000	1.0000	1.0000	0.9602	0.1042
Engineering maintenance and performance appraisal B10	0.0000	0.0000	1.0000	0.9464	0.1404

9.3.2.3 Evaluation Result Analysis

i. Comparative Analysis of the Indicator Layer.

According to Table 9.4, it can be observed that in terms of investment and construction costs, C1 performs the best, followed by C3, while C2 is relatively poor. In terms of operation and maintenance costs, C3 has lower costs due to lower maintenance costs for artificial wetlands. On the other hand, C1 incurs higher operation and maintenance costs due to the higher investment required for low-impact development technologies. In terms of municipal sewage cost savings, C2 demonstrates the best efficiency, followed by C3, as it can achieve deep purification of wastewater from sewage treatment plants while eliminating stormwater runoff. C2, on the other hand, is relatively weaker in this aspect.

By analyzing the pollutant removal efficiency indicators in Table 9.4, it can be seen that C1 performs well in removing SS, TP, and NH_4^+-N, while C3 has certain superiority in degrading organic pollutants. Additionally, the effluent quality of C1 and C3 is more stable, whereas C2, which relies mainly on detention facilities and lacks effective pollution removal through natural sedimentation alone, showing poorer and less stable pollutant removal performance. This indicates that C1, mainly based on LID, and C3, primarily reliant on artificial wetlands, have better operational effectiveness in reducing pollutants.

Table 9.4 Weighted data of the comprehensive evaluation index of the complete set of technology engineering application

Index layer	Weighted data			Maximum value	Minimum value
	C1	C2	C3		
Investment and construction costs B1	0.1023	0.0000	0.0992	0.1023	0.0000
Operation and maintenance costs B2	0.0000	0.0743	0.0934	0.0934	0.0000
Municipal sewage saves money on costs B3	0.0000	0.0901	0.0449	0.0901	0.0000
COD purifying rate B4	0.0592	0.0000	0.0897	0.0897	0.0000
SS purifying rate B5	0.0952	0.0000	0.0800	0.0952	0.0000
TP purifying rate B6	0.0956	0.0000	0.0810	0.0956	0.0000
NH3-N purifying rate B7	0.0549	0.0000	0.0893	0.0893	0.0000
Stability of the effluent water quality B8	0.0998	0.0000	0.0928	0.0998	0.0000
Technical specification and standard construction B9	0.0000	0.1042	0.1042	0.1042	0.0000
Engineering maintenance and performance appraisal B10	0.0000	0.0000	0.1404	0.1404	0.0000

Table 9.5 Positive and negative ideal solution of application evaluation system of complete technology engineering

Index layer	Positive ideal solution	Negative ideal solution
Investment and construction costs B1	0.1023	0.0000
Operation and maintenance costs B2	0.0934	0.0000
Municipal sewage saves money on costs B3	0.0901	0.0000
COD purifying rate B4	0.0897	0.0000
SS purifying rate B5	0.0952	0.0000
TP purifying rate B6	0.0956	0.0000
NH3-N purifying rate B7	0.0893	0.0000
Stability of the effluent water quality B8	0.0998	0.0000
Technical specification and standard construction B9	0.1042	0.0000
Engineering maintenance and performance appraisal B10	0.1404	0.0000

Table 9.6 The standard level distances and the total distances for the comprehensive evaluation of integrated technology engineering applications

Name	C1		C2		C3	
	D+	D−	D+	D−	D+	D−
Economic benefits A1	0.1296	0.1025	0.1039	0.1166	0.0458	0.1435
Ecological and environmental benefits A2	0.0458	0.1572	0.1849	0.0000	0.0200	0.1703
Long-term maintenance of the project A3	0.1749	0.1000	0.1723	0.1044	0.0000	0.1980
Total distance	0.2225	0.2124	0.2733	0.1565	0.0500	0.2978

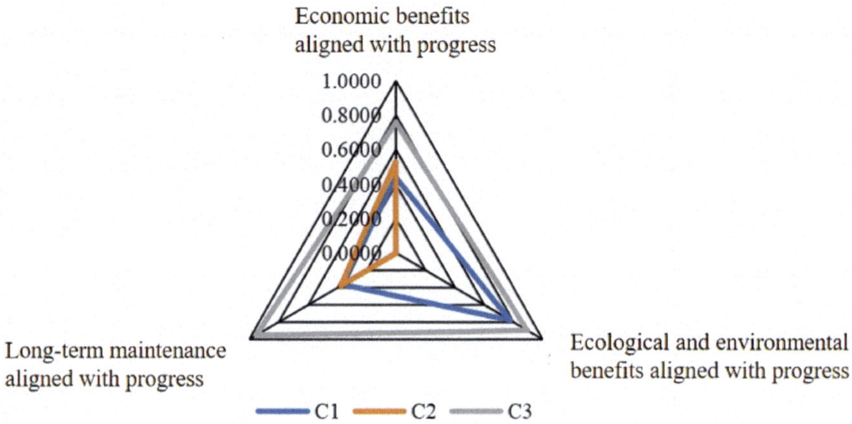

Fig. 9.3 Radar chart of the proximity/similarity of integrated technology engineering applications

Table 9.7 Evaluation results of closeness level of complete set of technical engineering

Name	C1	C2	C3
Economic benefits A1	0.4416	0.5288	0.7581
Ecological and environmental benefits A2	0.7744	0.0000	0.8949
Long-term maintenance of the project A3	0.3638	0.3773	0.9519
Overall closeness	0.4884	0.3641	0.8562

During the research and evaluation process, significant differences were also identified among these three typical engineering applications in terms of technical specifications, standard construction, engineering maintenance, and performance assessment. Among them, C3 has conducted a considerable amount of work in the aforementioned aspects, ensuring the stable operation of the engineering project. However,

C1 and C2 have undertaken relatively less work in terms of management. It is essential to carry out work in areas such as technical specifications, standard construction, engineering maintenance, and performance assessment. This is beneficial for the normal operation and maintenance of the engineering projects and represents necessary management measures that should be implemented in comprehensive technology engineering applications for urban stormwater runoff pollution control.

ii. Comparative Analysis of Criteria Levels

From Table 9.7, it can be seen that C3 performs best in terms of economic benefits, followed by C2 and then C1. In terms of ecological and environmental benefits, C3 performs best, followed by C1 and then C2. C3 also performs best in terms of long-term maintenance of the engineering project, followed by C2 and then C1.

In terms of economic benefits, C3 performs best mainly due to its highest weighted value for operation and maintenance costs, which is 0.0934, the highest among the three engineering applications. The weighted value for investment and construction costs is 0.0992, which is medium among the three. The weighted value for cost savings in municipal sewage treatment is 0.0449, also medium among the three. This indicates that C3 has clear advantages in terms of investment, operation and maintenance costs, and savings in sewage treatment costs. The relatively poor performance of C1 is mainly due to its lowest weighted value for operation and maintenance costs and cost savings in municipal sewage treatment. It shows that there is still space for improvement in terms of maintenance costs and cost savings in sewage treatment for C1.

In terms of ecological and environmental benefits, C3 performs best mainly because it has the highest weighted values for COD removal efficiency and NH_4^+-N removal efficiency among the three engineering applications, with values of 0.0897 and 0.0893, respectively. The weighted values for SS removal and TP removal efficiency are medium among the three. C2 performs relatively poorly because it relies solely on natural sedimentation, resulting in the lowest weighted values for COD, SS, TP, and NH_4^+-N removal efficiency, all of which are 0.0000.

In terms of long-term maintenance of the engineering project, C3 performs best mainly because it has the highest weighted values for technical specifications and standard construction, as well as engineering maintenance and performance assessment, among the three engineering applications, with values of 0.1042 and 0.1404, respectively. The weighted value for effluent water quality stability is medium among the three one. On the contrary, C1 performs relatively poorly mainly due to that it has the lowest weighted values for technical specifications and standard construction, as well as engineering maintenance and performance assessment, among the three one. This indicates that strengthening the technical standards, operation and maintenance, and performance evaluation of the engineering project is beneficial for its stable operation.

iii. Comprehensive Performance Analysis

According to Table 9.7, the proximity values for C1, C2, and C3 are 0.4884, 0.3641, and 0.8562, respectively. By ranking them, it can be concluded that C3 > C1 > C2.

The proximity values for economic benefits, ecological and environmental benefits, and long-term maintenance of C3 are 0.7581, 0.8949, and 0.9519, respectively, which are the highest among the three engineering applications. Therefore, C3 exhibits the highest level of comprehensive performance.

The proximity values for long-term maintenance of C1 and C2 are 0.3638 and 0.3773, respectively, which are very close. The overall performance of C1 versus C2 mainly depends on the proximity values for economic benefits and ecological and environmental benefits. Comparing the proximity values for economic benefits and ecological and environmental benefits of C1 and C2, it was find that C2 has a slightly higher proximity value for economic benefits (0.5288) compared to C1 (0.4416), but C1 has a significantly higher proximity value for ecological and environmental benefits (0.7744) compared to C2 (0.0000). Therefore, the overall performance of C1 is higher than that of C2.

In conclusion, among the three options, C3 exhibits the highest comprehensive performance, considering economic benefits, ecological and environmental benefits, and long-term maintenance.

iv. Analysis of Engineering Application Cases

C3 has the highest comprehensive proximity value of 0.8562, which is 1.7 times higher than the second-ranked C1. Therefore, it exhibits the best overall performance. C3 has several important advantages, including low investment and operation and maintenance costs, high pollutant removal efficiency, stable operation, and adherence to technical specifications and performance evaluation standards. The key advantages of C3 are as follows: ① the construction area of the project is small, resulting in low investment and construction costs. ② The operation and maintenance of the project mainly involve the removal of decaying plants in the wetland, which is a simple process [5]. This leads to lower operation and maintenance costs. ③ The project adopts a multi-stage wetland combination process, which enhances the natural reoxygenation capacity of the wetland and achieves a high reduction in pollutant load in the runoff [6]. As a result, the project achieves high removal efficiency for COD, NH_4^+-N, and TP.④ Artificial wetland technology has matured in China, and the project adheres to mature technical specifications and performance evaluation standards [7]. However, it should be noted that this project is designed for a smaller flow rate, thus resulting in lower cost savings for municipal sewage treatment.

C1 ranks second in terms of comprehensive closeness, with a comprehensive closeness of 0.4884. It has a lower overall performance level compared to typical artificial wetland projects but higher than C2. C1 suffers from issues such as high operation and maintenance costs, limited savings in municipal sewage costs, and a lack of technical standards and performance evaluation criteria. These problems are reflected in the following aspects: ① the project requires regular excavation to replace fill material [8], resulting in high operation and maintenance costs. ② The project handles a small volume of runoff, resulting in limited savings in municipal sewage costs. ③ The main LID measures used in the project are based on the low impact development concept from abroad, and such technologies are still in the early stages of development in China [9]. Therefore, corresponding technical standards and performance evaluation criteria have not yet been established.

C2 has a comprehensive closeness of only 0.3641, indicating poor overall performance. The fundamental reasons for this are as follows: ① the construction area of the detention tank is large, and the equipment is complex [10], resulting in high investment and construction costs. ② The primary method for removing pollution loads from runoff in the detention tank is through sedimentation and desilting [11], resulting in lower purification rates for COD, NH_3-N, and TP. ③ The detention tank has the advantage of collecting large volumes of runoff [12], but the project is subject to significant shock loading, leading to low stability in water quality. ④ Currently, the design and operation technology of detention tanks in China is still in the developmental stage, with the emphasis on establishing technical standards and specifications while lacking corresponding maintenance and performance evaluation criteria [13]. Therefore, C2 faces challenges such as high investment and construction costs and a lack of performance evaluation criteria, and further improvement is needed in pollutant removal rates and operational stability.

Based on the performance evaluation of the three complete technical projects mentioned above, C1 and C3 exhibit higher pollutant removal rates for stormwater runoff, while C2 shows lower removal rates. In terms of economic benefits, C1 has the lowest economic performance level. This is due to its high operation and maintenance costs as well as its weak runoff attenuation capacity, resulting in limited savings in municipal treatment costs. C2 ranks second in economic performance, benefiting from its low operation and maintenance costs and strong runoff attenuation capacity. C3 achieves the highest economic performance level due to its low investment and construction costs, low operation and maintenance costs, and certain runoff attenuation capacity.

In summary, C1 and C3 are suitable for areas with high environmental requirements but low runoff generation, such as residential communities, park green spaces, and municipal roads. C2, which focuses on detention facilities, is suitable for areas with large runoff generation, such as central urban areas with significant impervious surfaces. Additionally, C1 and C2 currently lack relevant technical standards and evaluation criteria. Therefore, special attention should be given to the maintenance of these two complete technologies and their applied projects in engineering practice to ensure their normal operation.

9.3.2.4 Engineering Improvement Suggestions

C1, showing high operation and maintenance costs, limited savings in municipal sewage costs, and a lack of technical standards and performance evaluation criteria, the following three recommendations are proposed: ① Using new, low-cost fillers with better adsorption performance to reduce excavation frequency and lower operation and maintenance costs [14]; ② Improving soil quality to increase the volume of runoff treated and boost municipal sewage savings [15]; ③ Conducting in-depth research on engineering technical specifications, standards construction, engineering maintenance, and performance evaluation to establish relevant technical specifications and performance evaluation standards.

C2, showing high investment and construction costs, lower pollutant removal efficiency, poor operational stability, and a lack of performance evaluation criteria, the following four recommendations are proposed: ① Conducting accurate diagnostic assessments of regional runoff and pollutant loads before construction and reasonably designing storage volume to reduce investment and construction costs [16]; ② Adding physical and chemical treatment equipment at the inlet and outlet of the storage facility to achieve efficient SS removal and partially eliminate COD, NH_4^+-N, and TP [17]; ③ Increasing the frequency of engineering monitoring to promptly respond and purify any abnormal water quality indicators to strengthen water quality stability; ④ Conducting in-depth research on engineering maintenance and performance evaluation to develop relevant performance evaluation standards.

C3, showing lower municipal sewage savings, is recommended that the project be combined with other low-impact development facility projects in the region to form an overall green infrastructure that can remove runoff pollution and absorb more runoff volume [18]. Rainwater collection and reuse projects should also be established to increase municipal sewage savings.

References

1. Zhao, Y., Li, C., An, C., et al. (2018). Analysis of operating costs of different urban sewage treatment plants. *Water Supply and Drainage, 54*(S2), 48–50. (in Chinese).
2. Huang, J. (2018). Cost model of COD and NH_3-N reduction in sewage treatment plant. *Chemical Engineering Design Communication, 44*(11), 207.
3. Zhao, L., Zhang, F., Dong, L., et al. (2013). Application of improved entropy weight topsis model in optimal selection of rural domestic sewage treatment. *Water Saving Irrigation*, (12), 52–54, 58. (in Chinese)
4. Zeng, Y. (2016). *Performance Evaluation of Higher Education Based on TOPSIS Method and Entropy Weight Method*. Jinan University. (in Chinese)
5. Zhang, H., Li, D., Zhang, J., et al. (2021). Application of combined constructed wetland in water quality up-to-standard projects for river assessment section. *Water Supply and Drainage, 57*(1), 49–53. (in Chinese).
6. Chai, H., Bao, Y., Lin, H., et al. (2013). Technology and measures for artificially enhanced natural reaeration of secondary rivers in mountainous city. *China Water & Wastewater, 29*(14), 9–12. (in Chinese).
7. Chang, Y., Wei, T., Ji, B., et al. (2019). Comparative analysis of relevant specifications/regulations of constructed wetlands in different regions of China. *China Water & Wastewater, 35*(8), 27–33. (in Chinese).
8. Duan, T., Cheng, G., Zheng, Y., et al. (2017). Advanced treatment of tail water from sewage plant by high-efficiency vertical flow constructed wetland-multistage ecological ponds. *Journal of Environmental Engineering, 11*(11), 5828–5835.
9. Wang, J., Che, W., & Yi, H. (2009). Low impact development for Urban Stormwater and Flood Control and Utilization. *China Water & Wastewater, 25*(14), 6–9, 16. (in Chinese)
10. Wang, J., Huang, T., Zhang, P., et al. (2015). Research progress on combined sewer overflow stormwater storage tank for pollution control. *Environmental Pollution and Control, 37*(08), 85–89, 95.

11. Shi, Z., & Hu, Y. (2008). Application of regulation reservoirs in combined sewerage systems. *Water & Wastewater, 7*, 43–45. (in Chinese).
12. Huang, J. X., Li, H. Z., Ye, J. F., et al. (2010). Recent research and development of storage tank in municipal drainage systems. *Environmental Science and Management, 35*(1), 115–118.
13. Yang, Z., Che, W., & Zhao, Y. (2020). General strategy and scientific decision-making of urban combined sewer overflow separation and CSO control. *China Water & Wastewater, 36*(14), 46–55. (in Chinese).
14. He, S., Wang, H., Zhang, Y., et al. (2014). Chemical/biological PRB technology for removing nitrate from groundwater. *China Water & Wastewater, 30*(3), 48–51.
15. Cai, X., Zhai, H., & Zhao, J. (2020). Improved methods and performance of a constructing sunken green space with low-permeability soil. *Water & Wastewater, 56*(S1), 696–701, 707.
16. Zhong, Y., Zi, T., & Zhen, X. (2021). Modeling research on performance improvement of stormwater treatment detention basins by implementing real-time control system. *Water & Wastewater, 57*(4), 144–150. (in Chinese).
17. Xu, G., Chen, C., Lin, W., et al. (2005). Study on pollution control of overflow from initial rainwater storage tank. *China Water & Wastewater, 8*, 19–22. (in Chinese).
18. Chang, X., Xu, Z., Zhao, G., et al. (2016). Urban rainfall-runoff simulations and assessment of low impact development facilities using SWMM model—A case study of Qinghe catchment in Beijing. *Journal of Hydroelectric Engineering, 35*(11), 84–93. (in Chinese).

Chapter 10
Performance Evaluation of the Engineering Application of Urban Rainfall Runoff Pollution Control in Clustered Areas

With the gradual prominence of rainfall-runoff pollution in China's urban and the development of related management technologies, a large number of individual technologies and their combination to form a complete set of technologies have been rapidly developed and applied in engineering. The urban area is between the urban area and a single residential area, business district, and other functional areas. It is relatively independent, with a specific scope, various functions, a relatively large area, and more types of subsurface areas. Usually, the factors influencing the runoff pollution load in this area are more complex, and it is often difficult to achieve the rainfall-runoff pollution control objectives with a single key technology and set of technologies. Instead, the integrated application of multiple key technologies and sets of technologies is required [1]. At present, in China, rainfall—runoff pollution control area projects have been implemented in Beijing, Shanghai, Jiaxing, Xi'an, Wuhan, Shenzhen, and other cities.. Through the construction of regional sponge facilities and the introduction of rainwater management concepts, the runoff pollution load in urban areas has been reduced to a greater extent, and at the same time, urban flooding has been alleviated to a certain extent. To comprehensively analyze the implementation effect of the runoff pollution control project in urban patches, this chapter constructs the evaluation index system and evaluation method for the application performance of the runoff pollution control project in urban patches and takes the Fengxi New City Area in Xixian New District, Xi'an, and Jiaxing Area as cases for performance evaluation. On this basis, improvement suggestions are proposed for the existing problems to provide a reference for the planning, design, construction, operation, and maintenance of rainfall-runoff pollution control projects in urbanized areas in China.

D. Sun and F. Qi, *Urban Rainfall Runoff Pollution Control Technology and Application*,
https://doi.org/10.1007/978-981-96-4174-1_10

10.1 Assessment Indicator System

10.1.1 Principles for Constructing the Assessment Indicator System

For details of the construction principles of specific indicators, refer to Sect. 8.1.1.1 in Chap. 8, and we will not repeat them here.

10.1.2 Construction of the Evaluation Indicator System

Engineering performance evaluation is often a complex, multi-dimensional system. To make a more accurate evaluation of the real situation of the implementation of engineering projects, the selection of assessment indicators should not only follows the above principles but also involve a reasonable division of the hierarchy level of the indicators, systematic categorization, and the formation of an indicator system structure with a rigorous logical framework.

For the performance evaluation of the project, the purpose, input, output, and impact of the project are conventional indicators. However, the rainfall-runoff pollution control project in urban areas involves more key technologies and sets of technologies. How to conduct standardized planning and design, construction operation and maintenance is the key to realizing the selection and configuration of multiple key technologies and sets of technologies. In addition, it is especially important to carry out the construction of a standardized management mechanism in the rainfall-runoff pollution control project in urban areas is an important indicator.

Therefore, this chapter takes the project purpose, inputs, outputs, process, and management factors as the starting point to construct the project performance assessment indicators, including economic benefits, ecological and environmental benefits, social benefits, project construction and operation, as well as the mechanism construction. The performance assessment indicator framework of the urban area management project is constructed from three aspects: target hierarchy, criteria hierarchy, and indicator hierarchy.

The first level is the target level, i.e. the overall assessment object "the performance of rainfall-runoff pollution control project in urban areas"; the second level is the criteria one, It is composed of five sub-systems condensed from the characteristics of various aspects of the urban area management project, namely, the economic benefit sub-system, the ecological and environmental benefit sub-system, the construction and operation sub-system, the social benefit sub-system and mechanism construction sub-system; Each subsystem can provide a reference direction for the selection of specific indicators in the third hierarchy according to its characteristics. The third level is the specific indicator one, which contains several specific indicators that can directly reflect the overall situation of each subsystem. The structure of the specific performance assessment indicator framework is shown in Fig. 10.1.

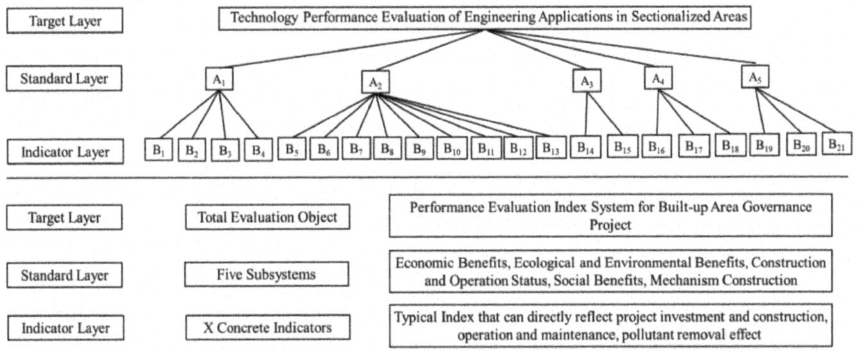

Fig. 10.1 Structure of the performance assessment indicator framework for rainfall runoff pollution control project in urban patchwork areas

10.1.3 Evaluation of Indicators

This chapter constructs assessment indicators for five criteria hierarchy, including economic benefits, ecological and environmental benefits, project construction and operation, social benefits, and mechanism construction. It finishes this from five aspects: the purpose, inputs, outputs, process, and management of rainfall-runoff pollution control projects in urban areas. Among them, economic benefits can reflect the economy of the project. Ecological and environmental benefits and social benefits can reflect the benefits brought to the environment and society after the project being constructed. The construction and operation status of the project can reflect the degree of stability of the project in its construction and operation. The mechanism construction can reflect the implementation of the project's operation and maintenance management system.

In the economic benefit subsystem, 'input cost' and 'output benefit' serve as the principal evaluative criteria reflecting the project's economic efficiency. The primary indicators for 'input cost' include 'investment in construction' and 'operation and maintenance expenses,' while 'output benefit' is primarily measured by the 'savings on water costs through reuse' and the 'reduction in municipal wastewater treatment expenditures.

In the ecological environment benefit subsystem, the primary goal of urban rainfall-runoff pollution control projects is to reduce runoff pollution, minimize total runoff volume, and improve the quality of urban aquatic and landscape environments. The most important aspect is the purification of key pollutants, such as SS, COD, TN, and TP. Additionally, it is essential to ensure a basic control rate of total runoff and enhance runoff volume control within urban areas. Moreover, through the reduction and purification of runoff pollution, the project aims to improve the comprehensive utilization of rainwater, facilitate wastewater regeneration, and increase the efficient use of unconventional water resources. After achieving reduced runoff pollution, total runoff control, and optimized use of non-conventional water resources, the improvement of the surrounding landscape environment becomes crucial. This is a key component of urban beautification and an important factor in evaluating the effectiveness of rainfall-runoff pollution control projects. Two key indicators reflecting the state of landscape habitats are the percentage of vegetation cover and the aesthetic quality of the view.

In the subsystem of engineering construction and operation status, both engineering construction and engineering operation are key elements that reflect the overall performance level of a project. Among them, the project quality pass rate serves as an indicator of construction quality and standards; while the stability of project operation is an indicator that characterizes the operation status and performance of the project.

The social benefit subsystem includes three indicators, including reducing the pressure of municipal wastewater treatment, alleviating urban flooding and recharging groundwater, reflecting the impact of the project on alleviating the pressure of urban wastewater treatment plants and guaranteeing water safety, and assessing the social benefits of the project to a certain extent.

In the mechanism construction subsystem, the control system and its implementation can provide a guarantee for the construction operation and maintenance of the project; the technical specifications and standards of construction are derived from the accumulated experience of long-term project operation, reflecting the typical characteristics of the project and its contribution to the standardization of the industry. Meanwhile, the performance evaluation and reward mechanism serves as both the technical and policy guarantee for the stable operation and efficient maintenance of the project, indirectly reflecting its ability to support ongoing operation and maintenance [3].

This chapter divides the performance evaluation indicators for rainfall-runoff pollution control projects in urban areas into three levels: target, criteria, and indicator levels. The criteria levels includes five aspects: economic benefits, ecological and environmental benefits, social benefits, project construction and operation status, and mechanism construction. For a detailed breakdown of the evaluation indicators system, please refer to Table 10.1, which illustrates the three-level evaluation system for assessing the performance of rainfall-runoff pollution control projects in urbanized areas.

a. Implications of Quantitative Indicators

The engineering performance evaluation index system determined in this chapter contains 13 quantitative evaluation indexes, including investment and construction cost, operation and maintenance cost, water saving cost for comprehensive utilization of rainwater, saving municipal wastewater treatment cost, COD purification performance, SS purification performance, TP purification performance, TN purification performance, total runoff control rate, utilization efficiency of in-situ rainwater treatment resource, utilization efficiency of rainwater and wastewater recycling, vegetation coverage rate, and engineering quality pass rate. Evaluation indicators, the specific meaning of each quantitative evaluation indicator is as follows.

Table 10.1 Three-level evaluation index system for performance assessment of rainfall-runoff pollution control project in urban patchwork areas

Target level	Standardized layer	Indicator level
Performance of Rainfall Runoff Pollution Control Project in Urban Patches	Economic benefit A_1	Investment construction cost B1, million
		O&M cost as a percentage B_2, %
		Costs of water savings from integrated rainwater utilization B_3, rmb
		Saving on municipal sewage treatment costs B_4, rmb
	Ecological and environmental benefits A_2	COD purification rate B_5, %
		SS purification rate B_6, %
		TP purification rate B_7, %
		TN purification rate B_8, %
		Total runoff control rate B_9, %
		Resource utilization rate for in-situ stormwater treatment B_{10}, %
		Stormwater reclamation utilization rate B_{11}, %
		percentage of vegetation cover B_{12}, %
		Aesthetics of the landscape, B_{13}
	Construction and operation status A_3	Project quality pass rate B_{14}, %
		Engineering Operational Stability, B_{15}
	social benefit A_4	Reduce pressure on municipal sewage treatment, B_{16}
		Mitigation of urban flooding, B_{17}
		Groundwater recharge, B_{18}
	institution-building A_5	Control systems and implementation, B_{19}
		Technical norms and standard building, B_{20}
		Performance appraisals and incentives, B_{21}

Note The indicator system determined in this study includes 8 qualitative indicators and 13 quantitative indicators, of which B1-B12 and B14 are quantitative; B13, and B15-B21 are qualitative indicators

i. Investment construction costs

Investment construction costs refer to all investment expenditures incurred by the project in capital construction activities that should be included in the the project's total cost. These costs include labor, machinery and equipment, and material. TA lower investment cost during the construction process indicates better overall project benefits.

ii. The operation and maintenance cost ratio

Operation and maintenance cost refers to the expenses incurred for the maintenance and management required to ensure the normal operation of the project after construction is completed. Since the operation and maintenance costs for the two urban rainfall-runoff pollution control projects cannot be directly measured, these costs are represented by the ratio of operation and maintenance costs to the total investment and construction costs in this study. In other words, the operation and maintenance costs are expressed as a percentage of the total investment and construction costs, as shown in Eq. (10.1).

$$\begin{aligned} &\text{O\&M cost as a percentage } (\%) \\ &= \frac{\text{Total post} - \text{project maintenance and management costs}}{\text{Total investment construction cost}} \times 100\% \end{aligned} \tag{10.1}$$

iii. Costs of water savings from integrated rainwater utilization

Urban runoff pollution control projects are designed to collect, purify, and comprehensively utilize runoff rainwater, thereby tapping into unconventional water resources and contributing to water conservation. In this context, the term "cost savings" refers to the economic value of the water resources conserved through the comprehensive utilization of rainwater in these projects.

iv. Saving municipal sewage treatment costs

Urban runoff pollution control projects successfully mitigate runoff volume and decrease pollution loads through functions such as rainwater infiltration, stagnation, storage, and purification. As a result, these projects reduce the amount of combined sewage into municipal wastewater treatment systems, thereby alleviating the pollution load on sewage treatment plants and generating cost savings for municipal sewage treatment.

v. SS, COD, TN and TP purification rate

The assessment of pollutant control effectiveness is a key metric for evaluating the performance of rainfall runoff pollution control projects within urban areas. This study focuses on several key pollutant indicators, namely, SS (suspended solids), COD (chemical oxygen demand), TN (total nitrogen), and TP (total phosphorus), along with their corresponding purification rates. The calculation method is presented in Eq. (10.2).

$$\text{Pollutant purification rate}(\%) = \frac{(\text{Pollutant concentration entering facility} - \text{pollutant concentration leaving facility})}{\text{Concentration of pollutants entering the facility}} \times 100\% \quad (10.2)$$

vi. Total runoff control rate

The total runoff control rate refers to the cumulative percentage of total rainfall that has been effectively managed within a specific project or facility. This management is achieved through both natural and engineered methods, such as infiltration, detention, and purification. The calculation of the total runoff control rate is provided in Eq. (10.3).

$$\text{Total runoff control rate}(\%) = \frac{\text{Cumulative controlled rainfall}}{\text{total rainfall}} \times 100\% \quad (10.3)$$

vii. Resource utilization rate of in-situ rainwater treatment

The utilization rate of on-site rainwater treatment resources refers to the ratio of the total volume of rainwater collected within an area—through methods such as in-situ purification, road watering, garden irrigation, municipal and industrial uses, cooling, landscape, and river recharge—to the total amount of rainfall received in the same area.

viii. Stormwater recycling rate

The utilization rate of recycled rainwater refers to the proportion of treated rainwater from sewage treatment plants that is used for purposes such as ecological water recharge in rivers, landscaping, municipal use, and industrial processes

ix. Percentage of vegetation cover

Vegetation cover can reflect the characteristics of a region's landscape structure; cover reflects the characteristics of a region's landscape structure. It is typically defined as the ratio of the land area covered by vegetation to the total land area of the region. The calculation method for this ratio is presented in Eq. (10.4).

$$\text{percentage of vegetation cover}(\%) = \frac{\text{Area covered by vegetation}}{\text{Area of land in the district}} \times 100\% \quad (10.4)$$

x. Engineering quality pass rate

Project quality pass rate refers to the proportion of completed unit works that are rated as qualified products during the acceptance appraisal within a certain period of time.

b. Meaning of Qualitative Indicators

This chapter identifies eight qualitative indicators in the engineering performance evaluation index system: aesthetics, stability of engineering operations, alleviation of municipal wastewater treatment pressure, reduction of urban flooding, groundwater recharge, control systems and implementation, development of technical norms and standards, and performance appraisal and incentive mechanisms. The specific meanings of the qualitative evaluation indicators are as follows.

i. Landscape aesthetics

Scenic aesthetics refers to the overall environmental quality of the runoff pollution control project in terms of sensory enjoyment. It is divided into three levels: natural beauty, landscape nature, and general scenic appeal.

ii. Stability of project operation

Project operation stability refers to the reliability and effectiveness of project's technology or equipment during actual operation, regardless of external conditions.

iii. Reducing pressure on municipal sewage treatment

The rainfall runoff pollution control project reduces the runoff volume, pollution load, and combined flow of sewage entering the treatment plant via the municipal pipe network. It comprehensively utilizes rainwater resources and alleviates the treatment pressure on municipal wastewater treatment plants.

iv. Alleviating urban flooding

The rainfall runoff pollution control project effectively mitigates urban flooding by reducing the runoff volume, implementing peak shaving and staggering, and employing measures such as source reduction, rainfall and sewage diversion, primary rainfall abandonment, storage of rainfall and sewage, terminal ecological management, and integrated plant-network coordination to enhance urban water security.

v. Groundwater recharge

The rainfall runoff pollution control project improves subsurface permeability, allowing rainwater to infiltrate into the ground and recharge groundwater. This indicator reflects the project's capacity to recharge groundwater to a certain extent.

vi. Control system and implementation, construction of technical norms and standards, performance evaluation and incentive mechanism

The implementation of management measures and the project's contribution to industry standardization and regulation directly highlight the model nature of urban runoff pollution treatment projects, while ensuring the effective construction, operation, and maintenance of the project.

10.1.4 Evaluation Criteria

By reviewing the existing literature on assessment criteria development at both domestic and international levels, field research on relevant projects, and consultation with experts in the field of urban rainfall runoff, this chapter effectively categorized the scoring levels for each qualitative indicator and established the evaluation criteria for each level. These criteria are presented in Table 10.2.

10.2 Assessment Methodology and Rating

10.2.1 Selection of Methodology

Comprehensively evaluating a multi-objective, multi-decision program involves considering numerous factors. One of the key challenges in this field is determining how to quantitatively analyze the priority and relative importance of these factors. For the performance evaluation of rainfall-runoff pollution control projects in urban areas, the critical issue is how to reasonably assign weights to each evaluation index and make an objective, and scientifically grounded judgment on the overall performance of the project. To address this challenge, it is essential to develop a performance evaluation model for rainfall-runoff pollution control projects in urban areas.

The performance evaluation of rainfall-runoff pollution control projects in urban areas involves a comprehensive assessment of boththe construction and operation operational effectiveness of large-scale runoff pollution control systems. Currently, there are many comprehensive evaluation methods for engineering performance evaluation, and the most common ones are the Delphi method, data envelopment method, hierarchical analysis method, fuzzy comprehensive evaluation method, entropy weight method, and TOPSIS method.

The Delphi method, also known as an expert scoring method, offers advantages such as simplicity and intuitive results. However, its evaluation process is often influenced by the the experience, personal preferences, and knowledge levels of the experts, leading to results that may lack systematic rigor and exhibit a degree of subjectivity. The data envelopment analysis (DEA) method, based on the theory of "relative efficiency", is widely used in efficiency evaluation. It has no specific requirements on the scale of the indicators and determines the weight of evaluation factors by analyzing observation data. While DEA provides a degree of objectivity, it only offers a judgment on the relative efficiency of the object under evaluation, without providing in-depth decision-making insights [5].

Hierarchical analysis method is a multi-objective decision-making method used more in the evaluation work at present, which has both subjective judgment and analysis by experts and precise calculation and reasoning by the mathematical model in the process of weight determination. However, it still can't eliminate the subjectivity

Table 10.2 Grading and scoring criteria for each qualitative indicator

Evaluation indicators	Marking standard	Value of a score
Aesthetics of the landscape	A. Natural and beautiful scenery	(6,9)
	B. Scenery Nature	(3,6)
	C. The view is mediocre	(1,3)
Engineering Operational Stability	A. Good stability of project operation	(6,9)
	B. Average engineering operational stability	(3,6)
	C. Poor engineering operational stability	(1,3)
Reduce pressure on municipal wastewater treatment	A. Reduces the pressure on municipal wastewater treatment to a large extent	(6,9)
	B. Municipal wastewater treatment can be appropriately relieved	(3,6)
	C. Slightly reduces pressure on municipal wastewater treatment	(1,3)
Mitigation of urban flooding	A. Alleviates urban flooding to a large extent	(6,9)
	B. Appropriate mitigation of urban flooding is possible	(3,6)
	C. Slightly relieves urban flooding	(1,3)
Groundwater recharge	A. Recharge groundwater to a large extent	(6,9)
	B. Recharge groundwater to some extent	(3,6)
	C. Slightly rechargeable groundwater	(1,3)
Control systems and implementation	A. Better control systems and implementation	(6,9)
	B. Control systems and implementation in general	(3,6)
	C. Poor control systems and implementation	(1,3)
Technical norms and standard building	A. Better construction of technical norms and standards	(6,9)
	B. Technical norms and standard-setting in general	(3,6)
	C. Poor construction of technical norms and standards	(1,3)
Performance appraisal and incentives	A. Improved performance appraisal and incentive mechanisms	(6,9)
	B. Performance appraisal and incentives in general	(3,6)
	C. Inadequate performance appraisal and incentive mechanisms	(1,3)

and randomness inherent in the evaluation process, and there are certain limitations regarding the number of evaluation samples.

The fuzzy comprehensive evaluation method is based on the fuzzy mathematical membership theory. This approach is simple, and easy to understand, through the precise mathematical theory model to deal with the fuzzy evaluation object. It presents embedded information in a way that reflects fuzzy data, leading to a more scientific and reasonable evaluation that is closer to the actual quantitative effect. However, its application requires that the number of evaluators be no less than 10, and the evaluators must have a deep understanding of the object being evaluated [6].

After the selection of comprehensive evaluation methods, this chapter finally adopts the evaluation model combining the entropy weight method and TOPSIS method to assess the performance evaluation of the rainfall-runoff pollution control project in the urban patchwork area. The evaluation method, based on the entropy weight-TOPSIS model, offers two key advantages. First, the entropy weight method derives the relative objectivity and reliability of the weights from the original data; second, the TOPSIS method is used to rank the advantages and disadvantages of the evaluated objects, enabling decision-making analysis [7].

This method is consistent with the performance evaluation method described in Chap. 9. Therefore, a detailed explanation of the basic principles and evaluation steps of the entropy weight method, the TOPSIS method, and their coupled approach is omitted.

10.2.2 Evaluation Levels

According to the evaluation method based on the Entropy Weight-TOPSIS model, the relative closeness calculated by Eq. (9.9) is used to determine the performance level of the evaluated subjects. A higher relative closeness indicates that the subject is closer to the ideal solution, signifying a higher performance level; conversely, a lower relative closeness suggests that the subject is further from the ideal solution indicating a lower performance level. Based on the research findings of established scholars, the relative closeness is categorized into four grade standards to represent the performance levels of urban rainfall-runoff pollution control pilot projects [8]. Detailed grade standards are presented in Table 10.3.

Table 10.3 Criteria for judging performance levels

Closeness	(0,0.3)	(0.3,0.6)	(0.6,0.8)	(0.8,1.0)
Performance level	Low	Medium	Favorable	Excellence

10.3 Case Study Project Overview

10.3.1 Runoff Pollution Control Project in Fengxi New Town, Xixian New Area

a. Basic Overview of the Developed Area of Fengxi New Town

The Xixian New Area is located at the junction of Xi'an and Xianyang cities. It is one of the earliest pilot areas in China for the construction of sponge cities. The area is composed of five distinct new towns, and Fengxi New Town is one of them. Positioned to the west of the Fenghe River and south of the Wei River, Fengxi New Town plays a key role in the Xixian New Area, covering a total area of 143 square kilometers.

Fengxi New Town is divided into three main areas based on the development layout, from north to south: the developed area, the core area, and the expansion area. The developed area is located south of the Wei River and north of the old Xibao Expressway. The core area is situated between the new and old Baomao Expressways, while the expansion area is located south of the new Baomao Expressway.

Fengxi New Town builds upon the existing infrastructure in the area and designates the construction zone for urban runoff pollution control projects as part of its core area. The planned land use within the core area includes the Science and Technology Innovation Harbor, the Information Industry Park, the Comprehensive Cultural and Educational Park, and the Service Core Area. The primary focus of the planned land use is urban construction, with an emphasis on creating a multi-level open space and green space framework. This framework centers around natural water systems, green corridors, circular parks, block parks, and road green spaces.

Based on the current topography and drainage network of the Xixian New Area, it is divided into five watershed hydrological zones: the Wei River Basin, Jing River Basin, Feng River Basin, Zao River Basin, and Xin River Basin. Fengxi New Town is primarily located within the Wei River Basin, Xin River Basin, and Feng River Basin.

Based on the GIS data and runoff simulation analysis of the current topography of Fengxi New Town, the watershed hydrological zones under natural terrain conditions are shown in Fig. 10.4. Using these natural terrain conditions and the corresponding watershed and sub-watershed divisions, the core area of Fengxi New Town is divided into 10 sub-watershed hydrological zones. These include two in the Xin River Basin: Xinhe District and Shahe District; four in the Wei River Basin: Weihe District 1, Weihe District 2, Weihe District 5, and Weihe District 8; three in the Feng River Basin: Fenghe District 1, Fenghe District 2, and Fenghe District 3; as well as the Green Corridor District, which spans both the Wei River and Feng River Basins.

The urban construction and management project in Fengxi New Town encompasses a total of six watershed hydrological zones, which include: Xinhe District (Watershed Zone 1, with a pilot area of 5.29 km^2), Weihe District 1 (Watershed Zone 2, with a pilot area of 6.17 km^2), Green Corridor District (Watershed Zone 3, with a pilot area of 6.45 km^2), Weihe District 2 (Watershed Zone 4, with a pilot area of 1.72 km^2), Fenghe District 2 (Watershed Zone 5, with a pilot area of 0.72 km^2), and Weihe District 8 (Watershed Zone 6, with a pilot area of 2.15 km^2). Details can be found in Fig. 10.5. This section focuses on evaluating the effectiveness of runoff pollution control projects within these six watershed hydrological zones in the urban construction area of Fengxi New Town.

b. Key Technologies and Engineering for Runoff Pollution Control in Urban Built-Up Areas

The land use in the built-up area before development was primarily agricultural fields and villages. After development, it mainly consists of parks, green spaces, commercial areas, industrial areas, and residential areas. The runoff pollution control project in the built-up area focuses on constructing a stormwater management system across four key aspects: controlling stormwater runoff from building complexes and municipal roads, implementing rainwater pipe networks and surface drainage systems, establishing end-of-pipe green corridors for storage and rainwater reuse. The construction concept is shown in Fig. 10.2.

The division of the six sub-watershed zones within the built-up area of Fengxi New Town is based on river and lake systems, as well as the natural topography, taking into account features such as green spaces, transportation networks, and the layout of major green infrastructure. To effectively control runoff pollution and improve water resource utilization, Fengxi New Town has implemented a series of runoff pollution control projects based on the principles of "source reduction, process control, endpoint regulation, and systematic management". The specific runoff pollution control projects and corresponding technologies used in each watershed zone are shown in Table 10.4.

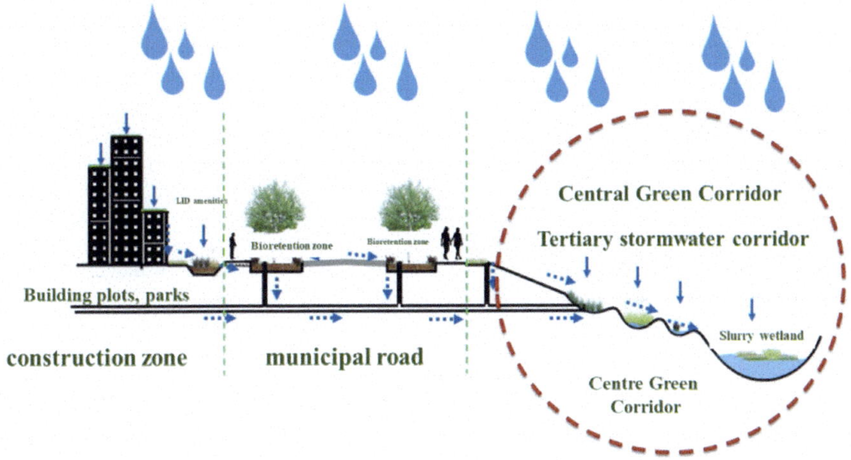

Fig. 10.2 Runoff pollution control concept in the built-up area

Table 10.4 Key technologies and treatment projects for runoff pollution control in each watershed zone of Fengxi New Town

Watershed Zone	project name	project type	Annual runoff total amount control rate,%	Rainwater storage volume,m^3	Main use of runoff pollution control technology
Watershed Zone 1 (New River Area)	Phase I of Jiaotong University Innovation Port	Overall development of the area	84	8218.6	Permeable and pollutant-reducing infiltration pipes, innovative rainwater inlet sediment interception facilities, infiltrated vegetative trenches, ecological parking lots, crushed stone infiltration trenches, stepped rain gardens, vegetation buffer zones and depressed green spaces for retention and pollutant reduction, composite media bio-retention facilities, water storage modules, and rainwater collection and reuse treatment facilities
	New Weisha Wetland Park	Park Green Space Category	88	1396.8	Surface flow constructed wetland, subsurface flow constructed wetland, permeable paving, ecological parking lot, ecological grass swale
Watershed Zone 2 (Wei River Zone 1)	Phase 1 of Tianfu Heyuan	Residential Community Category	84	3758.2	Infiltration rain garden, detention rain garden, depressed green space, vegetated shallow swale, raised planter, filter stone gabion, permeable brick paving, ecological parking lot, detention basin

(continued)

Table 10.4 (continued)

Watershed Zone	project name	project type	Annual runoff total amount control rate,%	Rainwater storage volume,m^3	Main use of runoff pollution control technology
	Phase 1 of Xinye Jiayuan	Residential Community Category	84	719.4	Stepped rain garden, permeable concrete paving, vegetated shallow swale, raised planter, ecological parking lot, rainwater downspout diversion
	Tianxiong West Road	Municipal Road Category	85	–	Permeable concrete paving, bio-retention facilities, vegetated swales
	Tianyuan road	Municipal Road Category	85	–	Permeable concrete paving, bio-retention facilities, vegetated swales
Watershed Zone 3 (Green Corridor Area)	South section of Qin Huang Avenue	Municipal Road Category	85	–	Prefabricated pollutant interception frame, window-type drainage outlet, ecological retention grass ditch, transmission grass ditch, sidewalk water passage culvert, permeable ceramic tile paving, dry pond
	Information 4th Road	Municipal Road Category	85	–	Permeable concrete paving, bio-retention facilities
	Data 2nd Road	Municipal Road Category	85	–	Permeable concrete paving, bio-retention facilities
	Data 6th Road	Municipal Road Category	85	–	Permeable concrete paving, bio-retention facilities, vegetated swales
	Kaiyuan Road (West Section)	Municipal Road Category	85	–	Permeable concrete paving, transmission vegetated swales, infiltrated vegetated swales, dry pond

(continued)

Table 10.4 (continued)

Watershed Zone	project name	project type	Annual runoff total amount control rate,%	Rainwater storage volume,m^3	Main use of runoff pollution control technology
	Feng wei avenue	Municipal Road Category	85	–	Permeable concrete paving, depressed green space, vegetated retention trench, aggregate storage ditch
	Xingyuan road	Municipal Road Category	85	–	Permeable concrete paving, vegetated retention trench, depressed green space, aggregate storage ditch
	Qinhuang Avenue Green Belt	Park Green Space Category	85	57,792	Drainage culvert, dry pond, transmission vegetated swale
	Phase 1 of Central Green Corridor	Park Green Space Category	87	4794.5	Rainwater corridor, constructed wetland, infiltration wetland, infiltration trench, filter bed, rainwater pumping station
Watershed Zone 4 (Wei River Zone 2)	West Section of Kangding Road	Municipal Road Category	85	–	Permeable concrete paving, conveyance-type vegetated swale, and bioretention facilities are three commonly used technologies for urban stormwater management
	Shangye Road	Municipal Road Category	85	–	Bioretention facilities, conveyance-type vegetated swales
	Tongde road	Municipal Road Category	85	–	Permeable concrete paving, bioretention facilities
	Tongde apartment	Residential Community Category	85	1378.6	Permeable brick paving, depressed green space, vegetated shallow trench, terraced rain garden
	Phase I of China Unicom	Residential Community Category	84	2285.2	Green roof, permeable paving, rain garden, depressed green space, disconnected downspouts

(continued)

Table 10.4 (continued)

Watershed Zone	project name	project type	Annual runoff total amount control rate,%	Rainwater storage volume,m^3	Main use of runoff pollution control technology
	Phase I of China Mobile	Residential Community Category	84	1581.6	Green roof, permeable paving, conveyance-type vegetated swale, depressed green space, gravel infiltration trench
	Western Cloud Valley	Residential Community Category	84	1236.2	Disconnected downspouts, permeable paving, green roof, elevated planting bed, eco-friendly parking lot
	Phase I of Tongde Jiayuan	Residential Community Category	84	2414	Rain garden, vegetated shallow trench, depressed green space, permeable concrete paving
	Phase I Headquarters economic Park	Residential Community Category	84	3048.7	Green roof, disconnected downspouts, eco-friendly parking lot, permeable paving
	Phase 1 of Ring Park	Park Green Space Category	90	2009.1	Infiltration pond, permeable paving, eco-friendly parking lot, depressed green space
Watershed Zone 5 (Feng River Zone 2)	Phase 3 of Ring Park	Park Green Space Category	88	2316.5	Infiltration pond, infiltration well, permeable paving, eco-friendly parking lot, ecological ditch
	Feng River Bank Treatment Project	Water System Category	85	2189.7	Ecological revetment, vegetated swale, infiltration ditch, stormwater pumping station
Watershed Zone 6 (Wei River Zone 8)	West Section of Tongyi Road	Municipal Road Category	85	–	Permeable concrete paving, bioretention facilities
	Yongping road	Municipal Road Category	85	–	Permeable concrete paving, bioretention facilities

(continued)

Table 10.4 (continued)

Watershed Zone	project name	project type	Annual runoff total amount control rate,%	Rainwater storage volume,m^3	Main use of runoff pollution control technology
	Xianyang Vocational and Technical College	Residential Community Category	84	6472.7	Artificial infiltration well, disconnected downspouts
	Phase 1 of Kangding Heyuan	Residential Community Category	84	1924.6	Depressed green space, rain garden, vegetated shallow trench, permeable concrete paving
	White Horse River Park	Park Green Space Category	90	1292.6	Infiltration pond, permeable paving, eco-friendly parking lot, ecological grass ditch

During the construction of the runoff pollution control project in Fengxi New Town, a variety of typical measures were implemented based on local conditions. These include a road runoff control system integrating facilities such as vegetated shallow trenches, stone-filled gabion filter strips, and overflow wells. The system meets is designed to support local plant growth and incorporates new substrate soils with good permeability in depressed green spaces, as well as newly constructed artificial infiltration wells. Figures 10.7, 10.8, 10.9, 10.10 provide an overview of the application of these facilities in the runoff pollution control project in Fengxi New Town (Figs. 10.3, 10.6).

Fig. 10.3 Vegetated shallow trench (left) and new substrate soil for depressed green space (right)

Fig. 10.4 Artificial infiltration well (left) and rain garden (right)

Fig. 10.5 Permeable asphalt pavement (left) and infiltration channel (right)

Fig. 10.6 Vertical greening on walls (left) and artificial wetland (right)

Fig. 10.7 Runoff Pollution Control Facilities and Projects in the Developed Area of Jiaxing City. **a** Green roof **b** Rainwater wetland **c** Wet stranded pool **d** Permeable ceramic pavement **d** Pollution control shore belt **e** A rainwater grate with a sewage interception hanging basket

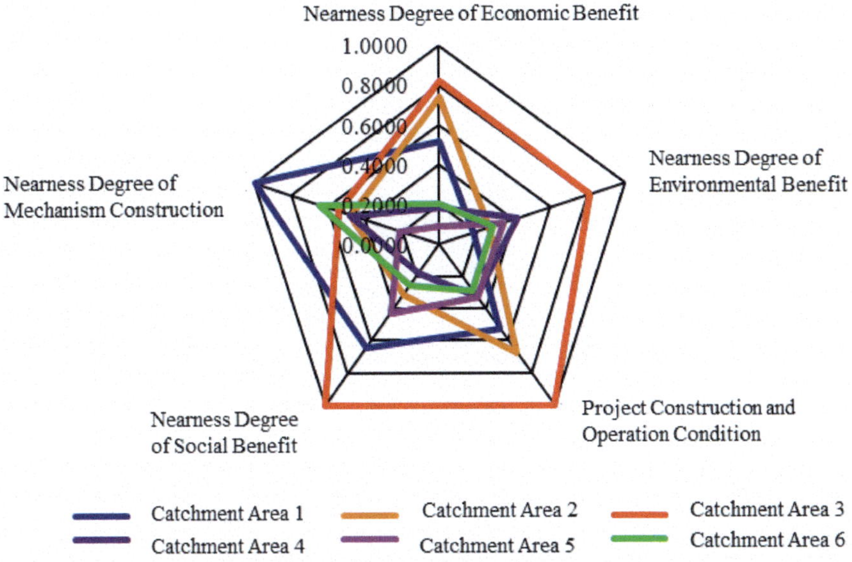

Fig. 10.8 Conformity of runoff pollution control projects in different catchment areas in Fengxi new city

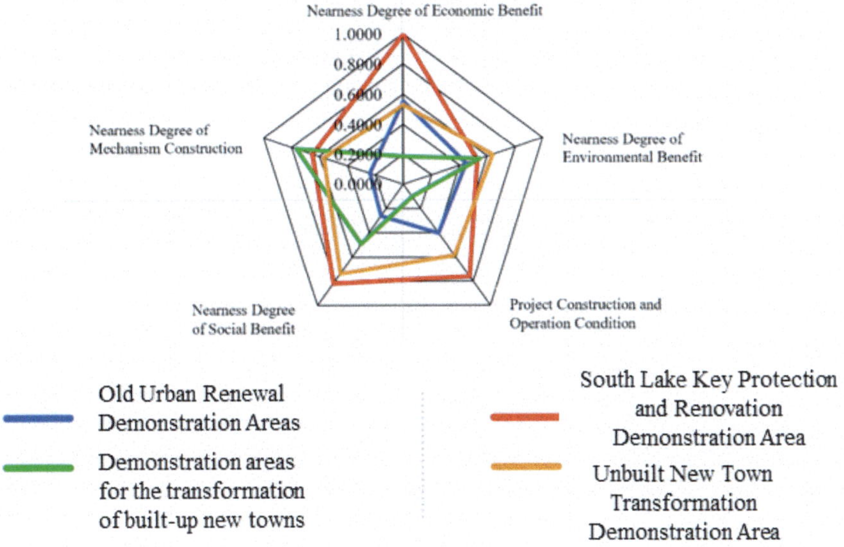

Fig. 10.9 Proximity of performance evaluation indicators for runoff pollution control projects in different catchment areas of the built-up area of Jiaxing city

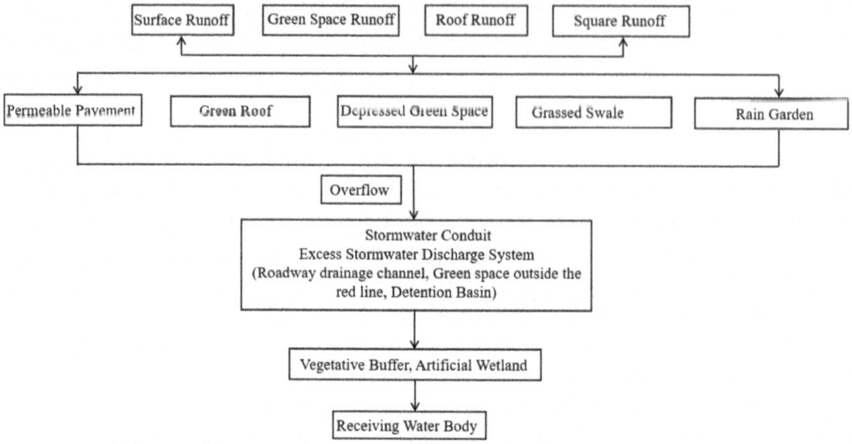

Fig. 10.10 Technical roadmap for runoff pollution control in the built-up area of Fengxi new city

c. Effectiveness of Urban Runoff Pollution Control Project

After the implementation of the runoff pollution control project, the annual total runoff control rate in the built-up area is 86.09%. The reductions in SS (suspended solids), COD (chemical oxygen demand), and TP (total phosphorus) loads are 67.31%, 72.22%, and 71.62% respectively, all exceeding the planned targets (SS: 60%, COD: 50%, TP: 40%). The control of total runoff volume and reduction of runoff pollutants across various catchment areas within the built-up area are summarized in Table 10.5.

10.3.2 Jiaxing City Runoff Pollution Control Project

a. General Overview of Jiaxing City's Built-Up Area

Jiaxing City is located in the northeast of Zhejiang Province, China, adjacent to Suzhou, Shanghai, and Hangzhou. It is also close to Taihu Lake and is one of the important cities in the Yangtze River Delta. When carrying out runoff pollution control and sponge city construction, Jiaxing determined a construction area surrounded by Xibanqiao Port, Changshuitang, Zuili Road, Fanggong Road, Furun Road, and Caihuajing. The entire sponge demonstration zone covers an area of 18.44 square kilometers. The construction area includes various typical blocks such as the old city, new city, Nanhu Scenic Area, and water source protection areas. In the construction process, various rainwater collection and utilization engineering technologies are utilized, which can effectively address various problems that may arise in the urban infrastructure construction related to urban rainfall runoff pollution control. It has significant demonstration significance.

Table 10.5 Construction and operational performance of runoff pollution control facilities in Fengxi new town's built-up area

Watershed	Annual runoff total amount control rate,%	SS reduction rate,%	COD reduction rate,%	COD reduction rate,%
Overall pilot	86.09	67.31	72.20	71.62
No.1 (Xin River 4#)	85.12	67.75	72.73	71.75
No.2 (Weihe River 1#)	85.96	62.34	68.06	66.86
No.3 (Green Corridor)	86.95	67.51	72.36	71.99
No.4 (Weihe River 2#)	87.05	75.60	79.33	79.14
No.5 (Feng River 2#)	86.97	74.40	78.24	77.90
No.6 (Weihe River 8#)	85.97	67.41	72.30	71.96
Planning objectives for built-up areas	85.00	60.00	50.00	40.00

Jiaxing City, located in the northeast of Zhejiang Province, China, is adjacent to Suzhou, Shanghai, and Hangzhou, and lies near Taihu Lake. As one of the key cities in the Yangtze River Delta, Jiaxing has undertaken runoff pollution control and sponge city construction initiatives in a designated area bounded by Xibanqiao Port, Changshuitang, Zuili Road, Fanggong Road, Furun Road, and Caihuajing. The sponge city demonstration zone covers an area of 18.44 square kilometers, which includes various representative districts such as the old city, new city, Nanhu Scenic Area, and water source protection zones. Throughout the construction process, a range of rainwater collection and utilization technologies are employed to address the challenges associated with urban infrastructure and urban rainfall runoff pollution control. This project serves as a significant demonstration for similar urban development and environmental management efforts.

Based on the characteristics of Jiaxing City's densely distributed urban river network, which lies within a plain river system, and considering the city's drainage pattern and control unit boundaries, watershed divisions have been delineated. The analysis takes into account factors such as the construction period, development density, greening rate, and the current status of municipal public facilities within the entire construction area. The runoff pollution control project in the entire urban area of Jiaxing is divided into four pilot areas, each with distinct characteristics: the old city transformation area, the key protected area of Nanhu Lake, the already built new city transformation area, and the future new city development area. This chapter focuses on the construction of rainfall runoff pollution control projects in the four watershed divisions within Jiaxing Sponge Demonstration Zone.

b. Key Technologies and Projects for Runoff Pollution Control in Urban Areas

In the construction of runoff pollution control projects in urban areas of Jiaxing, multiple pollution control measures with runoff detention, collection, and utilization have been adopted to address runoff pollution and flood issues caused by rainfall. The rainfall runoff pollution control projects and corresponding technologies used in each watershed division are presented in Table 10.6.

In the runoff pollution control process in Jiaxing, a range of typical measures have been implemented, tailored to local conditions. These measures include green roofs that serve multiple functions, such as providing viewing spaces, leisure areas, and rainwater storage; rain gardens integrated into the developed water system; and pollution control shorelines constructed along the riverbanks, as shown in Fig. 10.7.

c. The Effectiveness of Runoff Pollution Control Projects in the Developed Area

By implementing source control measures such as retention, filtration, and purification, combined with the interception and control provided by initial runoff detention ponds and storage tanks, the sediment (SS) reduction rate in the developed area has reached 45% to 60% (with a minimum target of 40%). The annual total runoff control rate has reached 78.2% (with a target control rate of 78%), as shown in Table 10.7.

10.4 Evaluation Results and Analysis

10.4.1 Determination of Indicator Weights

As described in Chap. 9, this section applies the entropy weighting method to determine the weights of various evaluation indicators. The data for quantitative evaluation indicators are sourced from the investigation of runoff pollution control projects in the surveyed cities. A consistent scoring standard for qualitative indicators has been established, with the details provided in Tables 10.8 and 10.9. During the quantification process of qualitative indicators, scoring levels and corresponding standards were developed based on field investigations of relevant projects and consultations with experts within the industry.

Expert rating forms were used to assign scores by subject-matter experts within the field, with the quantitative data for qualitative indicators derived from the average of these experts' scores. Specifically, for the runoff pollution control project in the built-up area of Fengxi New City in the Xixian New Area, 'n' experts were invited to participate in the scoring process. This group included engineering researchers (20%), engineering managers (40%), and engineering designers (40%). Similarly, for the runoff pollution control project in the built-up area of Jiaxing city, 'n' experts were similarly solicited, including engineering researchers (20%), engineering managers

Table 10.6 Overview of rainfall runoff control projects in various watershed divisions in Jiaxing

Watershed Zone	Project Name	Project type	Annual Runoff Total Control Rate (%)	The main adopted runoff pollution control technologies
Demonstration Zone for the Renovation of Old Urban Areas	Fucheng Primary School in Jiaxing City in English	Public buildings	80.00	Sunken Green Space. Vegetated Swale Terraced Rainwater. Garden Rainwater Wetland
	Xiuzhou apartment	Residential community	28.38	Multi-functional Tree Basin Integrated Retention. Detention. Filtration System Hydraulic Connection and Division of Impermeable Surfaces Initial Rainwater Runoff Drainage Multi-stage Filtration with Composite Media. Rainwater Runoff Reuse and Recirculation Filtration
	Gujing temple community	Residential community	61.19	Initial rainwater discharge Multi-functional tree basin Integrated retention. detention. Overflow purification. Multi-stage filtration with composite media. Rainwater downpipe disconnected
	Jiangnan building	Public buildings	80.00	Infiltration-type vegetated swale. Ecological parking lot Terraced rain garden. Composite media bio-retention facility. Water storage module and rainwater collection and reuse treatment facility. Multi-functional tree basin

(continued)

Table 10.6 (continued)

Watershed Zone	Project Name	Project type	Annual Runoff Total Control Rate (%)	The main adopted runoff pollution control technologies
	Jiannan apartment	Residential community	51.70	Multi-functional tree pit Integrated retention. detention, and overflow purification Hydraulic connection and division of impervious surfaces Initial discharge Multi-stage filtration with composite media Rainwater runoff reuse and recirculation filtration
	Ring Road (Qinjian Road-Xiucheng Bridge) east section	Municipal road class	80.00	Permeable concrete pavement. Window-type drain outlet. Initial rainwater discharge device. Rainwater inlet sediment efficient pollution interception facility
Demonstration Zone for Key Protection of the South Lake	South garden	Park green space class	66.10	Permeable pavement. Sunken green space. Rain garden. Ecological revetment
	Water Cube International Business Hotel	Residential community	66.10	Porous ceramic pavement. Multi-functional ecological tree pit. Rain garden. Rain downpipe disconnection. Sunken green space. Shallow vegetative swale
	Insulation factory dormitory	Residential community	53.10	Multi-functional tree pit. Infiltration-Permeation Vegetative Swale. Vegetated buffer and Sunken green space. Hydraulic connection separation of impervious surfaces. Rainwater runoff reuse cycling filtration

(continued)

Table 10.6 (continued)

Watershed Zone	Project Name	Project type	Annual Runoff Total Control Rate (%)	The main adopted runoff pollution control technologies
	Green space under Nanhu Bridge	Park green space	88.97	Rainwater inlet sediment efficient pollution interception facility. Infiltration-permeation vegetative swale. Vegetated buffer and sunken green space retention and pollutant reduction facilities. Rain garden
	The areas on both sides of Chengnan Road (Zhonghuan South Road—Zhenhe Ligang	Park green space category	67.90	Conveyance vegetated swale. Sunken green space. Subsurface drainage conduit. Permeable gravel trench. Rain garden. Multi-functional tree pit
	Zhonghuan South Road (Fanghezhou Road—Changshuitang)	Municipal road category	81.20	Permeable concrete pavement. Window-type drain outlet. Initial rainwater runoff disposal device. Rainwater inlet sediment efficient pollution interception facility
	Nanhu Citizen Square	Public building category	86.15	Permeable ceramic pavement. Permeable gravel trench. Multi-functional tree pit. Rain garden. Sunken green space
	Nanhu Parking Lot	Public building category	79.88	Porous permeable brick. Rainwater reuse facility. Vegetated swale. Permeable gravel trench
	Yanyu Community (Yanyu Garden)	Residential community category	81.19	Rain garden. Green roof. Rainwater downspout disconnect. Hydrological division of impervious surfaces. Multi-functional tree pit

(continued)

Table 10.6 (continued)

Watershed Zone	Project Name	Project type	Annual Runoff Total Control Rate (%)	The main adopted runoff pollution control technologies
Demonstration Zone for the Renovation of Newly Built Urban Areas	Boya Hotel	Residential community category	75.00	Multi-functional ecological tree pit. Rain garden. Disconnecting rainwater downspouts. Sunken green space. Permeable ceramic pavement
	Jiaxing Grand Theater	Public building category	82.70	Ecological silica sand filter stone. Granite mixed pavement. Drainage ditch. Sunken green space. Dry stream. Rain garden. Paper mulch shallow trench. Permeable parking lot. Permeable grass-embedded platform
	Fu Nan Garden area 1	Residential community category	75.00	Rain garden. Permeable vegetated swale. Vegetative buffer strip and sunken green space retention and reduction facility. Multi-functional ecological tree pit. Rainwater downspout disconnect
	Xindu Mingdi	Residential community category	80.00	Green roof. Rainwater downspout disconnect. Vegetative buffer strip and sunken green space retention and reduction facility. Composite media bio-retention facility. Water storage module and rainwater collection and reuse treatment facility
	Fu south park	Park green space category	80.90	Sunken green space. Rain garden. Rainwater wetland. Ecological parking lot

(continued)

Table 10.6 (continued)

Watershed Zone	Project Name	Project type	Annual Runoff Total Control Rate (%)	The main adopted runoff pollution control technologies
	Yuquan Road (Zhonghuan South Road—Youquan Road)	Municipal road category	85.50	Permeable concrete pavement. Window-type drainage outlet. Initial rainwater discharge device. Efficient sediment trapping facility for rainwater outlets
	The areas on both sides of Zhu'an Road (Yuquan Road—Nanhu Avenue)	Park green space category	80.00	Transportable vegetated swale. Ecological tree pit. Sunken green space. Drainage pipe. Rain garden
Demonstration Zone for the Construction of New Urban Areas	Funan 4th Phase	Residential community category	53.10	Terraced rain garden. Ecological parking lot. Sunken green space. Rainwater pond. Rainwater reuse facility
	Nanhu Avenue (Matang Jinggang - Longxinggang)	Municipal road category	85.00	Permeable ceramic pavement. Bio-retention facility. Window-type drainage outlet
	The areas on both sides of Nanhu Avenue (Matang Jinggang—Longxinggang)	Park green space category	85.00	Multifunctional ecological tree pit. New type of efficient sediment trapping facility for rainwater outlets. Infiltration and drainage vegetated swale. Permeable gravel trench. Vegetative buffer strip and sunken green space retention and reduction facility. Rain garden
	Nanhu Avenue (Longxinggang—Zuili Road)	Municipal road category	91.10	Permeable concrete pavement. Window-type drainage outlet. Initial rainwater discharge device. Efficient sediment trapping facility for rainwater outlets

(continued)

Table 10.6 (continued)

Watershed Zone	Project Name	Project type	Annual Runoff Total Control Rate (%)	The main adopted runoff pollution control technologies
	The areas on both sides of Nanhu Avenue (Longxinggang—Zuili Road)	Park green space category	91.10	Drainage pipe. Vegetated shallow ditch. Sunken green space. Transportable vegetated swale. Rain garden
	Central Park (North)	Park green space category	93.00	Terraced rainwater garden. Ecological parking lot. Sunken green space. Rainwater wetland
	Jiaxing Urban Planning and Design Institute Co., Ltd	Public building category	71.30	Green roof. Rainwater storage and detention facility. Ecological tree pit. Elevated planter bed. Disconnection of rainwater downspouts
	Botanical garden	Park green space category	93.00	Dry creek. Ecological drainage ditch. Permeable pavement. Rainwater pond. Pile revetment
	Fanggong Road—Furun Road	Municipal road category	72.98	Permeable concrete pavement. Window-type drainage outlet. Efficient sediment trapping facility for rainwater outlets
	The areas on both sides of Fanggong Road—Furun Road	Park green space category	66.10	Transportable vegetated swale. Sunken green space. Drainage pipe. Rain garden

(30%), and engineering designers (50%). The raw data from the relevant field surveys and expert scoring are presented in Tables 10.10 and 10.11.

In an evaluation framework with multiple indicators, the units and magnitudes of each indicator differ and cannot be directly compared. Therefore, standardizing the raw data for each indicator is the initial step in conducting subsequent engineering performance evaluations. Following the principles of the entropy weight-TOPSIS model, Eqs. (10.9–10.1) and (9.3) were applied to normalize the raw data found in Tables 10.10 and 10.11. The standardized data can be viewed in Tables 10.12 and 10.13.

Table 10.7 The construction and operational effectiveness of runoff pollution control facilities in the developed area of Jiaxing City

Watershed	Annual runoff total amount control rate, %	SS reduction rate, %	COD reduction rate, %	COD reduction rate, %
Overall pilot	78.20	53.00	57.18	43.86
Demonstration area of old urban area reconstruction	67.85	48.00	49.85	46.15
South Lake key protection demonstration area	76.71	45.00	55.77	46.43
New city reconstruction demonstration area has been built	82.12	62.00	60.24	42.86
No new city construction demonstration area has been built	85.26	57.00	62.88	40.00
Built-up area planning objectives	78.00	40.00	–	–

Following Eqs. (9.10) to (9.14), the weights of the various evaluation indicators were determined, resulting in the entropy values and entropy weights for the evaluation indicators of Fengxi New City and Jiaxing, respectively. The entropy weight for each indicator is presented in Tables 10.14 and 10.15.

10.4.2 Calculate Nearness Degree

As indicated by the evaluation steps of the entropy weight-TOPSIS model, a comprehensive assessment is conducted using TOPSIS, after determining the weights. The weighted data for the runoff pollution control projects in the built-up areas of Fengxi New City and Jiaxing, along with their positive and negative ideal solutions, can be calculated using Eqs. (10.9–10.4) to (9.6), as shown in Tables 10.16 and 10.17.

By integrating the data from Tables 10.16 and 10.17, the Euclidean distances for the overall catchment areas and the five criteria layers of Fengxi New City and Jiaxing are calculated respectively using Eqs. (9.7) and (9.8). The comprehensive relative closeness and the relative closeness for each of the five criteria layers in the

Table 10.8 Expert scoring table for runoff pollution control projects in various watershed areas of Fengxi new town

Evaluation indicators	Rating criteria	Score Range	Score					
			No.1 water Division	No.2 water Division	No.3 water Division	No.4 water Division	No.5 water Division	No.6 water Division
Scenic beauty rating	A. Beautiful natural scenery	(6,9)						
	B. Scenic natural beauty	(3,6)						
	C. Ordinary scenery	(1,3)						
Operational stability of the project	A. Good operational stability of the project	(6,9)						
	B. Moderate operational stability of the project	(3,6)						
	C. Poor operational stability of the project	(1,3)						
Alleviating the pressure on municipal wastewater treatment	A. Can significantly alleviate the pressure on municipal wastewater treatment	(6,9)						
	B. Can moderately alleviate the pressure on municipal wastewater treatment	(3,6)						
	C. Can slightly alleviate the pressure on municipal wastewater treatment	(1,3)						
Alleviating urban waterlogging	A. Can greatly alleviate urban waterlogging	(6,9)						
	B. Can moderately alleviate urban waterlogging	(3,6)						

(continued)

Table 10.8 (continued)

Evaluation indicators	Rating criteria	Score Range	Score					
			No.1 water Division	No.2 water Division	No.3 water Division	No.4 water Division	No.5 water Division	No.6 water Division
	C. Can slightly alleviate urban waterlogging	(1,3)						
Replenishing underground water	A. Can fill the groundwater to a large extent	(6,9)						
	B. Can replenish the groundwater to a certain extent	(3,6)						
	C. The groundwater can be slightly replenished	(1,3)						
Control system and implementation status	A. The control system and its implementation are good	(6,9)						
	B. The control system and implementation are general	(3,6)						
	C. The control system and its implementation are poor	(1,3)						
Technical specifications and standards development	A. Technical specifications and standard construction is good	(6,9)						
	B. Technical specifications and standard construction is general	(3,6)						
	C. Technical specifications and standard construction is poor	(1,3)						

(continued)

Table 10.8 (continued)

Evaluation indicators	Rating criteria	Score Range	Score					
			No.1 water Division	No.2 water Division	No.3 water Division	No.4 water Division	No.5 water Division	No.6 water Division
Performance assessment and reward mechanism	A. Performance appraisal and reward mechanisms have been improved	(6,9)						
	B. Performance appraisal and reward mechanisms are mediocre	(3,6)						
	C. Performance appraisal and reward mechanism are not perfect	(1,3)						

Table 10.9 The expert scoring table for runoff pollution control projects in various watershed subareas of Jiaxing city

Evaluating indicator	Scoring criteria	Score range	Value			
			Demonstration area for the renovation of old urban areas	Nanhu key protected demonstration Area	Demonstration area for the transformation of newly-built urban areas	Demonstration area for the construction of new urban areas
Scenic attractiveness	A. naturally beautiful scenery	(6,9)				
	B. natural scenery	(3,6)				
	C. ordinary scenery	(1,3)				
Stability of the project operation	A. excellent stability of the project operation	(6,9)				
	B. moderate stability of the project operation	(3,6)				
	C. poor operational stability of the project	(1,3)				
Relieving the pressure of municipal wastewater treatment	A. can largely alleviate the pressure on municipal sewage treatment	(6,9)				
	B. can appropriately alleviate the pressure on municipal sewage treatment	(3,6)				
	C. can slightly alleviate the pressure on municipal sewage treatment	(1,3)				
Mitigating urban Flooding	ACan largely alleviate urban flooding	(6,9)				

(continued)

Table 10.9 (continued)

Evaluating indicator	Scoring criteria	Score range	Value			
			Demonstration area for the renovation of old urban areas	Nanhu key protected demonstration Area	Demonstration area for the transformation of newly-built urban areas	Demonstration area for the construction of new urban areas
	B. can significantly alleviate urban waterlogging to a great extent	(3,6)				
	C: can slightly alleviate urban waterlogging	(1,3)				
Refill groundwater	A. can greatly replenish groundwater to a large extent	(6,9)				
	B. can replenish groundwater to a certain extent	(3,6)				
	C. can slightly replenish groundwater	(1,3)				
Regulatory system and implementation status	A. can slightly replenish groundwater	(6,9)				
	B. The management and implementation of control measures are average	(3,6)				
	C. The management and implementation of control measures are poor	(1,3)				

(continued)

Table 10.9 (continued)

Evaluating indicator	Scoring criteria	Score range	Value			
			Demonstration area for the renovation of old urban areas	Nanhu key protected demonstration Area	Demonstration area for the transformation of newly-built urban areas	Demonstration area for the construction of new urban areas
Construction of technical specifications and standards	A. The construction of technical specifications and standards is good	(6,9)				
	B. The construction of technical specifications and standards is average	(3,6)				
	C. The construction of technical specifications and standards is poor	(1,3)				
Performance assessment and incentive mechanism	A. The performance evaluation and reward mechanism is perfect	(6,9]				
	B. The performance evaluation and reward mechanism is average	(3,6)				
	C. The performance evaluation and reward mechanism is not perfect	(1,3)				

Table 10.10 The original data of representative expert evaluation for performance evaluation indicators of run-off pollution control project in Fengxi new city

Indicator level	Catchment Area 1	Catchment Area 2	Catchment Area 3	Catchment Area 4	Catchment Area 5	Catchment Area 6
Investment construction cost B1, million	8201.00	44,623.00	28,663.50	16,300.00	2894.20	14,214.00
O&M cost as a percentage B_2,%	7.00	8.00	7.00	7.00	6.00	7.00
Costs of water savings from integrated rainwater utilizationB_3,rmb	34,679.60	44,096.50	74,730.45	12,919.95	6152.13	17,161.44
Savings in municipal Wastewater treatment costs B_4,rmb	60,689.30	71,785.00	75,813.50	20,097.70	28,444.10	25,027.10
COD purification rate B_5,%	72.73	68.06	72.36	79.33	78.24	72.30
SS purification rate B_6,%	67.75	62.34	67.51	75.60	74.40	67.41
TP purification rate B_7,%	71.75	66.86	71.99	79.14	77.90	71.96
TN purification rate B_8,%	27.93	39.09	43.60	30.55	33.62	37.41
Total runoff control rate B_9,%	85.12	85.96	86.95	87.05	86.97	85.97
Resource utilization rate for rainwater treatment in place B_{10},%	9.00	10.00	17.00	12.00	9.00	9.00
Stormwater reclamation utilization rate B_{11},%	40.00	43.00	69.00	45.00	51.00	48.00
Percentage of vegetation cover B_{12},%	68.00	77.00	86.00	74.00	78.00	75.00
Landscape Aesthetics B_{13}	7.60	7.20	9.00	6.80	6.90	7.90
Project quality pass rate B_{14},%	92.00	95.00	95.00	93.00	91.00	93.00

(continued)

Table 10.10 (continued)

Indicator level	Catchment Area 1	Catchment Area 2	Catchment Area 3	Catchment Area 4	Catchment Area 5	Catchment Area 6
Engineering Operational Stability B_{15}	7.80	7.60	8.00	7.30	7.60	7.20
Reduce pressure on municipal sewage treatment B_{16}	7.90	7.80	8.20	7.10	6.90	7.40
Mitigating urban flooding B_{17}	8.20	7.70	8.80	8.00	7.90	8.10
Groundwater recharge B_{18}	8.30	7.30	8.80	7.20	8.40	7.00
Control systems and implementation B_{19}	8.40	8.40	8.20	8.40	8.20	8.30
Technical norms and standard building B_{20}	8.50	8.40	8.40	8.40	8.30	8.10
Performance appraisal and incentives B_{21}	8.00	7.60	8.00	7.60	7.70	8.00

catchment areas of both pilot cities are then computed using Eq. (9.9). The specific evaluation results can be found in Tables 10.18 and 10.19.

10.4.3 Evaluation Results and Improvement Suggestions

A. Runoff Pollution Control Project in Fengxi New City, Xixian New Area

I. Internal Analysis of Performance Indicators.

From the analysis of the results in Table 10.18 and Fig. 14, it is evident that the runoff pollution control projects in catchment areas 1 and 6 of Fengxi New City exhibit the highest levels of performance in terms of construction mechanisms. Catchment area 3 demonstrates high levels of performance across four aspects: economic benefits, ecological and environmental benefits, construction and operational status, and social benefits. In contrast, catchment areas 4 and 5 show lower performance levels across all five criteria layers. This indicates significant variations in performance levels across the six catchment areas in different domains. Such variations may be attributed to the number of runoff pollution control projects implemented in each catchment area and the variety of applied technological solutions.

Table 10.11 The original data table of representative expert evaluation for performance evaluation indicators of runoff pollution control project in the built-up area of Jiaxing city

Indicator level	Demonstration area for the transformation of old urban areas	South Lake Priority Protection Demonstration Area	Demonstration areas for the transformation of established new towns	Unbuilt New Town Construction Demonstration Area
Investment construction cost B1, million	17,392.99	69,140.98	30,800.84	3592.48
O&M cost as a percentage B_2,%	6.00	7.00	5.00	6.00
Costs of water savings from integrated rainwater utilizationB_3,rmb	3041.50	3300.00	1817.20	3141.60
Savings in municipal wastewater treatment costs B_4,rmb	2488.50	2700.00	1486.80	2570.40
COD purification rate B_5,%	49.85	55.77	60.24	62.88
SS purification rate B_6,%	48.00	45.00	62.00	57.00
TP purification rate B_7,%	46.15	46.43	42.86	40.00
TN purification rate B_8,%	75.59	69.11	66.96	72.32
Total runoff control rate B_9,%	67.85	76.71	82.12	85.26
Resource utilization rate for rainwater treatment in place B_{10},%	12.94	11.84	7.27	9.26
Stormwater reclamation utilization rate B_{11},%	36.70	38.60	37.80	38.80
percentage of vegetation cover B_{12},%	11.35	32.21	44.95	39.75
Landscape Aesthetics B_{13}	5.60	7.90	7.40	7.50

(continued)

Table 10.11 (continued)

Indicator level	Demonstration area for the transformation of old urban areas	South Lake Priority Protection Demonstration Area	Demonstration areas for the transformation of established new towns	Unbuilt New Town Construction Demonstration Area
Project quality pass rate B_{14},%	98.00	97.00	96.00	97.00
Engineering Operational Stability B_{15}	7.00	7.80	7.10	7.50
Reduce pressure on municipal sewage treatment B_{16}	7.10	7.20	6.80	7.40
Mitigating urban flooding B_{17}	7.00	7.90	7.70	7.60
Groundwater recharge B_{18}	4.70	6.00	5.70	5.50
Control systems and implementation B_{19}	7.50	7.60	7.80	7.80
Technical norms and standard building B_{20}	8.10	8.20	8.20	8.00
Performance appraisal and incentives B_{21}	7.10	7.50	7.30	7.40

Within catchment area 3, a variety of runoff pollution control projects and measures have been implemented, with the most notable being the Central Green Corridor artificial wetland. The Central Green Corridor integrates one-third of Fengxi New City's core area into a comprehensive sponge city entity, creating a striking addition to the landscape. In addition to its aesthetic value, the Green Corridor also serves dual purposes: rainwater storage and groundwater replenishment. It functions as the city's open space for stormwater management, combining both ecological and practical benefits. For Fengxi New City, the construction of the Central Green Corridor significantly reduces the risk for urban runoff pollution and pluvial flooding during the city's future development. By coordinating rainwater, reclaimed water, and surface water, the Corridor addresses ecological water replenishment needs, demonstrating the economic viability and long-term effectiveness of this ecological stormwater management solution. In contrast, catchment areas 4 and 5, due to lower level of construction and development, have fewer runoff pollution control projects, with most still under construction. As a result, these areas show generally lower performance across all five evaluation criteria.

Table 10.12 The standardized performance evaluation data table for runoff pollution control project in Fengxi new city

Indicator level	Catchment area 1	Catchment area 2	Catchment area 3	Catchment area 4	Catchment area 5	Catchment area 6
Investment construction cost B1, million	0.8728	0.0000	0.3825	0.6787	1.0000	0.7287
O&M cost as a percentage B_2,%	0.5000	0.0000	0.5000	0.5000	1.0000	0.5000
Costs of water savings from integrated rainwater utilizationB_3,rmb	0.4160	0.5533	1.0000	0.0987	0.0000	0.1605
Savings in municipal wastewater treatment costs B_4,rmb	0.7285	0.9277	1.0000	0.0000	0.1498	0.0885
COD purification rate B_5,%	0.4144	0.0000	0.3815	1.0000	0.9033	0.3762
SS purification rate B_6,%	0.4080	0.0000	0.3899	1.0000	0.9095	0.3824
TP purification rate B_7,%	0.3982	0.0000	0.4178	1.0000	0.8990	0.4153
TN purification rate B_8,%	0.0000	0.7122	1.0000	0.1672	0.3631	0.6050
Total runoff control rate B_9,%	0.0000	0.4352	0.9482	1.0000	0.9585	0.4404
Resource utilization rate for rainwater treatment in place B_{10},%	0.0000	0.2500	1.0000	0.3750	0.0000	0.0000
Stormwater reclamation utilization rate B_{11},%	0.0000	0.1034	1.0000	0.1724	0.3793	0.2759
Percentage of vegetation cover B_{12},%	0.0000	0.5000	1.0000	0.3333	0.5556	0.3889
Landscape Aesthetics B_{13}	0.3636	0.1818	1.0000	0.0000	0.0455	0.5000
Project quality pass rate B_{14},%	0.2500	1.0000	1.0000	0.5000	0.0000	0.5000

(continued)

Table 10.12 (continued)

Indicator level	Catchment area 1	Catchment area 2	Catchment area 3	Catchment area 4	Catchment area 5	Catchment area 6
Engineering Operational Stability B_{15}	0.7500	0.5000	1.0000	0.1250	0.5000	0.0000
Reduce pressure on municipal sewage treatment B_{16}	0.7692	0.6923	1.0000	0.1538	0.0000	0.3846
Mitigating urban flooding B_{17}	0.4545	0.0000	1.0000	0.2727	0.1818	0.3636
Groundwater recharge B_{18}	0.7222	0.1667	1.0000	0.1111	0.7778	0.0000
Control systems and implementation B_{19}	1.0000	1.0000	0.0000	1.0000	0.0000	0.5000
Technical norms and standard building B_{20}	1.0000	0.7500	0.7500	0.7500	0.5000	0.0000
Performance appraisal and incentives B_{21}	1.0000	0.0000	1.0000	0.0000	0.2500	1.0000

Among the five evaluation criteria, the performance level of construction mechanisms is higher than that of the other four criteria, while the overall economic benefit performance is relatively low. This evaluation result indicates that the the construction mechanisms of urban runoff pollution control projects have been notably effective. However, there are substantial variations in performance levels among different areas. The excessive focus on individual plot developments has led to lower overall effectiveness at the district level. Going forward, future construction should prioritize the integrated development of runoff pollution control projects across districts to prevent fragmentation and enhance overall performance.

ii. Comprehensive Performance Analysis

Analyzing the results from Table 10.18, the comprehensive the nearness degree of the six catchment areas exceeds 0.3. According to the performance evaluation standards in Table 10.3, the performance levels of all six areas are rated as medium or above. Notably, the performance of runoff pollution control projects in catchment area 3 is rated as "good". Overall, the performance of runoff pollution control projects in the built-up area of Fengxi New City in the Xixian New Area is above average. These evaluation results align with the actual local conditions.

iii. Improvement Recommendations

Table 10.13 The standardized performance evaluation data for runoff pollution control project in the built-up area of Jiaxing City

Indicator level	Demonstration area for the transformation of old urban areas	South Lake Priority Protection Demonstration Area	Demonstration areas for the transformation of established new towns	Unbuilt New Town Construction Demonstration Area
Investment construction cost B1, million	0.7895	0.0000	0.5849	1.0000
O&M cost as a percentage B_2,%	0.5000	0.0000	1.0000	0.5000
Costs of water savings from integrated rainwater utilizationB_3,rmb	0.8257	1.0000	0.0000	0.8932
Savings in municipal wastewater treatment costs B_4,rmb	0.8257	1.0000	0.0000	0.8932
COD purification rate B_5,%	0.0000	0.4543	0.7974	1.0000
SS purification rate B_6,%	0.1765	0.0000	1.0000	0.7059
TP purification rate B_7,%	0.9565	1.0000	0.4448	0.0000
TN purification rate B_8,%	1.0000	0.2491	0.0000	0.6211
Total runoff control rate B_9,%	0.0000	0.5089	0.8196	1.0000
Resource utilization rate for rainwater treatment in place B_{10},%	1.0000	0.8060	0.0000	0.3510
Stormwater reclamation utilization rate B_{11},%	0.0000	0.9048	0.5238	1.0000
Percentage of vegetation cover B_{12},%	0.0000	0.6208	1.0000	0.8452
Landscape Aesthetics B_{13}	0.0000	1.0000	0.7826	0.8261

(continued)

Table 10.13 (continued)

Indicator level	Demonstration area for the transformation of old urban areas	South Lake Priority Protection Demonstration Area	Demonstration areas for the transformation of established new towns	Unbuilt New Town Construction Demonstration Area
Project quality pass rate B_{14},%	1.0000	0.5000	0.0000	0.5000
Engineering Operational Stability B_{15}	0.0000	1.0000	0.1250	0.6250
Reduce pressure on municipal sewage treatment B_{16}	0.5000	0.6667	0.0000	1.0000
Mitigating urban flooding B_{17}	0.0000	1.0000	0.7778	0.6667
Groundwater recharge B_{18}	0.0000	1.0000	0.7692	0.6154
Control systems and implementation B_{19}	0.0000	0.3333	1.0000	1.0000
Technical norms and standard building B_{20}	0.5000	1.0000	1.0000	0.0000
Performance appraisal and incentives B_{21}	0.0000	1.0000	0.5000	0.7500

Based on the on-site investigation and performance evaluation of runoff pollution control projects in Fengxi New Town, the following recommendations are made for improving for runoff pollution control in the region.

Fengxi New Town is located in a northwest arid, water-deficient region with extensive collapsible loess soil that has good permeability (with an average permeability coefficient between 1×10^{-6} and 3.8×10^{-5} m/s). Given these conditions, source control technologies should be prioritized for planning runoff pollution control projects in conjunction with runoff storage and drainage measures such as process interception and excessive runoff discharge, as shown in Fig. 10.10.

Table 10.14 Entropy value and entropy weight of performance evaluation indicators for runoff pollution control project in Fengxi new city

Indicator level	Entropy value (physics)	Entropy weight
Investment construction cost B1 million	0.8739	0.0292
O&M cost as a percentage B_2%	0.8710	0.0299
Costs of water savings from integrated rainwater utilizationB_3 rmb	0.7514	0.0576
Savings in municipal wastewater treatment costs B_4 rmb	0.7473	0.0585
COD purification rate B_5%	0.8434	0.0363
SS purification rate B_6%	0.8440	0.0361
TP purification rate B_7%	0.8494	0.0349
TN purification rate B_8%	0.8217	0.0413
Total runoff control rate B_9%	0.8626	0.0318
Resource utilization rate for rainwater treatment in place B_{10}%	0.5163	0.1121
Stormwater reclamation utilization rate B_{11}%	0.7317	0.0622
Percentage of vegetation cover B_{12}%	0.8528	0.0341
Landscape Aesthetics B_{13}	0.7226	0.0643
Project quality pass rate B_{14}%	0.8364	0.0379
Engineering Operational Stability B_{15}	0.8163	0.0426
Reduce pressure on municipal sewage treatment B_{16}	0.8200	0.0417
Mitigating urban flooding B_{17}	0.7997	0.0464
Groundwater recharge B_{18}	0.7657	0.0543
Control systems and implementation B_{19}	0.7544	0.0569
Technical norms and standard building B_{20}	0.8856	0.0265
Performance appraisal and incentives B_{21}	0.7173	0.0655

Firstly, at the source locations where runoff is generated from urban buildings' rooftops, roads, plazas, parks, and green spaces, source reduction measures such as green roofs, permeable pavements, depressed green spaces, rain gardens, and grass swales can be implemented. These measures will help reduce runoff pollution, capture rainwater, and mitigate peak flow rates. Secondly, rainwater can be intercepted and stored through measures like detention ponds, regulation ponds, and artificial wetlands. Finally, these interception and storage measures can be combined with road drainage channels, green areas outside the red lines, and terminal water bodies (such as the Fenghe River and Weihe River) to form an excessive runoff discharge system. This system will manage overflow during heavy rainfall events when the pipe network's discharge capacity is exceeded, achieving peak flow reduction, rainwater storage, and promoting natural infiltration.

Table 10.15 The entropy value and entropy weight of performance evaluation indicators for runoff pollution control project in the built-up area of Jiaxing city

Indicator level	Entropy value (physics)	Entropy weight
Investment construction cost B1, million	0.7758	0.0429
O&M cost as a percentage B_2,%	0.7500	0.0479
Costs of water savings from integrated rainwater utilizationB_3,rmb	0.7902	0.0402
Savings in municipal wastewater treatment costs B_4,rmb	0.7902	0.0402
COD purification rate B_5,%	0.7582	0.0463
SS purification rate B_6,%	0.6678	0.0636
TP purification rate B_7,%	0.7529	0.0473
TN purification rate B_8,%	0.6992	0.0576
Total runoff control rate B_9,%	0.7667	0.0447
Resource utilization rate for rainwater treatment in place B_{10},%	0.7355	0.0506
Stormwater reclamation utilization rate B_{11},%	0.7676	0.0445
Percentage of vegetation cover B_{12},%	0.7792	0.0423
Landscape Aesthetics B_{13}	0.7884	0.0405
Project quality pass rate B_{14},%	0.7500	0.0479
Engineering Operational Stability B_{15}	0.6319	0.0705
Reduce pressure on municipal sewage treatment B_{16}	0.7631	0.0454
Mitigating urban flooding B_{17}	0.7822	0.0417
Groundwater recharge B_{18}	0.7783	0.0425
Control systems and implementation B_{19}	0.7244	0.0528
Technical norms and standard building B_{20}	0.7610	0.0458
Performance appraisal and incentives B_{21}	0.7652	0.0450

B. Runoff Pollution Control Project in the Built-Up Area of Jiaxing City

I. Internal Analysis of Performance Indicators

From Table 10.19 and Fig. 10.11, the evaluation results of runoff pollution control projects in the built-up areas of Jiaxing City show that the Nanhu Key Protection Demonstration Area has relatively high performance in terms of economic efficiency, engineering construction and operation, and social benefits. The newly constructed city transformation demonstration area has the highest performance in mechanism construction, while the undeveloped new city construction demonstration area ranks highest in ecological environmental benefits. In contrast, the old urban area transformation demonstration area falls short in all five criteria, with overall poor evaluation results.

Table 10.16 The weighted performance evaluation data table for runoff pollution control project in Fengxi new city

Indicator level	Data weighting						Maximum value	Minimum value
	Catchment Area 1	Catchment Area 2	Catchment Area 3	Catchment Area 4	Catchment Area 5	Catchment Area 6		
B_1	0.0255	0.0000	0.0112	0.0198	0.0292	0.0213	0.0000	0.0292
B_2	0.0149	0.0000	0.0149	0.0149	0.0299	0.0149	0.0000	0.0299
B_3	0.0240	0.0319	0.0576	0.0057	0.0000	0.0092	0.0576	0.0000
B_4	0.0427	0.0543	0.0585	0.0000	0.0088	0.0052	0.0585	0.0000
B_5	0.0150	0.0000	0.0138	0.0363	0.0328	0.0137	0.0363	0.0000
B_6	0.0147	0.0000	0.0141	0.0361	0.0329	0.0138	0.0361	0.0000
B_7	0.0139	0.0000	0.0146	0.0349	0.0314	0.0145	0.0349	0.0000
B_8	0.0000	0.0294	0.0413	0.0069	0.0150	0.0250	0.0413	0.0000
B_9	0.0000	0.0139	0.0302	0.0318	0.0305	0.0140	0.0318	0.0000
B_{10}	0.0000	0.0280	0.1121	0.0420	0.0000	0.0000	0.1121	0.0000
B_{11}	0.0000	0.0064	0.0622	0.0107	0.0236	0.0172	0.0622	0.0000
B_{12}	0.0000	0.0170	0.0341	0.0114	0.0189	0.0133	0.0341	0.0000
B_{13}	0.0234	0.0117	0.0643	0.0000	0.0029	0.0321	0.0643	0.0000
B_{14}	0.0095	0.0379	0.0379	0.0190	0.0000	0.0190	0.0379	0.0000
B_{15}	0.0319	0.0213	0.0426	0.0053	0.0213	0.0000	0.0426	0.0000
B_{16}	0.0321	0.0289	0.0417	0.0064	0.0000	0.0160	0.0417	0.0000
B_{17}	0.0211	0.0000	0.0464	0.0127	0.0084	0.0169	0.0464	0.0000
B_{18}	0.0392	0.0090	0.0543	0.0060	0.0422	0.0000	0.0543	0.0000
B_{19}	0.0569	0.0569	0.0000	0.0569	0.0000	0.0284	0.0569	0.0000
B_{20}	0.0265	0.0199	0.0199	0.0199	0.0133	0.0000	0.0265	0.0000
B_{21}	0.0655	0.0000	0.0655	0.0000	0.0164	0.0655	0.0655	0.0000

Table 10.17 The weighted performance evaluation data table for unoff pollution control project in the Built-up Area of Jiaxing City

Indicator level	Data weighting				Maximum value	Minimum value
	Old city transformation demonstration plot	South Lake Key Protection Demonstration Area	Demonstration areas for the transformation of established new towns	Unbuilt New Town Construction Demonstration Area		
B_1	0.0339	0.0000	0.0251	0.0429	0.0000	0.0429
B_2	0.0239	0.0000	0.0479	0.0239	0.0000	0.0479
B_3	0.0332	0.0402	0.0000	0.0359	0.0402	0.0000
B_4	0.0332	0.0402	0.0000	0.0359	0.0402	0.0000
B_5	0.0000	0.0210	0.0369	0.0463	0.0463	0.0000
B_6	0.0112	0.0000	0.0636	0.0449	0.0636	0.0000
B_7	0.0453	0.0473	0.0210	0.0000	0.0473	0.0000
B_8	0.0576	0.0143	0.0000	0.0358	0.0576	0.0000
B_9	0.0000	0.0227	0.0366	0.0447	0.0447	0.0000
B_{10}	0.0506	0.0408	0.0000	0.0178	0.0506	0.0000
B_{11}	0.0000	0.0403	0.0233	0.0445	0.0445	0.0000
B_{12}	0.0000	0.0262	0.0423	0.0357	0.0423	0.0000
B_{13}	0.0000	0.0405	0.0317	0.0335	0.0405	0.0000
B_{14}	0.0479	0.0239	0.0000	0.0239	0.0479	0.0000
B_{15}	0.0000	0.0705	0.0088	0.0441	0.0705	0.0000
B_{16}	0.0227	0.0302	0.0000	0.0454	0.0454	0.0000
B_{17}	0.0000	0.0417	0.0324	0.0278	0.0417	0.0000
B_{18}	0.0000	0.0425	0.0327	0.0261	0.0425	0.0000
B_{19}	0.0000	0.0176	0.0528	0.0528	0.0528	0.0000
B_{20}	0.0229	0.0458	0.0458	0.0000	0.0458	0.0000
B_{21}	0.0000	0.0450	0.0225	0.0337	0.0450	0.0000

It can be seen that there are differences in the performance levels of the four watersheds in different areas in Jiaxing City. Among them, the performance level of the Nanhu Key Protection Demonstration Area is relatively high on the 5 criteria levels. This is due to the extensive runoff pollution control projects implemented around Nanhu, which are part of the city's sponge city initiative. These efforts are complemented by programs such as the 'Twelfth Five-Year' Water Special Project, the River Chief System, and the 'Five Waters Governance,' all of which have contributed to positive outcomes.

Among the five evaluation criteria, the performance level of social benefits is higher than the other 4 criteria levels, while the overall performance level of economic benefits is relatively low. This indicates that Jiaxing City achieved significant results in social benefits construction in the construction of runoff pollution control projects. In comparison with the runoff pollution control project in Fengxi New City, although

Table 10.18 Evaluation results of the nearness degree level for runoff pollution control projects in Fengxi new city

Evaluation Object	Catchment Area 1	Catchment Area 2	Catchment Area 3	Catchment Area 4	Catchment Area 5	Catchment Area 6
Nearness Degree of Economic Benefit	0.5191	0.7435	0.8207	0.1842	0.0917	0.2068
Nearness Degree of Ecological and Environmental Benefits	0.1838	0.2664	0.8081	0.4181	0.3452	0.2865
Nearness Degree of Project Construction and Operation Condition	0.5231	0.6714	1.0000	0.3203	0.3286	0.2892
Nearness Degree of Social Benefits	0.6390	0.3141	1.0000	0.1834	0.4274	0.2582
Nearness Degree of Mechanism Construction	1.0000	0.4780	0.5444	0.4780	0.2163	0.6474
Nearness Degree of Comprehensive	0.4341	0.4210	0.7534	0.3790	0.3170	0.3501

Table 10.19 Assessment results of nearness degree level for runoff pollution control projects in the built-up area of Jiaxing city

Evaluation Object	Demonstration Area of Old Urban area reconstruction	South Lake key protection demonstration area	New city reconstruction demonstration area has been built	No new city construction demonstration area has been built
Nearness Degree of Economic Benefit	0.5560	1.0000	0.1852	0.5311
Nearness Degree of Environmental Benefit	0.4473	0.5246	0.5465	0.6385
Nearness Degree of Project Construction and Operation Condition	0.4045	0.7567	0.1014	0.5844
Nearness Degree of Social Benefit	0.2626	0.8153	0.4931	0.7343
Nearness Degree of Mechanism Construction	0.2387	0.6541	0.7655	0.5706
Nearness Degree of Comprehensive	0.4142	0.6436	0.4765	0.6062

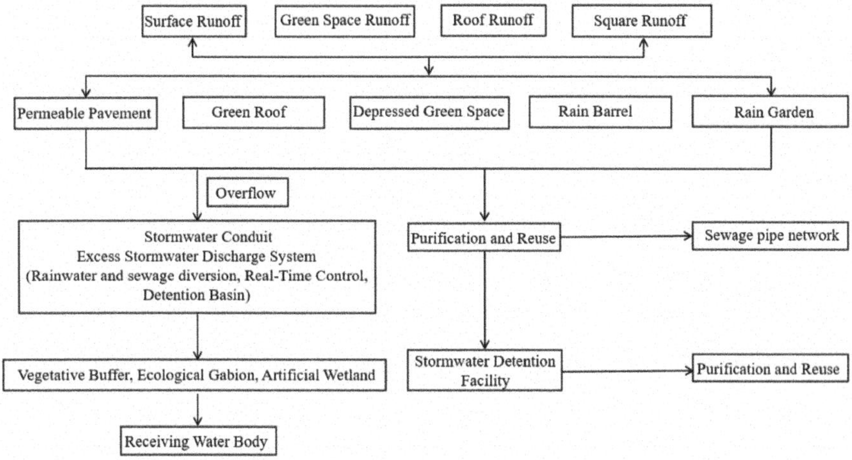

Fig. 10.11 Technical roadmap for runoff pollution control in the built-up area of Jiaxing city

the performance levels of each district in Jiaxing City are relatively similar, there are still some differences, and the overall effectiveness in the contiguous areas still needs to be improved.

ii. Comprehensive Performance Analysis

From the analysis results in Table 10.19, the comprehensive closeness of the four watershed areas is greater than 0.3. According to the performance evaluation criteria in Table 10.3, all four areas perform above average. Specifically, the runoff pollution control projects in the Nanhu Key Protection Area and the undeveloped new city construction area are rated as good. Overall, the performance of runoff pollution control projects in the built-up areas of Jiaxing City is above average.

iii. Improvement Suggestions

Based on the on-site investigation and performance evaluation of runoff pollution control projects in the built-up areas of Jiaxing City, the following suggestions are proposed for the management of runoff pollution in the area.

Jiaxing is a typical plain river network city with abundant rainfall resources, dense river networks, high groundwater levels, and poor soil permeability. These characteristic makes Jiaxing particularly suitable for measures focused on rainwater collection, storage, and utilization to control rainfall runoff pollution. According to the runoff generation points, the underlying surface of the built-up areas of Jiaxing can be divided into four categories: urban building roofs, urban roads, squares and green spaces, and urban water systems. For runoff generated from these surfaces, a combination of single or multiple runoff pollution control technologies can be applied to reduce pollution at the source, purify and utilize rainwater in the middle, and restore river water at the end, as shown in Fig. 10.11.

For urban building roof, environmentally friendly materials that minimize the impact on runoff quality should be prioritized. Where conditions allow, green roofs or rooftop gardens can be considered as runoff pollution control measures, particularly on suitable slopes.

At the same time, the roof runoff can be directed into the surrounding green areas via rainwater downpipes to promote infiltration of rainwater. Alternatively, grass swales and rainwater pipes can channel rainwater into rainwater collection modules such as rainwater tanks and rainwater irrigation for storage and reuse.

Regarding underlying surface, in addition to fulfilling basic transportation and pedestrian functions, permeable pavement designs can be adopted. Open curbs and hidden drainage channels can be installed along the road edges to facilitate runoff collection into surrounding pollution control facilities. In adjacent green belts, runoff pollution control measures such as grassed swales, ecological tree pits, depressed green spaces, and rain gardens can be implemented to retain and purify road runoff.

For square and green space underlying surface, green space serves as a natural urban permeable surface with excellent runoff purification and infiltration functions, playing a crucial role in controlling urban stormwater pollution. In the runoff pollution control project, it is important to fully utilize green spaces for runoff purification and infiltration. Additionally, in harmony with the terrain, small-scale runoff pollution control infrastructures such as grassed swales, gravel infiltration trenches, and rain gardens, can be installed around green spaces and squares to facilitate the collection and treatment of runoff rainwater from adjacent areas.

Urban water systems play a pivotal role in enhancing the urban ecological environment and are a vital part of the urban water cycle. Jiaxing city, with its dense river network and well-developed water systems, highlights the importance of river systems in managing rainfall runoff pollution. Before runoff enters the rivers, it is advisable to install facilities such as vegetative buffer zones, artificial wetlands, and ecological revetments to control runoff pollution. These measures help intercept rainfall runoff and its associated pollutants, protecting the urban aquatic ecosystem.

References

1. Liang, J. (2010). *Research on the evaluation methods of water pollution prevention and control technology in the pharmaceutical industry*. Hebei University of Science and Technology. (in Chinese).
2. Wang, P. (2014). *Research on the technology of overland flow interception for urban non-point source pollution*. Shenyang University. (in Chinese).
3. Qian, C. (2014). *Integration of common technologies and performance evaluation of river pollution control in southern Chinese cities*. East China Normal University. (in Chinese).
4. Chen, S. (2002). *Optimization and fuzzy identification theory and application in complex water resources systems*. Jilin University. (in Chinese).
5. Yu, J., Song, H., Wang, X., et al. (2010). Selection of comprehensive risk evaluation methods for urban gas pipelines. *Gas & Heat, 30*(9), 22–25. (in Chinese).
6. Yu, X., & Fu, D. (2004). Review of multi-indicator comprehensive evaluation methods. *Statistics & Decision, 11*, 119–121. (in Chinese).

7. Liu, Q., Han, H., Li, H., et al. (2017). Performance evaluation of sponge city construction based on the entropy weight TOPSIS method: A case study of Hebi City. *Henan Province. Yangtze River, 48*(14), 23–26. (in Chinese).
8. Lu, C., Wen, F., Yang, Q., et al. (2011). Performance evaluation and obstacle factor diagnosis of urban land use based on the improved TOPSIS method: A case study of Chongqing City. *Resources Science, 33*(3), 535–541. (in Chinese).

The manufacturer's authorised representative in the EU is Springer Nature Customer Service Centre GmbH, Europaplatz 3, 69115 Heidelberg, Germany. If you have any concerns regarding our products, please contact ProductSafety@springernature.com

Printed and bound by CPI Group (UK) Ltd, Croydon, CR0 4YY
01/10/2026
02225935-0018